Categorical Data Analysis and Multilevel Modeling Using R

Xing Liu

Eastern Connecticut State University

SAGE

Los Angeles | London | New Delhi
Singapore | Washington DC | Melbourne

*In memory of my grandparents, Liu, Liangsheng (*刘良生*) and Liu, Lishi (*刘李氏*).*
*To my parents, Canru Liu (*刘灿如*) and Suying Hou (*侯素英*), for their love and support.*
To Li, Lucy, Aileen, and Jerry.

SAGE

FOR INFORMATION:

SAGE Publications, Inc.
2455 Teller Road
Thousand Oaks, California 91320
E-mail: order@sagepub.com

SAGE Publications Ltd.
1 Oliver's Yard
55 City Road
London, EC1Y 1SP
United Kingdom

SAGE Publications India Pvt. Ltd.
B 1/I 1 Mohan Cooperative Industrial Area
Mathura Road, New Delhi 110 044
India

SAGE Publications Asia-Pacific Pte. Ltd.
18 Cross Street #10-10/11/12
China Square Central
Singapore 048423

Acquisitions Editor: Helen Salmon
Product Associate: Yumna Samie
Production Editor: Vijayakumar
Copy Editor: Calum Ross
Typesetter: TNQ Technologies
Proofreader: Benny Willy Stephen
Indexer: TNQ Technologies
Cover Designer: Candice Harman
Marketing Manager: Victoria Velasquez

Printed in the United States of America

ISBN 978-1-5443-2490-6

This book is printed on acid-free paper.

22 23 24 25 26 10 9 8 7 6 5 4 3 2 1

BRIEF CONTENTS

DETAILED CONTENTS

ACKNOWLEDGMENTS

Many people have contributed to my academic career and helped me during the writing of this book. Without them this book would not have been possible. First, I am grateful to Dr. Ann A. O'Connell, who was my statistics professor and major advisor for my Ph.D. degree at the University of Connecticut. I took her popular, well-known course on logistic regression and multilevel modeling and believed that was one of the most important courses in my graduate program. Her expertise in statistics and unique teaching pedagogy tremendously impacted my research and teaching career. I also would like to thank my other statistics and measurement professors, Drs. D. Betsy McCoach, Hariharan Swaminathan, and H. Jane Rogers. All those courses mattered. I am thankful to Dr. Scott Brown for his support while I worked for the Teachers for a New Era (TNE) Project at UConn. My data analysis and assessment experience there prepared me for the current position at Eastern. I am also grateful to Professor Hongwei Ye at Hohai University for inspiring and advising me to pursue graduate studies. His advice changed my life.

In addition, I would like to thank the following previously anonymous reviewers for their thoughtful reviews of the book. They provided extremely valuable comments and suggestions on an earlier version of the manuscript, which made the book better:

- Zack W. Almquist, University of Minnesota
- Yang Cao, University of North Carolina at Charlotte
- Jennifer Hayes Clark, University of Houston
- Tina Freiburger, University of Wisconsin–Milwaukee
- Alan Huebner, University of Notre Dame
- Ahmed Ibrahim, Johns Hopkins University
- Kyungkook Kang, University of Central Florida
- Brandon LeBeau, University of Iowa
- Man-Kit Lei, University of Georgia
- Julie Lorah, Indiana University
- Wendy L. Martinek, Binghamton University
- Amin Mousavi, University of Saskatchewan

- Ji Hoon Ryoo, University of Virginia
- Daniel S. Scheller, University of Texas at El Paso
- Richard Williams, University of Notre Dame

Furthermore, I would like to thank SAGE's editor, Helen Salmon, who has been very helpful and supportive throughout my writing of the two books. After my first book on ordinal logistic regression was published, Helen and I discussed the possibility of working on a book project on categorical data analysis at the AERA conference in 2016. After five years, I finally completed it. Thank you for your support and patience. I am also thankful to the Production Editor, Vijayakumar, for his help during the production stages.

A special thanks goes to the statisticians and programmers who contributed to the R community, including the R Development Core Team and the authors of the add-on packages, whose names with the related packages are listed in the Preface. In particular, I am thankful to Daniel Lüdecke for answering my questions related to the `ggeffects` package.

I also would like to thank my wonderful colleagues in the Education Department at Eastern Connecticut State University, where we have collaborated on various research projects. Special thanks to Drs. Hari Koirala, Sudha Swaminathan, and David Stoloff. Dr. Hari Koirala and I have worked collaboratively on many research grants, papers, and presentations for more than 12 years. We discussed research, shared ideas, and chatted about funny topics beyond math equations. He is a wonderful colleague, mentor, and friend. Dr. Sudha Swaminathan and I worked collaboratively on a national research grant, several projects, and the NCATE and CAEP accreditations. I thank her for the leadership to make the program better. Dr. David Stoloff hired me as a faculty when he was the department chair in 2007 and helped me in various aspects.

I am also grateful for the support from the administrators from Eastern Connecticut State University. I received the faculty development grant to purchase some of the references for this book project in 2017. I was also granted a sabbatical leave for fall 2020 so I could complete the draft for several book chapters. I would like to thank the Faculty Development Committee, the Sabbatical Leave Committee, and the administrators for their support.

Finally, very special thanks goes to my loving and caring wife, Li, for tolerating my writing on weekends and nights, and to my children, Lucy, Aileen, and Jerry. Lucy and Aileen often asked me, "Dad, have you finished your book yet?" When Jerry asked me for something, I always replied, "Wait a minute." I hope that I will have more playtime with him without waiting. Thanks to all of you who brought me joy and happiness.

ABOUT THE AUTHOR

Xing Liu, Ph.D., is a professor of educational research and assessment at Eastern Connecticut State University. He received his Ph.D. in measurement, evaluation, and assessment in the field of educational psychology from the University of Connecticut, Storrs. His interests include categorical data analysis, multilevel modeling, longitudinal data analysis, structural equation modeling, educational assessment, propensity score methods, data science, and Bayesian methods. He is the author of *Applied Ordinal Logistic Regression Using Stata: From Single-Level to Multilevel Modeling* (2016). His major publications focus on advanced statistical models. His articles have been recognized among the most popular papers published in the *Journal of Modern Applied Statistical Methods (JMASM)*. Dr. Liu is the recipient of the Excellence Award in Creativity/Scholarship at Eastern Connecticut State University.

PREFACE

PURPOSE

The purpose of this book is to provide the reader with a practical focus on regression techniques for analyzing categorical and count response variables using R, the free, open-source statistical software. It presents comprehensive coverage of generalized linear models for binary, ordinal, and nominal categorical response variables and count data in an easily accessible, practical way. It also covers generalized linear mixed models, or multilevel models for categorical and count response variables when data structure is nested or hierarchical. In addition, since Bayesian statistical methods have become popular in recent years, this book introduces Bayesian approaches to fit both generalized linear models and generalized linear mixed models. It provides a unified framework for both single-level and multilevel modeling of categorical and count response variables with both frequentist and Bayesian approaches with R. Emphasis is given to demonstrating how to conduct the analysis using R, how to interpret the models, and how to present the results for publication.

This book is built on and extends my Stata book (i.e., *Applied Ordinal Logistic Regression Using Stata*) with a broader range of regression techniques using R. It was inspired by colleagues and graduate students whom I encountered at the American Educational Research Association (AERA) conference. I was often asked the similar questions: Can R perform this type of analysis? Do you know if there are any packages in R that have a similar function to the program introduced in your paper? Their questions motivated me to work on an R book on categorical data analysis and multilevel modeling.

This book is designed as a text for graduate-level quantitative methods courses on applied categorical data analysis, or multilevel modeling for categorical response variables in education, social sciences, and behavioral sciences. Also, it is designed as a supplemental text for a course for applied Bayesian methods. In addition, it may be of interest to researchers in the areas of epidemiology, medicine, and public health, where categorical data are extensively analyzed. Finally, this book can be a reference for self-study for graduate students, applied researchers, and practitioners who are interested in categorical data analysis.

APPROACH OF THE BOOK

The book is organized in a structured format, which will facilitate readers to better understand regression models for categorical and count response variables and multilevel modeling, and apply them for their own research. Each chapter begins with a list of key objectives and concludes with a set of exercises. The concept of a new model is introduced first, followed by the research problem, questions, and variables in the data. R is then used to fit models with the example data. The R commands are explained and the output is interpreted. Finally, the guidelines for reporting the results are discussed and sample writings are provided. The R scripts are summarized at the end of each chapter, so readers can easily replicate the results. In addition, the end-of-chapter exercises provide readers with hand-on activities and help them to understand the model further.

A specific pedagogical approach used in this book is learning by doing. An effective way to learn how to conduct data analysis with R is to understand R code and practice. All of the R code is explained in the text and summarized both in the book and on the companion website. Readers can easily adapt the R code to meet the needs of their own research. The organization of the models, research questions and data, R commands, examples of model fitting, and exercises in sequence in each chapter provide readers with hands-on opportunities to practice R.

FEATURES OF THE BOOK

This book is practical. It provides both classical and Bayesian approaches in applying regression analysis and multilevel modeling for categorical and count response variables using R. It is nontechnical; the theory and mathematical computations are kept minimal since this book is intended for students, applied researchers, and practitioners in education, social sciences, and behavioral sciences. Although theories of the models are not the focus, the basic methodologies and the models equations are presented to facilitate the understanding of these models.

Readers are assumed to have basic knowledge of introductory statistics and linear regression models. They are also assumed to have limited or no previous knowledge in using R. The first two chapters introduce R basics and review introductory statistics, linear regression models, and the chi-square test, which lay the foundation for the remaining chapters.

ORGANIZATION OF THE BOOK

This book can be organized into four parts. The first part includes the initial two chapters, which introduce the basics of R and review introductory statistics with R. Readers with advanced knowledge in R and statistics may skip these two chapters and go directly to Chapter 3.

The second part covers Chapters 3–9, which present generalized linear models for binary, ordinal, nominal, and count response variables. It starts with the logistic regression model for binary response variables, then proceeds with several ordinal regression models (such as the proportional odds model, the partial proportional odds model and generalized ordinal logistic model, the continuation ratio model, the adjacent categories model, and the stereotype logistic regression model) and the multinomial logistic model for nominal response variables, and ends with Poisson regression models, negative binomial regression models, and zero-inflated models for count response data.

The third part includes Chapters 10–14, which discuss multilevel modeling for continuous, binary, ordinal, nominal, and count response variables. The fourth part covers Chapters 15 and 16, which focus on Bayesian techniques for generalized linear models and generalized linear mixed models, respectively. Chapter 15 introduces Bayesian inference and illustrates Bayesian logistic regression, ordinal regression, multinomial regression, Poisson regression, and negative binomial regression with the `rstanarm` and `brms` packages. Chapter 16 discusses Bayesian multilevel models for binary, ordinal, and count response variables with the `brms` and `MCMCglmm` packages. Please note that Bayesian multilevel multinomial models are introduced in Chapter 14.

Instructors may assign the first two chapters to students for review, or they may use the materials for an introductory class with a focus on a general introduction to R and data analysis. Chapter 3 can be assigned for a class on binary logistic regression models. Chapters 4–9 can be used as teaching materials for generalized linear models, and Chapters 10–14 can be used when introducing multilevel models for linear, binary, ordinal, nominal, and count response variables. In addition, Chapters 15 and 16 can be used as teaching materials for Bayesian generalized linear models and Bayesian generalized linear mixed models.

R, RSTUDIO, AND R PACKAGES

Why R? R is not only general-purpose statistical software but also a programming language environment. It is powerful, flexible, and freely available with rising popularity in various disciplines and research fields. R has extensive analytic capabilities and can be used to fit all the models covered in the book. I assume that readers are using R 4.1.0 and RStudio 1.4.1717. RStudio is a free, open-source integrated development environment (IDE) to make the use of R easier. Readers also need to install Rtools (version rtools40) before conducting Bayesian analysis with the `brms` package. I assume readers are using R on a Windows PC. If another operating system, such as macOS or Unix System, is used, readers may notice minor differences on screenshots. I use `typewriter font` (i.e., `Courier New`) for R commands, output, and variable names.

One strength of R is the extensibility with user-written or add-on packages. We often use more than one package to fit the same model, so readers can have their own choice. The following is a complete list of the add-on packages used in this book. Readers need to install them either by using the `install.packages()` function with the name of the package placed in quotation marks or by using the pull-down menu in RStudio or R. Other packages such as the `foreign`, `MASS`, `nlme`, and `nnet` packages are part of the R base distribution, so they do not need to be installed separately. If a package is removed from the CRAN repository, we can search and download it from the archive and then install it via the RStudio pull-down menu.

- `bbmle` (Bolker et al., 2021)
- `brms` (Bürkner, 2017)
- `car` (Fox & Weisberg, 2019)
- `dplyr` (Wickham et al., 2021)
- `ggeffects` (Lüdecke, 2018b)
- `ggplot2` (Wickham, 2016)
- `glmmTMB` (Brooks et al., 2017)
- `gmodels` (Warnes et al., 2018)
- `lme4` (Bates et al., 2015)
- `lmerTest` (Kuznetsova et al., 2017)
- `margins` (Leeper, 2021)
- `mclogit` (Elff, 2021)
- `MCMCglmm` (Hadfield, 2010)
- `mlogit` (Croissant, 2020)
- `mixor` (Archer et al., 2018)
- `nlme` (Pinheiro et al., 2021)
- `ordinal` (Christensen, 2019)
- `pastecs` (Grosjean et al., 2018)
- `pscl` (Jackman, 2020)
- `Rcmdr` (Fox, 2005)
- `rcompanion` (Mangiafico, 2021)
- `ResourceSelection` (Lele et al., 2019)
- `rstanarm` (Goodrich et al., 2020)
- `sjmisc` (Lüdecke, 2018a)
- `sjstats` (Lüdecke, 2021)

- `stargazer` (Hlavac, 2018)
- `texreg` (Leifeld, 2013)
- `tidyverse` (Wickham et al., 2019)
- `vcd` (Meyer et al., 2020)
- `VGAM` (Yee, 2010)

THE BOOK'S WEBSITE

A companion website for this book at **https://edge.sagepub.com/liu1e** contains resources for both instructors and students. The password-protected instructor site includes solutions to the end-of-chapter problems, and the datasets and R commands for the exercises. The open-access student site includes the datasets and R commands used in the book. The datasets are in the .dta format, one of the major data formats for the 2016 GSS. Other data formats, such as CSV, Excel, SAS, SPSS, and R, have also been created for students and instructors, and made available on the website.

1

R BASICS

OBJECTIVES OF THIS CHAPTER

This chapter begins with an introduction to the basics of R. The first section shows novice R users the interface, menus, and R toolbar. It introduces the RStudio and R Commander, the graphical user interfaces (GUIs), which make the use of R easier. It also introduces objects in R and R functions and arguments and shows how to install R and add-on packages, create script files, and import data. The second section shows readers several commonly used data types and various data management techniques, such as how to select cases and variables, create and recode variables, label categories for factors, and label variables. Further, it introduces `tidyverse` and how to use both the `dplyr` package in `tidyverse` and the `sjmisc` package for data management. The third section introduces basic graphic functions and the `ggplot2` package. After reading this chapter, you should be able to:

- Install, start, and exit R.
- Open existing data files.
- Enter commands, create a script file, and save output.
- Select cases and variables, create new variables, recode and label variables, and label values for categorical variables.
- Draw different types of graphs.

1.1 INTRODUCTION TO R

R is a programming language. It was developed by Ross Ihaka and Robert Gentleman in the 1990s based on the S programming language. After several years of development, R version 1.0.0 was officially released in 2000. R is now developed by the R Development

Core Team. It can be run on almost all operating systems across Windows, Mac, Linux, and Unix. R has different names. It is also called a program, a package, a system, or an environment. In this book, we treat R as a general-purpose statistical package and a programming language. Just like IBM SPSS, SAS, and Stata, R is also a general-purpose statistical tool for data management, data analysis, and graphing.

R is a powerful tool for data management, graphics, and data analyses. It is capable of conducting various statistical analyses, from basic statistical analyses to more complex models, such as generalized linear models, generalized additive models, multivariate analyses, time series, survival analysis, propensity score analysis, multilevel modeling, structural equation modeling, cluster analysis, machine learning, and Bayesian statistics. As an R user, you will be amazed at the capacities of R for statistical analysis. If I claim that R can do any statistical analysis, it may sound like I am exaggerating. However, if you name a modern statistical method, it is very likely that you can find an R package or function which has been developed for that method.

R has extensive programming capabilities. You can write packages with new functions which can be shared with other users in the R community. With the contributions of experts from various fields, new statistical techniques can be quickly implemented in R. For example, the user-written package `ordinal` was developed for ordinal regression models, the `VGAM` package was developed for generalized linear and additive models, and the `lme4` package was developed for mixed-effects models or multilevel models. We can install these user-written packages using the `install.packages()` function. After installation, these add-on packages can be executed in the same way as the base package in R. So, when functions of interest are unavailable in the base package in R, users can search online to see if they have been developed by other users.

R provides more than one solution for your statistical analysis. You may find multiple packages with similar functions for a statistical method. For example, you can use either the `clm()` function in the `ordinal` package or the `vglm()` function in the `VGAM` package to fit ordinal logistic regression models. There are also several packages developed for multinomial logistic regression models, such as the `vglm()` function in `VGAM`, the `multimon()` function in `nnet`, and the `mlogit()` function in `mlogit`. Therefore, you have a variety of choices to solve your problems. In addition, the results can be cross-validated if multiple packages are used for the same analysis.

R is an open-source programming language. This means that the source code is freely available to use. You can also modify the source code, create your own, and share it with your colleagues or in the R community.

R is free. It is free to download, install, and use. The benefit of having a free statistical package is self-evident.

1.1.1 Installing, Starting, and Exiting R

To install R on your computer, you need to download it from the Comprehensive R Archive Network (CRAN) at http://cran.r-project.org/. The CRAN is the online network for storing the R software, add-on packages, and documentation. Choose the link for the R installation file for Windows, Linux, or macOS operating systems. At the time of writing, the R version is R 4.1.0 (R Core Team, 2021). The installation of R is just as easy as the installation of any other program on these operating systems. For example, to install R on Windows, run the installation file by right clicking it and then select Run as Administrator. Then follow the steps to install it.

You can install 32-bit, 64-bit, or both versions on your computer. The advantage of the 64-bit version is that it can handle large data as long as your computer memory allows for it. Install 64-bit version if your computer supports 64-bit. To update R, you simply download and install the latest version following the procedures introduced above. You can either uninstall or keep the old version on your computer. In this book, we focus on R for Windows.

R can be started in two ways. First, you can run R by double-clicking the icon on the desktop. Second, you can start it by clicking the **Start** Menu on Windows, **All App**, and then **R**.

To exit R, you can either type the command `q()` and press **Enter** or use the pull-down menu. To use the menu, go to **File** and then click on **Exit**.

R is mainly a command-driven statistical package. To execute a command, you need to enter it on the command line and press the **Enter** key. Why do we still need to type commands in R instead of using point-and-click menus in other statistical packages? There is nothing wrong with using the graphic user interface (GUI) pull-down menus, but you may find that it will be more efficient to type commands. There are three reasons for this efficiency. First, it saves you time. For example, to run a simple regression analysis with a dependent variable `y` and an independent variable `x` in the `data1` dataset, you simply type:

```
lm(y ~ x, data = data1)
```

Second, it helps reproducible research. You can save all your commands to a script file so that you can replicate your analysis easily and share it with others. In addition, you can edit your script file for other analyses. For example, if we have three independent variables in a linear regression analysis, then we can modify the previous command as follows:

```
lm(y ~ x1 + x2 + x3, data = data1)
```

Third, it is extensible with add-on packages. You can use a variety of packages and functions in R which are not readily available on the GUI pull-down menus in other statistical packages. Due to R's extensive programming capacity, the number of add-on packages increases tremendously each year. You have more opportunities to find the right package to get your job done. In addition, you can write your own functions or edit the existing functions in R to make your work more efficient.

1.1.2 R at First Sight: R Console, Menus, and Toolbar

R Console

Once you start R, you will see the R Console windows (Figure 1.1).

In the R Console window, you need to type a command next to the > symbol and press the **Enter** key to execute it. You can copy a command from a Notepad and paste it here. You can also edit a command before execution. You can only execute one command at a time in this window, whereas you can run a series of commands via the script file which will be explained later in the chapter. R is interactive. After you execute a command, the output is displayed below. The more commands you type, the more results you get. The output can be copied and pasted into a text file or a Word document. It can also be saved into a text file, which will be explained in more detail next.

R Menus

R has seven pull-down main menus, including File, Edit, View, Misc, Packages, Windows, and Help. These menus provide basic tools and features that can be used by pointing and clicking.

File menu: The purpose of this menu is to help you with files. Options in this menu help you open and save a script file, load and save a workspace, load and save a history file, change directory, print results, and exit R.

FIGURE 1.1 ● R Console Window

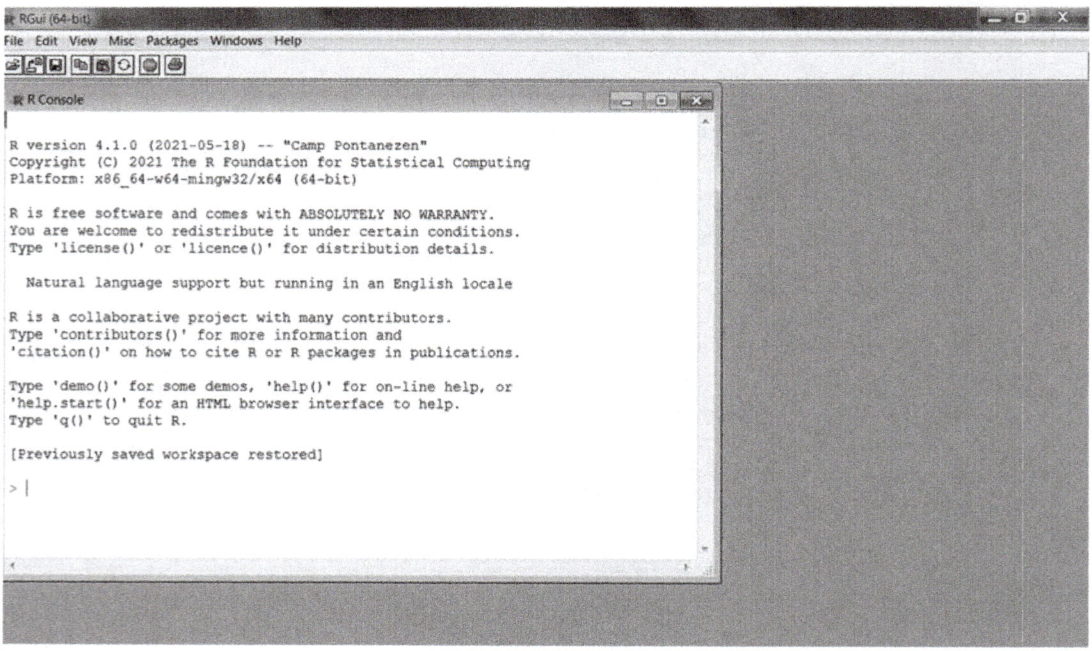

Edit menu: The Edit menu helps you copy texts from the R Console window and paste them in the same or another location. It also provides the Data Editor and the GUI Preference Editor.

View menu: The View menu shows you the tool bar and the status bar.

Misc menu: The Misc menu helps you stop the current computation, stop all computations, list objects, remove objects, and list search path.

Packages menu: The Packages menu helps you install, update, and load packages, select CRAN mirrors, and select repositories.

Windows menu: This menu helps you organize the R Console window. Several options include Cascade, Tile Horizontally, and Tile Vertically.

Help menu: This menu provides FAQ on R, R manuals in PDF, help on R functions, html help, search help, search for words in help list archives and documentation, and the links for R project and CRAN home pages, respectively.

Please note that R does not provide menus for data management, graphs, and statistics. If you need a GUI for such functions, you need to install an add-on package, the R Commander package (i.e., `Rcmdr`), which was developed by John Fox and his colleagues. An introduction to this package is provided in a later section.

R Toolbar

The toolbar, located below the main menus, comprises a set of icons. It helps you quickly access the most frequently used features. Familiarizing yourself with these icons will make the use of R more efficient.

These icons include Open Scripts, Load Workspace, Save Workspace, Copy, Paste, Copy and Paste, Stop Current Computation, and Print. Table 1.1 shows the icons of the toolbar, their titles, and their functions.

1.1.3 RStudio

RStudio (RStudio Team, 2020) is a free, open-source integrated development environment (IDE) for R. An IDE basically is a program that makes programming easier. RStudio includes an R console, a text editor, and tools for workspace, history, plots, packages, and help. RStudio not only makes R look fancy, but also provides a variety of tools which make R more user-friendly and more convenient.

You can download RStudio at https://rstudio.com/. At the website, go to Products, click on RStudio, and then choose the open-source edition. Follow the steps to install it on your

TABLE 1.1 ● Icons of the Toolbar, Their Titles, and Their Functions

Icon	Icon Title	Functions
	Open Scripts	Opens an existing script file (i.e., .R file)
	Load Workspace	Opens an existing workspace file (i.e, .RData file)
	Save Workspace	Saves the current workspace file (i.e, .RData file) to your computer
	Copy	Copies the selected text
	Paste	Pastes the copied text
	Copy and Paste	Copies and pastes the selected text to the current command line
	Stop Current Computation	Stops executing the current program
	Print	Prints your R output

computer. You need to install R first and then install RStudio. At the time of this writing, the RStudio version is RStudio 1.4.1717-3. Figure 1.2 shows the screenshot of RStudio.

- The upper left panel is the R script editor, an enhanced text-editor with highlighting. You can type R commands and create a script file.
- The lower left panel is the R console. The output is displayed after you execute R commands in the R script editor. You can also type and execute R commands in the R console.
- The upper right panel contains the workspace and history of commands. The workspace displays the existing and temporary datasets and other objects which you create. The history tab shows the history of R commands.
- The lower right panel shows the tabs for files, plots, packages, help, and viewer. The Files tab helps you see file directories and manage files; the Plots tab displays the graphs you create; the Packages tab displays the installed packages; the Help tab provides documentation and other help files or pages for R functions and packages; and the Viewer tab does not provide much information, but it can be useful for a particular package to display local web content.

FIGURE 1.2 ● Panels of RStudio

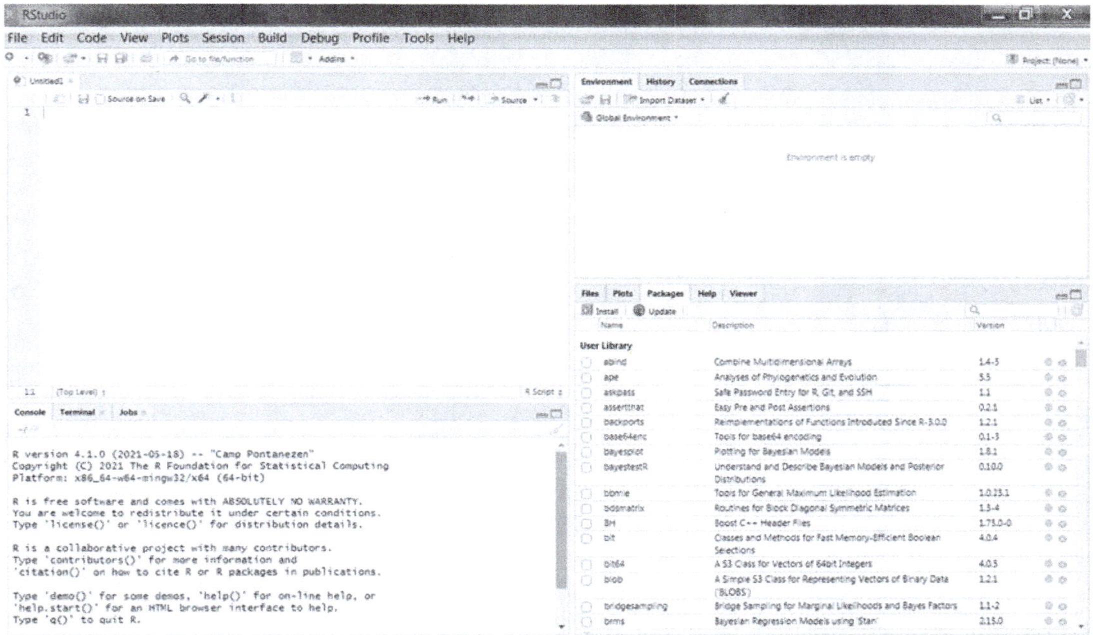

1.1.4 R Commander

R is a programming language, which does not have a well-built GUI system itself. The R Commander package (i.e., `Rcmdr`), which was developed by John Fox and his colleagues (Fox, 2005, 2017), provides a GUI for R. It is a menu system for reading data and conducting statistical analysis, which facilitates users to learn about the program. Users can use point-and-click menus to familiarize themselves with the features of importing data, recoding variables, making graphics, and choosing a variety of methods for statistical analysis. It is useful for novices or users who are familiar with other statistical packages such as IBM SPSS. It serves as a nice transitional tool to ease the anxiety of importing data and conducting data analysis without programming.

To install the `Rcmdr` package, use the following command.

```
install.packages("Rcmdr", dependencies = TRUE)
```

Choose the mirror close to you and then install it. To use the package, you need to load it using the `library(Rcmdr)` command. The screenshot is displayed in Figure 1.3.

The R Commander window provides a script file window at the top, an output window in the middle, and a messages window at the bottom. Just like IBM SPSS, you can load a dataset and then choose the statistical method of interest. You can also use menus to complete data management tasks, such as creating, deleting, and recoding variables, adding

FIGURE 1.3 ● Screenshot of R Commander

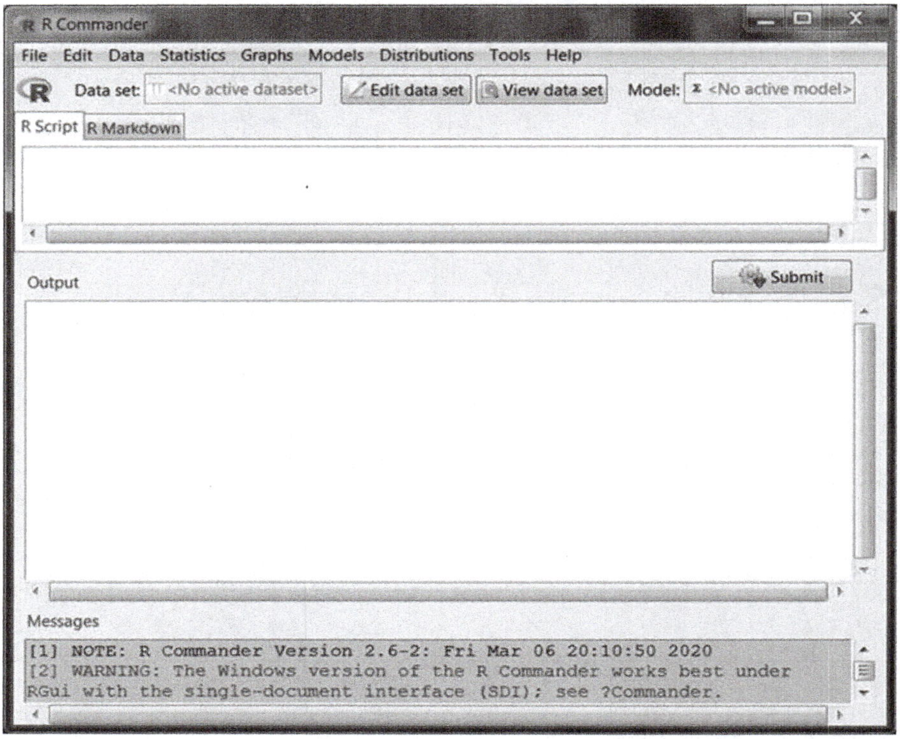

observations, and converting variables to factors. When you perform an analysis using the R Commander package, it displays the corresponding command in the script file window and the output in the output window. You can then save the command for future use.

There is nothing wrong with using the GUI pull-down menus to get the job done. As you become an experienced R user however, you will find that it will be more efficient to type commands and you will have a wide range of choices of packages.

1.1.5 R Base Package and Add-on Packages

The R base package is on your computer once R is installed. It contains many built-in functions for statistical analysis and graphing, such as `mean()` for means, `lm()` for linear regression models, and `glm()` for generalized linear models. The add-on packages are the user-written packages, also known as the third-party packages. They complement the base package by extending the capabilities of R. They are free and can easily be installed. They are normally stored on the CRAN.

You need to install the package of interest either by using the `install.packages()` function with the name of the package placed in quotation marks or by using the pull-down menu in RStudio or R. For example, the `ordinal` package is the user-written package for ordinal regression models. To install it by typing the command in R, you need to type: `install.packages("ordinal")`. Choose the mirror close to you and then follow steps to install it. If you want to install the package in RStudio, you type the `install.packages("ordinal")` command in the script editor window and run the command by highlighting it. You only need to install it once. To see the installed packages in RStudio, click on the **Packages** tab in the lower right panel.

You also need to load it after the installation by using the `library(ordinal)` command. It needs to be loaded again if you restart the R session. An error message will be displayed if you load a package which has not been installed. As of the writing, there are more than 15,000 packages available on the CRAN. The rapidly increasing add-on packages promote R's popularity.

1.1.6 Objects in R

R is an object-oriented language. In other words, everything in R can be treated as an object. Therefore, a data frame, a graph, a function, and a fitted model are all objects in R. For novices, this seems a confusing concept if no examples are provided. To clarify, let us see an example. We assign the value 5 to an object named x with the following command.

```
x <- 5
```

The assignment operator, <-, which is composed of a less than sign (<) immediately followed by a hyphen (-), is used to assign the value to the object. The value being assigned to the object is on the right-hand side of the assignment operator and the object name is on the left-hand side. We can also use the equal sign, =, as the assignment operator.

We can assign a linear model to an object named `mod` using the following command.

```
mod <- lm(y ~ x)
```

In each of the two examples, the object has a name and content. Typing the object name shows the content of the object. Although you can use any name you like, it is always good practice to have a meaningful and concise object name. Having a meaningful name helps you understand the object. Also, it is preferable to have a concise name rather than a lengthy one. An object name can be a combination of letters, numbers, and dots. There are several basic rules for naming objects.

1. Object names should not start with a number. They should start with a letter or dot. For example, 1.mod is not a correct object name, but mod.1 is.

2. Object names allow letters, numbers, underscore characters (_), and dots (.). You should avoid space and special characters such as @, #, $, %, and &. For example, you can have an object named `mod.1` or `mod_1`, but not `mod$1` since $ is an operator for accessing a variable.

3. R is case-sensitive. This rule also applies to object names, so the object `mod.1` is different from `Mod.1`.

1.1.7 Functions and Arguments in R

Functions in R are just like commands or procedures in other statistical packages. They are simply a set of instructions to perform a specific task. The form of an existing function in R is the function name with an argument or a set of arguments within the parentheses. In a programming language, arguments are inputs or parameters in a function, which are like options in other statistical packages. Let us see an example of a built-in function, `mean()`. To compute the mean of the variable x1, we use the `mean(x1)` command. In the command, `mean()` is the function for means and `x1` is the argument or input. Executing the command returns the output. When there are multiple arguments, they are separated by commas. For example, in the `mean(x1, na.rm = TRUE)` command, there are two arguments. One argument is the variable object and the other argument `na.rm = TRUE` is used to remove missing values. In a programming language, when executing a code in a function, we also say that we call a function. A call to that function just means that we execute that function. In the two examples above, we call the `mean()` function with the parentheses containing the arguments.

If you would like to create a new function, you can use the `function()` function. You need to assign the function a new name and have a body of statements. The body of a function is always enclosed by curly braces.

1.1.8 R Script Files

If you use the Console window to enter commands, you can enter one piece at a time. An R script file is a text file containing a list of R commands. R script files can help you put a list of commands in one file and execute them together as a batch. Using script-files helps you to organize commands, keep a record, and understand what you have done when you need them in the future. It is also helpful when you collaborate with other researchers on a research project. Your colleagues can simply replicate the analyses using the script files you provide, and they can modify them for new analyses. It will save you time when you need to replicate your statistical analyses. If you are an instructor, these script files are also helpful when preparing your instruction and for grading students' assignments.

To create a new script file in R, go to **File** and then click on **New Script**. You will see a new Untitled-R Editor window. To open an existed script in R, go to **File** and then click on **Open Script**. To run the script, you can highlight the commands, right click it and choose **Run line or selection**. You can also click on **Select all** and then run all the commands at once.

To save your script file in R, go to File and then click on Save as. Save the script with an ".R" extension.

RStudio has a highlighting text editor for script files, also known as a script editor. To create a new script file in RStudio, go to **File**, **New File**, and then click on **New Script**. One advantage to using the script editor in RStudio is that RStudio automatically loads the latest script when you open the program. Therefore, if you forget to save the script file in the last session, you can still do so. To run the script in RStudio, highlight the commands and click on **Run** at the upper right corner of the script editor.

In this book, all the R commands are saved as R scripts files in RStudio. The R scripts are provided at the end of each chapter. They are also available online to download so you can replicate all the analysis.

You should save the script file to the current working directory. A working directory is the place where R scripts, data files, output, and graphs are stored. Although it is not required to set a working directory to run R, it is cumbersome to type the full path with the file name to access that file each time. By setting the working directory, you can access the file folder directly without providing the full file path. Use the `setwd(" ")` function with the full path in the quotation marks to set the working directory. For example, the `setwd("C:/CDA")` command sets the existing folder named `CDA` in the `C` drive as the working directory. You need to create the folder first before setting is as the working directory. To check your current working directory, type the command `getwd()`.

A good habit when you create a script file is to have comments in the file. Comments in an R script file are lines beginning with a pound sign or a hash symbol (#). The comments may include time, project name, who wrote it, and for what purposes. The clearer your comments are, the better your documentation will be. The comments immediately following # in the script files will not be executed but will be displayed in your output.

1.1.9 How to Open an Existing Dataset via the Command Line or the Menus in RStudio

A dataset is called a data frame in R. A data frame is a rectangular-shaped dataset with variables in columns and observations in rows. The data frame in R is similar to the dataset in an Excel spreadsheet or statistical packages such as IBM SPSS, SAS, and Stata. R has other data structures, such as vectors, lists, matrices, and factors, which will be introduced in the following section.

If you work on a research project using R, the first question you may ask is "How can I open or import an existing IBM SPSS, SAS, Stata, or Excel dataset into R?" For novices, importing an existing dataset into R can be a daunting task or frustrating experience. Since R is a programming language, you cannot double click the dataset to open it. If you are not using the R Commander package (i.e., `Rcmdr`), you cannot open the dataset via the R pull-down menu, either. One solution is to use the `foreign` package which has functions to read SAS, SPSS, Stata, Systat, and other formats of data into R. The other solution is to open the dataset via the pull-down menus in RStudio. We will introduce both methods next.

The `foreign` package is now part of the R base installation, so you do not need to install it. However, if you would like to update it to the latest version, you can use the `install.packages("foreign")` command to install it. To use this package, we use the `library(foreign)` command to load it first.

Importing a Stata Dataset

In the following example, we would like to directly open the GSS:2016 dataset, which is a Stata data file with the .dta extension format. We use the `read.dta()` function with the file name enclosed in quotation marks. The basic syntax is as follows.

```
read.dta(file = " ")
```

To read a Stata 13 or higher dataset, use the `read.dta13()` function in the `readstata13` package. You need to install the `readstata13` package first by typing `install.packages("readstata13")` since it is a user-written package. After installation, load the package by typing `library(readstata13)`.

Importing a SPSS Dataset

To read a SPSS dataset with the .sav extension, use the `read.spss()` function.

Importing a SAS Dataset

To read a SAS permanent dataset, use the `read.ssd()` function; to read a SAS transport format, use the `read.xport()` function.

We can also use the `read_sas()` function in the `haven` package to open a SAS dataset. You need to install the `haven` package first by typing `install.packages("haven")` and then load the package by typing `library(haven)`.

Importing an Excel Dataset

To read an Excel dataset with the .csv extension, use the `read.table()` function or the `read.csv()` function.

To read an Excel dataset with the .xls extension, use the `read.xls()` function in the `gdata` package. You need to use `install.packages("gdata")` to install the package first and then load it with the `library(gdata)` command.

Importing a Delimited Text Dataset

To read an ASCII text dataset with the .txt extension, use the `read.table()` function.

R has its own data file format which has the .RData extension. You can use the `save()` function to save the data. For example, in the `save(data1, file = "data1.RData")` command, `data1` is the data object name and `file = "data1.RData"` specifies the data format.

Importing a Dataset Through the Menus in RStudio

In addition to typing the commands in the command line, it is also easy and handy to import an existing dataset through the menu system in RStudio with the following steps:

- First, go to **File** on the RStudio main menu and then click on **Import Dataset**. Select **From Text**, **From Excel**, **From SPSS**, **From SAS**, or **From Stata** according to your data file format and then click on it. You may be prompted to install the required packages. After you click on **Yes**, RStudio will automatically install them for you (Figure 1.4).
- Second, in the pop-up window, click on **Browse** to locate the dataset on your computer, select it, and click on **Open** (Figure 1.5).

FIGURE 1.4 ● Screenshot of Selecting a Dataset via RStudio

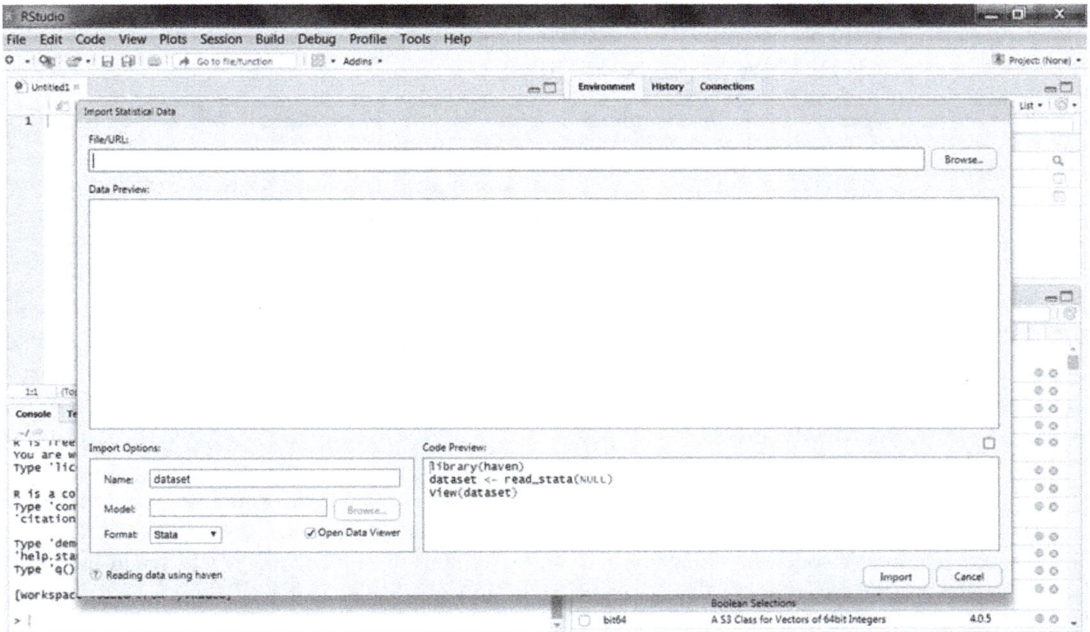

FIGURE 1.5 ● Screenshot of Opening a Dataset via RStudio

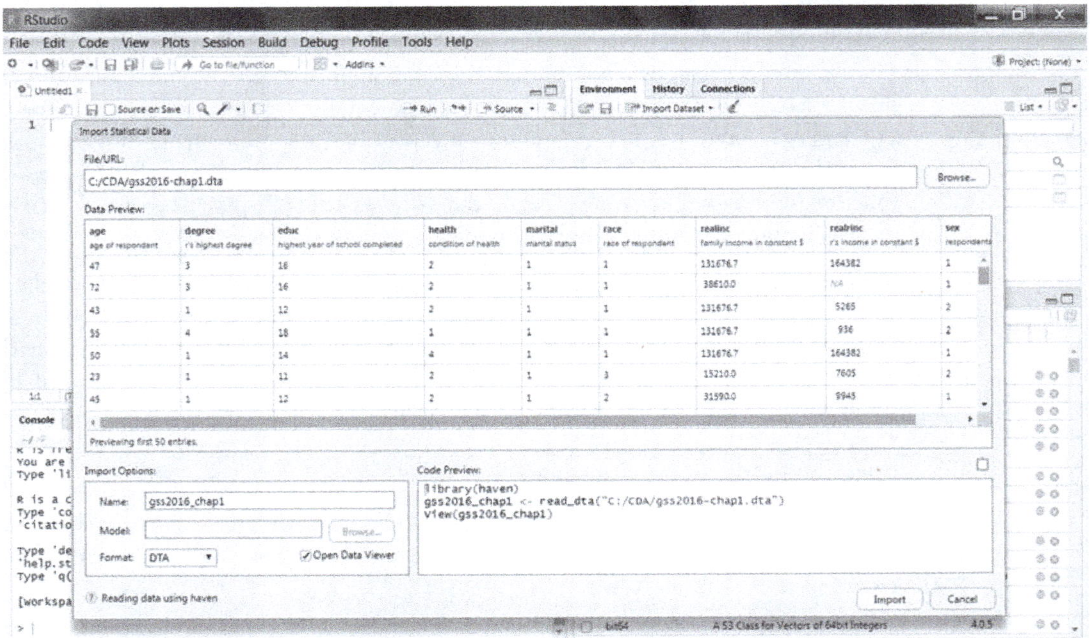

- Third, once previewing the dataset and clicking on **Import**, you will see the imported dataset shown in the upper left panel and the R command echoed in the lower left R console.

1.1.10 How to Save R Output

When you execute a command in the R Console window, R will display the output following the command. There are several ways to save the output. Three ways are introduced as follows. First, a simple way is to copy and paste it. You can copy the selected output as text and paste it into a Word document or a text file. If you paste it into a Word document, then you may specify the font as `Courier New` and set the font size to 9 or smaller to show the results properly. Second, we can use the `sink()` function to save the results in a plain text format. This function needs to be used before we conduct statistical analysis, so we can save all of the results at the beginning of your R session. For example, the `sink("filename.txt")` command saves the output to a file named `filename.txt` in the current working directory. If you are uncertain of your working directory, type `getwd()` in the **Command** window. You can use the `setwd(" ")` function with the path in the quotation marks to change the working directory. Please note that output will not be shown in the R console when we use the `sink()` function. Third, you may use the `knitr` package (Xie, 2015, 2021) to save the R code, the notes, the output, and graphs in one document which can be a Word document, a pdf file, a presentation, or a web page. This package is useful for reproducible research since the R code, the analysis, and the reporting are well documented and easily replicated.

1.1.11 What If I Have a Question? How Do I Get Help?

We have many questions when we use R. Thankfully, R provides rich resources for users in different ways.

First, R itself provides help with the documentation of packages and functions. You can use the `help()` function to display the documentation for a package or a function. For example, the `help(lm)` command provides the documentation for linear regression models. A shortcut is to type `?` followed by the name of a package or function. For example, using `?lm` is the same as using `help(lm)` to obtain the documentation for linear regression models.

Second, R has online help and search facilities, which provide a wide range of help. If you have a question for a particular command, you can either click on **Help** from the menu or type `help.search()` with the function name enclosed in quotation marks within the parentheses or type `??` followed by the function name. For example, if you are looking for the help files for the `glm()` function for generalized linear models, you could type `help.search("glm")`. The screenshot shown in Figure 1.6 shows the help page for the `glm()` function.

The same screenshot can be reached if you use the pull-down menu. Go to **Help** on the menu, click on **Search help**, and then enter `glm` in the box. You can see the search results after clicking on **OK**.

FIGURE 1.6 ● Screenshot of the `help.search("glm")` Command

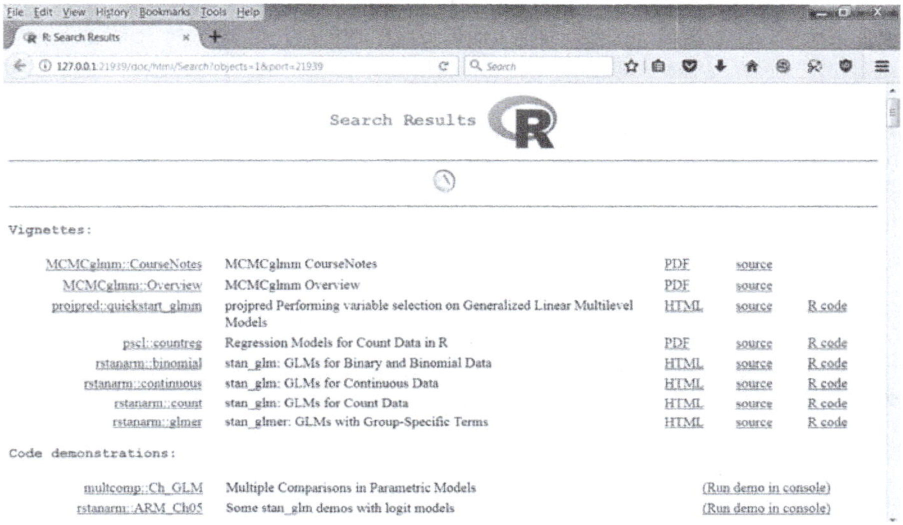

Third, if you would like to get a more detailed introduction to a package, use the `help(package = " ")` command with the package function name enclosed in quotation marks. To look for help files for a function in a package, include both the function name and the package name in parentheses following the `help()` function. For example, `help(vglm, package = "VGAM")` command displays the documentation for the `vglm()` function.

Fourth, to search the web for help, use the `RSiteSearch(" ")` function with the key word enclosed in quotation marks. This function provides a keyword search.

Fifth, go to http://www.r-project.org/ to access the R manuals. R has provided complete PDF documentation. Users can also have access to these manuals directly from the menu or the installation folder within R.

Finally, you can get help from the following two websites:

- R seek: Search engine for R topics at https://rseek.org
- https://stackoverflow.com provides a general discussion forum for R users.

1.2 R DATA STRUCTURES: VECTORS, MATRICES, DATA FRAMES, AND LISTS

R includes several commonly used data structures, such as vectors, factors, matrices, data frames, and lists. We briefly discuss each data structure or data type as follows.

1.2.1 Vector

Numeric Vector

A vector is a sequence of data elements of the same type. There are several different types of vectors, including numeric, character, complex, and logical vectors. In this section, we introduce two basic types of vectors, numeric and character vectors, which are a list of numeric values and characters or strings, respectively. We use the `c()` function, the concatenation function, to combine a list of numeric values or characters. For example, we would like to create a vector for a numeric variable named `age` using the function. The commands are as follows.

```
age <- c(47, 72, 43, 55, 50, 23, 45, 71, 86, 33)

age
```

In the first command, the `c()` function is used to combine the ages of 10 people and the created vector is assigned to an object named `age`. The second command is the vector name `age`. By typing the vector name, we get all the values of the vector. The output is as follows.

```
> # Use c() to create vectors
> age <- c(47, 72, 43, 55, 50, 23, 45, 71, 86, 33)
> age
  [1] 47 72 43 55 50 23 45 71 86 33
```

In the output, the first line shows the `c()` function with 10 values within the parentheses and the assigned object name `age`. The second line shows the vector name `age`. The third line shows the 10 values for `age`. `[1]` means that this line begins with the first value.

Character Vector

We can also use the `c()` function to combine a list of characters or strings for a character vector. For example, we want to create a vector for a character variable named `gender` by using the `c()` function. The characters are placed in quotation marks in the command as follows.

```
gender <-
c("male", "male", "female", "female", "male", "female", "male", "male", "female",
"female")
```

```
> gender <-
c("male", "male", "female", "female", "male", "female", "male", "male", "female", "female")
```

The command `gender` lists all the characters of the vector. The output is as follows.

```
> gender
 [1] "male"   "male"   "female" "female" "male"   "female" "male"   "male"
 [9] "female" "female"
```

In the output, `[1]` means that this first line begins with the first character value and `[9]` means that the second line begins with the ninth value.

Indexing a Vector

We can use indexing to access an individual element of a vector. Following the vector name, we use square brackets to place the position number. For example, the `age[5]` command references the fifth value of the `age` vector, which is `50`. The output is as follows.

```
> # Index a vector
> age[5]
[1] 50
```

In the second example, the `gender[3]` command references the third value of the `gender` vector, which is "`female`." The output is as follows.

```
> gender[3]
[1] "female"
```

Factor

A factor is a vector for categories or levels. In other words, a factor or a factor variable is a categorical variable with multiple levels. It includes categories or levels which are internally coded as integer values with labels for the corresponding categories. For example, the factor variable `health` has four categories with labels, including 1 = poor health, 2 = fair health, 3 = good health, and 4 = excellent health. We use the `factor()` function to create a factor or categorical variable. In this example, we use the following command.

```
health <- factor(health, levels = c(1, 2, 3, 4), labels = c("poor health", "fair
health", "good health", "excellent health"))
```

In the command, `health` is the variable name, `levels = c(1, 2, 3, 4)` specifies the four categories, and `labels = c("poor health", "fair health", "good health", "excellent health")` specifies the labels with the characters enclosed in the quotation marks. The `levels =` argument is optional and can be omitted.

A factor can be either a nominal or ordinal categorical variable. It is an ordinal variable when the categories are ordered. With the `ordered = TRUE` argument, we create an ordered factor as follows.

```
health <- factor(health, levels = c(1, 2, 3, 4), labels = c("poor health", "fair
health", "good health", "excellent health" ), ordered = TRUE)
```

1.2.2 Matrix

A matrix in R is a data structure with columns and rows in a two-dimensional rectangular layout with the same data type. Therefore, we cannot mix numeric and character data in a matrix. We use the `matrix()` function to create a matrix. For example, we use the `m1 <- matrix(1:8, nrow = 4, ncol = 2)` command to create the following 4 by 2 matrix which includes 4 rows and 2 columns. In the command, `1:8` represents the numbers from 1 to 8, which is a shortcut for the vector including the numbers from 1 to 8; `nrow = 4` specifies that the row number is 4; and `ncol = 2` specifies that the column number is 2. This matrix will be filled by columns since the `byrow=` argument is not specified. The resulting matrix is assigned to an object named `m1`.

```
> # Use matrix() to create matrices
> m1 <- matrix(1:8, nrow = 4, ncol = 2)
> m1
        [,1]    [,2]
[1,]       1       5
[2,]       2       6
[3,]       3       7
[4,]       4       8
```

If we would like the matrix to be filled by rows, we use the `byrow = TRUE` argument in the command. By running the `m2 <- matrix(1:8, nrow = 4, ncol = 2, byrow = TRUE )` command, we get the following output.

```
> m2 <- matrix(1:8, nrow = 4, ncol = 2, byrow = TRUE)
> m2
        [,1]    [,2]
[1,]       1       2
[2,]       3       4
[3,]       5       6
[4,]       7       8
```

Indexing a Matrix

We also use square brackets to reference elements in matrices. The row and column indices are enclosed in square brackets and are separated by a comma like this: [row, column]. For example, we use the `m1[2, 1]` command to access the element for the second row and the first column in the `m1` matrix. The output is as follows.

```
> # Index a matrix
> m1[2, 1]
[1] 2
```

The `m1[ , 2]` command accesses all the elements for the second column in the `m1` matrix. When the row index is not specified, all the rows in the matrix are selected. The output is as follows.

```
> m1[ , 2]
[1] 5 6 7 8
```

When the column index is not specified, all the columns are selected. The `m1[4, ]` command accesses all the elements for the fourth row in the `m1` matrix.

```
> m1[4, ]
[1] 4 8
```

1.2.3 Data Frame

As introduced in the previous section, a data frame in R is a rectangular-shaped dataset with variables in columns and observations in rows. It is also referred to as a dataset in other statistical packages. A data frame can include different types of variables or vectors, such as numeric variables, character variables, and factor variables. We use the `data.frame()` function to create data frames. When we work on research projects, in most situations we work on data frames.

In the following example, we use the `gss <- data.frame(age, gender)` command to create a data frame named `gss`. In the command, the two variables `age` and `gender` are separated by a comma.

```
> # Use data.frame() to create a data frame
> age <- c(47, 72, 43, 55, 50, 23, 45, 71, 86, 33)
> gender <-
c("male", "male", "female", "female", "male", "female", "male", "male", "female", "female")
> gss <- data.frame(age, gender)
> gss
     age   gender
1     47     male
2     72     male
3     43   female
4     55   female
5     50     male
6     23   female
7     45     male
8     71     male
9     86   female
10    33   female
```

Several functions such as `str()`, `dim()`, and `names()` are useful to inspect a data frame. In this example, we use the `str()` function to display the structure of the `gss` dataset. The output shows that there are 10 observations and two variables in the dataset. The names and types of the two variables are shown next.

```
> str(gss)
'data.frame': 10 obs. of 2 variables:
$ age : num 47 72 43 55 50 23 45 71 86 33
$ gender: Factor w/ 2 levels "female","male": 2 2 1 1 2 1 2 2 1 1
```

We can also use the `dim()` function to display the dimension of the dataset.

```
> dim(gss)
[1] 10 2
```

The output also shows that there are 10 observations and two variables in the dataset. We use the `names()` function to display the names of the variables in the dataset.

```
> names(gss)
[1] "age"    "gender"
```

Accessing a Variable in a Data Frame

After we create or import a data frame, variables in the data frame are not automatically available to us for analysis. We can use the dollar sign (`$`) to access a variable from a data frame by placing `$` between the name of the data frame and the variable of interest. For example, to access `age` in the `gss` data frame, we use `gss$age`. The output is as follows.

```
> # Reference a variable in a data frame
> gss$age
  [1] 47 72 43 55 50 23 45 71 86 33
```

Useful Functions for Descriptive Statistics

We can use several useful functions to conduct basic descriptive statistics. These functions include the `mean()` function, the `sd()` function, the `min()` function, the `max()` function, and the `length()` function to obtain the mean, standard deviation, maximum, minimum, and number of observations of a variable, respectively. In the following example, we use these functions to compute descriptive statistics for `age`.

```
> mean(gss$age)
[1] 52.5
> sd(gss$age)
[1] 19.10352
> min(gss$age)
[1] 23
> max(gss$age)
[1] 86
> length(gss$age)
[1] 10
```

To access a variable in a data frame, we can also use the `attach()` function to attach the data frame. In this way, we do not need to repeatedly type the name of the data frame followed by the `$` operator. If we attach more than one dataset with the same variable name, the variable in the first dataset will be masked or replaced by the second one or the newest one with the same name. Therefore, once completing the analysis on one dataset, it is a good practice to use the `detach()` function to detach it from the R session.

1.2.4 List

A list is simply a combination of objects. We can combine a vector, a matrix, and a data frame into a list. We use the `list()` function to create a list. In the following example, we use the `list.gss <- list(age, m1, gss)` command to create a list named `list.gss` with three components. In the command, there are three components separated by commas: `age` is the vector, `m1` is the matrix, and `gss` is the data frame. The output is as follows.

```
> # Use list() to create a list
> list.gss <- list(age, m1, gss)
> list.gss
[[1]]
 [1] 47 72 43 55 50 23 45 71 86 33

[[2]]
        [,1]    [,2]
[1,]       1       5
[2,]       2       6
[3,]       3       7
[4,]       4       8

[[3]]
      age    gender
1      47      male
2      72      male
3      43    female
4      55    female
5      50      male
6      23    female
7      45      male
8      71      male
9      86    female
10     33    female
```

We can add the names for the three components in the list. The output is as follows.

```
> list.gss2 <- list(age = age, matrix1 = m1, gssdata = gss)
> list.gss2
$`age`
  [1] 47 72 43 55 50 23 45 71 86 33


$matrix1
        [,1]    [,2]
[1,]       1       5
[2,]       2       6
[3,]       3       7
[4,]       4       8


$gssdata
      age    gender
1      47      male
2      72      male
3      43    female
4      55    female
5      50      male
6      23    female
7      45      male
8      71      male
9      86    female
10     33    female
```

1.3 DATA MANAGEMENT

Before conducting data analysis, we need to work on various data management tasks to make the data ready for analysis. This section introduces some commonly used functions for basic data management tasks, such as selecting variables and observations, creating a new variable, recoding a variable, dummy coding, reverse coding, labeling a variable, labeling values for a factor variable, dealing with missing values, combining data, reshaping data, and converting data types. We also briefly introduce the `dplyr` (Wickham et al., 2021) and `sjmisc` packages (Lüdecke, 2018a) and the pipe operator `%>%` for data management.

1.3.1 Selecting Variables

To select variables from a dataset, we first use the indexing method with square brackets following the name of the dataset. The basic syntax is `dataset[, c("var1", "var2", "var3")]`. In the command, we use the `c()` function to select variables with the variable names in quotation marks. In the square brackets, rows are not specified since all rows are selected.

The second method is to use the `subset()` function to select variables. The basic syntax of the function is `subset(dataset, select = c(var1, var2,` var3)). In the command, we first specify the data frame named dataset. Then we use the select = c(var1, var2, var3) argument to select variables of interest.

The third method is to the use the select() function in the dplyr package. You need to install the package first by typing install.packages("dplyr") and then load it with the library(dplyr) function. The basic syntax is select(dataset, var1, var2, var3). In the command, we first specify the dataset and then specify the three variables which are separated by commas. The quotation marks are not needed for each variable. Examples of using the dplyr package are introduced in Section 1.4.

1.3.2 Selecting Observations

To select observations from a dataset, we again use the indexing method with square brackets. For example, to select the first 100 observations in a dataset, we use the generic command, dataset[c(1:100),]. In the command, we use the c() function to select observations from 1 to 100. In the square brackets, following the comma, columns or variables are not specified since all variables are selected. We can also specify conditions using the indexing method. For example, in the dataset [age > 50 & gender = = "female",] command, we select the observations for age larger than 50 and select observations for females. The & sign means that both conditions are met. The double equal sign "==" is different from the single equal sign "=". The former sign is one of the logical operators.

Another method to select observations is to use the subset() function. The generic structure of the function is subset(dataset, age > 50 & gender = = "female"). In the command, we first specify the dataset. Then we use the age > 50 & gender = = "female" argument to select observations.

The third method is to the use the filter() function in the dplyr package. You need to install the package first by typing install.packages("dplyr") and then load it with the library(dplyr) function. The syntax is filter (dataset, age > 50 & gender = = "female").

1.3.3 Selecting Observations and Variables

To select observations and variables simultaneously from a dataset, we use the same three methods introduced above. For example, with the indexing method, the dataset[age > 50 & gender = = "female", c("var1", "var2", "var3")] command selects the observations for age larger than 50 and observations for females for the three variables. The same results can be obtained using the subset(dataset, age > 50 & gender = = "female", select = c(var1, var2, var3)) command. In the command, we first specify the data frame, and then specify the conditions for selecting observations, and finally specify the selected variables.

If we use the dplyr package, we need to first use filter(dataset, age > 50 & gender = = "female") to select observations and then use the select(da-taset, var1, var2, var3) command to select the three variables.

1.3.4 Creating a New Variable

When you work on a research project, you often need to create new variables, recode an existing variable to a new variable, or combine several variables into one variable. There are also situations when you need to label a variable or label values for a categorical variable. In this section, we will briefly introduce commands to fulfill these tasks.

The assignment operator, `<-` , is used to create a new variable.

```
var2 <- var1
```

This command creates a variable `var2`, which is the same as the original variable `var1`.

```
var3 <- var1 + var2
```

This command creates a variable `var3`, which is the sum of the two variables `var1` and `var2`.

Another method is to the use the `mutate()` function in the `dplyr` package. If we would like to create a variable which is a sum of the two other variables, the basic syntax is `mutate(dataset, var3 = var1 + var2)`. In the command, the first argument is `dataset` and the new variable `var3` equals the sum of `var1` and `var2`.

Now, let us see an example using the General Social Survey 2016 (GSS 2016) dataset. I would like to create a new variable `realinc1`, which is the variable `realinc/10000`. We first load the `foreign` package, import the data with the `read.dta()` function, and then attach the dataset. The `realinc1 <- realinc/10000` command is used to create the new variable `realinc1`.

```
> # Remove all objects
> rm(list = ls(all = TRUE))
# Import the Stata data with the foreign package
> library(foreign)
> chp1 <- read.dta("C:/CDA/gss2016-chap1.dta")
> attach(chp1)
> # Create a new variable
> realinc1 <- realinc/10000
```

The second method is to use the `mutate()` function in the `dplyr` package. We first load the package and then use the `chp1.n <- mutate(chp1, realinc2 = realinc/10000)` command. In this command, `chp1` is the dataset and `realinc2` is the new variable, which equals `realinc/10000`.

```
> # Create a new variable using mutate()
> library(dplyr)
> chp1.n <- mutate(chp1, realinc2 = realinc/10000)
```

1.3.5 Recoding a Variable

There are several ways to recode a variable. Three methods are introduced in this section. For example, we would like to create a new categorical variable `SES`, according to the family income `realinc`. The first method is to use the indexing method with conditions within square brackets ([ ]). We enter the following five commands:

```
> # Recode a variable
> chp1$SES <- NA
> chp1$SES[chp1$realinc >= 0 & chp1$realinc <= 11114] <- 1
> chp1$SES[chp1$realinc  >= 11115 & chp1$realinc <= 25739] <- 2
> chp1$SES[chp1$realinc  >= 25740 & chp1$realinc <= 38609] <- 3
> chp1$SES[chp1$realinc >= 38610] <- 4
> chp1$SES <- factor(chp1$SES)
```

The first command creates a new variable `SES` with `NA` for the missing values. The object `chp1$SES` refers to the variable `SES` in the dataset `chp1`. The second command creates the level 1 for `SES` if the value of `realinc` is equal to or greater than 0 and less than 11,114. The third command creates the level 2 for `SES` if the value of `realinc` is equal to or greater than 11,115 and less than 25,739. The fourth command creates the level 3 for `SES` if the value of `realinc` is equal to or greater than 25,740 and less than 38,609. The fifth command creates the level 4 for `SES` if the value of `realinc` is equal to or greater than 38,610.

We then use the `factor()` function to convert `SES` into a factor variable with the command `chp1$SES <- factor(chp1$SES)`. The `table()` function is used to display the frequencies for the four categories.

```
> # Create a factor
> chp1$SES <- factor(chp1$SES)
> table(chp1$SES)
  1      2      3      4
394    439    307    565
```

To convert `SES` to an ordered factor variable with labels, we use the command `chp1$SES <- factor(chp1$SES, labels = c("low SES", "low-middle SES", "upper-middle SES", "high SES"), ordered = TRUE)`. In the command, `labels = c("low SES", "low-middle SES", "upper-middle SES", "high SES")` specifies the labels and `ordered = TRUE` specifies that the variable is ordinal. The `table()` function is used to display the frequencies for the four categories.

```
> # Create a factor with labels
> chp1$SES <- factor(chp1$SES, labels = c("low SES", "low-middle SES", "upper-middle
SES", "high SES"), ordered = TRUE)
> table(chp1$SES)

        low SES     low-middle SES     upper-middle SES     high SES
            394                439                  307          565
```

The second method is to use the `recode()` function in the `car` package (Fox & Weisberg, 2019). You need to install the package first by typing `install.packages("car")` and then load it by using the `library(car)` function. Since the `car` package also contains a function with the same name as that in the `dplyr` package, we use `car::recode` to access the `recode()` function in the `car` package.

```
> # Recode a variable using recode() in the car package
> install.packages("car")
> library(car)
Loading required package: carData

Attaching package: 'car'

> chp1$SES2 <- car::recode(chp1$realinc, "0:11114 = 1; 11115:25739 = 2; 25740:
38609 = 3; 38610:hi = 4; else = NA")
```

In this command, `chp1$realinc` is the original variable `realinc` in the dataset `chp1` and `"0:11114 = 1; 11115:25739 = 2; 25740:38609 = 3; 38610:hi = 4; else = NA"` specifies how the ranges of the original values are coded to the new values. The recode specifications are separated by semicolons (;) and the argument is placed in the quotation marks. In the specifications, `hi` indicates the highest value and `else = NA` specifies the missing values. The recoded variable is named `chp1$SES2` which means `SES2` in the dataset `chp1`.

We again use the `factor()` function to convert `SES2` into a factor variable. The `table()` function is used to display the frequencies for the four categories.

```
> chp1$SES2 <- factor(chp1$SES2)
> table(chp1$SES2)

  1    2    3    4
394  439  307  565
```

We also convert `SES2` to an ordered factor variable with labels and use the `table()` function to display the frequencies for the four categories.

```
> chp1$SES2 <- factor(chp1$SES, labels = c("low SES", "low-middle SES", "upper-
middle SES", "high SES"), ordered = TRUE)
> table(chp1$SES2)

         low SES    low-middle SES    upper-middle SES      high SES
             394               439                 307           565
```

The third method is to use the `rec()` function in the `sjmisc` package. You need to install the package first by typing `install.packages("sjmisc")` and then load it by using the `library(sjmisc)` command. In the `rec()` function, the first argument can be either a data frame or a variable. If the first argument is a data frame, then the `rec()` function creates a new data frame; if the first argument is a variable, then the function creates a new variable. In the following two examples, we first demonstrate how to recode a variable and create a new variable. We also demonstrate how to recode that variable and then create a new data frame.

In the following command, `chp1$realinc` is the original variable `realinc` in the dataset `chp1`, so the function will create a new variable. The `rec = "0:11114 = 1; 11115: 25739 = 2; 25740:38609 = 3; 38610:hi = 4; else = NA"` argument specifies how the ranges of the original values are coded to the new values with the `rec=` argument. Just like the `recode()` function in the `car` package, in the `rec()` function, the recode specifications are separated by semicolons (;) and the argument is placed in quotation marks. In the specifications, `hi` indicates the highest value and `else = NA` specifies the missing values. The recoded variable is named `chp1$SES3` which means `SES2` in the dataset `chp1`. The `table()` function displays the frequencies, and the results are the same as those using the first two methods.

```
> # Recode a variable using rec () in the sjmisc package: method 1
> # Install sjmisc by using install.packages("sjmisc")
> library(sjmisc)
> chp1$SES3 <- rec(chp1$realinc, rec = "0:11114 = 1; 11115: 25739 = 2; 25740:38609
= 3; 38610:hi = 4; else = NA")
> table(chp1$SES3)

  1    2    3    4
394  439  307  565
```

In the next command, with everything else the same, the first argument is the data frame `chp1` and the second argument is the variable, so the function will create a new data frame. This data frame named `chp1.re` contains the recoded variable `realinc_r` only. The new variable name `realinc_r` is automatically assigned and is the combination of the original name `realinc` and the suffix `_r`. We use the `add_columns()` function in the `sjmisc` package to add the new data frame to the original data frame. In the `chp1 <- add_columns(chp1.re, chp1, replace = FALSE)` command, `chp1.re` is the new data frame and `chp1` is the original data frame. The `replace = FALSE` argument specifies that both data frames are kept. The combined dataset is named `chp1`. The `table()` function displays the same frequencies as above.

```
> # Recode a variable using rec() in the sjmisc package: method 2
> chp1.re <- rec(chp1, realinc, rec = "0:11114 = 1; 11115:
  25739 = 2; 25740:38609 = 3; 38610:hi = 4; else = NA", append = FALSE)
> chp1 <- add_columns(chp1.re, chp1, replace = FALSE)
> table(chp1.re$realinc_r)

  1     2     3     4
394   439   307   565
```

1.3.6 Creating a Dummy or Binary Variable

We can also use the `rec()` function to create a dummy or binary variable. For example, let us create a variable `education` with a value of 1 for respondents' highest year of school completed greater than 13.79 years, and a value of 0 otherwise. The `table()` function is used to display the frequencies of the two categories.

```
> # Create a binary or dummy variable: method 1
> chp1$education <- rec(chp1$educ, rec = "min:13.79=0; 14:hi=1; else = NA")
> table(chp1$education)

  0     1
929   944
```

Another method to create a dummy variable is to use the `dicho()` function in the `sjmisc` package. In the `chp1$education2 <- dicho(chp1$educ, dich.by = 13.79, append = TRUE)` command, `chp1$educ` is the variable, the `dich.by = 13.79` argument splits the variable into two groups by the value of 13.79, and the `append = TRUE` argument adds the new variable to the data frame. The `table()` function displays the same frequencies as those above.

```
> # Create a binary or dummy variable: method 2 using dicho() in sjmisc
> chp1$education2 <- dicho(chp1$educ, dich.by = 13.79, append = TRUE)
> table(chp1$education2)

  0     1
929   944
```

1.3.7 Reverse Coding with `rec()`

You can use the `rec()` function to reverse categories of a variable. For example, a survey item uses a four-point scale of 1–4, with 1 = excellent and 4 = poor. You want to reverse the order and define poor to be 1 and strongly excellent to be 4. In the following example, we first recode the character labels to the numeric labels with excellent = 1 and poor = 4. The `table()` function displays the frequencies.

```
> # Reverse coding with rec()
> table(health)
health
excellent       good      fair      poor       dk       iap       na
      414        914       427       118        0         0        0
> chp1$health.n <- rec(chp1$health, rec = "excellent=1; good=2; fair=3; poor=4;
else = NA")
> table(chp1$health.n)


  1     2     3     4
414   914   427   118
```

For simplicity, we can also use the `rec = "rev"` argument in the `rec()` function for reverse coding. In the example, with the `rec = "rev"` argument, we reverse the order of the categories of the variable named `health.rev`. The `table()` function displays the frequencies. As we can see, the categories are reversed in the frequency table.

```
> chp1$health.rev <- rec(chp1$health.n, rec = "rev")
> table(chp1$health.rev)


  1     2     3     4
118   427   914   414
```

1.3.8 Labeling Values for Factor Variables

We can use the `factor()` function to recode a numeric variable to a factor variable. As introduced above, we can also use the `rec()` function in the `sjmisc` package to recode the values to the categories and then label them. The purpose of labeling values is to define the numeric values of a categorical variable. This will make your analysis easier and interpretation clearer. Once you label values of a categorical variable, the label will appear in your output when you conduct an analysis. In the following example, we use the `str()` function to display the structure of the recoded variable `health.rev`.

```
> str(chp1$health.rev)
  num [1:1873] 3 3 3 4 1 3 3 3 1 4 ...
```

The output shows that `health.rev` is numeric with 1,873 observations. We can use the `factor()` function to convert the numeric variable to a factor with labels. In the command, `chp1$health.rev` refers to the variable `health.rev` in the dataset, `chp1`, `levels = c(1, 2, 3, 4)` specifies the values of the four categories, and `labels = c("poor", "fair", "good", "excellent")` specifies the labels. The resulting output by the `table()` function displays the frequencies of the four categories with labels.

```
> # Label a factor
> chp1$health.rev <- factor(chp1$health.rev, levels = c(1, 2, 3, 4), labels =
c("poor", "fair", "good", "excellent"))
> table(chp1$health.rev)

         poor       fair       good    excellent
          118        427        914          414
```

1.3.9 Labeling a Variable

Labeling shows the meaning of a variable. To label a variable, the first method is to use the `label()` function in the `Hmisc` package. You need to install the package first by typing `install.packages("Hmisc")` and then load it using the `library(Hmisc)` function. The basic syntax is `label(varname) <- "label text"`.

For example, if you want to label a variable `health.rev` with the text `recoded health status`, type the following command.

```
label(health.rev) <- "recoded health status"
```

1.3.10 The `row_sums()` and `row_means()` Functions in the `sjmisc` Package

We can use the `row_sums()` function in `sjmisc` package to create a new variable, which is a summation of several items in a survey. For example, if we would like to create a sum score for five variables, we use the following command:

```
row_sums(dataset, var1:var5, n = 5)
```

In the command, `dataset` is the name of the data frame, `var1:var5` specifies the five variables, and `n=5` specifies the minimum number of non-missing values per row. In this example, we specify `n = 5` since there are no missing values in these five variables. This command creates a variable with a total row score of `var1`, `var2`, `var3`, `var4`, and `var5`.

We can also use the `row_means()` function in the `sjmisc` package to create a variable with a row mean score of a set of variables. The basic syntax is `row_means(dataset, var1:var5, n=5)`.

1.3.11 How to Deal With Missing Values When Recoding Variables

Missing data need to be coded as NA so R can recognize them correctly. If datasets from other software packages are imported into R, missing data need to be coded correctly since the datasets may include user-defined missing values, such as 999, na, and iap. We can use the `rec()` function in the `sjmisc` package to recode these user-defined missing values to NA.

If we want to exclude missing values in a data frame, we can use the `no.omit()` function with the name of the data frame enclosed in parentheses and create a new data frame like this: `data.new <- no.omit(dataname)`.

1.3.12 Other Useful Data Management Functions

The following commands will be briefly introduced, but examples using real data will be omitted here due to space limitations:

1. Combining data
   - The `rbind()` function can be used to add cases to the existing variables. When we have two datasets containing the same variables with the same variable types, this command can be applied to combine different cases into one dataset.
   - The `cbind()` can be used to add variables from two datasets without a common unique identification variable. If the two datasets have a common identification variable, we can use the `merge()` function. The generic structure of the `merge()` function is `merge(data1, data2, by = "ID", all = TRUE, sort = TRUE)`. In the command, `data1` and `data2` are the two datasets, `by = "ID"` specifies the sorting ID, `all = TRUE` keeps the unmatched cases from both datasets, and `sort = TRUE` specifies that the ID variable should be sorted.
2. Reshaping data
   The `reshape` package (Wickham, 2007) is useful when we reorganize data into different forms. You need to install the package first by typing `install.packages("reshape")` and then load it with the `library(reshape)` function. The `melt()` function reorganizes the dataset in the long form, whereas the `cast()` function transforms the dataset in the wide form. For example, in longitudinal data analysis, a person-level

dataset is in the wide form, a multivariate layout with one record per individual; on the other hand, a person-period dataset is in the long form with multiple records for each individual, representing each time-point for data collection.

3. Converting variable types
   A variable can be coded in either a numeric or a string format. A numeric variable deals with numbers, whereas a character or a string variable contains text data. The `as.numeric()` function can be used to convert a character variable into a numeric variable, whereas the `as.character()` function can be used to convert a numeric variable into a character variable. The `factor()` function can be used to convert a numeric variable or a character variable into a factor variable.

1.4 DATA MANAGEMENT WITH THE `tidyverse` AND `sjmisc` PACKAGES

The `tidyverse` (Wickham et al., 2019) is not one package, but a collection of several packages that make importing, cleaning, exploring, managing, and visualizing data in R much easier. It includes the following major packages:

- `readr` for importing datasets
- `tibble` for creating data frames
- `dplyr` for data management or data munging
- `tidyr` for creating tidy data
- `ggplot2` for visualizing data
- `purr` for functional programming
- `stringr` for string management
- `forcat` for handling factors

You can use the `install.packages("tidyverse")` command to install all the packages above and then use the `library(tidyverse)` command to load them. In this chapter, we mainly focus on the `dplyr` and `ggplot2` packages in `tidyverse`.

In the previous section, we discussed several useful functions in the `dplyr` and `sjmisc` packages for data management. Specifically, in the `dplyr` package, we introduced the `filter()` function for selecting observations, the `select()` function for selecting variables by their names, the `mutate()` function for creating new variables; in the `sjmisc` package, we introduced the `rec()` function for recoding variables, the `add_columns()` function for adding created variables to a data frame, and the `dicho()` function for creating dummy variables; we also introduced how to use the `row_sums()` function and `row_means()` function to create a sum score and a mean score for a set of variables, respectively.

In the `dplyr` package (Wickham et al., 2021), there are other useful functions. For example, we can also use the `arrange()` function to sort a variable or a set of variables in a data frame, use the `group_by()` function to group data by a factor or categorical variable, and use the `summarize()` function to summarize data. Further, we can use the pipe operator `%>%` to follow each function which you would like to pipe. The pipe `%>%` means "then" when we execute these functions sequentially. In other words, you run a function and then conduct the second one by specify `%>%` between these two functions or steps. You continue to run functions as a chain with `%>%` until you complete the work. The strength of using pipes is to execute a series of functions in sequence without creating temporary data frames. A summary of the major functions in the `dplyr` package is as follows.

- `filter()` function for selecting observations
- `select()` function for selecting variables
- `mutate()` function for creating new variables
- `arrange()` function for sorting variables
- `group_by()` function for grouping data by a factor variable
- `summarize()` function for summarizing data

Please note that the `summarize()` function can be also written as `summarise()`. In addition, the pipe operator `%>%` is a part of the `magrittr` package, which is automatically loaded when you load the `dplyr` package or `tidyverse`.

In the `sjmisc` package (Lüdecke, 2018a), there are other useful functions for data management, which are summarized below.

- `rec()` function for recoding variables
- `add_columns()` function for adding variables as columns to a data frame
- `dicho()` function for creating dummy variables
- `row_sums()` function for creating a sum score for a set of variables
- `row_means()` function for creating a mean score for a set of variables
- `to_factor()` function for recoding a variable into a factor variable with labels
- `var_rename()` function for renaming variables
- `to_long()` function for transforming a data frame from wide format to long format

- descr() function for descriptive statistics
- frq() function for creating a frequency table for factor variables

Since all the functions in the sjmisc package work together with the dplyr package, using both packages makes the data management tasks easier. In addition, the pipe operator %>% works with both packages, which makes the commands more concise.

In the first example, we would like to select the two variables, sex and age, from the data frame chp1. We then would like to split the data by the grouping variable sex and conduct descriptive statistics of age by the two groups of sex. We use the select() and the group_by() function in the dplyr package and the descr() function in the sjmisc package with pipes. This single-line chp1 %>% select(sex,age) %>% group_by(sex) %>% descr(age) command tells R to run it as a chain. It reads "load dataset chp1, then select sex and age, then group the data by sex, and then describe age." The output is as follows.

```
> # Data management using the dplyr and sjmisc packages with the pipe operator %>%
> library(dplyr)
> library(sjmisc)
> chp1 %>% select(sex,age) %>% group_by(sex) %>% descr(age)


## Basic descriptive statistics


Grouped by: male


var     type label     n NA.prc  mean    sd   se  md trimmed       range skew
age integer    age   831      0 48.62 17.08 0.59  48   48.09 71 (18-89) 0.18

Grouped by: female



var     type label     n NA.prc   mean    sd   se  md trimmed      range skew
age integer    age  1042      0  51.02 17.67 0.55  52   50.62 71 (18-89) 0.12
```

In the second example, we would like to select the two variables, sex and health, from the data frame chp1. We then also would like to split the data by the grouping variable sex and create a frequency table of health grouped by sex. We use the select() and the group_by() function in the dplyr package and the frq() function in the sjmisc package with pipes. The chp1 %>% select(sex, health) %>% group_by(sex) %>% frq(health) command tells R to load the dataset, select sex and age, group the data by sex, and then create the frequency table for health. The output is as follows.

```
> chp1 %>% select(sex,health) %>% group_by(sex) %>% frq(health)

health <categorical>
# grouped by: male
# total N=831  valid N=831  mean=2.15  sd=0.83

      Value      |     N     |    Raw %     |    Valid %    |     Cum. %
-------------------------------------------------------------------------
excellent        |    180    |    21.66     |     21.66     |     21.66
good             |    402    |    48.38     |     48.38     |     70.04
fair             |    197    |    23.71     |     23.71     |     93.74
poor             |     52    |     6.26     |      6.26     |    100.00
Dk               |      0    |     0.00     |      0.00     |    100.00
Iap              |      0    |     0.00     |      0.00     |    100.00
Na               |      0    |     0.00     |      0.00     |    100.00
<NA>             |      0    |     0.00     |      <NA>     |      <NA>


health <categorical>
# grouped by: female
# total N=1042  valid N=1042  mean=2.12  sd=0.83

Value          |     N     |    Raw %     |    Valid %    |     Cum. %
-----------------------------------------------------------------------
excellent      |    234    |    22.46     |     22.46     |     22.46
good           |    512    |    49.14     |     49.14     |     71.59
fair           |    230    |    22.07     |     22.07     |     93.67
poor           |     66    |     6.33     |      6.33     |    100.00
Dk             |      0    |     0.00     |      0.00     |    100.00
Iap            |      0    |     0.00     |      0.00     |    100.00
Na             |      0    |     0.00     |      0.00     |    100.00
<NA>           |      0    |     0.00     |      <NA>     |      <NA>
```

1.5 GRAPHS

R is powerful in drawing various graphs. In this section, some basic functions in R will be introduced. We will focus on the `hist()` function for histograms, the `barplot()` function for bar charts, the `boxplot()` function for box plots, and the `plot()` function for scatterplots. In addition, the `ggplot2` package will be introduced.

For each type of graph, R offers rich options, which may seem complicated. You can save the commands into a script file so that the graphs can be easily reproduced or edited.

In the following sections, I will show you how to create basic graphs, such as histograms, bar charts, box plots, and scatterplots.

1.5.1 Histograms

The histogram is one of the most frequently used graphs, and is used to present in a visual manner a frequency distribution of data. In a histogram, scores appear on a horizontal scale and frequency counts are displayed on a vertical scale. Histograms are normally used for continuous variables. They can also be used for ordinal variables if their underlying traits are continuous.

The function for histograms is `hist()`. For example, to see the distribution of the variable `age` using the `gss` data file, simply enter the `hist(age)` command. The histogram is shown in Figure 1.7.

```
> # Create a histogram
> hist(age)
```

We can customize the graph by adding more arguments or parameters to the `hist()` function. For example, we can use the `main=` argument to add a title for the graph, use the `xlab` and `ylab` arguments to provide labels for the x axis and y axis, respectively, use the `xlim` and `ylim` arguments to specify ranges of the axes, and use the `col=` argument to choose colors. We can also use the `breaks=`

FIGURE 1.7 ● Histogram of Age

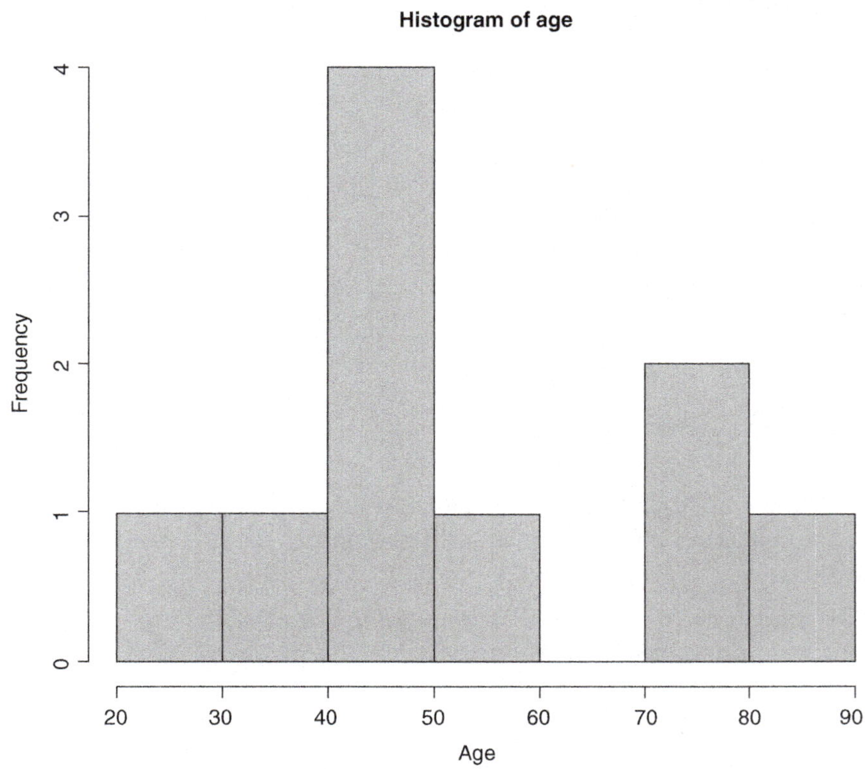

FIGURE 1.8 ● Histogram of Age in GSS

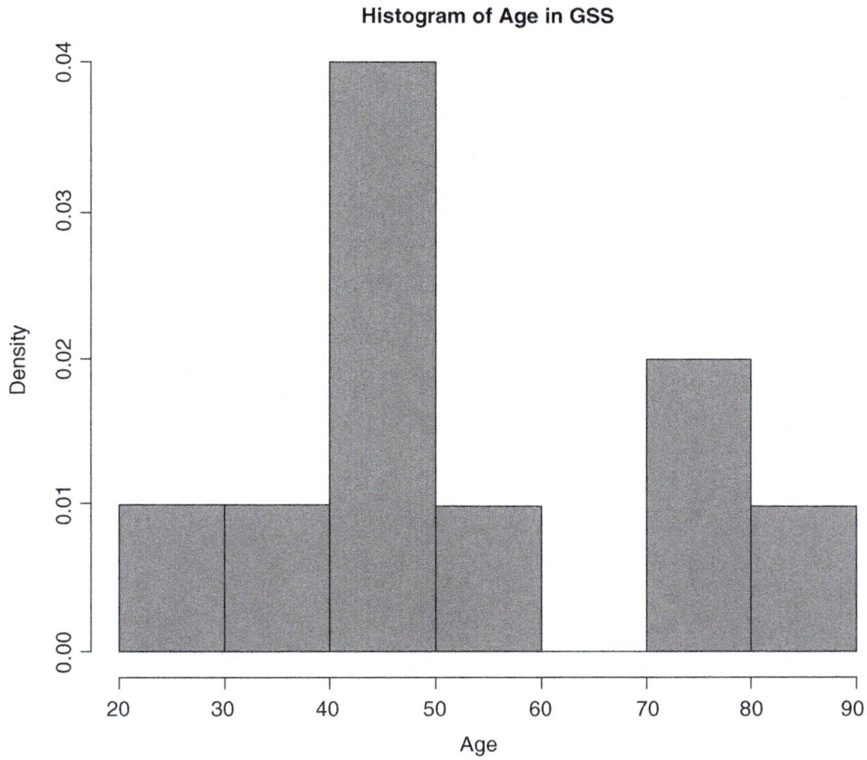

argument to specify the number of cells and use the `freq = FALSE` argument to request the percentages instead of the frequencies. The following is a command with customized arguments: `hist(age, main = "Histogram of Age in GSS", xlab = "Age", xlim = c(20,90), col = "lightblue", freq = FALSE)`. The updated histogram is shown in Figure 1.8.

```
> # Create a histogram with customized arguments
> hist(age, main="Histogram of Age in GSS",
+  xlab="Age", xlim=c(20,90), col="lightblue", freq=FALSE)
```

1.5.2 Bar Charts

Bar charts are normally used to display frequencies for categorical variables. The `barplot()` function is used for bar charts. For example, to draw a bar chart for a categorical variable `gender`, type the following command: `barplot(table(gender))`. In the command, we use the `table()` function to get the frequencies since `gender` is a factor or categorical variable. We will then see the following output and the graph shown in Figure 1.9.

FIGURE 1.9 ● Bar Chart of Gender

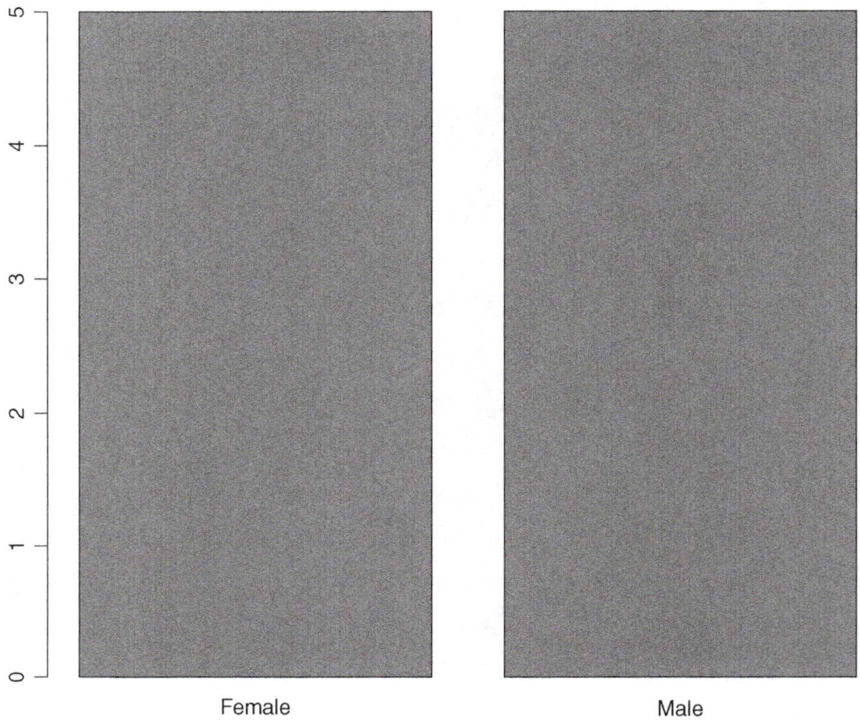


```
> # Create a bar plot
> barplot(table(gender))
```

We can also customize the bar chart by using the `main=` argument to add a title for the graph, using the `xlab=` and `ylab=` arguments to provide labels for x axis and y axis, respectively, using the `names.arg=` argument to name bars, and using `col=` to specify colors.

1.5.3 Box Plots

Box plots are useful for displaying a distribution and to identify outliers for a continuous variable. They display the 25th and 75th percentiles, median, whiskers, and outliers. In a box plot, the lower and upper ends of the box indicate the 25th and 75th percentiles, respectively. The width of the box indicates the interquartile range. The horizontal line in the box is the median, which is the 50th percentile. The lines below and above the box are whiskers, which indicate the spread of your data. Observations beyond the whiskers are shown as dots, which are outliers.

FIGURE 1.10 ● Box Plot of Age

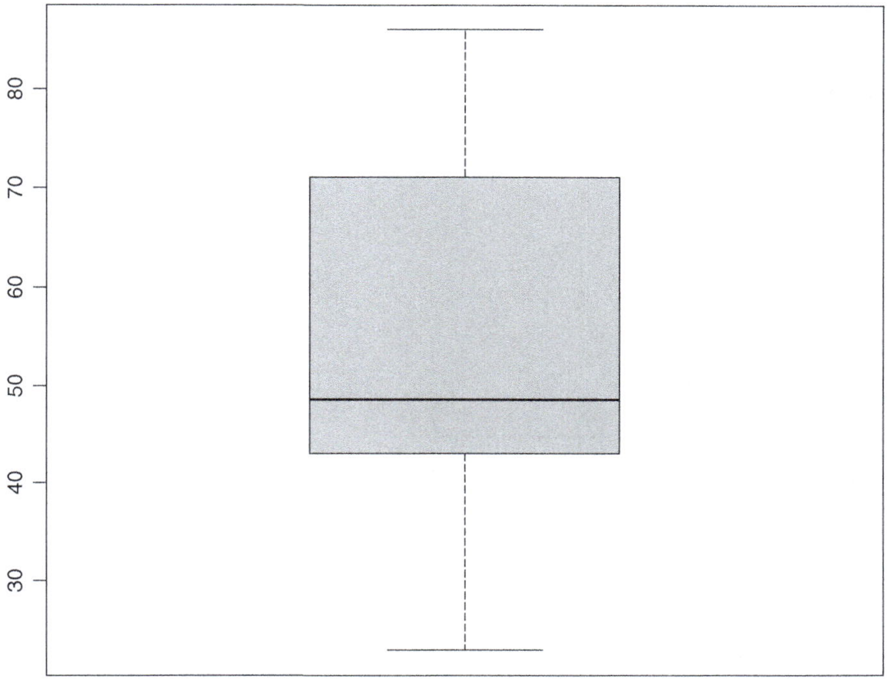

We use the `boxplot(age)` command to draw a box plot for `age` (Figure 1.10).

```
> # Create a box plot
> boxplot(age)
```

We can also customize the box plot by using the `main=` argument to add a title, using the `xlab=` and `ylab=` arguments to provide labels for the x axis and y axis, respectively, using `col=` to specify colors, and using the `horizontal = TRUE` argument to make the graph horizontally.

To display several box plots in one graph, we use the `boxplot()` function with the model equation for the variables enclosed in parentheses. For example, the `box-plot(age ~ gender)` command produces the boxplots for `age` by the grouping variable `gender` (Figure 1.11).

```
> # Create a box plot grouped by gender
> boxplot(age ~ gender)
```

FIGURE 1.11 ● Box Plots of Age by Gender

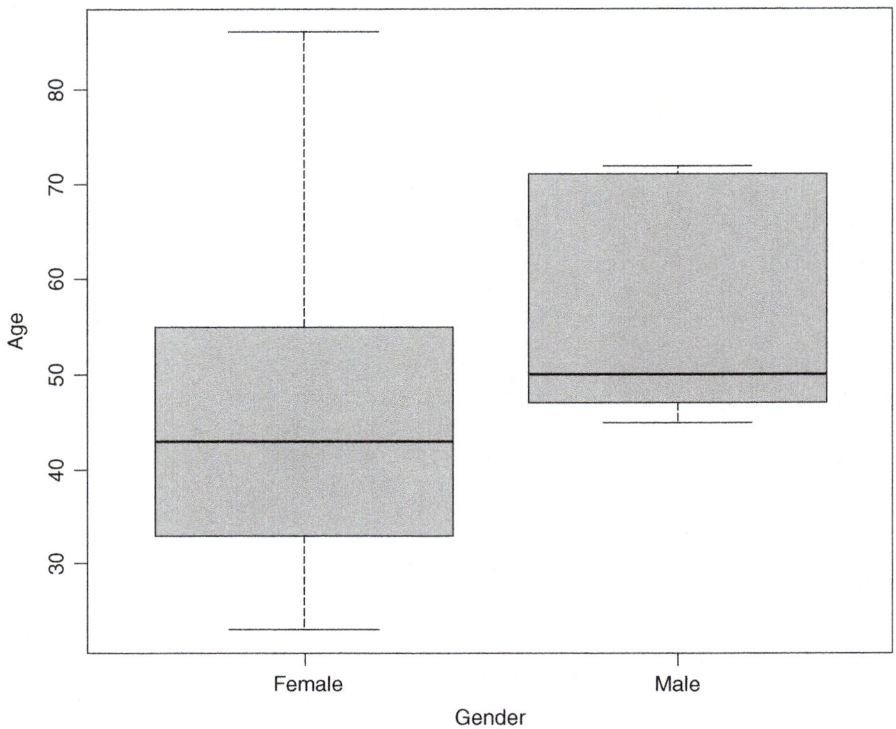

1.5.4 Scatterplots

Scatterplots are used to show a relationship between two variables. They are two-dimensional graphs, with the x axis displaying values of one variable and the y axis displaying values for another variable.

To see a scatterplot of two variables `var1` and `var2`, type the following command:

```
plot(var1, var2)
```

This command tells R to draw a two-way scatterplot for `var1` and `var2`.

For example, we would like to draw a scatterplot for the pre- and post-math tests. We first use the `c()` function to create two variables and then use the `data.frame()` function to create a data frame named `math`. After attaching the dataset, we use the `plot()` function to draw the graph.

```
> # Create a data frame for a scatterplot
> pretest <- c(80, 90, 68, 88, 75, 71, 83, 75, 95, 86, 77, 90, 97, 94, 89)
> posttest <- c(83, 95, 70, 89, 80, 73, 85, 78, 94, 89, 77, 92, 98, 95, 90)
> math <- data.frame(pretest,posttest)
> # Create a scatterplot
> plot(pretest, posttest)
```

The graph is displayed in Figure 1.12.

We can also customize the box plot by using the `main=` argument to add a title, using the `sub=` argument to add a subtitle, using the `xlab=` and `ylab=` arguments to provide labels for x axis and y axis, respectively, and using `col=` to specify colors.

FIGURE 1.12 ● Scatterplot of the Pretest and Posttest of Math Achievement

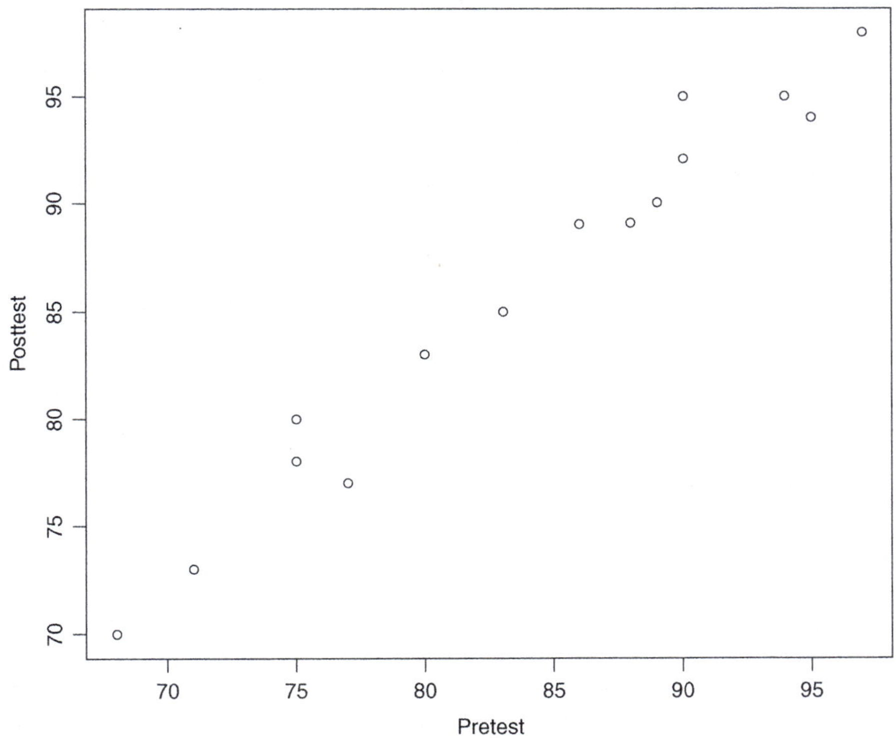

We can also use the `scatterplot(var1, var2)` function in the `car` package to draw scatterplots. An example is not provided here.

1.5.5 Scatterplots with `ggplots2`

The `ggplot2` package (Wickham, 2016), developed by Hadley Wickham, is a popular package for R graphing with a unified framework. It is powerful in providing a variety of types of graphs. It is flexible since you have a variety of options for customizing color, height, size, and shape of graphs. You can use the `install.packages("ggplot2")` command to install the package and then use the `library(ggplot2)` command to load it. This package is a part of `tidyverse`, so if you have installed `tidyverse`, you do not need to reinstall `ggplot2`.

The `ggplot()` function in the `ggplot2` package includes three major elements: data, aesthetics (aes), and geometry (geometric objects).

1. Data: we need to specify a data frame.
2. Aesthetics (aes): this argument specifies the x and y lines, colors, point size, and point shape.
3. Geoms: this argument specifies many types of graphs. For example, we can specify histogram (`geom_histogram`), line (`geom_line`), box plot (`geom_boxplot`), bar (`geom_bar`), point (`geom_point`), and many other types of graphs.

In the following example, we use the `ggplot(math, aes(x=pretest, y=posttest)) + geom_point()` command to draw the scatterplot for the two variables, `pretest` and `posttest`. In the command, `math` is the dataset, `aes(x=pretest, y=posttest))` specifies the x and y variables, and `geom_point()` specifies that the type of graph is a scatterplot with the point geom. The R code is shown below, and the graph is displayed in Figure 1.13.

```
> # Create a scatterplot using ggplot() in ggplot2
> library(ggplot2)
> ggplot(math, aes(x=pretest, y=posttest)) + geom_point()
```

In addition to the point plot, we can also make a line plot with `geom_line()`. Further, to add a regression line to the graph, we use the `ggplot(math, aes(x=pretest, y=posttest)) + geom_point() + stat_smooth(method=lm)` command. In the command, the `stat_smooth(method=lm)` function specifies that the method of linear regression is used to fit the regression line. The `stat_smooth()`

FIGURE 1.13 ● Scatterplot with the Point Goem in ggplot2

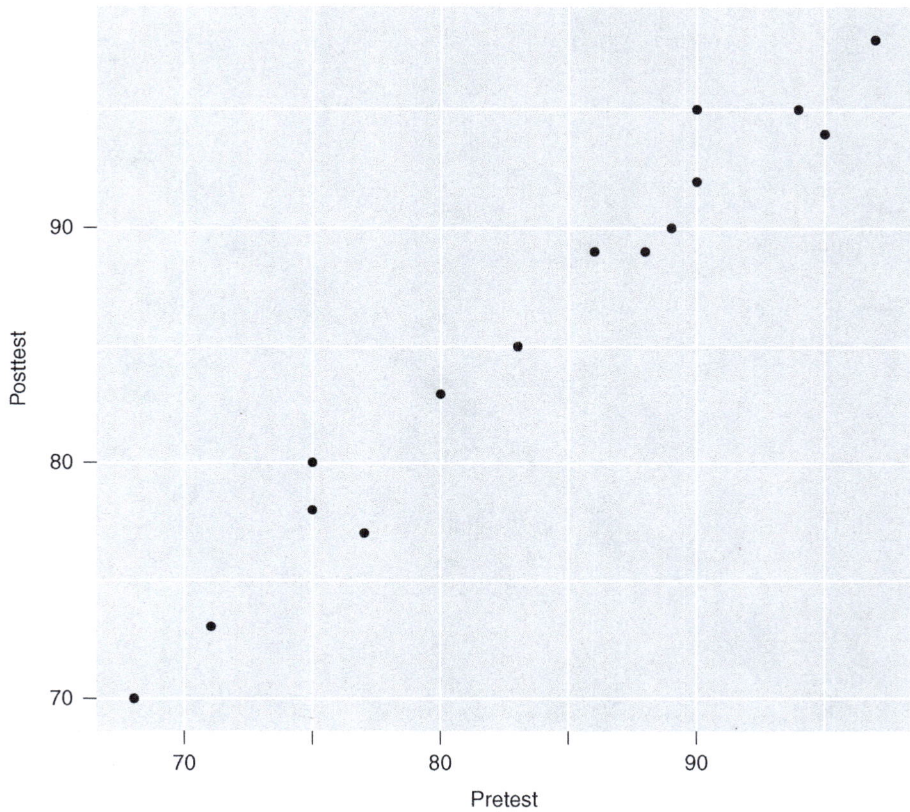

function also produces a 95% confidence interval for the regression line. The R code is shown below, and the resulting graph is displayed in Figure 1.14.

```
> ggplot(math, aes(x=pretest, y=posttest)) + geom_point() + stat_smooth(method=lm)
```

1.5.6 How to Save Graphs

R does not automatically save the graphs you create. To save your graph if you are running RStudio, you can either save it from the plot panel or enter the command. For example, if you would like to use the menu in the plot panel in RStudio, go to **Plots** and click on **Export**. You can choose to **Save as Image** or **Save as PDF**. Or, if you

FIGURE 1.14 ● Scatterplot with the Regression Line

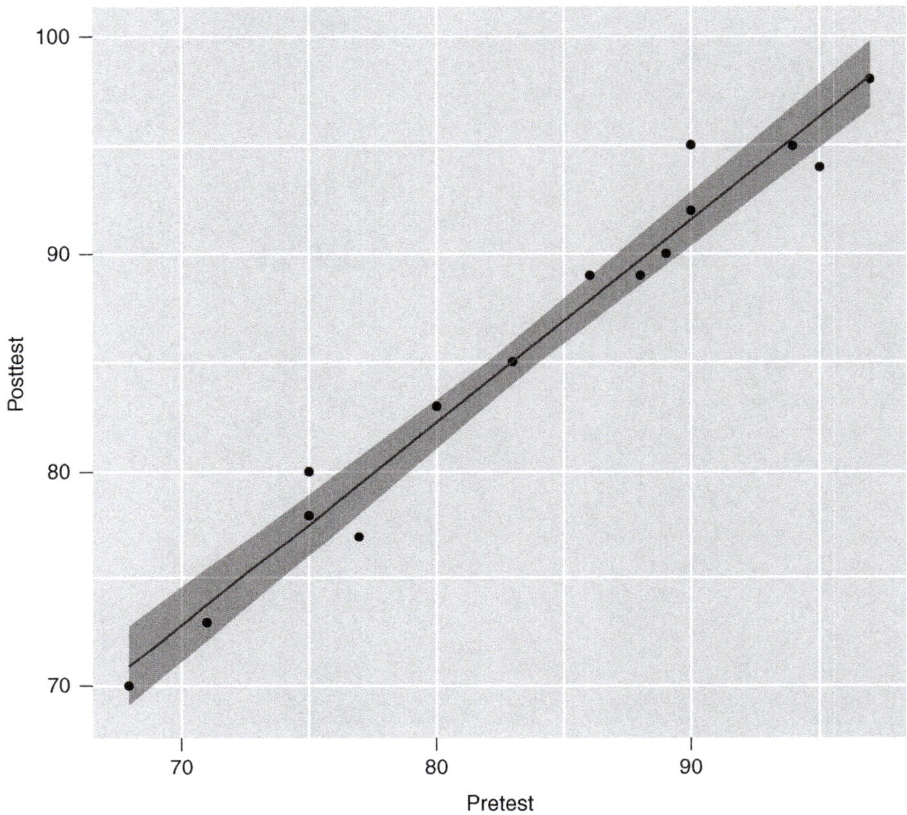

would like to save a graph created by the `ggplot()` function, use the `ggsave()` function to save the file.

The created graph can be copied and pasted into a Word document, but this method would be tedious if you needed to save a lot of graphs all at one time.

1.6 SUMMARY OF R COMMANDS IN THIS CHAPTER

```
# Chapter 1 R Script

# The following user-written packages need to be installed first by using install.packages(" ") and then
by loading it with library()
# library(dplyr)
# library(sjmisc)
# library(car)
# library(ggplot2)

# Section 1.2
# Use c() to create vectors
age <- c(47, 72, 43, 55, 50, 23, 45, 71, 86, 33)
age
gender <-
c("male", "male", "female", "female", "male", "female", "male", "male", "female", "female")
gender
# Index a vector
age[5]
gender[3]

# Use matrix() to create matrices
m1 <- matrix(1:8, nrow=4, ncol=2)
m1
m2 <- matrix(1:8, nrow=4, ncol=2, byrow=TRUE)
m2
# Index a matrix
m1[2, 1]
m1[4, ]
m1[ ,2]

# Use data.frame() to create a data frame
gss <- data.frame(age, gender)
gss
str(gss)
dim(gss)
names(gss)

# Reference a variable in a data frame
gss$age

# Several functions for descriptive statistics
mean(gss$age)
sd(gss$age)
min(gss$age)
max(gss$age)
length(gss$age)

# Use list() to create a list
list.gss <- list(age, m1, gss)
list.gss
list.gss2 <- list(age=age, matrix1=m1, gssdata=gss)
list.gss2
```

```
# Remove all objects
rm(list = ls(all=TRUE))


# Section 1.3
library(foreign)
chp1 <- read.dta("C:/CDA/gss2016-chap1.dta")
attach(chp1)
str(chp1)


# Create a new variable
realinc1 <- realinc/10000
# Create a new variable using mutate()
library(dplyr)
chp1.n <- mutate(chp1, realinc2=realinc/10000)
mean(realinc1, na.rm=TRUE)
mean(chp1.n$realinc2, na.rm=TRUE)

# Recode a variable
chp1$SES <- NA
chp1$SES[chp1$realinc >=0 & chp1$realinc <= 11114] <- 1
chp1$SES[chp1$realinc  >= 11115 & chp1$realinc <= 25739] <- 2
chp1$SES[chp1$realinc  >= 25740 & chp1$realinc <= 38609] <- 3
chp1$SES[chp1$realinc >= 38610] <- 4


# Create a factor
chp1$SES <- factor(chp1$SES)
table(chp1$SES)


# Create a factor with labels
chp1$SES <- factor(chp1$SES, labels = c("low SES", "low-middle SES", "upper-middle SES", "high SES"),
ordered = TRUE)
table(chp1$SES)


# Recode a variable using recode() in the car package
# install.packages("car")
library(car)
chp1$SES2 <- car::recode(chp1$realinc, "0:11114 = 1; 11115: 25739 = 2; 25740:38609 = 3; 38610:hi = 4;
else = NA")
chp1$SES2 <- factor(chp1$SES2)
chp1$SES2 <- factor(chp1$SES, labels = c("low SES", "low-middle SES", "upper-middle SES", "high SES"),
ordered = TRUE)
table(chp1$SES2)


# Recode a variable using rec () in the sjmisc package: method 1
library(sjmisc)
chp1$SES3 <- rec(chp1$realinc, rec = "0:11114 = 1; 11115: 25739 = 2; 25740:38609 = 3; 38610:hi = 4;
else = NA")
table(chp1$SES3)


# Recode a variable using rec () in the sjmisc package: method 2
chp1.re <- rec(chp1, realinc, rec = "0:11114 = 1; 11115:25739 = 2; 25740:38609 = 3; 38610:hi = 4; else
= NA", append = FALSE)
chp1 <- add_columns(chp1.re, chp1, replace = FALSE)
table(chp1.re$realinc_r)


# Create a binary or dummy variable: method 1
chp1$education <- rec(chp1$educ, rec = "min:13.79=0; 14:hi=1; else = NA")
table(chp1$education)
```

```
# Create a binary or dummy variable: method 2 using dicho() in sjmisc
chp1$education2 <- dicho(chp1$educ, dich.by = 13.79, append = TRUE)
table(chp1$education2)

# Reverse coding with rec()
table(health)
chp1$health.n <- rec(chp1$health, rec = "excellent=1; good=2; fair=3; poor=4; else = NA")
table(chp1$health.n)
chp1$health.rev <- rec(chp1$health.n, rec = "rev")
table(chp1$health.rev)
str(chp1$health.rev)

# Label a factor
chp1$health.rev <- factor(chp1$health.rev, levels=c(1, 2, 3, 4), labels = c("poor", "fair", "good",
"excellent"))
table(chp1$health.rev)
table(chp1$sex)

# Section 1.4
# Data management using the dplyr and sjmisc packages with the pipe operator %>%
library(dplyr)
library(sjmisc)
chp1 %>% select(sex,age) %>% group_by(sex) %>% descr(age)
chp1 %>% select(sex,health) %>% group_by(sex) %>% frq(health)

# Section 1.5
age <- c(47, 72, 43, 55, 50, 23, 45, 71, 86, 33)
gender <-
c("male", "male", "female", "female", "male", "female", "male", "male", "female", "female")
gss <- data.frame(age, gender)

# Create a histogram
hist(age)
# Create a histogram with customized arguments
hist(age, main="Histogram of Age in GSS",
    xlab="Age", xlim=c(20,90), col="lightblue", freq=FALSE)

# Create a bar plot
barplot(table(gender))

# Create a box plot
boxplot(age)
# Create a box plot grouped by gender
boxplot(age ~ gender)

# Create a data frame for a scatterplot
pretest <- c(80, 90, 68, 88, 75, 71, 83, 75, 95, 86, 77, 90, 97, 94, 89)
posttest <- c(83, 95, 70, 89, 80, 73, 85, 78, 94, 89, 77, 92, 98, 95, 90)
math <- data.frame(pretest,posttest)

# Create a scatterplot
plot(pretest, posttest)
# Create a scatterplot using ggplot() in ggplot2
library(ggplot2)
ggplot(math, aes(x=pretest, y=posttest)) + geom_point()
ggplot(math, aes(x=pretest, y=posttest)) + geom_line()
ggplot(math, aes(x=pretest, y=posttest)) + geom_point() + stat_smooth(method=lm)
```

Glossary

Data frame A rectangular-shaped dataset in R with variables in columns and observations in rows, also referred to as a dataset in other statistical packages.

Factor A categorical variable or a factor variable with multiple levels.

Functions A set of instructions to perform a specific task in R, just like commands or procedures in other statistical packages.

`ggplot2` An add-on package for R graphing with a unified framework, also a part of `tidyverse`.

Matrix A data structure in R with columns and rows in a two-dimensional rectangular layout with the same data type.

R A programming language and a general-purpose statistical tool for data management, data analysis, and graphing.

R Commander (i.e., `Rcmdr`) An add-on package providing a graphic user interface (GUI) system for R.

R script file A text file containing a list of R commands.

RStudio A free, open-source integrated development environment (IDE) for R that makes programming easier.

`sjmisc` An add-on package for data transformation and management.

`tidyverse` A collection of several add-on packages that make importing, cleaning, exploring, managing, and visualizing data in R easier.

Vector A sequence of data elements of the same type.

Exercises

Use the GSS 2016 data available at **https://edge.sagepub.com/liu1e** for the following problems.

1. Find the variable `happy` and recode it to a new variable `happyrev`. Recode the values of 1, 2, and 3 in `happy` into the values of 3, 2, and 1, respectively, for the new variable.
2. Label the new variable `happyrev`, `happiness`, and then label its values 1 "not too happy", 2 "pretty happy", and 3 "very happy".
3. Produce a histogram for `educ`.
4. Draw a scatterplot to explore the relationship between `educ` and `coninc`.

2

REVIEW OF BASIC STATISTICS

OBJECTIVES OF THIS CHAPTER

This chapter reviews descriptive statistics, simple linear regression, multiple linear regression, and the chi-square test using R. When introducing each type of inferential statistics, an example of research design is provided followed by the research questions, the R command, and the output. The R commands are explained and the output is interpreted in detail. In addition, a sample of reporting the results for each analysis is provided. Finally, the commands for creating publication-quality tables using R are introduced, and the guidelines for reporting results are discussed. This chapter focuses on conducting basic statistical analyses using R, as well as on interpreting and presenting the results. After reading this chapter, you should be able to:

- Conduct analysis of descriptive statistics for continuous and categorical variables.
- Conduct simple linear regression and multiple regression.
- Conduct the chi-square test.
- Interpret R output for these analyses.
- Make publication-quality tables using R.
- Write results for research reports.

This section reviews the basic statistics covered in most introductory statistics courses. It shows you how to use R to perform basic statistics analyses in descriptive statistics, simple linear regression, multiple regression, and the chi-square test.

2.1 UNDERSTAND YOUR DATA USING DESCRIPTIVE STATISTICS

Descriptive statistics should never be overlooked. Whenever you conduct statistical analysis, you first need to understand your data. The descriptive statistics analysis helps you describe your data by showing you types of data, graphs of the distribution of variables, and various statistical indices, such as the central tendency and variability of your data. For continuous variables, you can draw graphs, such as histograms, box plots, and stem-and-leaf plots to show frequency distributions of your data. You can also create a scatterplot to check the relationship between two variables.

The central tendency of a variable tells you the central values of a distribution. Common measures of central tendency include the mean, median, and mode. The mean of a variable is the arithmetic average of scores, the median is the middle score when all values of a variable are ordered, and the mode is the score that is most frequently occurring. In addition to the central tendency of a distribution, we also need to understand its variability. Measures of variability show the spread of a variable. They include the range, variance, and standard deviation. The range is simply the difference between the highest and lowest scores. Variance is the average summed square of each score from the mean. To compute the variance of a set of scores, we need to follow three steps. First, subtract each score from the mean. Second, square each deviation score from the first step and sum them together to get the total sum of squares. Third, get the average sum of squares by dividing the total sum of squares by the number of scores. The standard deviation is just the square root of the variance.

2.2 DESCRIPTIVE STATISTICS FOR CONTINUOUS VARIABLES USING R

2.2.1 The `summary()` Function

The `summary()` function can be used for descriptive statistics analysis. It provides the basic descriptive statistics, such as the minimum, the first quartile, the mean, the median, the third quartile, and the maximum values for a continuous variable and frequencies for a categorical variable. Before running a descriptive analysis, you may want to know whether the missing values are properly coded, and how they are coded. Although you may browse the data, you may not be able to identify all types of missing values. A better way is to use the `summary()` function.

```
> # Import GSS 2016 Stata data file
> library(foreign)
> chp2 <- read.dta("C:/CDA/gss2016-chap1.dta")
> attach(chp2)

> summary(chp2)
      age                 degree                educ              health
 Min.   :18.00   high school    :940   Min.   : 0.00   excellent:414
 1st Qu.:35.00   bachelor       :358   1st Qu.:12.00   good     :914
 Median :50.00   graduate       :220   Median :14.00   fair     :427
 Mean   :49.95   lt high school :216   Mean   :13.79   poor     :118
 3rd Qu.:63.00   junior college:139   3rd Qu.:16.00   dk       :  0
 Max.   :89.00   dk             :  0   Max.   :20.00   iap      :  0
                  (Other)       :  0                   na       :  0
         marital           race            realinc          realrinc
 married      :819   white:1374   Min.   :   234   Min.   :   234
 widowed      :170   black: 315   1st Qu.: 11115   1st Qu.:  8775
 divorced     :325   other: 184   Median : 25740   Median : 17550
 separated    : 65   iap  :   0   Mean   : 32484   Mean   : 24140
 never married:494                3rd Qu.: 38610   3rd Qu.: 31590
 na           :  0                Max.   :131677   Max.   :164382
                                  NA's   :168      NA's   :798
     sex
 male  : 831
 female:1042
```

The output shows the minimum, the first quartile, the mean, the median, the third quartile, and the maximum values for a continuous variable and frequencies for a categorical variable. For example, let us look at the first variable `age`, which is the age of the respondent. It is numeric with a range from 18 to 89.

We can also use the `head(chp2)` command to see the first six cases of all the variables in the dataset.

```
> head(chp2)
  age      degree educ    health marital  race  realinc realrinc    sex
1  47    bachelor   16      good married white 131676.7   164382   male
2  72    bachelor   16      good married white  38610.0       NA   male
3  43 high school   12      good married white 131676.7     5265 female
4  55    graduate   18 excellent married white 131676.7      936 female
5  50 high school   14      poor married white 131676.7   164382   male
6  23 high school   11      good married other  15210.0     7605 female
```

The `summary(chp2)` command provides descriptive statistics for all the variables in the dataset named `chp2`. If you want to get descriptive statistics for a particular variable, you just enter the variable name within the parentheses of the function. Let us take a look at the summary of the variable `age`, which is the age of a respondent to the General Social Survey 2016 (GSS 2016) and run an analysis of descriptive statistics with the `summary(age)` command.

```
> summary(age)
  Min.  1st Qu.  Median    Mean  3rd Qu.    Max.
 18.00    35.00   50.00   49.95    63.00   89.00
```

The R output displays the minimum, the first quartile, the mean, the median, the third quartile, and the maximum values of the variable. The mean age is 49.95 years. The minimum age is 18 years old, and the maximum is 89 years old.

We can also use the `str(age)` command to look at the structure of the variable `age`. The output shows that the class of the variable `age` is integer and there are 1,873 observations.

```
> str(age)
 int [1:1873] 47 72 43 55 50 23 45 71 86 33 ...
```

We can use the `mean()` function, `sd()` function, `max()` function, `min()` function, and `length()` function to obtain the mean, standard deviation, maximum, minimum, and number of observations of the variable `age`, respectively. The `na.rm=TRUE` argument is used to delete the missing values.

```
> mean(age, na.rm=TRUE)
[1] 49.95462
> sd(age, na.rm=TRUE)
[1] 17.44434
> max(age, na.rm=TRUE)
[1] 89
> min(age, na.rm=TRUE)
[1] 18
> length(age)
[1] 1873
```

2.2.2 The `tapply()` Function for Grouped Summaries

In R, if you are interested in descriptive statistics of a variable by another grouping variable, we can use the `tapply()` function to split the data file and apply a function. In this example, we would like to see the descriptive statistics for male and female respondents, respectively. We first use the `mean <- tapply(age, sex, mean, na.rm=TRUE)` command to obtain the mean for the variable `age` for each category of sex. In the tapply() function, age is the continuous variable in which we are interested, sex is the grouping variable with two categories, and mean is the function. The na.rm=TRUE argument is used to remove the missing values. The computed means for the two groups are named the object mean. Then we repeat the process to obtain the standard deviations (sd), maximums (max), minimums (min), and numbers of observations (length) of age for males and female separately. Finally, we use the cbind(mean, sd, max, min, length) command to combine the results.

```
> # Descriptive statistics by group: method 1
> mean <- tapply(age, sex, mean, na.rm=TRUE)
> sd <- tapply(age, sex, sd, na.rm=TRUE)
> max <- tapply(age, sex, max, na.rm=TRUE)
> min <- tapply(age, sex, min, na.rm=TRUE)
> length <- tapply(age, sex, length)
> cbind(mean, sd, max, min, length)
            mean        sd  max  min  length
male    48.62214  17.07903   89   18     831
female  51.01727  17.66650   89   18    1042
```

The R output shows the descriptive statistics of age for males and female separately. Of 831 male respondents, the mean of the age is 48.622 years and the standard deviation is 17.079 years. Of 1,042 female respondents, the average age is 51.017 years and the standard deviation is 17.667 years.

2.2.3 The group_by() Function and the summarize() Function for Grouped Summaries

Another option is to use the group_by() function and the summarize() function in the dplyr package (Wickham et al., 2021). If not installed, you need to install the package first by typing install.packages("dplyr") and then load it by using the library(dplyr) function. In the following example, we first use the gender <- group_by(chp2, sex) function to create a grouping variable and name it as an object gender. In the group_by() function, chp2 is the data frame and sex is the grouping variable. We then use the summarize() function to compute the descriptive statistics for age for the two categories of the variable sex. The following is the output.

```
> # Descriptive statistics by group: method 2
> library(dplyr)
> gender <- group_by(chp2, sex)
> summarize(gender, mean(age, na.rm=TRUE), sd(age, na.rm=TRUE), max(age, na.rm=TRUE), min(age,
na.rm=TRUE))
# A tibble: 2 x 5
   sex     `mean(age, na.rm ~  `sd(age, na.rm =~  `max(age, na.rm~  `min(age, na.rm~
   <fct>                <dbl>              <dbl>             <int>             <int>
1  male                  48.6               17.1                89                18
2  female                51.0               17.7                89                18
```

2.3.4 The `group_by()` Function and the `descr()` Function for Grouped Summaries

The third option is to use the `group_by()` function in the `dplyr` package and the `descr()` function in the `sjmisc` package (Lüdecke, 2018a). Since both the `dplyr` and `sjmisc` packages have been loaded, we do not need to load them again. This single-line `chp2 %>% group_by(sex) %>% descr(age)` command tells R to run it as a chain. We first take the data frame `chp2`, then split the data by the grouping variable `sex`, and finally compute the descriptive statistics of `age` for the two categories of the variable `sex`. In this command, the pipe operator `%>%` means "then," which helps us to run a series of functions in sequence.

```
> # Descriptive statistics by group: method 3
> # library(sjmisc)
> # library(dplyr)
> chp2 %>% group_by(sex) %>% descr(age)

## Basic descriptive statistics

Grouped by: male

var     type label   n NA.prc  mean    sd   se md trimmed        range skew
age  Integer   age 831      0 48.62 17.08 0.59 48   48.09 71 (18-89)  0.18

Grouped by: female

var     type label    n NA.prc  mean    sd   se md trimmed        range skew
age  integer   age 1042      0 51.02 17.67 0.55 52   50.62 71 (18-89)  0.12
```

2.2.5 Descriptive Statistics for Multiple Variables With `stat.desc()`

The previous examples focus on the descriptive statistics for one continuous variable. If you would like to see the descriptive statistics for more than one variable, simply use the `stat.desc()` function in the `pastecs` package (Grosjean et al., 2018) and list the variables in the `c()` function. You need to install the package first by typing `install.packages("pastecs")` and then load the package with the `library(pastecs)` command. In the following example, we use the `stat.desc(chp2[, c("age", "educ")])` command to compute the descriptive statistics for the two variables `age` and `educ`. The following results are displayed.

```
> # Descriptive statistics for two variables
> # Install pastecs by using install.packages()
> library(pastecs)
> stat.desc(chp2[, c("age", "educ")])
                        age          educ
nbr.val        1.873000e+03  1.873000e+03
nbr.null       0.000000e+00  2.000000e+00
nbr.na         0.000000e+00  0.000000e+00
min            1.800000e+01  0.000000e+00
max            8.900000e+01  2.000000e+01
range          7.100000e+01  2.000000e+01
sum            9.356500e+04  2.582100e+04
median         5.000000e+01  1.400000e+01
mean           4.995462e+01  1.378590e+01
SE.mean        4.030749e-01  6.953498e-02
CI.mean.0.95   7.905233e-01  1.363742e-01
var            3.043051e+02  9.056169e+00
std.dev        1.744434e+01  3.009347e+00
coef.var       3.492038e-01  2.182916e-01
```

The output shows that the means of `age` and `educ` are 49.955 and 13.786, respectively. The variable `educ` is the highest year of school completed. The smallest value is 0, whereas the largest value is 20. The variance is 9.056, and the standard deviation of 3.009.

2.2.6 Descriptive Statistics for Multiple Variables With `descr()`

Another option is to use the `descr()` function in the `sjmisc` package. You need to install the package first by typing `install.packages("sjmisc")` and then load it with the `library(sjmisc)` command. In the `descr(chp2, age, educ)` command, `chp2` is the data frame and `age` and `educ` are the two selected variables in the data.

```
> # Descriptive statistics for two variables: method 2
> library(sjmisc)
> descr(chp2, age, educ)

## Basic descriptive statistics

 var      type  label     n  NA.prc   mean     sd    se  md  trimmed       range   skew
 age   integer    age  1873       0  49.95  17.44  0.40  50    49.48  71 (18-89)   0.15
educ   integer   educ  1873       0  13.79   3.01  0.07  14    13.80   20 (0-20)  -0.20
```

The output shows the variable name, type, label, total number, percentage of missing data, mean, standard deviation, standard error, mode, range, and skewness. For example, the mean age is 49.95 years and the standard deviation is 17.44 years. The range is 71 years. The skewness is .15, which is less than 1. It indicates that the distribution is almost symmetric. We can use the Shapiro–Wilk test with the `shapiro.test()` function to test normality.

2.2.7 The `group_by()` Function and the `descr()` Function for Grouped Summaries of Multiple Variables

If we would like to see the mean of these two continuous variables across the categories of a third variable, we can use the `chp2 %>% group_by(sex) %>% select(age, educ) %>% descr()` command. The following example displays each mean of `age` and `educ` for males and females separately.

```
# Descriptive statistics for multiple variables by group
> chp2 %>% group_by(sex) %>% select(age, educ) %>% descr()

## Basic descriptive statistics

Grouped by: male

 var    type   label    n  NA.prc   mean    sd    se  md  trimmed       range
 age  integer    age  831       0  48.62 17.08  0.59  48    48.09  71 (18-89)
educ  integer   educ  831       0  13.74  2.99  0.10  14    13.76   20 (0-20)
  skew
  0.18
 -0.29

Grouped by: female

 var    type   label     n  NA.prc   mean    sd    se  md  trimmed       range
 age  integer    age  1042       0  51.02 17.67  0.55  52    50.62  71 (18-89)
educ  integer   educ  1042       0  13.82  3.03  0.09  14    13.83   20 (0-20)
  skew
  0.12
 -0.13
```

The R output shows the descriptive statistics of `age` and `educ` for males and female separately. In the first table, the mean, standard deviation, minimum, and maximum of the variable `age` for males are 48.62, 17.08, 18, and 89, respectively. In addition, the mean, standard deviation, minimum, and maximum of the variable `educ` for males are 13.74, 2.99, 0, and 20, respectively. The descriptive statistics of `age` and `educ` for female are displayed in the second table.

2.3 FREQUENCY DISTRIBUTION FOR CATEGORICAL VARIABLES USING R

For categorical variables, such as gender or ethnicity, it does not make much sense to get summary statistics, such as the mean and the standard deviation. Instead we normally do a frequency analysis to get the frequency of each value for categorical variables. The `table()` function is used for creating frequency tables. The following example shows you how to create a frequency table for the nominal variable `degree` from the GSS 2016 dataset. Before we start, let us take a look at the structure of this variable with the `str(degree)` command first.

```
> str(degree)
 Factor w/ 8 levels "lt high school",..: 4 4 2 5 2 2 2 3 2 2 ...
```

The output shows that `degree` is a factor variable with eight levels. To see detailed levels and frequency for each level, we use the `table()` function below.

2.3.1 The `table()` Function for a Single Categorical Variable

Next, let us look at the frequency table of the variable `degree` using the `table(degree)` command. We can also use the `table(marital)` command for the other categorical variable, `marital`.

```
> # Frequency table for a factor or categorical variable
> table(degree)
degree
lt high school  high school  junior college  bachelor  graduate
           216          940             139       358       220
            dk          iap              na
             0            0               0
> table(marital)
marital
      married   widowed   divorced   separated never married
          819       170        325          65           494
           na
            0
```

The `table(degree)` command displays the frequency for each level or category of the variable `degree`. The output shows eight categories. Since "iap" (i.e., inapplicable), "dk" (i.e., don't know), and "na" (i.e., no answer) represent the missing values in the Stata dataset for this chapter, there are five actual levels ranging from "lt high school" to "graduate."

2.3.2 The `frq()` Function in the `sjmisc` Package for a Single Categorical Variable

To see more detailed information such as percentages and cumulative percentages in a frequency table, we use the `frq()` function in the `sjmisc` package. The package needs to be loaded first. In the `frq(chp2, degree)` command, `chp2` is the data frame and `degree` is the categorical variable. We also use the same `frq()` function for `marital`.

```
# Frequency table with frq() in sjmisc
> # library(sjmisc)
> frq(chp2, degree)

degree <categorical>
# total N=1873 valid N=1873 mean=2.69 sd=1.24

Value           |   N |  Raw % |  Valid % |  Cum. %
---------------------------------------------------
lt high school  | 216 |  11.53 |    11.53 |   11.53
high school     | 940 |  50.19 |    50.19 |   61.72
junior college  | 139 |   7.42 |     7.42 |   69.14
bachelor        | 358 |  19.11 |    19.11 |   88.25
graduate        | 220 |  11.75 |    11.75 |  100.00
dk              |   0 |   0.00 |     0.00 |  100.00
iap             |   0 |   0.00 |     0.00 |  100.00
na              |   0 |   0.00 |     0.00 |  100.00
<NA>            |   0 |   0.00 |     <NA> |    <NA>


> frq(chp2, marital)

marital <categorical>
# total N=1873 valid N=1873 mean=2.60 sd=1.66

Value           |   N |  Raw % |  Valid % |  Cum. %
---------------------------------------------------
married         | 819 |  43.73 |    43.73 |   43.73
widowed         | 170 |   9.08 |     9.08 |   52.80
divorced        | 325 |  17.35 |    17.35 |   70.15
separated       |  65 |   3.47 |     3.47 |   73.63
never married   | 494 |  26.37 |    26.37 |  100.00
na              |   0 |   0.00 |     0.00 |  100.00
<NA>            |   0 |   0.00 |     <NA> |    <NA>
```

Interpreting R Output

The output shows the number of observations, valid observations with no missing values, mean, and standard deviation on the top of the frequency table. In the first table, the first column shows the levels of the degree, the labels of the categories. The frequency column (labeled "`frq`") shows the number of respondents who reported their highest degrees. The `Raw %` column shows that 11.53% of respondents have a degree less than high school, 50.19% have a high-school degree, 7.42% have a junior college degree, 19.11% have a bachelor's degree, and 11.75% have a graduate degree. The percentages from the missing values which are coded as "iap," "dk," "na," and "NA" are 0%. The `Valid %` column shows the percentages for data with no missing values. The last column labeled `Cum. %` provides the cumulative percentage for each category. For example, the cumulative percent of having a junior college degree or less is 69.14%, which equals the percent of respondents having less than a high-school degree (11.53%) plus the percent having a high-school degree (50.19%), plus the percent having a junior college degree (7.42%). The other frequency table for the variable `marital` can be interpreted in the same way.

2.3.3 The `table()` Function for a Two-Way Table

If we want to get a two-way cross-tabulation table for two categorical variables, we use the `table()` function with the two variables which are separated by a comma within the parentheses. For example, the command `tab <- table(degree, race)` tells R to create a two-way table of frequency counts for two nominal variables `degree` and `race` and assign `tab` as the object name. The `ftable(tab)` command also produces the same frequency table. To get the marginal totals of the rows and columns of the table, we use the `addmargins (tab)` command.

```
> # Cross-tabulation
> tab <- table(degree, race)
> summary(tab)
Number of cases in table: 1873
Number of factors: 2
Test for independence of all factors:
    Chisq = NaN, df = 21, p-value = NA
    Chi-squared approximation may be incorrect
> tab
                      race
degree                 white  black  other  iap
  lt high school         132     40     44    0
  high school            683    186     71    0
  junior college         104     23     12    0
  bachelor               286     44     28    0
  graduate               169     22     29    0
  dk                       0      0      0    0
  iap                      0      0      0    0
  na                       0      0      0    0
```

```
> ftable(tab)
                   race
degree              white  black  other  iap
  lt high school      132     40     44    0
  high school         683    186     71    0
  junior college      104     23     12    0
  bachelor            286     44     28    0
  graduate            169     22     29    0
  dk                    0      0      0    0
  iap                   0      0      0    0
  na                    0      0      0    0

> addmargins(tab)
                   race
degree              white  black  other  iap   Sum
  lt high school      132     40     44    0   216
  high school         683    186     71    0   940
  junior college      104     23     12    0   139
  bachelor            286     44     28    0   358
  graduate            169     22     29    0   220
  dk                    0      0      0    0     0
  iap                   0      0      0    0     0
  na                    0      0      0    0     0
  Sum                1374    315    184    0  1873
```

Interpreting R Output

The output produced by the `table(degree, race)` command or the `ftable (tab)` command displays a two-way cross-tabulation table, where the first column lists the categories for the variable `degree` and the first row lists the categories for the variable `race`. The variable `degree` is the row variable since its categories are across the rows of the table. The variable `race` is the column variable since its categories are across the top of the table. Each cell shows a relative frequency of subjects in each subgroup. For example, there were 176 White respondents who did not have a high-school degree.

The table produced by the `addmargins(tab)` command displays the marginal totals of the rows and columns. The last column is the row total. It shows the total frequency for each of the five degree levels. The last row is the column total. It provides the total frequency for each of the three categories of the variable `race`. The row and column totals are also called marginal totals or frequencies.

2.3.4 The `CrossTable()` Function in the `gmodels` Package

To get the relative frequency of each cell within its row and column, we use the `CrossTable()` function in the `gmodels` package (Warnes et al., 2018). We need to install the package first by typing `install.packages("gmodels")` and then load it by using the `library(gmodels)` command. In the following example, we use the `CrossTable(degree, race, digits=2)` command to create a frequency table. In the function, `degree` and `race` are the two categorical variables and the argument `digits=2` specifies the number of decimals.

```
> # Cross-tabulation with CrossTable() in gmodels
> library(gmodels)
> CrossTable(degree, race, digits=2)

  Cell Contents
|-------------------------|
|                       N |
| Chi-square contribution |
|           N / Row Total |
|           N / Col Total |
|         N / Table Total |
|-------------------------|


Total Observations in Table:  1873


               | race
        degree |     white |     black |     other | Row Total |
---------------|-----------|-----------|-----------|-----------|
lt high school |       132 |        40 |        44 |       216 |
               |      4.42 |      0.37 |     24.46 |           |
               |      0.61 |      0.19 |      0.20 |      0.12 |
               |      0.10 |      0.13 |      0.24 |           |
               |      0.07 |      0.02 |      0.02 |           |
---------------|-----------|-----------|-----------|-----------|
   high school |       683 |       186 |        71 |       940 |
               |      0.06 |      4.93 |      4.93 |           |
               |      0.73 |      0.20 |      0.08 |      0.50 |
               |      0.50 |      0.59 |      0.39 |           |
               |      0.36 |      0.10 |      0.04 |           |
---------------|-----------|-----------|-----------|-----------|
junior college |       104 |        23 |        12 |       139 |
               |      0.04 |      0.01 |      0.20 |           |
               |      0.75 |      0.17 |      0.09 |      0.07 |
               |      0.08 |      0.07 |      0.07 |           |
               |      0.06 |      0.01 |      0.01 |           |
---------------|-----------|-----------|-----------|-----------|
      bachelor |       286 |        44 |        28 |       358 |
               |      2.08 |      4.36 |      1.46 |           |
               |      0.80 |      0.12 |      0.08 |      0.19 |
               |      0.21 |      0.14 |      0.15 |           |
               |      0.15 |      0.02 |      0.01 |           |
---------------|-----------|-----------|-----------|-----------|
      graduate |       169 |        22 |        29 |       220 |
               |      0.36 |      6.08 |      2.53 |           |
               |      0.77 |      0.10 |      0.13 |      0.12 |
               |      0.12 |      0.07 |      0.16 |           |
               |      0.09 |      0.01 |      0.02 |           |
---------------|-----------|-----------|-----------|-----------|
  Column Total |      1374 |       315 |       184 |      1873 |
               |      0.73 |      0.17 |      0.10 |           |
---------------|-----------|-----------|-----------|-----------|

> chisq.test(degree, race)

   Pearson's Chi-squared test

data:  degree and race
X-squared = 56.286, df = 8, p-value = 2.481e-09
```

Interpreting R Output

In the output, each category of the first variable `degree` takes one row, and each category of the second variable `race` takes one column. The row percentage displays the relative frequency of a cell within each row (i.e., each degree level), and the column percentage displays the relative frequency of a cell with each category of race. For example, 683 White respondents had a high-school degree, which was 73% of the 940 respondents who had a high-school degree (683/940 = 73%). Its column percentage was 50%. This means 50% of the 1,374 White respondents had a high-school degree (683/1,374 = 50%).

2.4 SIMPLE LINEAR REGRESSION

2.4.1 Simple Linear Regression: An Introduction

Regression is used when we predict a dependent variable from an independent variable or multiple independent variables. When there is only one independent variable, the regression is simple linear regression; when there are two or more independent variables, it is called multiple linear regression. The independent variable is the explanatory or predictor variable, whereas the dependent variable is the outcome or response variable, the one we are interested in predicting from the predictor variable. In linear regression, the dependent variable is continuous and the independent variable(s) can be either continuous or categorical. The simple linear regression can be expressed as follows:

$$Y = \beta_0 + \beta_1 X_1 + e \tag{2.1}$$

where Y is the observed value of the dependent variable, X_1 is the value of the independent variable, β_0 is the intercept, β_1 is the regression coefficient that is the slope of the regression line, and e is the error term, also known as the residual. The error term e is the difference between the observed value and the predicted value of the dependent variable. It is assumed to have a normal distribution with a mean of 0 and a constant variance at every value of the independent variable. The values of e are mutually independent from each other.

The simple linear regression can also be expressed as a sample equation, which is the predicted regression equation:

$$\hat{Y} = b_0 + b_1 X_1 \tag{2.2}$$

where $\hat{Y}$ is the predicted or fitted value of the dependent variable, b_0 is the intercept, and b_1 is the regression coefficient. The estimated intercept and coefficient can also be expressed as $\hat{\beta}_0$ and $\hat{\beta}_1$, respectively.

To estimate the intercept β_0 and the regression coefficient β_1, the method of ordinary least squares (OLS) is used, which minimizes the sum of squared residuals. Since the residual e is the difference between the observed value Y and predicted value $\hat{Y}$, of the dependent variable, the estimated intercept and regression coefficients are the values when the total amount of the squared errors is as small as possible.

In the following example, we are interested in how well annual family income can be predicted by a respondent's individual income using the GSS 2016 dataset. The dependent variable is annual family income, and the independent variable is the respondent's annual income.

Research Questions:

1. Can an individual's income predict the family income?
2. How well can the annual family income be predicted from an individual's income?

2.4.2 The `lm()` Function and Extractor Functions

The `lm()` function is used for linear regression analysis. The model formula in `lm()` specifies the dependent variable and the predictor variable(s), which are separated by the tilde (~). For example, in simple linear regression, there is only one dependent variable and one independent variable, so the model syntax in the function is `y ~ x`, where `y` is the dependent variable and `x` is the independent variable. The command `lm(y ~ x, data = data1)` tells R to run a simple regression analysis predicting the dependent variable `y` with an independent variable `x`. The data argument is `data = data1`. When there is more than one predictor variable in the formula, they are connected by plus (+) symbols. For more details on how to use this command, type `help(lm)` in the command prompt.

Once a model is fitted, we can use extractor functions to extract specific results. For example, we can use the `summary()` function to display the summary results of the fitted model, use the `coef()` function to extract the coefficients, use the `confint()` function to request the confidence intervals of the coefficients, and use the `anova()` function to request the ANOVA table. Other useful extractor functions include the `fitted()` function for creating the fitted values, the `residuals()` function for the residual values, the `predict()` function for the predicted values of an outcome variable at specific values of a predictor variable or multiple predictor variables, the `AIC()` function for the AIC statistic, and the `plot()` function for diagnostic plots. These are the generic functions and most of them can be applied to other models introduced in the book.

In the following example, the command `slm <-lm(realinc1 ~ realrinc1)` tells R to predict the dependent variable `realinc1` from the independent variable `realrinc1`. The fitted model is named `slm`. The output is shown by the `summary(slm)` command.

```
> # Simple linear regression
> realinc1 <- realinc/10000
> realrinc1 <- realrinc/10000
> slm <- lm(realinc1 ~ realrinc1)
> summary(slm)

Call:
lm(formula = realinc1 ~ realrinc1)

Residuals:
    Min       1Q   Median       3Q      Max
-1.8654  -1.4220  -0.7374   0.5563  11.2236

Coefficients:
            Estimate  Std. Error  t value   Pr(>|t|)
(Intercept)  1.87035    0.08842     21.15   <2e-16 ***
realrinc1    0.78711    0.02284     34.46   <2e-16 ***

---
Signif. codes: 0 '***' 0.001 '**' 0.01 '*' 0.05 '.' 0.1 ' ' 1

Residual standard error: 2.211 on 1045 degrees of freedom
  (826 observations deleted due to missingness)
Multiple R-squared: 0.5319, Adjusted R-squared: 0.5315
F-statistic: 1188 on 1 and 1045 DF, p-value: < 2.2e-16
```

2.4.3 Interpreting R Output: The Coefficients Table

Two new variables `realinc1` and `realrinc1` are created so that the family income and the respondent's income are on a scale of $10,000. With this transformation, the R output would show the results properly. In the R output for the regression analysis, the first part is the call, which shows the R command for the model. The second part shows the residuals. The minimum, first quarter, median, third quarter, and maximum values of the residuals are shown here. The third part shows the coefficients table including the parameter estimates for the predictor variable and the intercept, their standard errors, the t statistics, and the associated p values. The fourth part shows the residual standard error and the degrees of freedom. The fifth part shows the multiple R^2 and the adjusted multiple R^2. Finally, the sixth part shows the F statistic, the degrees of freedom, and the associate p value.

First, let us take a look at the regression table. In this model, we have one independent variable, or the predictor, and the constant. The null hypothesis for the t test is that the coefficient of the predictor variable is zero, and the alternative hypothesis is that the coefficient of the predictor variable is significantly different from zero. In this example, $\beta_1 = .787$ is the regression coefficient. The t statistic tests whether the regression coefficient of the independent variable, respondent's income, is significantly different from zero. In other words, it tests whether the effect of the independent variable on the dependent variable is significant.

The t statistic equals the parameter estimate divided by its standard error. For the predictor variable `realrinc1`, $t = .787/.023 = 34.46$. The associated p value, `Pr(>|t|)` < `.001`, which means that the probability of having a t value larger than the absolute value of the observed t value of 34.46 is less than .001 if the null hypothesis is true, so we reject the null hypothesis. The rejection of the null hypothesis indicates that the regression coefficient of the independent variable, respondent's income, is significantly different from zero. Therefore, the predictor variable, respondent's income (`realrinc1`) is a significant predictor of the dependent variable, family income (`realinc1`).

In the model, the estimated constant (β_0) is shown as `(Intercept)`, = 1.870. This is the mean family income when the value of the independent variable equals zero.

The estimated regression coefficient $\beta_1 = .787$. It can be interpreted as follows: For a one-unit increase in the respondent's annual income, the dependent variable, the family income, is increased by a value of .787.

Substituting the values of the constant and regression coefficient into the equation of the estimated regression model:

$$\hat{Y} = b_0 + b_1X_1$$

We get:

$$\hat{Y} = 1.870 + .787X_1$$

When $X_1 = 0$, the predicted outcome variable = 1.870, which is the constant.

Next, the output reports the residual standard error and the degrees of freedom. The residual standard error = 2.211, which is the square root of the mean squares of variance for the `Residual`. The degrees of freedom for the residual $= n - 1 - k = 1{,}047 - 1 - 1 = 1{,}045$ since $n = 1{,}047$ and the number of predictors $k = 1$.

2.4.4 Interpreting R Output: The Multiple R^2 and the F Statistic

Further, the multiple R^2 and the adjusted multiple R^2 are reported. R^2 is called the coefficient of determination, indicating the strength of the prediction. It is the proportion of error variance explained by the predictor. $R^2 = .532$, which indicates that 53.2% of the variance in the family income is explained by the respondent's income. The adjusted R^2 takes the sample size and the number of predictors into consideration, and it is a less biased estimate of the population R^2. Adjusted $R^2 = .352$.

Finally, the output shows the F statistic, the degrees of freedom, and the associate p value. The F statistic tests whether the overall model with one predictor in this example can significantly predict the dependent variable.

Null hypothesis: The overall model with one predictor in this example can significantly predict the dependent variable.

Alternative hypothesis: The overall model with one predictor in this example cannot significantly predict the dependent variable.

$F(1, 1{,}045) = 1187.50$, $p < .001$, which indicates that the model with one predictor, respondent's income, is significantly different from zero.

2.4.5 The ANOVA Table

We use the `anova()` function to request the ANOVA table, which was discussed in the one-way ANOVA section.

```
> anova(slm)
Analysis of Variance Table


Response: realincl
             Df  Sum Sq  Mean Sq  F value        Pr(>F)
realrincl     1  5805.0   5805.0   1187.5  < 2.2e-16 ***
Residuals  1045  5108.4      4.9


---
Signif. codes: 0 '***' 0.001 '**' 0.01 '*' 0.05 '.' 0.1 ' ' 1
```

The source of variance includes three components: the variance for the model, the variance for the residuals, and the total variance. The total sum of squares of the variance (SST) is partitioned into the model sum of squares (SSM) and the residual sum of squares (SSR). Since this model has only one independent variable or predictor variable, the SSM is the sum squares of the predictor variable `realrincl`. The Analysis of Variance table displays the variance for `realrincl` and the residuals, so we can compute the SST as follows.

$$\text{SST} = \text{SSM} + \text{SSR} = 5,805 + 5,108.4 = 10,913.4$$

The `Df` column lists the degrees of freedom for their respective variances. The total degrees of freedom $= n - 1 = 1{,}047 - 1 = 1{,}046$. The degrees of freedom for the model equal the number of predictors $k = 1$. The degrees of freedom for the residual $= n - 1 - k = 1{,}047 - 2 = 1{,}045$.

The `Mean Sq` column lists the mean squares of the variance, which is the ratio of the sum squares and the corresponding degrees of freedom. For example, MS (model) = 5,805.0/1 = 5,805.0.

The F statistic equals the ratio of the mean squares of variance for the model (i.e., `realrincl`) to the mean squares of variance for the `Residuals`. F = MSM/MSR = 5,805.0/4.9 = 1,187.50. $F(1, 1{,}045) = 1{,}187.50$.

2.4.6 The coef() Function and the confint() Function

The regression coefficient and the constant can also be obtained using the `coef(slm)` command and their confident intervals can be obtained with `confint(slm)`.

```
> coef(slm)
(Intercept)  realrincl
  1.8703542  0.7871115
> confint(slm)
                  2.5 %     97.5 %
(Intercept)  1.6968568  2.0438516
realrincl    0.7422919  0.8319312
```

2.4.7 Effect Size With the eta_sq() Function

We can also compute the effect size using use the `eta_sq()` function in the `sjstats` package (Lüdecke, 2021). You need to install the package first by typing `install.packages("sjstats")` and then load it with `library(sjstats)`. In the following example, we use the `eta_sq(slm)` command to compute the eta square (η^2).

```
> # Install sjstats using install.packages()
> library(sjstats)
> eta_sq(slm)
     term etasq
1 realrincl 0.532
```

In the output, the overall model $\eta^2 = .532$, which is the same as R^2 in the output of the `summary(slm)` command above.

2.4.8 Reporting the Results

A simple regression analysis was conducted to investigate whether an individual's annual income was a predictor of the family income and how accurate the prediction was: $F(1, 1{,}045) = 1{,}187.50$, $p < .001$, which indicated that the model with one predictor, respondent's income, was significantly different from zero. The regression coefficient $\beta = .787$, $p < .001$, which indicates that there was a significant effect of respondent's income on the dependent variable, family income. For a one-unit increase in the respondent's annual income, the family income was increased by a factor of .787.

2.5 MULTIPLE LINEAR REGRESSION

2.5.1 Multiple Linear Regression: An Introduction

Multiple linear regression is simply an extension of the simple linear regression when there are two or more independent variables. It is used to predict a continuous dependent variable from a combination of predictors that can be either continuous or categorical variables. Similar to simple regression, multiple linear regression can be expressed as follows:

$$Y = \beta_0 + \beta_1 X_1 + \beta_2 X_2 + \ldots + \beta_k X_k + e$$

where Y is the continuous dependent variable; X_1, X_2, ..., and X_k are a set of independent variables; β_0 is the intercept; and β_1, β_2, and β_k are the regression coefficients for predictors. As with simple linear regression, the intercept and regression coefficients are estimated using the OLS method.

In the following example, we are interested in how well the family income can be predicted by a combination of three independent variables using the GSS 2016 dataset. The dependent variable is the annual family income, and the independent variables are the respondent's annual income, the highest years of education completed, and age.

Research Question: How accurately can the family income be predicted from a set of three independent variables, the respondent's annual income, the highest years of education completed, and age?

2.5.2 The `lm()` Function

As with simple regression, the function for multiple regression is still the same `lm()` function. The model formula in `lm()` specifies the dependent variable and the predictor variable(s), which are separated by the tilde (~). When there are multiple predictor variables, they are connected by plus (+) symbols. For more details on how to use this command, type `help(lm)` in the command prompt.

In the following example, the command `mlm <- lm(realinc1 ~ realrinc1 + educ + age)` tells R to predict the dependent variable `realinc1` from the three independent variables `realrinc1`, `educ`, and `age`. The fitted model is named `mlm`. The output is shown by the `summary(mlm)` command.

```
# Multiple linear regression
> mlm <- lm(realinc1 ~ realrinc1 + educ + age, data=chp2)
> summary(mlm)

Call:
lm(formula = realinc1 ~ realrinc1 + educ + age, data = chp2)

Residuals:
    Min       1Q   Median      3Q      Max
-3.3887  -1.2862  -0.6401  0.5646  11.6320
```

```
Coefficients:
             Estimate  Std. Error  t value    Pr(>|t|)
(Intercept)  -1.219610   0.394154   -3.094   0.002026 **
realrincl     0.733981   0.023196   31.642    < 2e-16 ***
educ          0.168181   0.023660    7.108   2.17e-12 ***
age           0.018453   0.004787    3.855   0.000123 ***

______
Signif. codes: 0 '***' 0.001 '**' 0.01 '*' 0.05 '.' 0.1 ' ' 1

Residual standard error: 2.147 on 1043 degrees of freedom
  (826 observations deleted due to missingness)
Multiple R-squared: 0.5594, Adjusted R-squared: 0.5581
F-statistic: 441.4 on 3 and 1043 DF, p-value: < 2.2e-16
```

2.5.3 Interpreting R Output: The Coefficients Table

In the R output for the regression analysis, the first part is the call, which shows the R command for the model. The second part shows the residuals. The minimum, first quarter, median, third quarter, and maximum values of the residuals are shown here. The third part shows the coefficients table including the parameter estimates for the predictor variable and the intercept, their standard errors, the t statistics, and the associated p values. The fourth part shows the residual standard error and the degrees of freedom. The fifth part shows the multiple R^2 and the adjusted multiple R^2. Finally, the sixth part shows the F statistic, the degrees of freedom, and the associated p value.

First, in the regression model table, we can see the estimated regression coefficients, their standard errors, the t values, and the associated p values for the constant and the three predictor variables. The 95% confident intervals can be obtained with the `confint(mlm)` command, which will be introduced later in this section.

The t statistic in the regression table tests whether each regression coefficient of the three independent variables is significantly different from zero, controlling for the other independent variables. It is the ratio of the estimated regression coefficient to its standard error.

The regression coefficient for the first predictor, respondent's annual income (`realrincl`), $\beta = .734$. The t value equals the ratio of the regression coefficient to its standard error: $t = .7340/.0232 = 31.642$.

Under the heading, `P(>|t|)`, $p <$ `2e-16`. This means that the probability of having a t value larger than the absolute value of 31.64 is close to zero if the null hypothesis is true. Since $p < .001$, we can conclude that the regression coefficient of the independent variable, respondent's income, is significantly different from zero when holding the other two predictors constant. Therefore, the respondent's income is a significant predictor of the dependent variable, family income.

The regression coefficient $\beta_1 = .734$. This can be interpreted as follows: For a one-unit increase in the respondent's annual income, the dependent variable, the family income, is increased by a factor of .734 when holding the other two predictors constant.

The regression coefficient for the second predictor, the highest years of education completed (`educ`), $\beta_2 = .168$, $t = .1682/.0237 = 7.10$.

Under the heading, `P(>|t|)`, $p <$ `2e-16`. This means that the probability of having a t value larger than the absolute value of 7.10 is almost zero if the null hypothesis is true. Since $p < .001$, we can conclude that the regression coefficient of the independent variable, the highest years of education completed, is significantly different from zero when holding the other two predictors constant. Therefore, the predictor, the highest years of education completed, is a significant predictor of the dependent variable, family income.

The regression coefficient $\beta_2 = .168$, which means that for a one-unit increase in the highest years of education completed, the dependent variable, the family income, is increased by a factor of .168 when holding the other two predictors constant.

The regression coefficient for the third predictor age (`age`), $\beta_3 = .018$. $t = .0185/.0048 = 3.854$.

Under the heading, `P(>|t|)`, $p < .001$. We can conclude that the regression coefficient of the independent variable, age, is significantly different from zero when controlling for the other two predictors. Therefore, the predictor, age, is a significant predictor of the dependent variable, family income.

In the model, the constant (β_0), shown as `(Intercept)`, $= -1.220$. It is also known as the intercept of the model and is the mean family income when the values of the independent variables equal zero.

Substituting the values of the constant and regression coefficients into the equation of the estimated regression model:

$$\hat{Y} = \beta_0 + \beta_1 X_1 + \beta_2 X_2 + \beta_3 X_3$$

We get:

$$\hat{Y} = -1.220 + 0.734X_1 + 0.168X_2 + 0.018X_3$$

Next, the output reports the residual standard error and the degrees of freedom. The residual standard error $= 2.147$, which is the square root of the mean squares for the residual (i.e., mean square error). The degrees of freedom for the residual $= n - 1 - k = 1{,}047 - 1 - 3 = 1{,}043$ since $n = 1{,}047$ and the number of predictors $k = 3$.

2.5.4 Interpreting R Output: The Multiple R^2 and the F Statistic

Further, the multiple R^2 and the adjusted multiple R^2 are reported. R^2 is the coefficient of multiple determination, which tells us the strength of the prediction with a set of predictors. It is the proportion of error variance explained by the model with all the predictors. $R^2 = .559$, which indicates that 55.9% of the variance in the family income is explained using a fitted model with the three predictors overall. The adjusted R^2 takes the sample size and the number of predictors into consideration, and it is a less biased estimate of the population R^2. Adjusted $R^2 = .558$.

Finally, the output shows the F statistic, the degrees of freedom, and the associate p value. The F statistic in the multiple regression tests whether the overall model with all the predictors can significantly predict the dependent variable.

Null hypothesis: The regression coefficients of all three independent variables are equal to zero ($\beta_1 = \beta_2 = \beta_3 = 0$).

In other words, all three independent variables are not significant predictors of the dependent variable, family income.

Alternative hypothesis: At least one of the three regression coefficients of the independent variables is different from zero, controlling for the others. (At least one β_j is not equal to 0.)

In other words, at least one of the three independent variables is a significant predictor of the dependent variable, family income.

$F(3, 1{,}043) = 441.4$, $p < .001$, which indicates that the overall model with three predictors, respondent's income, the highest years of education completed, and age, is significant. In other words, at least one independent variable significantly predicts the dependent variable.

2.5.5 The `coef()` Function and the `confint()` Function

The regression coefficient and the constant can also be obtained using the `coef(mlm)` command and their confident intervals can be obtained with the `confint(mlm)` command.

```
> coef(mlm)
(Intercept)  realrincl       educ         age
-1.2196105   0.7339815  0.1681815   0.0184535

> confint(mlm)
                    2.5 %       97.5 %
(Intercept)  -1.993036447  -0.44618454
realrincl     0.688464467   0.77949850
educ          0.121755538   0.21460741
age           0.009059685   0.02784731
```

A 95% confidence interval for a regression coefficient means that we are 95% confident that this interval contains the true coefficient with repeated samples. The 95% confidence interval for the predictor `realrincl` is [.688, .779]. It can be interpreted as follows: For a one-unit change in the respondent's annual income, we are 95% confident that the change in the dependent variable, the family income, is between .668 and .779 when controlling for the other two variables.

The 95% confidence interval for the predictor `educ` is [.122, .215]. This means that for a one-unit change in the highest years of education completed, we are 95% confident that the change in the dependent variable, the family income, is between .122 and .215 when holding the other two predictors constant.

The 95% confidence interval for the predictor `age` can be interpreted in a similar way.

2.5.6 The ANOVA Table

We use the `anova()` function to request the ANOVA table, which was discussed in the one-way ANOVA section. The command `anova(mlm)` requests the ANOVA table for the multiple regression model.

```
> anova(mlm)
Analysis of Variance Table

Response: realincl
            Df  Sum Sq  Mean Sq   F value      Pr(>F)
realrincl    1  5805.0   5805.0  1259.157   < 2.2e-16 ***
educ         1   231.4    231.4    50.188   2.567e-12 ***
age          1    68.5     68.5    14.859    0.000123 ***
Residuals 1043  4808.5      4.6

---
Signif. codes:  0 '***' 0.001 '**' 0.01 '*' 0.05 '.' 0.1 ' ' 1
```

Just as explained for simple regression, in the ANOVA table, the `Sum Sq` column displays the sum of squares for each predictor variable and the sum of squares for the residuals (also called the sum of squared errors). The total sum of squares can be obtained by adding all the sum of squares. The `Df` column displays the respective degrees of freedom related to each type of sum of squares. The `Mean Sq` column shows the mean squares of the variance, which is the ratio of the sum squares to the corresponding degrees of freedom. The `F value` column shows the *F* statistics. Finally, the `Pr(>F)` column displays the associated *p* values.

2.5.7 Effect Size With the `eta_sq()` Function

We again use the `eta_sq()` function in the `sjstats` package to compute the effect size. Since we have loaded the package in the previous section for simple linear regression, you do not need to load it again. We can also use the `anova_stats()` function in the `sjstats` package to compute more effect size statistics. In the following example, we use the `eta_sq(mlm, ci.lvl = .95)` command to compute the eta square (η^2). In the `eta_sq()` function, `mlm` is the model object and the `ci.lvl = .95` argument requests the 95% confidence interval.

```
> eta_sq(mlm, ci.lvl = .95)
        term  etasq  conf.low  conf.high
1  realrincl  0.532     0.466      0.598
2       educ  0.021     0.010      0.033
3        age  0.006     0.001      0.012
```

The η^2 for respondent's annual income (`realrincl`) is .532, which indicates that 53.2% of the variance in the family income is explained by this predictor variable. The eta squares (η^2) for the other two predictor variables `educ` and `age` can be interpreted in the same way. The same results can be obtained with the `effectsize::eta_squared(mlm, partial = FALSE)` command.

We can also use the `eta_sq(mlm, partial=TRUE, ci.lvl = .95)` command to compute the partial η^2. The output is as follows.

```
> eta_sq(mlm, partial=TRUE, ci.lvl = .95)
        term  partial.etasq  conf.low  conf.high
1  realrincl          0.547     0.510      0.580
2       educ          0.046     0.024      0.073
3        age          0.014     0.003      0.031
```

In the output, the partial η^2 for respondent's annual income (`realreincl`) is .547, which indicates that 54.7% of the variance in the family income is explained by this predictor variable while excluding variance explained by the other two predictor variables. The partial eta squares (η^2) for the other two predictor variables `educ` and `age` can be interpreted in the same way.

2.5.8 Computing the Predicted Values With the `ggpredict()` Function in `ggeffects`

The `ggpredict()` function in the `ggeffects` package (Lüdecke, 2018b) is used to compute the predicted outcome variable if we know the specified values of a predictor variable or multiple predictor variables. This command is particularly useful for generalized linear models that will be introduced in the following chapters. The `ggeffects` package needs to be installed first by typing `install.packages("ggeffects")`. We then load it with `library(ggeffects)`. In the first example, we compute the predicted values of the outcome variable when the predictor variable `educ` is specified at the values of 12, 14, and 16 and the other two predictors are held at their means. Please note when predictor variables are non-numeric or categorical, they are held at the reference level or category. The command is as follows: `mlm.educ <- ggpredict(mlm, terms="educ[12, 14, 16]")`. In the `ggpredict()` function, `mlm` is the model object and the `terms="educ[12, 14, 16]"` option specifies the predictor variable `educ` at the values of 12, 14, and 16. When there are more than one variable, the `terms` option can specify up to four variables, including the second to fourth grouping variables. In this example, we include only one variable in the `terms` option. The output is assigned to an object named `mlm.educ`. Either the `as.data.frame()` function or the `sqrt(diag(vcov()))` function can be used to request the standard errors of the predicted values.

```
> # Predicted values with ggpredict() in ggeffects
> # Install ggeffects using install.packages()
> library(ggeffects)
> mlm.educ <-ggpredict(mlm, terms="educ[12, 14, 16]")
> mlm.educ

# Predicted values of realinc1


educ  |  Predicted  |          95% CI
-----------------------------------------

  12  |       3.41  |  [3.25, 3.58]
  14  |       3.75  |  [3.62, 3.88]
  16  |       4.09  |  [3.93, 4.24]



Adjusted for:
* realrinc1 =  2.46
*        age = 44.00
> as.data.frame(mlm.educ)
    x  predicted    std.error   conf.low   conf.high  group
1  12   3.413709  0.08459629  3.247904   3.579515      1
2  14   3.750072  0.06672711  3.619290   3.880855      1
3  16   4.086435  0.07890954  3.931775   4.241095      1

> sqrt(diag(vcov(mlm.educ)))
          1           2           3
0.08459629  0.06672711  0.07890954

> plot(mlm.educ)
Loading required namespace: ggplot2
```

In the output, when `educ` equals 8, 13, and 16 and the two predicator variables `realreinc1` and `age` are held at their means, 2.46, and 44.00, respectively, the mean predicted outcomes are 3.41, 3.75, and 4.09, respectively. The same values can be obtained if we substitute the specified values of `educ` and the means of the other two predictor variables into the multiple regression equation provided earlier. Figure 2.1 shows the predicted values of the outcome variable for `educ` at 12, 14, and 16 when the other two predictor variables are held at their means.

FIGURE 2.1 ● Predicted Values of the Outcome Variable for educ at 12, 14, and 16 With Others Fixed at Their Means

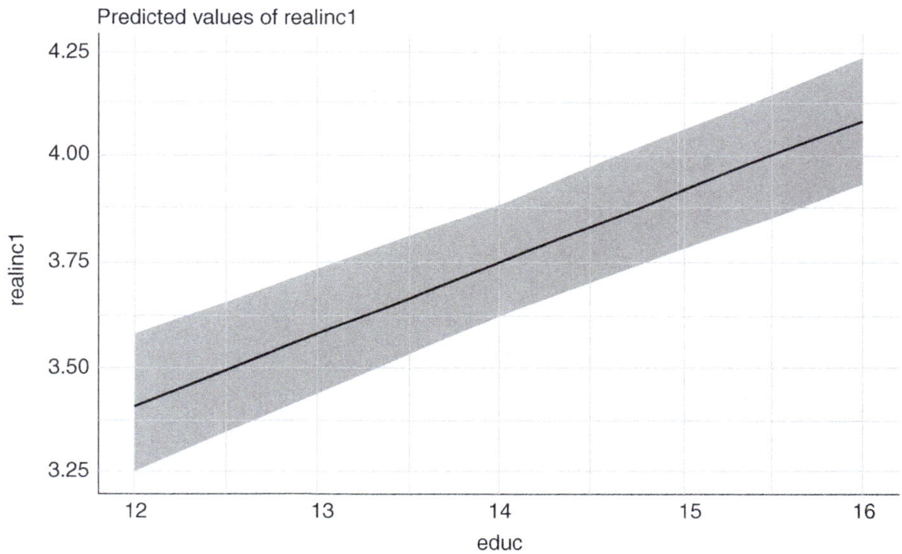

In the second example, we compute the predicted values of the outcome variable when the three predictor variables are specified at their mean values and at one standard deviation below and above the mean values with the `meansd` option. The command is as follows: `mlm.m <- ggpredict(mlm, terms=c("educ[meansd]", "realrinc1[meansd]", "age[meansd]"))`. In the `ggpredict()` function, `mlm` is the model object and the `terms= c("educ[meansd]", "realrinc1[meansd]", "age[meansd]")` option specifies the mean value and at one standard deviation below and above the mean value for each predictor variable. In this example, we use the `c()` function to include the three variables in the `terms` option. The output is assigned to an object named `mlm.m`.

```
> mlm.m <- ggpredict(mlm, terms=c("educ[meansd]", "realrinc1[meansd]", "age
[meansd]"))
> mlm.m

# Predicted values of realinc1
# realrinc1 = -0.54
#       age = 31.1

 educ  | Predicted  |        95% CI
-----------------------------------
11.30  |      0.86  | [0.62, 1.10]
14.20  |      1.35  | [1.12, 1.57]
17.10  |      1.83  | [1.55, 2.11]

# realrinc1 = 2.46
#       age = 31.1

 educ  | Predicted  |        95% CI
-----------------------------------
11.30  |      3.06  | [2.83, 3.29]
14.20  |      3.55  | [3.36, 3.73]
17.10  |      4.04  | [3.81, 4.26]

# realrinc1 = 5.45
#       age = 31.1

 educ  | Predicted  |        95% CI
-----------------------------------
11.30  |      5.25  | [4.97, 5.54]
14.20  |      5.74  | [5.51, 5.98]
17.10  |      6.23  | [5.98, 6.48]

# realrinc1 = -0.54
#       age = 45

 educ  | Predicted  |        95% CI
-----------------------------------
11.30  |      1.11  | [0.91, 1.32]
14.20  |      1.60  | [1.41, 1.79]
17.10  |      2.09  | [1.84, 2.34]

# realrinc1 = 2.46
#       age = 45

 educ  | Predicted  |        95% CI
-----------------------------------
11.30  |      3.32  | [3.13, 3.50]
14.20  |      3.80  | [3.67, 3.93]
17.10  |      4.29  | [4.11, 4.48]
```

```
# realrincl = 5.45
#       age = 45

 educ  | Predicted  |          95% CI
---------------------------------------
11.30  |       5.51 |  [5.26, 5.76]
14.20  |       6.00 |  [5.81, 6.19]
17.10  |       6.49 |  [6.28, 6.69]


# realrincl = -0.54
#       age = 59

 educ  | Predicted  |          95% CI
---------------------------------------
11.30  |       1.37 |  [1.12, 1.62]
14.20  |       1.86 |  [1.63, 2.10]
17.10  |       2.35 |  [2.06, 2.64]


# realrincl = 2.46
#       age = 59

 educ  | Predicted  |          95% CI
---------------------------------------
11.30  |       3.58 |  [3.35, 3.80]
14.20  |       4.06 |  [3.88, 4.25]
17.10  |       4.55 |  [4.32, 4.78]


# realrincl = 5.45
#       age = 59

 educ  | Predicted  |          95% CI
---------------------------------------
11.30  |       5.77 |  [5.49, 6.05]
14.20  |       6.26 |  [6.03, 6.48]
17.10  |       6.75 |  [6.50, 6.99]
```

Since each predictor variable is specified at the three values, the mean value and at one standard deviation below and above the mean value, the output displays nine tables for the combinations of the three variables. The mean for `educ` is 14.20 and the values for one standard deviation below and above the mean value are 11.30 and 17.10, respectively. Similarly, the mean for `realrincl` is 2.46 and the values for one standard deviation below and above the mean value are −.54 and 5.45, respectively;

the mean for `age` is 45.00 and the values for one standard deviation below and above the mean value are 31.10 and 59.00, respectively.

Let us take a look at the first table in the output. When `educ` = 11.30, `realrinc1` = −.54, and `age` = 31.10, the estimated margin or the mean predicted outcome ($\hat{Y}$) is .86. When `educ` = 14.20, `realrinc1` = −.54, and `age` = 31.10, the estimated margin or the mean predicted outcome ($\hat{Y}$) is 1.35. When `educ` = 17.10, `realrinc1` = −.54, and `age` = 31.10, the estimated margin or the mean predicted outcome ($\hat{Y}$) is 1.83.

When all three predictor variables are held at their means (i.e., `educ` = 14.20, `realrinc1` = 2.46, and `age` = 45.00), the mean predicted outcome ($\hat{Y}$) is 3.80. The same result is displayed in the fifth table in the output.

2.5.9 Reporting the Results

A multiple regression analysis was conducted to predict the annual family income using three predictors, the respondent's annual income, the highest years of education completed, and age. $F(3, 1{,}043) = 441.4$, $p < .001$, which indicated that the overall model with the three predictors was significant. $R^2 = .559$, which indicated that 55.9% of the variance in the family income was explained by the fitted model with the three predictors overall. Adjusted $R^2 = .558$.

The respondent's income was a significant predictor of the dependent variable, family income ($\beta = .734$, $t = 31.642$, $p < .001$). For a one-unit increase in the respondent's annual income, the dependent variable, the family income, was increased by a factor of .734 when holding the other two predictors constant.

The predictor, the highest years of education completed, was also a significant predictor of the dependent variable, family income ($\beta = .168$, $t = 7.108$, $p < .001$). For a one-unit increase in the highest years of education completed, the dependent variable, the family income, was increased by a factor of .168 when holding the other two predictors constant.

In addition, age was a significant predictor of the dependent variable, family income ($\beta = .018$, $t = 3.855$, $p < .001$). For a one-unit increase in age, the family income was increased by a factor of .018 when holding the other two predictors constant.

2.6 CHI-SQUARE TEST

2.6.1 The Chi-Square Test: An Introduction

The chi-square test of independence is used to investigate the relationship between two categorical variables. Each categorical variable has two or more levels/categories. A two-way contingency table can be constructed with one variable as the row variable and the other as the column variable. The rows list different levels of the row variable, and the columns represent categories of the column variable. For example, a 4×5 contingency table shows frequencies for a row variable with four levels and a column variable with five levels.

A simple case for the chi-square test is a two-by-two frequency table, which includes two categorical variables with two levels for each category. In it the rows represent two categories of one variable and the columns represent two categories of the other. Each cell where a row and column intersects tells the frequency number of participants that fall into each subgroup.

In the following example, we are interested in whether two categorical variables, health status and marital status, are related using the GSS 2016 dataset. One variable, health status, has four levels, including poor, fair, good, and excellent. The other variable, marital status, has five levels: married, widowed, divorced, separated, and never married.

Research Question: Is there a relationship between the two categorical variables, health status and marital status? Or, in other words, are health status and marital status independent of each other?

> *Null hypothesis*: There is a relationship between the two categorical variables, health status and marital status.
>
> *Alternative hypothesis*: There is no relationship between the two categorical variables, health status and marital status.

2.6.2 The `CrossTable()` Function in the `gmodels` Package

To get the relative frequency of each cell within its row and column, we use the `CrossTable()` function in the `gmodels` package. Since the package has been installed earlier in this chapter, we just need to load it with `library(gmodels)`. If it has been loaded, you do not need to load it again. In the following example, we use the `CrossTable(health, marital, digits=2)` command to create a frequency table. In the function, `health` and `marital` are the two categorical variables and the argument `digits=2` specifies the number of decimals.

```
> # Cross-tabulation with CrossTable() in gmodels
> library(gmodels)
> CrossTable(health, marital, digits=2)
   Cell Contents
|  ------------------------- |
|                          N |
|  Chi-square contribution   |
|              N / Row Total |
|              N / Col Total |
|            N / Table Total |
|  ------------------------- |

Total Observations in Table:  1873

             | marital
      health |    married |    widowed |   divorced |  separated | never married |  Row Total |
------------ | ---------- | ---------- | ---------- | ---------- | ------------- | ---------- |
   excellent |        196 |         35 |         60 |          7 |           116 |        414 |
             |       1.24 |       0.18 |       1.95 |       3.78 |          0.42 |            |
             |       0.47 |       0.08 |       0.14 |       0.02 |          0.28 |       0.22 |
             |       0.24 |       0.21 |       0.18 |       0.11 |          0.23 |            |
             |       0.10 |       0.02 |       0.03 |       0.00 |          0.06 |            |
------------ | ---------- | ---------- | ---------- | ---------- | ------------- | ---------- |
        good |        430 |         76 |        162 |         32 |           214 |        914 |
             |       2.30 |       0.58 |       0.07 |       0.00 |          3.04 |            |
             |       0.47 |       0.08 |       0.18 |       0.04 |          0.23 |       0.49 |
             |       0.53 |       0.45 |       0.50 |       0.49 |          0.43 |            |
             |       0.23 |       0.04 |       0.09 |       0.02 |          0.11 |            |
------------ | ---------- | ---------- | ---------- | ---------- | ------------- | ---------- |
        fair |        160 |         46 |         65 |         21 |           135 |        427 |
             |       3.82 |       1.35 |       1.12 |       2.58 |          4.45 |            |
             |       0.37 |       0.11 |       0.15 |       0.05 |          0.32 |       0.23 |
             |       0.20 |       0.27 |       0.20 |       0.32 |          0.27 |            |
             |       0.09 |       0.02 |       0.03 |       0.01 |          0.07 |            |
------------ | ---------- | ---------- | ---------- | ---------- | ------------- | ---------- |
        poor |         33 |         13 |         38 |          5 |            29 |        118 |
             |       6.70 |       0.49 |      15.00 |       0.20 |          0.14 |            |
             |       0.28 |       0.11 |       0.32 |       0.04 |          0.25 |       0.06 |
             |       0.04 |       0.08 |       0.12 |       0.08 |          0.06 |            |
             |       0.02 |       0.01 |       0.02 |       0.00 |          0.02 |            |
------------ | ---------- | ---------- | ---------- | ---------- | ------------- | ---------- |
Column Total |        819 |        170 |        325 |         65 |           494 |       1873 |
             |       0.44 |       0.09 |       0.17 |       0.03 |          0.26 |            |
------------ | ---------- | ---------- | ---------- | ---------- | ------------- | ---------- |
```

Interpreting R Output

The output provides a two-way contingency table. The row variable, health status, has four levels, and the column variable, marital status, has five levels. The last row shows the column totals (819, 170, 325, 65, and 494), and the last column displays the row totals (414, 914, 427, and 118). The row and column totals are also called the marginal totals according to the places where they are located. The total number of observations is shown at the lower right bottom of the table (n = 1,873). The row percentage and column percentage are also displayed in the table.

2.6.3 The `chisq.test()` Function

We use the `chisq.test()` function to conduct the chi-square test for the two categorical variables `health` and `marital`. The command is `chisq.test(health, marital)` and the resulting output is displayed as follows.

```
> chisq.test(health, marital)

        Pearson's Chi-squared test

data: health and marital
X-squared = 49.423, df = 12, p-value = 1.764e-06

Warning message:
In chisq.test(health, marital) : Chi-squared approximation may be incorrect
```

The Pearson $\chi^2 = 49.423$ with the degrees of freedom of 12. The number of degrees of freedom in the χ^2 test $= (r - 1) \times (c - 1) = 3 \times 4 = 12$ since r is the number of rows and c is the number of columns.

$\chi^2_{(12)} = 49.423$, $p < .001$, which indicates that there is a significant relationship between health status and marital status.

2.6.4 Cramér's *V*

There are several ways to compute the Cramér's *V*. We first use the `xtab_statistics()` function in the `sjstats` package. The `sjstats` package needs to be installed first by typing `install.packages("sjstats")`. We then load it with `library(sjstats)`.

Before we use the `xtab_statistics()` function, we need to work on the data frame to correctly code the missing values as "NA." The current data frame includes missing values coded as "iap" (i.e., inapplicable), "dk" (i.e., don't know), and "na" (i.e., no answer). These three types of missing values need to be coded as "NA" in R. We first load the data frame `chp2.bs`, then we select the two factor variables `health` and `marital` by using the `dplyr::select(health, marital)` command.

In the command, `dplyr::select` means that we use the `select ()` function from the `dplyr` package with `::`, the double colons. Finally, we recoded the three types of missing values to "NA" by using the `mutate ()` function in the `dplyr` package and the `rec ()` function in the `sjmisc` packages. Since both the `dplyr` and `sjmisc` packages have been loaded in the previous section, we do not need to load them again. You need to load them if you have not done so. The command is as follows: `dplyr::mutate(health_re = rec(health, rec = "iap=NA; dk=NA; na=NA; else=copy"), marital_re = rec(marital, rec = "na=NA; else=copy"))`. The recoded two variables are named `health_re` and `marital_re`, respectively. The pipe operator `%>%` is used to run a series of functions in sequence. The new data frame is named as the object `new`.

```
> # Recode user-defined missing values to NA
> new<-chp2 %>%
+  dplyr::select(health, marital) %>%
+  dplyr::mutate(health_re = rec(health, rec = "iap=NA; dk=NA; na=NA; else=copy"),
marital_re = rec(marital, rec = "na=NA; else=copy") )

One or more of the old values are recoded into identical new values.
Please check if you correctly specified the recode-pattern,
else separate multiple values with comma, e.g. rec="a,b,c=1; d,e,f=2".
```

To check if the two variables are correctly recoded, we use the `table (new$health_re, new$marital_re)` command to create a two-way cross-tabulation table.

```
> table(new$health_re, new$marital_re)


            divorced  married  never married  separated  widowed
excellent         60      196            116          7       35
fair              65      160            135         21       46
good             162      430            214         32       76
poor              38       33             29          5       13
```

Since "iap," "dk," and "na" are not shown in the cross-tabulation table, these missing values are coded correctly.

We use the `xtab_statistics()` function in the `sjstats` package to compute the Cramér's *V*. In the `xtab_statistics(new, health_re, marital_re)` command, `new` is the data frame and `health_re` and `marital_re` are the two recoded factor variables.

```
# Compute chi-square test statistic and Cramer's V
> library(sjstats)
> xtab_statistics(new, health_re, marital_re)

# Measure of Association for Contingency Tables
     (using Fisher's Exact Test)

 Chi-squared: 49.4229
  Cramer's V : 0.0938
     p-value : <0.001
```

In the output, Cramér's $V = .094$. It also shows that the Pearson $\chi^2 = 49.423$ and $p < .001$.

The second method to compute the Cramér's V is to use the `cramer()` function in the `sjstats` package. In the `cramer()` function, `health_re ~ marital_re` is the model equation and `data=new` is the data argument. The command and the resulting output are as follows.

```
> cramer(health_re ~ marital_re, data=new)
[1] 0.09378524
```

The third method to compute the Cramér's V is to use the `assocstats()` function in the `vcd` package (Meyer et al., 2020). The `vcd` package needs to be installed first by typing `install.packages("vcd")`. We then load it with `library(vcd)`. We first create a two-way table by using the `tab <- table(new$health_re, new$marital_re)` command and the table is assigned to an object named `tab`. Then we use the `summary(assocstats(tab))` command to display the statistics of association between the two factor variables.

```
> # Install vcd by using install.packages()
> library(vcd)
Loading required package: grid
> tab <- table(new$health_re, new$marital_re)
> summary(assocstats(tab))

Number of cases in table: 1873
Number of factors: 2
Test for independence of all factors:
  Chisq = 49.42, df = 12, p-value = 1.764e-06
  Chi-squared approximation may be incorrect
                X^2 df  P(> X^2)
Likelihood Ratio 47.973 12 3.1603e-06
Pearson          49.423 12 1.7639e-06

Phi-Coefficient      :  NA
Contingency Coeff.   :  0.16
Cramer's V           :  0.094
```

The output displays the likelihood ratio χ^2 test statistic, the Pearson χ^2 test statistic, the contingency coefficient, and the Cramér's V. The results are the same as those above.

We can also compute the Cramér's V manually by following the equation

$$\text{Cramér's } V = \sqrt{\frac{\chi^2}{N*(k-1)}}$$

where χ^2 is the Pearson chi-square statistic, N is the sample size, and k is the smaller number of categories of the two variables. In this example, $\chi^2 = 49.423$ and $N = 1{,}873$. Since the numbers of categories for `health_re` and `marital_re` are 4 and 5, respectively, the smaller number of categories is 4. $k - 1 = 3$. We compute the Cramér's V as follows.

```
> # Compute the Cramer's V
> sqrt(49.423/1873/3)
[1] 0.09378536
```

Cramér's V indicates the strength of association between two categorical variables. Its size lies between 0 and 1. In the output, Cramér's $V = .094$, which indicates that the relationship between two variables is weak.

In the output, we can easily see that each cell displays the row and column percentages in addition to the frequency. The row percentage equals the number of counts in a cell divided by its corresponding row total at the margin (i.e., the marginal total).

2.6.5 Follow-Up Chi-Square Test With the `chisq.test()` Function

The Pearson χ^2 test discussed earlier indicated that there is a significant relationship between health status and marital status. It is an omnibus test for the overall model. If we are interested in the relationship between the variables for subcategories, then we can conduct follow-up tests. For example, we can examine the health status between married and widowed. The research question would be as follows: Do the married and the widowed differ among the levels of health status?

Since we know the frequency distribution from the earlier output, we can use the `matrix()` function to create a 4 by 2 matrix with the command `matrix(c(196, 35, 160, 46, 430, 76, 33, 13), nrow=4, byrow=T)` and the matrix is assigned to an object named `list`. We then conduct a chi-square test with the `chisq.test(list)` command.

```
> # Create a matrix and conduct a chi-square test
> list <- matrix(c(196, 35, 160, 46, 430, 76, 33, 13), nrow=4, byrow=T)
> chisq.test(list)
```

```
	Pearson's Chi-squared test

data: list
X-squared = 10.133, df = 3, p-value = 0.01747
```

$\chi^2_{(3)} = 10.133, p < .05$, which indicates that there is a significant relationship between health status and marital status and that the married have better health status levels than the widowed.

Other follow-up comparisons can be done in a similar way. To control the Type I error due to multiple comparisons, we need to adjust the significant level. To do this, we can use the Bonferroni adjustment.

2.6.6 Reporting the Results

A Pearson chi-square test was conducted to investigate the relationship between two categorical variables, health status and marital status, using the GSS 2016 dataset. $\chi^2_{(12)} = 49.423$, $p < .001$, which indicated that there was a significant relationship between health status and marital status. Cramér's *V* was used to indicate the strength of association between two categorical variables. Cramér's $V = .094$, which indicated that the relationship between two variables was weak.

2.7 MAKING PUBLICATION-QUALITY TABLES USING R

Once you have conducted statistical analyses and interpreted results, the final step of the research process is to report the results and submit your manuscript for publication. While writing the manuscript, you may find that you need to summarize your research findings and display them in tables. You might wonder whether there are any tools to reduce your workload, or automatically combine results from the fitted models and generate a single publication-style regression table.

R has several user-written packages to accomplish this job with no hassle. We use the `stargazer` package (Hlavac, 2018) and the `texreg` package (Leifeld, 2013) to produce regression tables throughout this book since they work with most of the models covered here.

To install these two add-on packages, you can use the `install.packages()` function and then load it with `library()` function. For example, if you would like to install `stargazer`, type `install.packages("stargazer")`, choose a mirror close to you, and then install the package. To use it, you need to load it with the `library(stargazer)` command.

The help files of these two packages provide various examples with explanations on how to use them. The following is an example of using the `stargazer()` function in the `stargazer` package to make a table for the results of two regression models. After fitting the single-predictor model `slm` and the multiple-predictor model `mlm` introduced in the previous section, we use the `stargazer(slm, mlm, type="text", align=TRUE, out="chp2.lrmod.txt")` command to create a table. In the `stargazer()` function, we first specify `slm` and `mlm`, the two model objects to be presented, and then the type of the table with the `type="text"`

```
> library(stargazer)

Please cite as:

 Hlavac, Marek (2018). stargazer: Well-Formatted Regression and Summary Statistics Tables.
 R package version 5.2.2. https://CRAN.R-project.org/package=stargazer

> stargazer(slm, mlm, type="text", align=TRUE, out="chp2.lrmod.txt")

==================================================================
                                  Dependent variable:
                    ----------------------------------------------
                                       realincl
                          (1)                          (2)
------------------------------------------------------------------
realrincl               0.787***                     0.734***
                        (0.023)                      (0.023)
educ                                                 0.168***
                                                     (0.024)
age                                                  0.018***
                                                     (0.005)
Constant                1.870***                     -1.220***
                        (0.088)                      (0.394)
------------------------------------------------------------------
Observations             1,047                        1,047
R2                       0.532                        0.559
Adjusted R2              0.531                        0.558
Residual Std. Error   2.211 (df = 1045)           2.147 (df = 1043)
F Statistic       1,187.513*** (df = 1; 1045)  441.401*** (df = 3; 1043)
==================================================================
Note:                                  *p<0.1; **p<0.05; ***p<0.01

> stargazer(slm, mlm, type="html", align=TRUE, out="chp2.lrmod.htm")
```

argument. The `align=TRUE` argument requests to align the results of the two models. The `out="chp2.lrmod.txt"` argument saves the output named `chp2.lrmod.txt`.

Table 2.1 shows the results of two regression models using the `stargazer()` function.

TABLE 2.1 ● Results of Two Regression Models: An Example

	Dependent variable:	
	realinc1	
	(1)	(2)
realrinc1	0.787***	0.734***
	(0.023)	(0.023)
educ		0.168***
		(0.024)
age		0.018***
		(0.005)
Constant	1.870***	−1.220***
	(0.088)	(0.394)
Observations	1,047	1,047
R^2	0.532	0.559
Adjusted R^2	0.531	0.558
Residual Std. Error	2.211 (df = 1,045)	2.147 (df = 1,043)
F Statistic	1,187.513*** (df = 1; 1,045)	441.401*** (df = 3; 1,043)

Note:
*$p < 0.1$
**$p < 0.05$
***$p < 0.01$

A similar table can be created if you use the `screenreg()` and `htmlreg()` functions from the `texreg` package. You need to install `texreg` first by typing `install.packages("texreg")` if you haven't already done so and then load the package by typing `library(texreg)`. To create a plain text table similar to Table 2.1, we use the `screenreg(list(slm, mlm))` command. In the `screenreg()` function, we specify the two model objects to be presented with the `list()` function. We can also use the `htmlreg()` function to create the same regression table and save it to a Microsoft Word file named `chap2.doc` with the following command: `htmlreg(list(slm, mlm), file="chap2.doc", doctype=TRUE, html.tag=TRUE, head.tag=TRUE)`. To save space, the created table is omitted here.

Both the `stargazer` package and the `texreg` package can be of good help when creating a regression table since you do not need to start from scratch. They are particularly useful when you fit a series of models and need to summarize parameter estimates from each model. Please note that different fields or journals have different requirements for the table format. You still need to do some editing to the created tables in the manuscript submitted for publication.

2.8 GENERAL GUIDELINES FOR REPORTING RESULTS

Once the data analysis is complete, the next step is to present and interpret the results, which normally are included in the results section of a research report. When reporting the results of your statistical analyses, a general rule is to provide sufficient information for readers to understand your analyses and the findings of your study. What are the major elements that should be included in the results section? The answer varies since different disciplines and journals may have their own reporting requirements, and one statistic commonly reported in one field may not be needed by another field. We provide the following general guidelines for reporting results.

First, describe the analyses you have conducted, explain the variables with descriptive statistics, and state what research questions have been addressed.

Second, when reporting the results of a statistical test, provide the value of the test, the degrees of freedom, and the associated p value, followed with an explanation of the meaning of your findings. You may also need to form a conclusion about whether the test is significant. If the test is significant, provide the effect size if available. The reporting of the effect sizes was recommended in APA (2020), and various measures of effect size for continuous and categorical outcome variables and their estimation methods were introduced in Kline (2013).

Third, when summarizing numerical information in tables, make sure they can be easily interpreted. The labels, categories, and numbers in the tables should be concise and clear, so that readers can understand them without much effort. Complex tables may confuse readers with excessive information.

Fourth, tables and written text should be complementary to each other. If the results can be summarized in a sentence, you do not need a table. On the other hand, if you have many categorical variables and you need report frequencies for all categories, it will be tedious to report them in paragraph after paragraph in written text, and it will be boring for readers to read your description. Instead, a table containing these frequencies is sufficient to help readers quickly understand this information.

Fifth, after the results of statistical tests are presented, you also need to interpret them. Readers are more interested in the meaning of the statistical results than in the technical information related to the tests. The numerical information is important since it is the evidence supporting your conclusion. To help readers understand these statistics, you need to interpret them in a clear manner.

Finally, when summarizing the results, also keep the audience in mind. If you submit your article to a journal for publication, reviewers are interested in reading the statistics, tables, and graphs. You might receive feedback from them asking you to provide more statistics. However, if your readers have limited backgrounds in statistics, you will need to explain your results in plain English in addition to the statistics you provide in the text.

2.9 SUMMARY OF R COMMANDS IN THIS CHAPTER

```
# Chap 2 R Script

# Remove all objects
rm(list = ls(all=TRUE))
# Set a working directory; here we create a folder named CDA in the C drive (C:) first
setwd("C:/CDA")

# The following user-written packages need to be installed first by using
install.packages(" ") and then by loading it with library()
# library(dplyr)      # It is already installed for Chapter 1
# library(sjmisc)     # It is already installed for Chapter 1
# library(gmodels)
# library(ggeffects)
# library(pastecs)
# library(stargazer)
# library(sjstats)
# library(vcd)

# Import GSS 2016 Stata data file
library(foreign)
chp2 <- read.dta("C:/CDA/gss2016-chap1.dta")
attach(chp2)
summary(chp2)
head(chp2)
str(age)
summary(age)
mean(age, na.rm=TRUE)
sd(age, na.rm=TRUE)
max(age, na.rm=TRUE)
min(age, na.rm=TRUE)
length(age)
```

```
# Descriptive statistics by group: method 1
mean <- tapply(age, sex, mean, na.rm=TRUE)
sd <- tapply(age, sex, sd, na.rm=TRUE)
max <- tapply(age, sex, max, na.rm=TRUE)
min <- tapply(age, sex, min, na.rm=TRUE)
length <- tapply(age, sex, length)
cbind(mean, sd, max, min, length)

# Descriptive statistics by group: method 2
library(dplyr)
gender <- group_by(chp2, sex)
summarize(gender, mean(age, na.rm=TRUE), sd(age, na.rm=TRUE), max(age, na.rm=TRUE),
min(age, na.rm=TRUE))

# Descriptive statistics by group: method 3
library(sjmisc)
chp2 %>% group_by(sex) %>% descr(age)

# Descriptive statistics for two variables
library(pastecs)
stat.desc(chp2[, c("age","educ")])

# Descriptive statistics for two variables: method 2
#library(sjmisc)
descr(chp2, age, educ)

# Descriptive statistics for multiple variables by group
chp2 %>% group_by(sex) %>% select(age, educ) %>% descr()

# Frequency table for a factor or categorical variable
str(degree)
table(degree)
table(marital)
# Frequency table with frq() in sjmisc
#library(sjmisc)
frq(chp2, degree)
frq(chp2, marital)

# Cross-tabulation
tab <- table(degree, race)
summary(tab)
tab
ftable(tab)
addmargins(tab)
prop.table(tab)

# Cross-tabulation with CrossTable() in gmodels
library(gmodels)
CrossTable(degree, race, digits=2)
chisq.test(degree, race)

# Simple linear regression
realinc1 <- realinc/10000
```

```
realrinc1 <- realrinc/10000
slm <- lm(realinc1 ~ realrinc1)
summary(slm)
anova(slm)
coef(slm)
confint(slm)
library(sjstats)
eta_sq(slm)

# Multiple linear regression
mlm <- lm(realinc1 ~ realrinc1 + educ + age, data=chp2)
summary(mlm)
anova(mlm)
coef(mlm)
confint(mlm)
eta_sq(mlm, ci.lvl = .95)
eta_sq(mlm, partial=TRUE, ci.lvl = .95)
anova_stats(mlm, digits = 3)

# Predicted values with ggpredict() in ggeffects
library(ggeffects)
mlm.educ <- ggpredict(mlm, terms="educ[12, 14, 16]")
mlm.educ
as.data.frame(mlm.educ)
sqrt(diag(vcov(mlm.educ)))
plot(mlm.educ)

mlm.m  <-  ggpredict(mlm,  terms=c("educ[meansd]",  "realrinc1[meansd]",  "age
[meansd]"))
mlm.m

# Create a results table with stargazer()
library(stargazer)
stargazer(slm, mlm, type="text", align=TRUE, out="chp2.lrmod.txt")
stargazer(slm, mlm, type="html", align=TRUE, out="chp2.lrmod.htm")

# Cross-tabulation with CrossTable() in gmodels
library(gmodels)
CrossTable(health, marital, digits=2)
chisq.test(health, marital)

# Recode user-defined missing values to NA
new <- chp2 %>%
 dplyr::select(health, marital) %>%
 dplyr::mutate(health_re = rec(health, rec = "iap=NA; dk=NA; na=NA; else=copy"),
marital_re = rec(marital, rec = "na=NA; else=copy" ) )

table(new$health_re, new$marital_re)

# Compute chi-square test statistic and Cramer's V
library(sjstats)
xtab_statistics(new, health_re, marital_re)
cramer(health_re ~ marital_re, data=new)
```

```
library(vcd)
tab <- table(new$health_re, new$marital_re)
summary(assocstats(tab))

# Compute the Cramer's V
sqrt(49.423/1873/3)

# Create a matrix and conduct a chi-square test
list <- matrix(c(196, 35, 160, 46, 430, 76, 33, 13), nrow=4, byrow=T)
chisq.test(list)

 detach(chp2)
```

Glossary

Descriptive statistics helps you describe your data by showing types of data, graphs of the distribution of variables, and various statistical indices, such as the central tendency and variability of your data.

Frequency analysis provides the frequency of each value for categorical variables.

Multiple linear regression is used when we predict a dependent variable from two or more independent variables.

Simple linear regression is used when we predict a dependent variable from an independent variable.

The chi-square test of independence is used to investigate the relationship between two categorical variables.

Exercises

Use the GSS 2016 data available at **https://edge.sagepub.com/liu1e** for the following problems.

1. Run a descriptive statistics analysis for `coninc` and interpret the results.
2. Conduct a frequency analysis for `happy`. What percentages of respondents are very happy?
3. Make a two-way table for `degree` and `class`.
4. Run a chi-square test to investigate the relationship between `degree` and `class`.
5. Conduct a multiple regression analysis to estimate `tvhours` from the two predictor variables `educ` and `age`.
   a. Write a research question.
   b. Find the *F* statistic from the output and interpret whether the overall model is statistically significant.
   c. Which predictor variables are significant? Interpret the coefficients for the two predictor variables.
   d. Produce a table for the regression output using `stargazer`.
   e. Write a concise report to summarize the results.

3

LOGISTIC REGRESSION FOR BINARY DATA

OBJECTIVES OF THIS CHAPTER

This chapter introduces logistic regression models for binary data. It introduces concepts of odds, odds ratio, and goodness-of-fit statistics of the model; describes how to test the significance of predictors; and shows how to interpret parameter estimates. Following the description of data, two logistic regression models with the `glm()` function are illustrated with step-by-step instructions. R commands and output are explained in detail. In addition, probit regression with R is also introduced. The focus of this chapter is on fitting binary logistic regression models using R, as well as on interpreting and presenting the results. After reading this chapter, you should be able to:

- Determine when a logistic regression model is used.
- Conduct logistic regression using R.
- Interpret the output.
- Interpret the model in terms of odds ratios.
- Interpret the marginal effects
- Compute and plot the predicted probabilities.
- Compare models using the likelihood ratio test and other fit statistics.
- Present the results in publication-quality tables.
- Write the results for publication.

3.1 LOGISTIC REGRESSION MODELS: AN INTRODUCTION

In multiple linear regression, a set of independent variables is used to predict a continuous outcome variable. In this chapter, logistic regression models are introduced for the analysis of dichotomous response variables with two categories.

Research examples include binary response variables such as life satisfaction (Hoffmann, 2016), having a fracture or not (Hosmer et al., 2013), gun ownership (Kaufman, 2019), computer use (Menard, 2010), and skipping school or not (Smithson & Merkle, 2014). These binary response variables have two categories: having an event or not having an event. They are normally coded as 1 for having an event and 0 for not having an event.

We can use multiple linear regression as a preliminary method to estimate a binary outcome variable. The method is referred to as the linear probability model. Due to its simplicity in computation and easy interpretation as with the linear regression model, the linear probability model is still popular in economics. However, this model has several drawbacks because using multiple regression with the ordinary least squares method to estimate a binary response variable violates the assumptions of normality, homoscedasticity, and the linear relationship between the dependent variable and independent variables. First, in multiple linear regression, we assume that the dependent variable has a normal distribution. When the dependent variable is binary, it has a Bernoulli distribution, which is not a normal distribution. Second, the homoscedasticity assumption suggests that the error variances of the dependent variable are the same across each value of the independent variables. This assumption is violated because the error variances depend on the values of the predictor variables and vary across them, which is referred to as heteroscedasticity. Heteroscedasticity may lead to biased standard errors of the estimates and invalid statistical inference. Third, multiple linear regression also assumes that the relationship between the dependent variable and independent variables is linear. When the dependent variable is dichotomous, the relationship between this variable and an independent variable is a nonlinear, S-shaped curve, which can be easily seen if we draw a scatterplot showing the relationship between the dichotomous dependent variable and the independent variable. In addition, we may see the predicted values of the dependent variable are out of the range of 0 and 1. In other words, the estimated values of the outcome variable may be negative (i.e., smaller than 0) or larger than 1. These values are nonsensical since they are the predicted probabilities from the independent variables. Therefore, we should use the linear probability model with caution. If we do use this model, the robust standard error approach should be used to adjust the standard errors to obtain unbiased standard errors of the estimates.

3.1.1 Why Do We Need a Logistic Transformation?

As explained previously, if we use linear regression to estimate binary data, heteroscedasticity (i.e., unequal error variance) is present in the model and the predicted value may have values less than 0 or larger than 1. To estimate the probability of success for having an event, a logistic transformation logit(p) or a probit transformation probit(p)

can be used. A regression model with the logistic transformation is called the logistic regression model, whereas a model with the probit transformation is called the probit regression model. Both models produce similar results and use of either model is down to the choice of researchers. This chapter introduces the binary logistic regression models first and then the probit regression models.

The form of the simple logistic regression model can be expressed as follows:

$$\text{logit}(p) = \alpha + \beta X$$

where p is the probability when the outcome variable equals 1, $P(Y = 1)$; logit(p) is the logistic transformation of the probability of success or of an event occurrence; and on the right side, α is the intercept and β is the logit regression coefficient.

This equation looks like that for simple linear regression. The noticeable difference is the logistic transformation on the left side of logistic regression. Instead of directly estimating the dependent variable, we estimate the logistic transformation (i.e., logit) of the probability of a success, which is also known as the logarithm of the odds or "log odds." Odds are the ratio of the probability of success to the probability of failure. The transformation between probabilities and odds will be introduced in detail in the next section. Therefore, in simple logistic regression, we estimate the relationship between an independent variable and the binary outcome variable on a scale of the logit or log odds. In other words, the relationship between the predictor variable X and the logit of the outcome (i.e., the logit transformation of the probability when the outcome variable $Y = 1$) is linear. The logit of the outcome logit(p) can be easily transformed back to the probability of the outcome $P(Y = 1)$ since the logit is the natural logarithm of the odds or log odds.

Since logit(p) is ln(odds), which is expressed as $\ln\frac{p}{1-p}$, the form of the simple logistic regression can also be rewritten as:

$$\ln\frac{p}{1-p} = \alpha + \beta X \tag{3.1}$$

where ln is the natural logarithm. For simplicity, we read "ln(odds)" log odds or log of the odds.

When the probability (p) varies from 0 to 1, the log odds or logit will vary from negative infinity to positive infinity.

In multiple logistic regression, we have more than one predictor variable. The following is the form of the multiple logistic regression model:

$$\ln\frac{p(x)}{1-p(x)} = \alpha + \beta_1 X_1 + \beta_2 X_2 + \ldots + \beta_p X_p \tag{3.2}$$

where X_1, X_2, ..., X_p are the predictor variables and β_1, β_2, ..., β_p are the logit coefficients of these predictors. This equation can be also expressed as:

$$\text{logit}[p(x)] = \alpha + \beta_1 X_1 + \beta_2 X_2 + \ldots + \beta_p X_p \tag{3.3}$$

3.1.2 Probabilities, Odds, and Odds Ratios

Just as the mean and standard deviation are the cornerstone for the descriptive statistics and linear models, probabilities, odds, and odds ratios are the key concepts for logistic regression models. Probability is the chance of success or of having an event occur. It can be expressed as a proportion. It equals the frequency of success or of having an event divided by the total number of observations or events. Probabilities range from 0 to 1. When probability is 0, it means that the chance of success is 0 or that there is no chance an event will occur. When probability is 1, it means that the chance of success is certain.

If p is the probability of success or having an event, then the probability of failure or of not having an event is $1 - p$ since these two probabilities are complementary. The sum of all possible probabilities equals 1.

The odds are the ratio of two probabilities, the probability of success or of having an event (p) to the probability of failure or of not having an event ($1 - p$). Since $p = P(Y = 1)$ and $1 - p = P(Y = 0)$, the odds can be expressed as follows.

$$\text{Odds} = \frac{P(Y = 1)}{P(Y = 0)} = \frac{p}{1 - p}$$

Since the probability p varies from 0 to 1, the odds vary from 0 to positive infinity. With the increase of the probability of success or of having an event, the odds also get larger. Let us see three examples:

$$\text{When } p = 0.1,\ \text{odds} = 0.1/(1 - 0.1) = 0.11.$$

$$\text{When } p = 0.5,\ \text{odds} = 0.5/(1 - 0.5) = 1.00.$$

$$\text{When } p = 0.1,\ \text{odds} = 0.1/(1 - 0.1) = 9.00.$$

The preceding examples focus on the odds of having an event. What are the odds of not having that event? When the odds of having an event = 99, the odds of not having an event = 1/99 = 0.01 since they are just the multiplicative odds or the reciprocal of the odds of having an event. This is an important property of odds. It is useful when we compare different categories for the odds in various ordinal logistic regression models in the following chapters.

The examples show us how to compute odds if we know the probability of success or of having an event. But if we know the odds of success or of having an event, can we compute the probability of success or of having an event? This backward transformation is also easy. Since odds $= \frac{p}{1-p}$, with a simple transformation, we get $p = \frac{\text{odds}}{1 + \text{odds}}$. In other words, the probability of success or of having an event equals the odds divided by (1 + odds).

For example, the command `table(healthy, wrkfull)` produces a two-way cross-tabulation table of two binary variables, the health status and working full-time.

The health status has two categories with 1 = good health and 0 = not good health; the working full-time is coded as full-time and not.

```
> tab <- table(healthy, wrkfull)
> tab
           wrkfull
healthy      0     1
      0    363   182
      1    635   693

> ftable(tab)
        wrkfull    0      1
healthy
0                 363    182
1                 635    693

> addmargins(tab)
         wrkfull
healthy          0      1     Sum
      0        363    182     545
      1        635    693    1328
    Sum        998    875    1873
```

Since the probability of success or of having an event is a ratio of the frequency of success or of having an event to the total observations, the probability of having good health *P*(healthy = 1) = 1,328/1,873 = .709. The probability of not having good health *P*(healthy = 0) = 545/1,873 = .291. Therefore, overall, the odds of having good health = *P*(healthy = 1)/*P*(healthy = 0) = .709/.291 = 2.436.

Next, let us compute the odds of having good health for not working full-time.

To compute the probability of having good health for people who are currently not working full-time, we look at the ratio of the frequency of people with good health who are not working full-time to the total frequency of people who are not working full-time: *P*(healthy =1 | not full-time work) = 635/998 = .636.

The probability of having poor health for people who do not work full-time *P*(healthy = 0 | not full-time work) = 363/998 = .364.

For people who do not work full-time, the odds of having good health = *P*(healthy = 1 | not full-time work)/*P*(healthy = 0 | not full-time work) = .636/.364= 1.747.

For full-time work, the odds of having good health = *P*(healthy =1 | full-time work)/ *P*(healthy = 0 | full-time work). Since *P*(healthy = 1 | full-time work) = 693/875 = .792 and *P*(healthy = 0 | full-time work) = 182/875 = .208, the odds of having good health for the people who work full-time = .792/(1 − .792) = 3.808.

3.1.3 Transformation Among Probabilities, Odds, and Log Odds in Logistic Regression

Forward Transformation

If we know the probability of success or of having an event, the odds of success are just the probability of success divided by the probability of failure. By taking the natural logarithm of odds, we get log odds or logit. Table 3.1 presents the forward transformation from probabilities to odds to log odds.

TABLE 3.1 ● Forward Transformation From Probabilities to Odds to Log Odds

Transformation	Probabilities	Odds	Logit or Log Odds
Forward	p	$\frac{p}{1-p}$	$\ln\frac{p}{1-p}$

Backward Transformation

If we know the logit, since it is the natural logarithm of the odds or log odds, then the odds are the antilogarithm. In other words, the odds of success can be obtained by taking the exponential of the logit. To transform the odds back to the probability of success, we divide the odds by 1 plus the odds. The backward transformation is displayed in Table 3.2.

TABLE 3.2 ● Backward Transformation From Log Odds to Odds to Probabilities

Transformation	Logit or Log Odds	Odds	Probabilities
Backward	Logit(p) or $\ln\frac{p}{1-p}$	Odds = exp (logit)	$p = \frac{\text{odds}}{1+\text{odds}}$

Odds Ratio

What is an odds ratio (OR)? It is just a ratio of two odds. Since the odds of having good health for the people who work full-time are 3.808 and the odds of having good health for not working full-time are 1.747, the ratio of the odds for working full-time to the odds for not working full-time = 3.808/1.747 = 2.180. In other words, the odds of having good health for working full-time are 2.180 times the odds for not working full-time.

When OR > 1, the odds of success or of having an event for one group are larger than the odds for the other group. For example, OR = 2 indicates the odds of success for one group are two times the odds for the other group.

When OR < 1, the odds in one group are less than the odds in the other group. For example, OR = 0.5 indicates that the odds for one group are 0.5 times the odds for the other group. In other words, the odds for the second group are twice the odds for the first group. When OR is less than 1, we can take the inverse of it and make it more interpretable.

When OR = 1, the odds for one group are the same as the ones for the other group.

3.1.4 Bernoulli Distributions, the Likelihood Function, and Maximum Likelihood Estimation

Bernoulli Distributions

When a discrete random variable is a binary outcome variable with 1 for having an event or success and 0 for not having an event or failure, it follows a Bernoulli distribution. When $Y = 1$, the probability of success or of having an event, $p = P(Y = 1)$; when $Y = 0$, the probability of failure or of not having an event, $1 - p = P(Y = 0)$. The probability function, also called the probability mass function (PMF), for Bernoulli distributions is expressed as:

$$P(Y = y) = p^y(1 - p)^{1-y} \tag{3.4}$$

where y can be either 1 or 0.

For each observation, we add subscripts to the equation as follows.

$$P(Y = y_i) = p_i^{y_i}(1 - p_i)^{1-y_i} \tag{3.5}$$

where y_i can be either 1 or 0.

With the Bernoulli probability function or PMF, we can obtain the probability at each value of the discrete random variable. For all the observations in the sample, you can obtain the joint probability function by multiplying the probability function of each independent observation as follows.

$$P(Y = y_1, y_2, \ldots, y_n) = \prod_{i=1}^{n} p_i^{y_i}(1 - p_i)^{1-y_i} \tag{3.6}$$

where $\prod$ is the product term for the joint probability.

The Likelihood Function

By reparametrizing the function above, we obtain the likelihood function.

$$L(p; y) = \prod_{i=1}^{n} p_i^{y_i}(1 - p_i)^{1-y_i} \tag{3.7}$$

where $L(p; y)$ is the likelihood function, which is the function of the unknown parameter p given the observed data y. The likelihood function is not the same as the probability function, although both equations on the right-hand side look the same. In the likelihood function, the parameter is unknown and needs to be estimated, whereas in the probability function we use the known parameter to obtain probabilities at different values of the discrete random variable. The likelihood function is often expressed as the form of the log likelihood function by taking the natural log of the function in Equation 3.7, so we can work on addition rather than multiplication of the function of individual observations.

The log likelihood function for the Bernoulli distribution can be expressed as:

$$\ln[L(p;\ y)] = \sum_{i=1}^{n} \{y_i \ln p_i + (1 - y_i)\ln(1 - p_i)\} \tag{3.8}$$

Since $(1 - y_i)\ln(1 - p_i) = \ln(1 - p_i) - y_i\ln(1 - p_i)$, the equation above can be rewritten as:

$$\ln[L(p;y)] = \sum_{i=1}^{n} \{y_i \ln p_i + \ln(1 - p_i) - y_i\ln(1 - p_i)\} = \sum_{i=1}^{n} \left\{y_i \ln\frac{p_i}{1 - p_i} + \ln(1 - p_i)\right\} \tag{3.8b}$$

If we define $l(p;\ y) = \ln[L(p;\ y)]$, then the log likelihood function is expressed as:

$$l(p;y) = \sum_{i=1}^{n} \left\{y_i \ln\frac{p_i}{1 - p_i} + \ln(1 - p_i)\right\} \tag{3.9}$$

Maximum Likelihood Estimation

We use maximum likelihood estimation to estimate the unknown parameter by maximizing the likelihood in the function above. In other words, we look for the value of the parameter p which makes the log likelihood value l the largest. If we find that estimated value of the parameter p, then that value is the maximum likelihood estimate that maximizes the likelihood of observing the data.

Let us see an example of the sample data with 10 Bernoulli trials. The set of the sample data includes the value of each observation y_i, which is either 1 or 0

$$\{1,\ 0,\ 0,\ 1,\ 0,\ 1,\ 0,\ 1,\ 1,\ 1\}$$

The likelihood function is:

$$L(p;y) = p(1 - p)(1 - p)p(1 - p)p(1 - p)ppp = p^6(1 - p)^4$$

The log likelihood function is:

$$l(p;y) = \ln\left[p^6(1 - p)^4\right] = 6\ln p + 4\ln(1 - p)$$

To use maximum likelihood estimation to estimate the unknown parameter p, we take the derivative of the log likelihood function $l(p; y)$ with respect to p, set it to 0, and find the solution to the equation.

$$\frac{d}{dp}l = \frac{d}{dp}[6\ln p + 4\ln(1-p)] = \frac{6}{p} - \frac{4}{1-p} = 0$$

Therefore, we obtain the estimated p or $\hat{p} = \frac{6}{10} = \frac{3}{5}$.

The $\hat{p}$ value is the maximum likelihood estimate of the parameter p. With $\hat{p} = \frac{3}{5}$, the log likelihood function $l(p; y)$ has its maximum value. In other words, among all the possible values for the parameter p, the estimated $p = \frac{3}{5}$ is the maximum likelihood estimate which makes the sample data with 10 Bernoulli trials above to be the most likely.

Similarly, in simple logistic regression, we use maximum likelihood estimation to estimate the parameter, the logit coefficient β. We take the first derivative of the log likelihood function $l(\beta; y, x)$ with respect to the coefficient β, set it to 0, and solve. To estimate the standard error of the coefficient, we take the second derivative of the log likelihood function with respect to the coefficient to obtain the Hessian matrix. The variance–covariance matrix of the coefficient is just the negative inverse Hessian matrix. Recall that in simple logistic regression, we estimate the relationship between an independent variable and the binary outcome variable on a scale of the logit or log odds and the probability of the outcome $P(Y = 1)$ is conditional on the values of the predictor variable. Maximum likelihood estimation of the coefficient is far more complex than the example demonstrated above. Technical detail of maximum likelihood estimation is beyond the scope of this book; interested readers should refer to Hilbe (2009) for more detail.

The Binomial Distribution

The Bernoulli distribution is a special case of the binomial distribution with one trial for each individual. When we count the number of successes for the total trials, the discrete random variable follows a binomial distribution. With k successes in n trials with the success probability p, the binomial distribution is expressed as:

$$P(Y = k) = \binom{n}{k} p^k (1-p)^{n-k} \tag{3.10}$$

where $\binom{n}{k}$ is the binomial coefficient and is read as "n choose k." It gives us the number of ways we have k successes in n trials.

$$\binom{n}{k} = \frac{n!}{(n-k)!k!}$$

where $n!$ is n factorial or the factorial of n. $n! = n*(n - 1) \ldots 2*1$.

The log likelihood function for the binomial distribution is expressed as:

$$l(p;\ y;\ n) = \sum_{i=1}^{n} \left\{ y_i \ln \frac{p_i}{1-p_i} + n_i \ln(1-p_i) + \ln \binom{n}{k} \right\} \tag{3.11}$$

Since $\ln\binom{n}{k}$ is a constant, when $n = 1$, the log likelihood function for the binomial distribution is the same as the log likelihood function for the Bernoulli distribution; so is the estimate by maximum likelihood estimation.

3.1.5 Goodness-of-Fit Statistics

To assess whether a model fits the data well, we normally look at several measures of fit statistics rather than at a single measure. The deviance, log likelihood ratio test, pseudo R^2, AIC, and BIC statistics are introduced next.

Deviance

Deviance is one of the goodness-of-fit statistics used in the logistic regression. It compares the currently fitted model and the saturated model. It is defined as −2 times the difference in log likelihood between these two models, which can be expressed as −2(log likelihood of the current model − log likelihood of the saturated model). The saturated model is the model that fits the data perfectly.

In binary logistic regression, the saturated model estimates one parameter for each observation so the number of parameters equals the number of sample size. Its likelihood of fitting the data is 1, so the log likelihood of the saturated model equals ln1, which is 0. The expression of deviance can be simplified to −2 log likelihood of the current model. It is often abbreviated as −2LL.

Deviance shows how well a model fits the data compared with a saturated model that fits the data perfectly. If the discrepancy in log likelihood between the fitted model and the saturated model is small, then this model has a good fit. On the contrary, if the deviance is large, then the fitted model has a poor fit. Therefore, smaller deviance means a better fit. In a linear regression model, we minimize the error variance and would like to see the sum of squared residuals as small as possible. Similarly, in a logistic regression model, we would like to minimize the deviance and would like to see the discrepancy in log likelihood between the fitted model and the saturated model as small as possible.

Model Comparisons Using the Deviance Difference or Likelihood Ratio Test

Deviance is often used to compare nested models. Models are nested when more constraints can be put on parameters in one model than in the other. One model is called the reduced model, which contains less parameters, and the other is called the full model, within which the reduced model is nested but has more parameters. In the logistic regression, the reduced model has fewer variables than the full model, and the former model is a subset of the latter. The difference in deviance between nested models follows a chi-square distribution. The degrees of freedom of the distribution equal the difference in the number of parameters between these two models. The difference in deviance is often expressed as G = Deviance for the reduced model − Deviance for the full model or as follows: $G = D_{\text{Reduced}} - D_{\text{Full}}$.

This test is also known as the likelihood ratio test since the difference in deviance is the difference in −2LL, which can be expressed as a ratio of likelihood in logarithm.

$$G = -2\ln\frac{\text{Likelihood}_{\text{Reduced}}}{\text{Likelihood}_{\text{Full}}} = -2\text{LL}_{\text{Reduced}} - (-2\text{LL}_{\text{Full}}) = D_{\text{Reduced}} - D_{\text{Full}} \tag{3.12}$$

where $D_{\text{Reduced}} = -2\text{LL}_{\text{Reduced}}$ and $D_{\text{Full}} = -2\text{LL}_{\text{Full}}$.

In simple logistic regression with only one independent variable, the likelihood ratio test compares the deviance between the null model with only the intercept (D_0) and the model with one independent variable (D_m). The likelihood ratio test, $G = -2\text{LL}_0 - (-2\text{LL}_m) = D_0 - D_m$. If the likelihood ratio chi-square test is significant, then we reject the null hypothesis and conclude that the model with one independent variable fits the data better than the model with only the intercept (null model).

Predictor Selections Using the Likelihood Ratio Test

The likelihood ratio chi-square test can be used to test whether a predictor variable contributes to the model by comparing the models with and without the variable. A significant test means that the added variable contributes to the model. It is also useful for model developing. We can build a series of nested models from a simple model with one predictor to more complex models with multiple predictors. The likelihood ratio chi-square test can be used to decide which model fits the data better and whether the predictor variables should be kept in or removed from the model.

Pseudo R^2

In linear regression models, the coefficient of determination, R^2, is the index for the model fit. It is a ratio of the variance explained by the model to the total variance. It indicates how much variance in the dependent variable is accounted for by an independent variable or a set of independent variables. Analogous to R^2 in the linear regression, several pseudo R^2 are used in logistic regression (Hardin & Hilbe, 2018; Long & Freese, 2014; Menard, 2010). However, the interpretation of these pseudo R^2 in logistic regression is different from the interpretation of R^2 in the linear regression. In addition, there is no consensus as to which of the pseudo R^2 measures is the best. Table 3.3 displays three major pseudo R^2 measures and their formulas.

1. The Likelihood Ratio R^2
   The likelihood ratio R^2, written as R_L^2, is also known as McFadden's R^2, or the McFadden R^2. It is the reduction in deviance from the fitted model (D_m) to the null model that only contains the intercept (D_0). It is expressed as:

$$R_{\mathrm{L}}^2 = 1 - \frac{D_{\mathrm{m}}}{D_0} = 1 - \frac{-2\mathrm{LL_m}}{-2\mathrm{LL_0}} \quad (3.13)$$

when model deviance equals -2 times log likelihood ($D = -2\mathrm{LL}$), the likelihood ratio R^2 can also be the reduction in log likelihood between the fitted model and the null model. It is expressed as:

$$R_{\mathrm{L}}^2 = 1 - \frac{-2\mathrm{LL_m}}{-2\mathrm{LL_0}} = 1 - \frac{\mathrm{LL_m}}{\mathrm{LL_0}} \quad (3.14)$$

2. Cox and Snell's R^2
   Cox and Snell's R^2 or the Cox and Snell R^2, written as R_{ML}^2, is also known as the maximum likelihood R^2. It is based on the likelihood function of the fitted model (L_{m}) and the null model, which only contains the intercept (L_0). It is expressed as:

$$R_{\mathrm{ML}}^2 \left(\frac{L_0}{L_{\mathrm{m}}}\right)^{2/n} \quad (3.15)$$

where n is the total number of observations.

3. Nagelkerke's R^2
   This is also called Cragg and Uhler's R^2. It is an adjustment to the Cox and Snell R^2 by dividing the maximum value of the Cox and Snell R^2. It is expressed as:

$$R_{\mathrm{N}}^2 = R_{\mathrm{ML}}^2 / \text{maximum } R_{\mathrm{ML}}^2 \quad (3.16)$$

4. Other Pseudo R^2 Measures
   Other pseudo R^2 measures include the adjusted McFadden's R^2, McKelvey and Zavoina's R^2, Efron's R^2, Tjur's R^2, and the Count and adjusted Count R^2. Refer to Hardin and Hilbe (2018) and Long and Freese (2014) for detailed introductions to pseudo R^2 measures.

TABLE 3.3 ● Three Major Pseudo R^2 Measures

Pseudo R^2	Formulas
Likelihood ratio R^2 (McFadden's R^2)	$R_{\mathrm{L}}^2 = 1 - \frac{D_m}{D_0} = 1 - \frac{-2\mathrm{LL_m}}{-2\mathrm{LL_0}}$
Cox and Snell's R^2 (maximum likelihood R^2)	$R_{\mathrm{ML}}^2 \left(\frac{L_0}{L_m}\right)^{2/n}$
Nagelkerke's R^2 (Cragg and Uhler's R^2)	$R_{\mathrm{N}}^2 = R_{\mathrm{ML}}^2/\text{maximum } R_{\mathrm{ML}}^2 = \left\{1-\left(\frac{L_0}{L_m}\right)\right\}^{2/n} / (1-L_0)^{2/n}$

Information Criteria Indices: AIC and BIC

The likelihood ratio tests are used for comparisons between nested models. But how do you compare models if they are not nested (e.g., if two models contain different sets of independent variables or if one model has more missing values so the sample size is different between the two models)? To compare non-nested models, the Akaike information criterion (AIC) (Akaike, 1974) and the Bayesian information criterion (BIC) (Schwarz, 1978) statistics are commonly used. Both statistics are based on the deviance statistics and can be seen as an adjustment to the deviance. Just like the likelihood ratio test, they can also be used to compare nested models. Table 3.4 presents formulas for the AIC and BIC statistics.

The AIC adjusts or penalizes the deviance by the number of predictors:

$$\text{AIC} = -2(\text{LL}_{\text{m}} - k) = D_{\text{m}} + 2k \qquad (3.17)$$

where k is the number of parameters (the number of independent variables plus the intercept) and -2LL_{m} or D_{m} is the deviance of the fitted model.

Remember that we would like to see smaller deviance; this is still true for the AIC statistic. A smaller AIC means a better fit of the model.

The BIC adjusts the deviance by its degrees of freedom and the sample size:

$$\text{BIC} = -2\text{LL}_{\text{m}} + \ln(n) \times k = D_{\text{m}} + \ln(n) \times k \qquad (3.18)$$

where k is the number of parameters (the number of independent variables plus the intercept), n is the number of the sample size, and D_{m} is the deviance of the fitted model. Similar to the AIC statistic, we also prefer a smaller BIC statistic.

The AIC and BIC can be used to compare both nested and non-nested models. When comparing nested models, the BIC often leads to the selection of more parsimonious models, which are simpler models with few parameters. The BIC statistic has a larger penalty term in the form than that in the AIC statistic, so if the focus is to choose a more parsimonious model, the BIC is normally preferred.

TABLE 3.4 ● Information Criteria Indices: AIC and BIC Statistics

Information Criteria Indices	Formulas
AIC	$-2(\text{LL}_{\text{m}} - k) = D_{\text{m}} + 2k$
BIC	$-2\text{LL}_{\text{m}} + \ln(n) \times k = D_{\text{m}} + \ln(n) \times k$

3.1.6 Testing Significance of Predictors

To test the statistical significance of each predictor, the Wald test is used. It is computed as a ratio of the parameter estimate for each predictor variable in the model to its corresponding standard error.

$$\text{Wald } z = \frac{\hat{\beta}}{\text{SE}(\hat{\beta})} \tag{3.19}$$

where $\hat{\beta}$ is the estimated logit coefficient and $\text{SE}(\hat{\beta})$ is the standard error. Statistical software packages report either the univariate Wald z statistic, which follows a standard normal distribution, or a Wald chi-square test statistic, z^2, which follows a chi-square distribution. The null hypothesis is that the coefficient of each predictor variable is 0. Rejection of the null hypothesis indicates that the effect of a particular predictor variable is significant.

$$\text{Confidence intervals for the parameter } \beta = \hat{\beta} \pm z{*}\text{SE}(\hat{\beta})$$

where $\hat{\beta}$ is the estimated logit coefficient for a predictor, z is the z-score from a normal distribution for the chosen confidence interval, and $\text{SE}(\hat{\beta})$ is the standard error. For the 95% confidence intervals, $z = 1.96$. Therefore, 95% confidence intervals for the parameter $\beta = \hat{\beta} \pm 1.96{*}\text{SE}(\hat{\beta})$.

Another way to test the significance of predictors is the likelihood ratio test, which compares -2LL of different models. With only one additional predictor added to the existing model, the difference in deviance between the nested models has a chi-square distribution with one degree of freedom (i.e., the degree of freedom equals the difference between the number of predictors). A significant likelihood ratio test means that the variable is a significant predictor in the model. For a univariate test of a single predictor, the results of the Wald test and the likelihood ratio test are equivalent. The likelihood ratio test can also be used to test the contribution of two or more variables when they are added to the current model. A significant likelihood ratio test for the nested models means that a set of variables overall makes a significant contribution to the model.

3.1.7 Interpretation of Model Parameter Estimates in Logistic Regression

Probabilities, Odds, and Odds Ratios in Logistic Regression

In Section 3.1.3, the transformation from probabilities to odds and to odds ratios was introduced. Let us take a look at their transformation in logistic regression. In the simple logistic regression, $\ln\frac{p}{1-p} = \alpha + \beta X$, where p is the probability when $Y = 1$. The estimated coefficient is the logit coefficient, which is the coefficient on the scale of logit or log odds. In the simple logistic regression, the outcome variable is dichotomous with values of 1 and 0. We estimate the relationship between the predictor variable and the logit function of the probability that $Y = 1$, the log odds.

Exponentiating both sides of the equation, we get the odds of success or of having an event:

$$\text{Odds}(Y = 1) = \exp(\alpha + \beta X)$$

If the independent variable X is a categorical variable with the values of 0 and 1, what are the odds of success or of having an event?

When $X = 0$, odds $(Y = 1) = \exp(\alpha)$, which is the exponentiated intercept.

When $X = 1$, odds $(Y = 1) = \exp(\alpha + \beta)$, which is the exponentiated sum of intercept and logit coefficient.

The odds ratio of the group 1 ($X = 1$) to the group 2 ($X = 0$):

$$\text{OR} = \frac{\exp(\alpha + \beta)}{\exp(\alpha)} = \frac{\exp(\alpha) \times \exp(\beta)}{\exp(\alpha)} = \exp(\beta) \tag{3.20}$$

This is the simple case for a one-unit increase in an independent variable (e.g., from 0 to 1 in the previous example); the change in the odds is the odds ratio, which is the exponentiated logit coefficient. When the independent variable is continuous, for a one-unit increase from any value of x to the value of $(x + 1)$, the change in the odds is still the exponentiated logit coefficient. Since the 95% confidence intervals for the parameter $\beta = \hat{\beta} \pm 1.96 * \text{SE}(\hat{\beta})$, by exponentiating this form we obtain the corresponding confidence intervals for the odds ratio, $\exp[\hat{\beta} \pm 1.96 * \text{SE}(\hat{\beta})]$.

When the logit coefficient is positive, it indicates the relationship between the predictor variable and the logit function of the probability is positive. By exponentiating the logit coefficient, we get the odds ratio, which is larger than 1. This means that the odds of success or of having an event increases for a one-unit increase in the predictor variable.

When the logit coefficient is negative, it indicates that the relationship between the predictor variable and the logit function is negative. The exponentiated coefficient, the odds ratio, is less than 1. This means that the odds of success or of having an event decreases for a one-unit increase in the predictor variable.

When the logit coefficient equals 0, the odds ratio equals 1. This indicates that there is no relationship between the predictor and the odds of success.

Interpreting an Odds Ratio as a Percentage Change in Odds

Another way of interpreting odds ratios is the percentage change in odds. It can be calculated by using (odds ratio − 1) × 100%. A positive percentage change in odds indicates there is an increase in the odds, whereas a negative percentage change corresponds to a decrease in the odds. A zero-percentage change indicates no change in odds at all. In other words, the predictor variable does not influence the odds of success.

For example, if an OR for a predictor variable equals 1.2, then the percentage change in the odds can be computed as follows: $(1.2 - 1) \times 100\% = 20\%$. This indicates that each one-unit increase in the predictor variable corresponds to an increase of 20% in the odds of success.

In another example, if OR = .80, then the percentage change in the odds is $(0.80 - 1) \times 100\% = -20\%$. Since the percentage change is negative, it indicates that for each one-unit increase in the predictor variable, there is a decrease of 20% in the odds of success.

Interpreting Coefficients in Terms of Predicted Probabilities

In addition to odds ratios, we can interpret logistic regression models in terms of predicted probabilities or predictive margins. Recall that the logit is the natural logarithm of the odds, or log odds, and the odds of success is obtained by taking the exponential of the logit. By using the backward transformation method introduced previously, we can easily transform the odds back to the probability of success. The predicted probability can be calculated at specified values of a set of predictor variables. When a combination of the values of the predictor variable changes, the predicted probabilities are different.

Interpreting Marginal Effects as Changes in Predicted Probabilities

A marginal effect is a change in a response variable related to the change in an independent variable. Mathematically, it is the partial derivative of the regression equation with respect to an individual predictor variable in the model. In a simple linear regression model, a marginal effect is the same as the regression coefficient. In logistic regression, a marginal effect is a change in the predicted probability with a change in the values of a predictor variable. When a predictor variable is binary with the values of 1 and 0, the marginal effect is a discrete change in the predicted probability; when a predictor variable is continuous, the marginal effect is an instantaneous rate of change in the predicted probability, which is a rate of change at a specific value of the predictor variable rather than the two values for a binary predictor variable. Therefore, a marginal effect for a continuous predictor variable in logistic regression is often an approximate change in the predicted probability with a one-unit change, or strictly speaking, it is a change in the predicted probability with a small change in that predictor variable. An average marginal effect (AME) is the mean value of the marginal effects at each value of a predictor variable. Marginal effects can be computed at means or at specified values of predictor variables, which are referred to as marginal effects at means (MEM) or marginal effects at representative values (MER), respectively.

We can also compute the predicted probabilities for a particular variable at given values. These predicted probabilities are also called predictive margins or adjusted probabilities. Changes in the predicted probabilities can be used to compute the marginal effects. For example, when a predictor variable is binary with two categories, the difference between the predicted probabilities is the marginal effect of that variable.

3.2 RESEARCH EXAMPLE AND DESCRIPTION OF THE DATA AND SAMPLE

Research Problem and Questions: In this example, we are interested in investigating whether the binary response variable, health status, can be predicted by four predictor variables, marital status, years of education, age, and gender. The research question is as follows: Which predictor variables are associated with the likelihood of having good health? In other words, can health status be predicted by the four preceding variables?

Description of the Data and Sample: The data for the following analyses were the General Social Survey 2016 (GSS 2016). The following are the variables:

- `healthy`: the recoded variable of health (health status) with 1 = good health and 0 = not good health
- `wrkfull`: working full-time with 1 = full-time and 0 = not full-time
- `maritals`: the recoded variable of marital (marital status) with 1 = currently married and 0 = not currently married
- `female`: recoded variable of sex with 1 = female and 0 = male
- `educ`: the highest education.

3.3 GENERALIZED LINEAR MODELS AND THE `glm()` FUNCTION

3.3.1 Generalized Linear Models and the `glm()` Function: An Introduction

Generalized linear models extend the linear regression model when the response variable is a noncontinuous variable, such as a binary, an ordinal, a nominal, or a count variable. When the response variable is continuous, the linear regression model is a special case of generalized linear models. In generalized linear models, the response variable follows a distribution from the exponential family, such as a normal distribution for a continuous response variable, a binomial distribution for a binary variable, and a Poisson distribution for a count variable. There are three common components in the framework of generalized linear models: the random component, the systematic component, and the link function.

First, a probability distribution of the response variable Y is the random component of the model. For example, the binomial distribution of a binary response variable is the random component in a logistic regression model.

Second, a generalized linear model also includes the systematic component, a linear combination of the predictor variables, which is referred to as the linear predictor with the form $\alpha + \beta_1X_1 + \beta_2X_2 + \ldots + \beta_pX_p$. If we define the linear predictor as η (*eta*), the systematic component can be expressed as the form: $\eta = \alpha + \beta_1X_1 + \beta_2X_2 + \ldots + \beta_pX_p$.

Third, a generalized linear model has a component called the link function to connect the linear predictor and the expected value $E(Y)$ or the conditional mean μ (mu) of the response variable. The link function is written as $g(\mu)$, which links the linear predictor η and the conditional mean μ of the response variable. For example, the identity link, $g(\mu) = \mu$, is used for linear regression models, the logit link, $g(\mu) = \ln(\mu/1-\mu)$, is used for logistic regression models, and the natural log link, $g(\mu) = \ln(\mu)$, is used for Poisson regression and negative binomial regression models. These link functions are the most commonly used canonical links which connect the linear predictor η with the natural parameter, the conditional mean μ. Other link functions, such as the probit function and the complementary log-log, can also be used in models for binary, ordinal, and nominal response variables. Table 3.5 provides the types of outcome variables, the probability distributions, the link functions with names, and the corresponding models which are covered in this book.

TABLE 3.5 ● Outcome Variables, Distributions, Link Functions, Function Names, and the Corresponding Models

Outcome Variable	Distribution	Link Function	Link Function Name	Model
Continuous	Gaussian	μ	identity	Linear regression
Binary	Bernoulli or Binomial	$\ln(\mu/1-\mu)$	logit	Logistic regression
Ordinal	Multinomial	$\ln\left(\frac{P(Y\leq j)}{P(Y>j)}\right)$	logit	Ordinal logistic regression (proportional odds model)
Ordinal	Multinomial	$\ln\left(\frac{P(Y=j)}{P(Y>j)}\right)$	logit	Ordinal logistic regression (continuation ratio odds model)
Ordinal	Multinomial	$\ln\left(\frac{P(Y=j+1)}{P(Y=j)}\right)$	logit	Ordinal logistic regression (adjacent categories model)
Nominal	Multinomial	$\ln\left(\frac{P(Y=j)}{P(Y=J)}\right)$	logit	Multinomial logistic regression
Count	Poisson	$\ln(\mu)$	natural log	Poisson regression
Count	Negative binomial	$\ln\left(\frac{\alpha\mu}{1+\alpha\mu}\right)$ (canonical form)	natural log	Negative binomial regression (canonical)
Count	Negative binomial	$\ln(\mu)$	natural log	Negative binomial (NB-2)

3.3.2 The `glm()` Function

The `glm()` function is normally used to fit generalized linear models where a binary logistic regression model belongs. The model formula syntax of `glm()` is similar to that of `lm()` introduced in the last chapter but includes the additional `family=` argument for the probability distribution of the outcome variable and the `link=` argument for the link function. The model formula in `glm()` specifies the dependent variable and the predictor variable(s), which are separated by the tilde (~). When there are multiple predictor variables in the formula, they are connected by plus (+) symbols.

In addition to the model formula, next, we also need to specify the probability distribution of the outcome variable with the `family=` argument. For example, the `family = binomial` argument is used for a binary outcome variable in logistic regression and `family = poisson` is specified for a count variable in Poisson regression. Other families include Gaussian, Gamma, inverse Gaussian, quasi, quasi-Poisson, and quasi-binomial. We also need to specify the link function associated with the `family` argument. For example, `(link = "logit")` stands for the logit link and `(link = "log")` is used for the log link. Several link functions can be used in the `binomial` family. In this chapter, we will only focus on the logit and probit links. Please also note that the current `glm()` function cannot be used to fit models for ordinal and nominal response variables, so other packages will be introduced in the following chapters.

Most extractor functions for the `lm()` function can also be used for the `glm()` function. We can use the `summary()` function to display the summary results of the fitted model, use the `coef()` function to extract the coefficients, use the `confint()` function to request the confidence intervals of the coefficients, and use the `anova()` function to conduct the likelihood ratio test for model comparison. In addition, we can the `exp(coef())` function to obtain the odds ratios. Other useful extractor functions include the `fitted()` function for creating the fitted values, the `residuals()` function for the residual values, the `predict()` function for the predicted values of an outcome variable, the `AIC()` function for the AIC statistic, and the `plot()` function for diagnostic plots.

3.4 SIMPLE LOGISTIC REGRESSION USING R

3.4.1 Simple Logistic Regression: R Syntax

In simple logistic regression, there is only one binary dependent variable with values of 1 and 0 and one independent variable, which can be either categorical or continuous. The command `glm(y ~ x, family = binomial(link = "logit"))` tells R to fit a simple logistic regression model predicting the binary dependent variable *y* with an independent variable *x* by specifying the binomial family and the logit link function. For more details on how to use this command, type `help(glm)` in the command prompt.

In the following example, the command `glm(healthy ~ wrkfull, data = chp3.lr, family = binomial(link = "logit"))` tells R to predict the dependent variable `healthy` from the independent variable `wrkfull`. The fitted model is named `LR.1`. The output is shown by the `summary(LR.1)` command.

```
> # Simple logistic regression
> LR.1 <- glm(healthy ~ wrkfull, data = chp3.lr, family = binomial(link = "logit"))
> summary(LR.1)


Call:
glm(formula = healthy ~ wrkfull, family = binomial(link = "logit"),
    data = chp2.lr)
```

```
Deviance Residuals:
    Min         1Q    Median        3Q       Max
-1.7721    -1.4222    0.6829    0.9509    0.9509

Coefficients:
                Estimate   Std. Error   z value    Pr(>|z|)
(Intercept)       0.5592       0.0658     8.499    < 2e-16 ***
wrkfull           0.7778       0.1061     7.328   2.34e-13 ***

---
Signif. codes: 0 '***' 0.001 '**' 0.01 '*' 0.05 '.' 0.1 ' ' 1

(Dispersion parameter for binomial family taken to be 1)

    Null  deviance: 2258.9  on  1872  degrees of freedom
Residual  deviance: 2203.2  on  1871  degrees of freedom
AIC: 2207.2

Number of Fisher Scoring iterations: 4
```

3.4.2 Interpreting R Output

In the R output for simple logistic regression, the first part is the call, which shows the R syntax for the model. The second part shows the deviance residuals, which are similar to the residuals in a linear regression model. The minimum, first quarter, median, third quarter, and maximum values of the deviance residuals are shown here. The third part shows the coefficients table including the parameter estimates for the predictor variable and the intercept, their standard errors, the Wald z statistics, and the associated p values. The null hypothesis for the Wald test is that the coefficient of the predictor variable is 0, and the alternative hypothesis is that the coefficient of the predictor variable is significantly different from 0.

The Wald z statistic equals the parameter estimate divided by its standard error. For the predictor variable `wrkfull`, Wald $z = .778/.106 = 7.328$. Some other statistical software packages report the Wald chi-square test statistic, which is the squared Wald z statistic. The associated p value, `Pr(>|z|) < .001`, so we rejected the null hypothesis. The rejection of the null hypothesis indicates that the predictor variable `wrkfull` is a significant predictor of the dependent variable `healthy`.

Finally, the fourth part of the output shows the fit statistics including the null deviance, the residual deviance, and the AIC. The null deviance is the deviance for the null model with the intercept only. The residual deviance is the deviance for the fitted model, which is defined as −2(log likelihood of the current model − log likelihood of the saturated model). The difference between the residual deviance and the null deviance can be used to evaluate the significance of the fitted model. Please note that the residual deviance is not the same as the deviance residual. The deviance residual is the residual for each case, whereas the residual deviance is the total of the deviance residuals multiplied by 2.

3.4.3 Interpreting the Coefficients

The logit coefficients can also be obtained using `coef(LR.1)` and their confident intervals can be obtained with `confint(LR.1)`.

```
> coef(LR.1)
(Intercept)      wrkfull
  0.5592222    0.7778011

> confint(LR.1)
Waiting for profiling to be done...
                  2.5 %       97.5 %
(Intercept)   0.4309678    0.6889906
wrkfull       0.5709823    0.9872410
```

Since the estimated intercept and coefficient are .559 and .778, respectively, the simple logistic regression model could be expressed as:

$$\text{logit}(p) = .559 + .778X$$

The regression coefficient of the `wrkfull` predictor is .778, which is also called the logit coefficient since we estimate the relationship between the predictor variable and the logit of the probability when the outcome variable `healthy` takes the value of 1.

The `confint(LR.2)` command produces the profile confidence intervals which are based on the chi-square distribution of the likelihood ratio test statistic. We can also use the `confint.default(LR.2)` command to obtain the standard confidence intervals, which are based on the Wald test statistic.

```
> confint.default(LR.1)
                  2.5 %       97.5 %
(Intercept)   0.4302568    0.6881876
wrkfull       0.5697574    0.9858449
```

The results of the profile confidence intervals and the standard confidence intervals look similar. The profile confidence intervals are preferred when the sample size in the model is small. We will use the `confint()` function throughout the book.

3.4.4 Interpreting the Odds Ratio

When $X = 0$, the estimated logit(p) = .559. Since logit is log odds, exponentiating .559 gives us the odds of having good health. The odds of having good health for people who do not work full-time are exp(.559) = 1.749.

When $X = 1$, the estimated logit(p) = .559 + .778 = 1.337. So the odds of having good health for people who work full-time are exp(1.337) = 3.808.

Odds ratio = 3.808/1.749 = 2.177, which is the ratio of odds of having good health for working full-time to the odds for not working full-time.

If we directly exponentiate the logit coefficient exp(.778) = 2.177, then we get the same odds ratio as earlier. You get the following output displaying the odds ratio if you run the `exp(coef(LR.1))` command. The corresponding confidence intervals can be obtained using the `exp(confint(LR.1))` command.

```
> exp(coef(LR.1))
(Intercept)     wrkfull
   1.749311    2.176681

> exp(confint(LR.1))
Waiting for profiling to be done...
                  2.5 %      97.5 %
(Intercept)    1.538746    1.991704
wrkfull        1.770005    2.683820
```

Odds ratio = 2.177. It is interpreted as follows: For each one-unit increase in the predictor variable, the odds of having good health increase by a factor of 2.177. Since the predictor variable `wrkfull` is also a binary variable, it can also be interpreted as the odds of having good health for the people who work full-time are 2.177 times the odds of having good health for those who do not work full-time.

3.4.5 Interpreting the Pseudo R^2

We use the `nagelkerke()` function in the `rcompanion` package (Mangiafico, 2021) to obtain the pseudo R^2 statistics. The `rcompanion` package needs to be installed first by typing `install.packages("rcompanion")`. We then load it with `library(rcompanion)`. The output produced by the `nagelkerke(LR.1)` command is displayed as follows.

```
> # PseudoR2
> library(rcompanion)
> nagelkerke(LR.1)
$`Models`

Model: "glm, healthy ~ wrkfull, binomial(link = \"logit\"), chp3.lr"
```

```
Null: "glm, healthy ~ 1, binomial(link = \"logit\"), chp3.lr"

$Pseudo.R.squared.for.model.vs.null
                              Pseudo.R.squared
McFadden                             0.0246663
Cox and Snell (ML)                   0.0293107
Nagelkerke (Cragg and Uhler)         0.0418352

$Likelihood.ratio.test
Df.diff   LogLik.diff   Chisq      p.value
     -1        -27.86   55.72   8.3585e-14

$Number.of.observations

Model   :   1873
Null    :   1873

$Messages
[1] "Note: For models fit with REML, these statistics are based on refitting with ML"

$Warnings
[1] "None"
```

McFadden's $R^2 = 0.0247$. It is the likelihood ratio R^2, which is normally written as R_L^2.

$$R_L^2 = 1 - \frac{-2LL_m}{-2LL_0} = 1 - (2,203.2/2,258.9) = 0.0247$$

```
> 1-2203.2/2258.9
[1] 0.02465802
```

The same results can be computed using the equations for the pseudo R^2 introduced in the previous section. In the R syntax below, we first fit a null model with the intercept only and name the model object `LR.0`. We also create the following objects. `LLM` is the log-likelihood value for the single-predictor model and `LL0` is the log-likelihood value for the null model. In addition, `McFadden` is the object name for the McFadden R^2, `CS` for the Cox and Snell R^2, and `NG` for the Nagelkerke R^2.

```
> LR.0 <- glm(healthy ~ 1, data = chp3.lr, family = binomial(link = "logit"))
> LLM <- logLik(LR.1)
> LL0 <- logLik(LR.0)
> McFadden <- 1-(LLM/LL0)
> McFadden
'log Lik.' 0.02466634 (df=2)
> CS <- 1-exp(2*(LL0-LLM)/1873)
> CS
'log Lik.' 0.02931066 (df=1)
> NG <- CS/(1-exp(2*LL0/1873))
> NG
'log Lik.' 0.0418352 (df=1)
```

The results of the three pseudo R^2 are the same as those produced by the `nagelkerke()` function.

3.4.6 AIC and BIC Statistics

AIC is reported by `summary(LR.1)`. The `AIC(LR.1)` command also reports the AIC statistic. The `BIC(LR.1)` command provides the BIC statistic.

```
> AIC(LR.1)
[1] 2207.209
> BIC(LR.1)
[1] 2218.28
```

In the output, AIC = 2,207.209. The same value can be easily derived if we use the following equation:

$$\text{AIC} = -2(\text{LL}_\text{m} - k) = (D_\text{m} + 2k) = 2,203.2 + 2 \times 2 = 2,207.2$$

BIC in the output is 2,218.28. It can be calculated using this equation:

$$\text{BIC} = D_\text{m} + \ln(n) \times k = 2,203.2 + \ln(1,873) \times 2 = 2,218.27$$

The AIC and BIC statistics themselves do not indicate whether the model fits the data well. They are useful for model comparison purposes, particularly when comparing non-nested models. More details will be provided in the multiple logistic regression section.

3.4.7 Testing the Overall Model Using the Likelihood Ratio Test

To test if the overall model is significant, we use the `anova(LR.1, update(LR.1, ~1), test = "Chisq")` command. The `anova()` function compares the deviance statistics of the fitted model `LR.1` and the null model using the chi-square test.

```
> LR.1 <- glm(healthy ~ wrkfull, data = chp3.lr, family = binomial(link = "logit"))
> anova(LR.1, update(LR.1, ~1), test = "Chisq")
Analysis of Deviance Table

Model 1: healthy ~ wrkfull
Model 2: healthy ~ 1
     Resid.Df    Resid. Dev    Df    Deviance     Pr(>Chi)
1        1871        2203.2
2        1872        2258.9    -1      -55.72    8.358e-14 ***

---
Signif. codes: 0 '***' 0.001 '**' 0.01 '*' 0.05 '.' 0.1 ' ' 1
```

`Deviance` = `55.72`, which is the likelihood ratio chi-square test statistic and can be written as $LR\ \chi^2_{(1)} = 55.72$. It is the difference in the -2 log likelihood (-2LL) between the current model, which contains the one predictor and the intercept, and the null or empty model, which contains only the intercept. The deviance of the current model was 2,203.2 and that of the null model was 2,258.9. Therefore, the difference in deviance was 55.72, with 1 degree of freedom, which is the difference in the number of predictors between these two models. Please note that the deviance is negative in the output since there is a decrease in the residual deviance from the null model to the one-predictor model.

The null hypothesis of the likelihood ratio chi-square test is that the logit coefficient of the predictor variable `wrkfull` is not significant or that the predictor variable does not contribute to the model.

The alternative hypothesis is that the coefficient of the predictor is significant in the model or that it significantly contributes to the model.

The associated p value with the likelihood ratio chi-square test `Prob(>chi) < 0.0001` indicates that the null hypothesis is rejected. Therefore, the overall model with one predictor is significant.

3.5 MULTIPLE LOGISTIC REGRESSION USING R

The multiple logistic regression model is simply an extension of the simple logistic regression model when there are two or more predictor variables in the model. The following equation is for the multiple logistic regression model:

$$\ln\frac{p(x)}{1-p(x)} = \alpha + \beta_1 X_1 + \beta_2 X_2 + \ldots + \beta_p X_p \qquad (3.21)$$

where X_1, X_2, …, X_p are the predictor variables and β_1, β_2, …, β_p are the logit coefficients of these predictors.

The multiple logistic regression estimates the logit of the probability or log odds when the outcome variable equals 1 given a set of predictors, which can be either categorical or continuous. The odds of success or of having an event in multiple regression is the ratio of probability of success to the probability of failure given a set of predictor variables.

3.5.1 Interpretation of Model Parameter Estimates and Odds Ratios in Multiple Logistic Regression

The estimated regression coefficients are the logit coefficients, which are the coefficients on the scale of logit or log odds. The logit coefficient can be interpreted as the change in the predicted logit or the log odds for a one-unit increase in the predictor variable when holding other predictors constant or controlling for the effects of other predictors.

The exponentiated logit coefficients are the odds ratios. The odds ratio of each predictor can be interpreted as the change in the odds for a one-unit change in the dependent variable when holding other predictors constant.

When an odds ratio is larger than 1, the odds of success or of having an event increase for a one-unit increase in the predictor variable when controlling for other predictors.

When an odds ratio is less than 1, the odds of success or of having an event decrease for a one-unit increase in the predictor variable when controlling for other predictors.

When an odds ratio equals 1, there is no relationship between the predictor and the odds of success when holding other variables constant.

3.5.2 Model Fitting Based on the Likelihood Ratio Test and Information Criteria Statistics

Just like linear regression models, models in logistic regression can be fitted from a simple logistic regression model to a more complex model to the full model. A baseline model can be the starting point for model building. A series of nested models can be compared by using the likelihood ratio test or the deviance difference test. Models are nested when one model is a special case of the other. As explained in the Goodness-of-Fit Statistics section, the difference in deviance between nested models has an approximately chi-square distribution with the degrees of freedom equal to the difference in the number of parameters between these two models. It is often expressed as $G =$ deviance for the reduced model $-$ deviance for the full model or as $D_{\text{Reduced}} - D_{\text{Full}}$. If the likelihood ratio chi-square test or the deviance test is significant, then we reject the null hypothesis and conclude that one model fits the data better than the other model.

The result of the likelihood ratio test is valid when the competing models are fitted on the same data. In real data analysis, it is common to see that the sample size varies among the nested models. If there is a great difference between the sample sizes of fitted models, for example, if some variables have more missing values, then the results are questionable. Under this circumstance, two information criteria measures, the AIC and BIC statistics, are more appropriate for model comparisons. They are commonly used to compare non-nested models. The smaller the AIC and BIC statistics, the better fit the model.

3.5.3 The `glm()` Function for Multiple Logistic Regression

We still use the `glm()` function for multiple logistic regression. In the model formula, the dependent variable `healthy` and the four predictor variables are separated by the tilde (~). The four predictor variables include `wrkfull`, `maritals`, `female`, and `educ`, which are connected by plus (+) symbols. We also specify the data and the family arguments.

In the following example, the `glm(healthy ~ wrkfull + maritals + female + educ, data = chp3.lr, family = binomial(link = "logit"))` command tells R to predict the dependent variable `healthy` from the four independent variables. The fitted model is named `LR.2`. The following output is shown by the `summary(LR.2)` command.

```
> # Multiple logistic regression
> LR.2 <- glm(healthy ~ wrkfull + maritals + female + educ, data = chp3.lr, family =
binomial(link = "logit"))
> summary(LR.2)

Call:
glm(formula = healthy ~ wrkfull + maritals + female + educ, family = binomial(link =
"logit"),
  data = chp3.lr)


Deviance Residuals:
    Min        1Q    Median       3Q       Max
-2.3201   -1.1877    0.6439   0.8652    1.7620



Coefficients:
              Estimate   Std. Error   z value   Pr(>|z|)
(Intercept)   -1.82016      0.26294    -6.922   4.45e-12 ***
wrkfull        0.65505      0.11112     5.895   3.74e-09 ***
maritals       0.36738      0.10943     3.357   0.000787 ***
female         0.18217      0.10807     1.686   0.091868 .
educ           0.16184      0.01881     8.605    < 2e-16 ***

---
Signif. codes: 0 '***' 0.001 '**' 0.01 '*' 0.05 '.' 0.1 ' ' 1


(Dispersion parameter for binomial family taken to be 1)


    Null deviance: 2258.9  on 1872  degrees of freedom
Residual deviance: 2103.4  on 1868  degrees of freedom
AIC: 2113.4

Number of Fisher Scoring iterations: 4
```

3.5.4 Interpreting R Output

In the R output for multiple logistic regression, the first part is the call, which shows the R syntax for the model. The second part shows the minimum, first quarter, median, third quarter, and maximum value of the deviance residuals. The third part shows the coefficients table including the parameter estimates for the four predictor variables and the intercept, their standard errors, the Wald z statistics, and the associated p values.

For the predictor variable `wrkfull`, Wald $z = 5.895$. The associated p value, `Pr(>|z|) < .001`, so we rejected the null hypothesis. The rejection of the null hypothesis indicates that the predictor variable `wrkfull` is a significant predictor of the dependent variable `healthy`.

For the predictor variable `maritals`, Wald $z = 3.357$. The associated p value, `Pr(>|z|) < .001`, so we rejected the null hypothesis. For the predictor variable `educ`, the Wald $z = 8.605$. The associated p value, `Pr(>|z|) < .001`, so we also reject the null hypothesis. Therefore, `maritals` and `educ` are significant predictors of the dependent variable.

For the other new predictor variable `female`, the Wald $z = 1.686$. The associated p value `Pr(>|z|) = .092`, so we fail to reject the null hypothesis and conclude that there is no significant effect of female on the outcome variable. In other words, whether a person is a female or male does not significantly predict whether that person has good health.

Finally, the fourth part of the output shows the fit statistics including the null deviance, the residual deviance, and the AIC.

3.5.5 Interpreting the Coefficients

The logit coefficients can also be obtained using the `coef(LR.2)` command and their confident intervals can be obtained with the `confint(LR.2)` command.

```
> coef(LR.2)
(Intercept)     wrkfull    maritals      female        educ
-1.8201572    0.6550481   0.3673780   0.1821725   0.1618404

> confint(LR.2)
Waiting for profiling to be done...
                  2.5 %       97.5 %
(Intercept) -2.34090516   -1.3094672
wrkfull      0.43826244    0.8740228
maritals     0.15362395    0.5827676
female      -0.02966614    0.3941360
educ         0.12540105    0.1991806
```

The regression coefficients of the predictors in multiple logistic regression are the logit coefficients since they are on the scale of the logit or log odds. They are the partial effects of the predictor variables on the logit or log odds of the outcome variable (having good health). Each logit coefficient can be interpreted as the change in the logit or log odds of having good health for each one-unit increase in the predictor when controlling for the effects of other variables. A positive logit coefficient indicates there is an increase in the logit coefficient for a one-unit increase in the predictor, whereas a negative logit coefficient indicates that there is a decrease in the logit coefficient for a one-unit increase in the predictor when controlling for other predictors.

3.5.6 Interpreting the Odd Ratios

You get the following output displaying the odds ratios if you run the `exp(coef(LR.2))` command. The corresponding confidence intervals can be obtained using the `exp(confint(LR.2))` command.

```
> exp(coef(LR.2))
(Intercept)      wrkfull     maritals       female         educ
  0.1620003    1.9252351    1.4439437    1.1998212    1.1756726

> exp(confint(LR.2))
Waiting for profiling to be done...
                  2.5 %      97.5 %
(Intercept)  0.09624049   0.2699638
wrkfull      1.55001164   2.3965322
maritals     1.16605232   1.7909884
female       0.97076958   1.4831023
educ         1.13360300   1.2204023

> cbind(exp(coef(LR.2)), exp(confint(LR.2)))
Waiting for profiling to be done...
                                2.5 %      97.5 %
(Intercept)  0.1620003   0.09624049   0.2699638
wrkfull      1.9252351   1.55001164   2.3965322
maritals     1.4439437   1.16605232   1.7909884
female       1.1998212   0.97076958   1.4831023
educ         1.1756726   1.13360300   1.2204023
```

For `wrkfull`, odds ratio = 1.925. The odds of having good health for the people who work full-time are 1.925 times the odds of having good health for those who do not work full-time.

For `maritals`, odds ratio = 1.444, which is larger than 1. This indicates that the odds of having good health for the married were 1.444 times the odds for the unmarried.

For `educ`, odds ratio = 1.176, for each one-unit increase in education, the odds of having good health increase by a factor of 1.176. Another interpretation is the percentage change in odds. It can be calculated by using (odds ratio − 1) × 100%. In this example, (1.176 − 1) × 100% = 17.6%. For each one-unit increase in education, there is an increase of 17.6% in the odds of having good health.

For `female`, odds ratio = 1.200, $p > .05$, which is not significant. When an odds ratio takes a value of 1, it means that there is no significant effect of the predictor variable on the odds of success when holding other variables constant. Therefore, being a female or not does not significantly influence whether that person has good health.

3.5.7 Interpreting the Pseudo R^2

The pseudo R^2 statistics for the multiple logistic regression model are obtained using the `nagelkerke(LR.2)` command. The output is displayed as follows.

```
> # PseudoR2(LR.2)
> nagelkerke(LR.2)
$`Models`

Model: "glm, healthy ~ wrkfull + maritals + female + educ, binomial(link = \"logit\"), chp3.lr"
Null: "glm, healthy ~ 1, binomial(link = \"logit\"), chp3.lr"


$Pseudo.R.squared.for.model.vs.null
                                    Pseudo.R.squared
McFadden                                   0.0688652
Cox and Snell (ML)                         0.0796992
Nagelkerke (Cragg and Uhler)               0.1137550


$Likelihood.ratio.test
Df.diff    LogLik.diff     Chisq        p.value
     -4        -77.781    155.56    1.3082e-32

$Number.of.observations


Model     :     1873
Null      :     1873

$Messages
[1] "Note: For models fit with REML, these statistics are based on refitting with ML"


$Warnings
[1] "None"
```

The McFadden R^2 = 0.069, the Cox Snell R^2 = 0.080, and the Nagelkerke R^2 = 0.114.

These three pseudo R^2 statistics are larger than those in the previous simple logistic regression model. (McFadden's R^2 = .025 for the one-predictor model.) It seems that the full model fits better than the simple logistic regression model, but we also need to look at other fit statistics.

Please note that a pseudo R^2 measure should not be interpreted as variance explained by the predictors in the model as its analog in multiple linear regression. When it is used for model comparisons, instead of depending on this single statistic, it should be used together with other fit statistics.

3.5.8 AIC and BIC Statistics

The AIC and BIC statistics are obtained using `AIC(LR.2)` and `BIC(LR.2)`, respectively.

```
> AIC(LR.2)
[1] 2113.367
> BIC(LR.2)
[1] 2141.044
```

To compare the AIC and BIC statistics of the two fitted model, we use `AIC(LR.1, LR.2)` and `BIC(LR.1, LR.2)`.

```
> AIC(LR.1, LR.2)
        df       AIC
LR.1     2   2207.209
LR.2     5   2113.367

> BIC(LR.1, LR.2)
        df       BIC
LR.1     2   2218.280
LR.2     5   2141.044
```

The AIC and BIC statistics decrease by 93.842 and 77.236, respectively. Since smaller AIC and BIC statistics means a better model fit, the results support that the multiple logistic regression model fits the data better.

3.5.9 Hosmer–Lemeshow Goodness-of-Fit Statistic

We use the `hoslem.test()` function in the `ResourceSelection` package (Lele et al., 2019) to obtain the Hosmer–Lemeshow goodness-of-fit statistic. The `ResourceSelection` package needs to be installed first by typing `install.packages("ResourceSelection")`. We then load it with `library(ResourceSelection)`. The output produced by the `hoslem.test(healthy, fitted(LR.2), g = 10)` syntax is displayed as follows. The syntax includes the outcome variable, `healthy`, the estimated probabilities, `fitted(LR.2)`, and the number of groups, `g = 10`.

```
> # Hosmer-Lemeshow goodness-of-fit test
> library(ResourceSelection)
> hoslem.test(healthy, fitted(LR.2), g = 10)

Hosmer and Lemeshow goodness of fit (GOF) test

data: healthy, fitted(LR.2)
X-squared = 7.9332, df = 8, p-value = 0.44

> hlt2<-hoslem.test(healthy, fitted(LR.2), g = 10)
> cbind(hlt2$observed,hlt2$expected)
                  y0    y1      yhat0     yhat1
[0.212,0.53]     109    82  102.48990   88.5101
(0.53,0.575]      79   111   82.67738  107.3226
(0.575,0.652]     79   131   79.11126  130.8887
(0.652,0.685]     60   132   62.82160  129.1784
(0.685,0.723]     47   135   52.67628  129.3237
(0.723,0.758]     54   131   46.16049  138.8395
(0.758,0.79]      40   144   39.57522  144.4248
(0.79,0.833]      29   176   37.20245  167.7975
(0.833,0.857]     23   127   22.74877  127.2512
(0.857,0.932]     25   159   19.53663  164.4634
```

In the output, the Hosmer–Lemeshow chi-square test has a value of 7.933, with the degrees of freedom equal to 8. The associated *p* value is .44, which is not significant. Therefore, the model fits the data well.

To see the estimated probabilities by the 10 groups, we use the `cbind(hlt2$observed, hlt2$expected)` command. The first column lists the estimated probabilities for these groups, which are ordered in sequence. The second column `y0` lists the number of observations for the outcome variable with the value of 0, and the third column `y1` is the number of observations for the outcome variable with the value of 1. The last two columns are `yhat0` and `yhat1`, respectively, which are the expected number of frequencies for no event and the event occurring. Within each group, the frequencies of the observed cases are compared with the expected frequencies, and we would like to see small discrepancies between them. This test follows a chi-square distribution with the degrees of freedom equal to the number of groups −2. A nonsignificant *p* value indicates that the model fits the data well since there is no significant difference between the observed and expected data.

3.5.10 Testing the Overall Model Using the Likelihood Ratio Test

To test if the overall model is significant, we use the `anova(LR.2, update(LR.2, ~1), test = "Chisq")` command. The `anova()` function compares the deviance statistics of the fitted model `LR.2` and the null model using the chi-square test.

```
> LR.2 <- glm(healthy ~ wrkfull + maritals + female + educ, data = chp3.lr, family = binomial(link =
"logit"))
> anova(LR.2, update(LR.2, ~1), test = "Chisq")
Analysis of Deviance Table

Model 1: healthy ~ wrkfull + maritals + female + educ
Model 2: healthy ~ 1
     Resid. Df    Resid. Dev    Df    Deviance      Pr(>Chi)
1         1868        2103.4
2         1872        2258.9    -4     -155.56    < 2.2e-16 ***

---
Signif. codes: 0 '***' 0.001 '**' 0.01 '*' 0.05 '.' 0.1 ' ' 1
```

The deviance of the current model was 2,103.4 and that of the null model was 2,258.9. Therefore, the difference in deviance was 155.56 with 4 degrees of freedom. The deviance difference is the likelihood ratio chi-square test statistic. It is the difference in the -2 log likelihood (-2LL) between the current model, which contains the four predictors and the intercept, and the null or empty model, which contains only the intercept.

The associated p value with the likelihood ratio chi-square test `Prob(>chi) < 0.0001` indicates that the null hypothesis is rejected. Therefore, the overall model with the four predictors is significant.

3.5.11 Model Comparison Using the Likelihood Ratio Test

To identify which model fits the data better, the likelihood ratio test or the deviance difference test can be used. Recall that this test compares the reduced model, which contains less parameters, and the full model, which contains all parameters. The difference in deviance is often expressed as G = Deviance for the reduced model − Deviance for the full model or as $D_{\text{Reduced}} - D_{\text{Full}}$. The difference in deviance between nested models has a chi-square distribution with the degrees of freedom equal to the difference in the number of parameters between these two models.

The `anova()` function is used for the likelihood ratio test or the deviance difference test. Next, we compare the simple logistic regression model and the multiple logistic regression model with the `anova(LR.1, LR.2, test = "Chisq")` command.

```
> anova(LR.1, LR.2, test = "Chisq")
Analysis of Deviance Table

Model 1: healthy ~ wrkfull
Model 2: healthy ~ wrkfull + maritals + female + educ
     Resid. Df    Resid. Dev    Df    Deviance      Pr(>Chi)
1         1871        2203.2
2         1868        2103.4     3      99.842    < 2.2e-16 ***

---
Signif. codes: 0 '***' 0.001 '**' 0.01 '*' 0.05 '.' 0.1 ' ' 1
```

The likelihood ratio chi-square test $\chi^2_{(3)} = 99.842$, $p < .001$. The same result can be obtained if you compute it using the following equation:

$$G = D_{\text{Reduced}} - D_{\text{Full}} = 2203.2 - 2103.4 = 99.80,\ \text{df} = 1871 - 1868 = 3$$

3.5.12 Interpreting the Marginal Effects in Logistic Regression

We load the `margins` package (Leeper, 2021) with the `library(margins)` command and compute the average marginal effects with the `margins(LR.2)` command. This package needs to be installed first with the `install.packages("margins")` command. The results are named `LR.2`, and the summary of the results is obtained with `summary(marg2)` as follows.

```
> library(margins)
> marg2<-margins(LR.2)
> summary(marg2)
  factor      AME      SE       z       p     lower   upper
    educ   0.0307  0.0033  9.2173  0.0000   0.0241  0.0372
  female   0.0345  0.0204  1.6896  0.0911  -0.0055  0.0745
maritals   0.0696  0.0205  3.3897  0.0007   0.0293  0.1098
 wrkfull   0.1241  0.0204  6.0787  0.0000   0.0841  0.1641
```

The average marginal effect for `educ` is .031. Since `educ` is continuous, its marginal effect is an instantaneous rate of change in the probability of having good health with a small change in education. The result indicates that on average the predicted probability of having good health increases by 3.1% for a one-unit increase in education when holding all other predictors constant.

The average marginal effect for `maritals` is .070. Since `maritals` is binary, the marginal effect is a discrete change when its value moves from 0 to 1. This indicates that on average the predicted probability of having good health for the married is 7% higher than that for the unmarried when holding the other predictors constant. The other two predictor variables are also binary, so the interpretation is similar.

3.5.13 Computing the Predicted Probabilities With the `predict()` Function

We can use the `predict()` function to compute the predicted probabilities for a particular variable at given values. For example, we would like to compute the predicted probabilities for `educ` at the specified values of 12, 14, and 16 when holding the other predictor variables at their means. We first create a data frame with the `data.frame()` function and then apply the `predict()` function. In the `data.frame()` function, `educ = c(12, 14, 16)` specifies the values of `educ`; `maritals = rep(mean(maritals), 3)` repeats the mean of `maritals` three times; `wrkfull = rep(mean(wrkfull), 3)` repeats the mean of wrkfull three times; and female = rep(mean(female), 3) repeats the mean of female three times. The created data frame is assigned to an object named newdf.

```
> newdf <- data.frame(educ = c(12,14,16),
+                     maritals=rep(mean(maritals), 3),
+                     wrkfull=rep(mean(wrkfull), 3),
+                     female=rep(mean(female), 3))
> newdf
     educ    maritals    wrkfull      female
1      12   0.4372664   0.467165   0.5563267
2      14   0.4372664   0.467165   0.5563267
3      16   0.4372664   0.467165   0.5563267
```

In the predict() function, we first specify the model object LR.2 and then the newdata = newdf argument, followed by the type = "response" argument for the predicted probabilities. The predicted probabilities labeled pred.prob are provided in the data frame named newdf.

```
> newdf[,c('pred.prob')] <- predict(LR.2, newdata=newdf, type="response")
> newdf
  educ    maritals    wrkfull      female   pred.prob
1   12   0.4372664   0.467165   0.5563267   0.6659520
2   14   0.4372664   0.467165   0.5563267   0.7337268
3   16   0.4372664   0.467165   0.5563267   0.7920447
```

The same results can be obtained by using the ggpredict() function in the ggeffects package in the following section.

3.5.14 Computing the Predicted Probabilities With the ggpredict() Function in the ggeffects Package

The ggpredict() function in the ggeffects package (Lüdecke, 2018b) is used to compute the predicted probabilities when the outcome variable $Y = 1$ for predictor variables at specified values. The ggeffects package needs to be installed first by typing install.packages("ggeffects"). We then load it with library (ggeffects). In the first example, we compute the predicted probabilities of being healthy (i.e., $Y = 1$) for educ at the values of 12, 14, and 16 when holding the other three predictor variables constant at their means. Please note that the predictor variables need to be numeric so that they can be held constant at their means. The syntax is as follows: margins <- ggpredict(LR.2, terms = "educ[12, 14, 16]"). In the ggpredict() function, LR.2 is the fitted model and the terms = "educ[12, 14, 16]" option specifies the predictor variable educ at the values of 12, 14, and 16. When there are more than one variable, the terms option can specify up to four variables, including the second to fourth grouping variables. In this example, we include only one variable in the `terms` option. The output is assigned to an object named `margins`. To request the standard errors of the predicted probabilities, we can use either the `as.data.frame()` or the `sqrt(diag(vcov()))` function. The `vcov()` function computes the variance–covariance matrix, so the standard errors can be obtained by taking the square root of the variances, which are on the diagonal of the matrix.

```
> library(ggeffects)
> margins <- ggpredict(LR.2, terms = "educ[12, 14, 16]")
> margins

# Predicted values of healthy

educ    |    Predicted    |          95% CI
-------------------------------------------
  12    |         0.67    |    [0.64, 0.69]
  14    |         0.73    |    [0.71, 0.75]
  16    |         0.79    |    [0.77, 0.82]


Adjusted for:
*  wrkfull = 0.47
* maritals = 0.44
*   female = 0.56
> as.data.frame(margins)
     x   predicted    std.error     conf.low    conf.high    group
1   12   0.6659520   0.05899403    0.6397564    0.6911624        1
2   14   0.7337268   0.05581622    0.7118151    0.7545469        1
3   16   0.7920447   0.07470175    0.7668988    0.8151319        1

> sqrt(diag(vcov(margins)))
         1            2            3
0.05899403   0.05581622   0.07470175

> plot(margins)
```

When `educ` equals 12, 14, and 16, and the other three predictor variables are held at their means, the predicted probabilities of being healthy (i.e., $Y = 1$) are .666, .734, and .792, respectively. The results are the same as those obtained in the last section. They are plotted using the `plot(margins)` command. Figure 3.1 shows the predicted probabilities of being healthy (i.e., $Y = 1$) for `educ` at 12, 14, and 16. With an increase in years of education, the probabilities of having good health ($Y = 1$) increase.

We can also compute the predicted probabilities for a continuous variable at given values by different groups. In the third example, we compute the predicted probabilities of being healthy (i.e., $Y = 1$) for `educ` at the values of 12, 14, and 16 by the two groups in `wrkfull` when holding other variables at their means. The syntax is as

FIGURE 3.1 ● Estimated Probabilities When $Y = 1$ for educ at 12, 14, and 16 With Others Fixed at Their Means

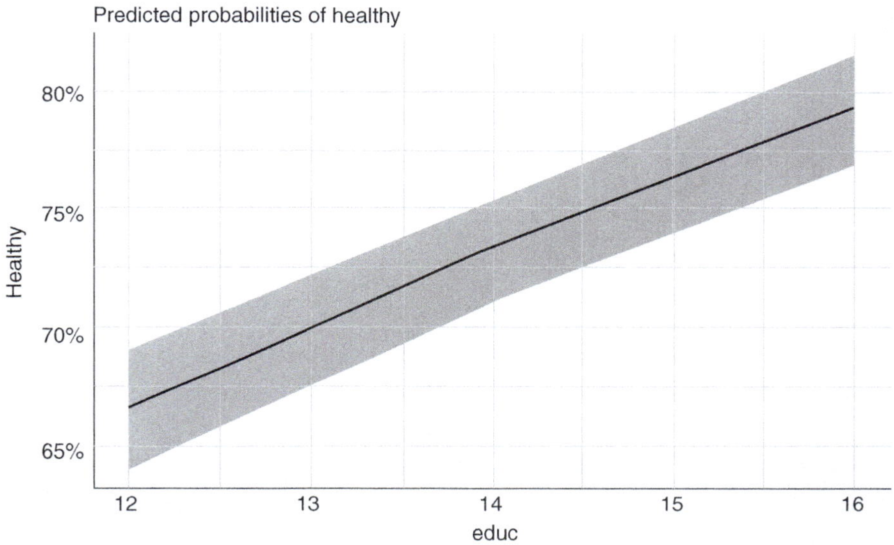

follows: `margins.ew <- ggpredict(LR.2, terms = c("educ[12, 14, 16]", "wrkfull"))`. In the `ggpredict()` function, the `terms = c("educ [12, 14, 16]", "wrkfull")` option specifies both `educ` and `wrkfull`, with the latter as the grouping variable. The output is assigned to an object named `margins.ew` and is plotted with the `plot(margins.ew)` command.

```
> margins.ew <- ggpredict(LR.2, terms = c("educ[12, 14, 16]", "wrkfull"))
> margins.ew

# Predicted values of healthy

# wrkfull = 0

educ    |   Predicted   |         95% CI
----------------------------------------
  12    |        0.59   |   [0.56, 0.63]
  14    |        0.67   |   [0.64, 0.70]
  16    |        0.74   |   [0.70, 0.77]
```

```
# wrkfull = 1

educ     |   Predicted     |          95% CI
----------------------------------------
  12     |          0.74   |   [0.70, 0.77]
  14     |          0.80   |   [0.77, 0.82]
  16     |          0.84   |   [0.82, 0.87]


Adjusted for:
* maritals = 0.44
*   female = 0.56

> plot(margins.ew)
```

FIGURE 3.2 ● Predicted Probabilities of Being Healthy for `educ` at 12, 14, and 16 by `wrkfull`

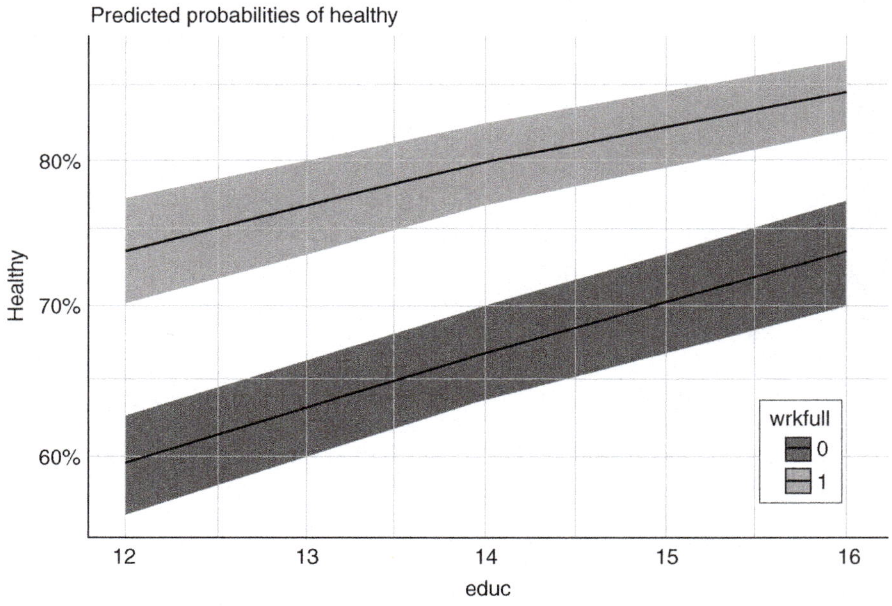

As shown in the graph, the predicted probabilities of having good health ($Y = 1$) increase with an increase in years of education and the probability for the people who work full-time is higher than the probability for those not working full-time.

3.6 PROBIT REGRESSION USING R

In addition to the logistic regression models introduced above, probit regression is also a popular method for analyzing binary response variables. The probit model assumes a standard cumulative normal distribution with a mean of 0 and a standard deviation of 1. While the logit link function is used for logistic regression models, the probit link function is used for the probit regression models. The probit link, or probit transformation, can be expressed as the inverse of the cumulative density function (cdf) of the standard normal distribution $\Phi^{-1}(\pi)$, where Φ is the cdf for the standard normal. The model can be expressed as:

$$\Phi^{-1}(\pi) = \alpha + \beta_1 X_1 + \beta_2 X_2 + \ldots + \beta_p X_p \tag{3.22}$$

where X_1, X_2, ..., X_p are the predictor variables and β_1, β_2, ..., β_p are the probit coefficients of these predictors. By taking the inverse on both sides of the equation, we get the following equation:

$$\pi(Y = 1|x_1, x_2 \ldots x_p) = \Phi\left(\alpha + \beta_1 X_1 + \beta_2 X_2 + \ldots + \beta_p X_p\right) \tag{3.23}$$

where Φ (.) denotes the cdf for the standard normal and a linear combination of the intercept and a set of predictors that are specified within the parentheses. The probit coefficients are *z* scores or standard normal scores.

3.6.1 Interpretation of Model Parameter Estimates in Probit Regression

First, the estimated regression coefficients are the probit coefficients, which can be interpreted as the change in the predicted probit for a one-unit increase in the predictor variable when holding the other predictors constant. Second, since probit regression is based on the cumulative normal distribution Φ, the regression coefficient can also be interpreted as the change in the *z* scores for each one-unit increase in the predictor variable. Recall that the change in *z* scores is the number of standard deviations from the mean of a variable. Third, as with logistic regression, we can estimate marginal effects in probit regression. A marginal effect can be interpreted as the change in the predicted probability for a one-unit change in a predictor variable. Fourth, we can interpret probit coefficients in terms of predicted probabilities at representative values of a set of predictor variables.

3.6.2 The `glm()` Function for Multiple Probit Regression

In the following example, the `glm(healthy ~ wrkfull + maritals + female + educ, data = chp3.lr, family = binomial(link = "probit"))` command tells R to predict the dependent variable `healthy` from the four independent variables with the probit link function. The fitted model is named `PR.2`. The following output is shown by the `summary(PR.2)` command.

```
> # Probit regression
> PR.2 <- glm(healthy ~ wrkfull + maritals + female + educ, data = chp3.lr, family =
binomial(link = "probit"))
> summary(PR.2)

Call:
glm(formula = healthy ~ wrkfull + maritals + female + educ, family = binomial(link =
"probit"),
  data = chp2.lr)

Deviance Residuals:
    Min        1Q    Median       3Q      Max
-2.3716   -1.2015    0.6466   0.8735   1.7227


Coefficients:
             Estimate   Std. Error   z value   Pr(>|z|)
(Intercept)  -1.04382     0.15487    -6.740    1.58e-11 ***
wrkfull       0.38772     0.06535     5.933    2.98e-09 ***
maritals      0.21948     0.06463     3.396    0.000684 ***
female        0.10578     0.06423     1.647    0.099553 .
educ          0.09425     0.01096     8.599     < 2e-16 ***

---
Signif. codes: 0 '***' 0.001 '**' 0.01 '*' 0.05 '.' 0.1 ' ' 1

(Dispersion parameter for binomial family taken to be 1)

    Null deviance: 2258.9  on 1872  degrees of freedom
Residual deviance: 2105.0  on 1868  degrees of freedom
AIC: 2115

Number of Fisher Scoring iterations: 4
```

The probit coefficients can also be obtained using the `coef(PR.2)` command and their confident intervals can be obtained with the `confint(PR.2)` command. The output is omitted here.

3.6.3 Interpreting Probit Coefficients in R Output

The coefficients table displays the probit coefficients for the four predictor variables and the intercept, their standard errors, the Wald z statistics, and the associated p values.

For the predictor variable `wrkfull`, the probit coefficient $\beta = .388$, which is positive. This indicates that working full-time increases the probability of having good health when holding all the other predictors constant. It can also be interpreted that working full-time increases the z score by .388. The probit coefficients of the other three predictors can be interpreted in a similar way.

We can use the marigns() function in the margins package to compute marginal effects and the ggpredict() function in the ggeffects package to compute the predicted probabilities. The output is omitted here.

3.7 MAKING PUBLICATION-QUALITY TABLES

3.7.1 Presenting the Results Using the stargazer Package

We can use the stargazer package (Hlavac, 2018) to make a table containing the results of the fitted models with the glm() function. After fitting the single-predictor model LR.1 and the multiple-predictor model LR.2, we use the syntax as follows: stargazer(LR.1, LR.2, type = "text", align = TRUE, out = "lr2mod.txt"). In the stargazer() function, we first specify the two model objects to be presented and then the type of table. The option type = "text" specifies the table type and the align = TRUE option aligns the results of the two models. The out = "lr2mod.txt" argument saves the output named lr2mod.txt.

```
> library(stargazer)
> stargazer(LR.1, LR.2, type = "text", align = TRUE, out = "lr2mod.txt")

====================================
                    Dependent variable:
                 -------------------------
                          healthy
                      (1)              (2)
-------------------------------------------
wrkfull            0.778***         0.655***
                   (0.106)          (0.111)

maritals                            0.367***
                                    (0.109)

female                               0.182*
                                    (0.108)

educ                                0.162***
                                    (0.019)

Constant           0.559***        -1.820***
                   (0.066)          (0.263)

-------------------------------------------
Observations        1,873            1,873
Log Likelihood   -1,101.605       -1,051.684
Akaike Inf. Crit. 2,207.209        2,113.367
====================================
Note:             *p<0.1; **p<0.05; ***p<0.01
```

We can also create the table in the HTML format and copy it into Microsoft Word. The syntax is as follows: `stargazer(LR.1, LR.2, type = "html", align = TRUE, out = "lr2mod.htm")`. It produces Table 3.6, as shown here in its original format, presenting the results of both the single-predictor and multiple-predictor logistic regression models.

TABLE 3.6 ● Results of the Logistic Regression Models: Single-Predictor and Multiple-Predictor Models (Shown in Original Format Generated by R)

	Dependent variable:	
	Healthy	
	(1)	(2)
Wrkfull	0.778***	0.655***
	(0.106)	(0.111)
Maritals		0.367***
		(0.109)
Female		0.182*
		(0.108)
Educ		0.162***
		(0.019)
Constant	0.559***	−1.820***
	(0.066)	(0.263)
Observations	1,873	1,873
Log Likelihood	−1,101.605	−1,051.684
Akaike Inf. Crit.	2,207.209	2,113.367

Note:
*$p < 0.1$
**$p < 0.05$
***$p < 0.01$

3.8 REPORTING THE RESULTS

When you summarize the results of logistic regression models, it is good practice to describe the statistical method you use for data analysis, the dependent variable and the independent variables in the models, and your research hypothesis or the purpose of your study. Also, report the likelihood ratio test statistic, the degrees of freedom, and the associated p value. Based on the likelihood ratio test result, you may discuss whether the fitted model fits better than the null model containing only the intercept. If more than one model is fitted, then the deviance statistics for each model may be presented, and the results of the likelihood ratio test comparing deviance statistics between nested models may be provided.

In the body of the text, you may interpret the parameter estimates or the odds ratios for the predictor variables.

You could have a table containing all the parameter estimates, their standard errors, *p* values, and odds ratios. A summary of major fit statistics also might be included in the table, which could include the likelihood ratio test statistic with its degrees of freedom; a couple of pseudo R^2 values, such as the likelihood ratio R^2 and Cox and Snell's R^2; the deviance statistic; and the AIC and BIC statistics. If more than one model is fitted, then the results of all the competing models from the simple model to the full model can be presented in the table. For detailed guidelines and recommendations on reporting the results of logistic regression, see Peng et al. (2002). In addition, O'Connell and Amico (2010) provided a list of key elements of logistic regression that should be addressed when judging a manuscript from a reviewer's perspective. Huck (2012) also provided examples on reporting the results of logistic regression from published articles. The following is an example of summarizing results from the multiple logistic regression model.

The multiple logistic regression analysis was conducted to estimate the probability of having good health from four predictor variables. The dependent variable was having good health or not, and the independent variables were working full-time, marital status, years of education, and gender. The likelihood ratio chi-square test statistic for the four-predictor model $LR\ \chi^2_{(4)} = 155.56$, $p < .001$, indicated that the overall model with all four predictors was significant. When comparing the multiple logistic regression model with the simple logistic regression model, the likelihood ratio chi-square test $\chi^2_{(3)} = 99.842$, $p < .001$, indicated the full model had a better fit. Table 3.6 presents the logit coefficients and standard errors of the full model. The results are interpreted in terms of odds ratios.

For `wrkfull`, odds ratio = 1.925. The odds of having good health for the people who work full-time are 1.925 times the odds of having good health for those who do not work full-time.

For `maritals`, odds ratio = 1.444, which is larger than 1. This indicates that the odds of having good health for the married were 1.444 times the odds for the unmarried.

For `educ`, odds ratio = 1.176, for each one-unit increase in education, the odds of having good health increase by a factor of 1.176.

For `female`, odds ratio = 1.200, $p > .05$, which is not significant. This indicates that being a female or not does not significantly influence whether that person has good health.

3.9 SUMMARY OF R COMMANDS IN THIS CHAPTER

```
# Chap 3 R Script
# Remove all objects
rm(list = ls(all = TRUE))

# The following user-written packages need to be installed first by using
install.packages(" ") and then by loading it with library()
# library(foreign)
# library(rcompanion)
# library(ResourceSelection)
# library(margins)
# library(ggeffects)    # It is already installed for Chapter 2
# library(stargazer)    # It is already installed for Chapter 2

# Import GSS 2016 Stata data file
library(foreign)
chp3.lr <- read.dta("C:/CDA/gss2016.dta")
chp3.lr$educ <- as.numeric(chp3.lr$educ)
chp3.lr$wrkfull <- as.numeric(chp3.lr$wrkfull)
chp3.lr$maritals <- as.numeric(chp3.lr$maritals)
attach(chp3.lr)
head(chp3.lr)
table(healthy)
table(wrkfull)

# Cross tabulation
tab <- table(healthy, wrkfull)
summary(tab)
tab
ftable(tab)
addmargins(tab)

# Simple logistic regression
LR.1 <- glm(healthy ~ wrkfull, data = chp3.lr, family = binomial(link = "logit"))
summary(LR.1)
coef(LR.1)
confint(LR.1)
confint.default(LR.1)
exp(coef(LR.1))
exp(confint(LR.1))

library(rcompanion)
nagelkerke(LR.1)
1-2203.2/2258.9
LR.0 <- glm(healthy ~ 1, data = chp3.lr, family = binomial(link = "logit"))
LLM <- logLik(LR.1)
LL0 <- logLik(LR.0)
McFadden <- 1-(LLM/LL0)
```

```
McFadden
CS <- 1-exp(2*(LL0-LLM)/1873)
CS
NG <- CS/(1-exp(2*LL0/1873))
NG

AIC(LR.1)
BIC(LR.1)

anova(LR.1, update(LR.1, ~1), test="Chisq")

# Multiple logistic regression
LR.2 <- glm(healthy ~ wrkfull + maritals + female + educ, data = chp3.lr, family =
binomial(link = "logit"))
summary(LR.2)
coef(LR.2)
confint(LR.2)
exp(coef(LR.2))
exp(confint(LR.2))
cbind(exp(coef(LR.2)), exp(confint(LR.2)))

nagelkerke(LR.2)
1-2103.4/2258.9
AIC(LR.2)
BIC(LR.2)
AIC(LR.1, LR.2)
BIC(LR.1, LR.2)

# Hosmer-Lemeshow goodness-of-fit test
library(ResourceSelection)
hoslem.test(healthy, fitted(LR.2), g = 10)
hlt2 <- hoslem.test(healthy, fitted(LR.2), g = 10)
cbind(hlt2$observed, hlt2$expected)

anova(LR.2, update(LR.2, ~1), test="Chisq")

# Model comparison using the log likelihood ratio test
anova(LR.1, LR.2, test = "Chisq")

# Marginal effects
library(margins)
marg2 <- margins(LR.2)
summary(marg2)

# Predicted probabilities
newdf <- data.frame(educ=c(12,14,16),
                    maritals=rep(mean(maritals), 3),
                    wrkfull=rep(mean(wrkfull), 3),
                    female=rep(mean(female), 3))
newdf
newdf[,c('pred.prob')] <- predict(LR.2, newdata=newdf, type="response")
newdf
```

```
# Predicted probabilities with ggpredict() in ggeffects
library(ggeffects)
margins <- ggpredict(LR.2, terms="educ[12, 14, 16]")
margins
as.data.frame(margins)
sqrt(diag(vcov(margins)))
plot(margins)

margins.ew <- ggpredict(LR.2, terms = c("educ[12, 14, 16]", "wrkfull"))
margins.ew
plot(margins.ew)

# Probit regression
PR.2 <- glm(healthy ~ wrkfull + maritals + female + educ, data = chp3.lr, family =
binomial(link = "probit"))
summary(PR.2)

# Making tables to display the results
library(stargazer)
stargazer(LR.1, LR.2, type="text", align=TRUE, out="lr2mod.txt")
stargazer(LR.1, LR.2, type="html", align=TRUE, out="lr2mod.htm")

detach(chp3.lr)
```

Glossary

A Bernoulli distribution is a distribution for a binary variable with 1 for having an event or success and 0 for not having an event or failure. It is a special case of the binomial distribution with one trial for each individual.

A logit coefficient is the regression coefficient in the logit scale in logistic regression models.

A marginal effect is a change in a response variable related to the change in an independent variable. In logistic regression, it is a change in the predicted probability with a change in the values of a predictor variable.

A probit regression model is used with the probit transformation of a binary outcome variable.

An odds ratio (OR) is a ratio of two odds.

Deviance compares the currently fitted model and the saturated model. It is defined as –2 times the difference in log likelihood between these two models.

Generalized linear models extend the linear regression model when the response variable follows a distribution from the exponential family. They include three common components: the random component, the systematic component, and the link function.

Logistic regression is a regression method used to predict a binary outcome variable with the logistic transformation.

The AIC penalizes the deviance by the number of predictors.

The BIC penalizes the deviance by its degrees of freedom and the sample size.

The likelihood ratio test is the difference in deviance between nested models. It is the difference in –2LL, which can be expressed as a ratio of likelihood in logarithm.

The linear probability model is used to estimate a binary outcome variable with the linear regression method.

The odds are the ratio of two probabilities, the probability of success or of having an event (p) to the probability of failure or of not having an event ($1 - p$).

Exercises

Use the GSS 2016 data available at **https://edge.sagepub.com/liu1e** for the following problems.

1. Conduct a logistic regression analysis to examine the relationship between the outcome variable `gunlaw` and three predictor variables from the three predictor variables `sex`, `educ`, and `age`. Before conducting the analysis, recode `gunlaw` into a new variable named `gun` so that 1 = favoring gun permit and 0 = opposing gun permit.
2. Conduct the likelihood ratio test of the overall model and interpret it.

3. Compute the deviance statistic for the model.
4. List three measures of pseudo R^2 and the AIC statistic.
5. Identify the logit coefficient, the Wald z test, and the 95% confidence interval for the predictor variable `educ`.
6. Computer the odds ratios for `educ` and `age`.
7. Make a publication-quality table containing the estimated logit coefficients.
8. Write a report to summarize the results from the output.

4

PROPORTIONAL ODDS MODELS FOR ORDINAL RESPONSE VARIABLES

OBJECTIVES OF THIS CHAPTER

This chapter introduces proportional odds models for ordinal response variables. It starts with an introduction to the model followed by a discussion of the odds and odds ratios in the model, goodness-of-fit statistics of the model, the proportional odds assumption, and how to interpret parameter estimates. After a description of the data, the proportional odds models with the `clm()` function in the `ordinal` package and the `vglm()` function in `VGAM` are illustrated with step-by-step instructions. R commands and output are explained in detail. The chapter focuses on fitting proportional odds models using R, as well as on interpreting and presenting the results. After reading this chapter, you should be able to:

- Identify when a proportional model is used.
- Conduct proportional odds models, and test the assumption using R.
- Interpret the output.
- Interpret the model in terms of odds ratios.
- Compute and plot the predicted probabilities.
- Compare models using the likelihood ratio test and other fit statistics.
- Present results in publication-quality tables using R.
- Write the results for publication.

4.1 PROPORTIONAL ODDS MODELS: AN INTRODUCTION

In the last chapter, we focused on binary logistic regression models when the outcome variable is dichotomous with values of 1 and 0. In your research, you may often encounter ordinal outcome variables, which are categorical variables with ranks or orders, for example, student's socioeconomic status ordered from low to high; children's proficiency in early reading scored from level 0–5; and a response scale of a survey instrument with five levels, ordered from strongly disagree to strongly agree.

Research examples for ordinal response variables in the literature include a three-level response scale for an item related to astrology in the 2006 General Social Survey (Agresti, 2010), the deprivation level with three categories (not deprived, mildly deprived, or severely deprived) (Borooah, 2002), student persistence through high school with three levels (i.e., dropped out, still in school but behind peers, and persisted) (Heck et al., 2012), four-category drug user type (Menard, 2010), children's literacy proficiency level (O'Connell, 2006), and four-category level of severity of illness (Rabe-Hesketh & Skrondal, 2012).

In the preceding examples, the outcome variables of interest are ordinal response variables with more than two categories. In this chapter and the following two chapters, we will focus on various logistic regression models for ordinal response variables, which are referred to as ordinal logistic regression models when they are broadly defined. The first model, which is introduced in this chapter, is the proportional odds (PO) model. The PO model, which is also called the cumulative odds model, is one of the most commonly used models for the analysis of ordinal response data. It is so popular that sometimes we just call it the ordinal regression model or the ordered logit model. The PO model is a generalization of a binary logistic regression model when the response variable has more than two ordinal categories. It is used to estimate the odds of being at or below a particular level of the response variable. For example, if there are J levels of ordinal outcomes, then the model makes $J-1$ predictions, each estimating the odds of being at or below the jth level of the outcome variable, which are referred to as the cumulative odds. The odd ratios for each predictor variable are assumed to be the same across all categories, which is referred to as the PO assumption, or the parallel lines assumption. This model can also estimate the odds of being above a particular level of the ordinal response variable as well, because below and above a particular category are just two opposite directions.

The PO model can be expressed in the logit form as follows:

$$\text{logit}[\pi_j(x)] = \ln\left(\frac{\pi_j(x)}{1-\pi_j(x)}\right) = \alpha_j + (-\beta_1 X_1 - \beta_2 X_2 - \ldots - \beta_p X_p) \quad (4.1)$$

where $\pi_j(x) = P(Y \leq j \mid x_1, x_2, \ldots, x_p)$, which is the probability of being at or below category j given a set of predictors. $j = 1, 2, \ldots, J-1$. α_j are the cut points, and β_1,

$\beta_2, \ldots, \beta_p$ are the logit coefficients. To estimate the ln(odds) of being at or below the j-th category, the PO model can be rewritten as:

$$\begin{aligned}\text{logit}\left[P(Y \leq j|x_1, x_2, \ldots, x_p)\right] &= \left(\frac{P(Y \leq j|x_1, x_2, \ldots, x_p)}{P(Y > j|x_1, x_2, \ldots, x_p)}\right) \\ &= \alpha_j + (-\beta_1 X_1 - \beta_2 X_2 - \ldots - \beta_p X_p)\end{aligned} \tag{4.2}$$

To understand the PO model for an ordinal response variable, we can think of it as several binary logistic regression models that are estimated simultaneously. The outcome variables of these binary models are dichotomized from the ordinal outcome variable comparing outcomes at or below a category ($Y \leq$ cat. j) and above that category ($Y >$ cat. j). Therefore, each binary logistic regression estimates odds of being at or below a category (coded as 1) versus above that category (coded as 0). Although each logistic model has a different intercept, the estimated logit coefficients are constrained to be equal. In other words, the regression lines are parallel, or the odds are proportional across the categories. Therefore, for each predictor variable, we only need to estimate one regression coefficient rather than multiple coefficients. This constraint is the proportional odds assumption or the parallel lines assumption.

The `clm()` function from the `ordinal` package (Christensen, 2019) in R uses Equation 4.2 to express the PO model where there are negative signs before the logit coefficients in the linear predictor, whereas the `vglm()` function in the `VGAM` package (Yee, 2010) uses a different parameterization with positive signs before the logit coefficients.

Although this chapter focuses on the commonly used logit link function for ordinal logistic regression models, the probit link can be used to fit ordinal probit models, where the cumulative probability of being at or below a particular category can be expressed as the cumulative standard normal distribution function. See Chapter 3 on the discussion of the probit model. For more information on ordinal regression models, refer to Agresti (2010, 2013, 2015, 2019), Ananth and Kleinbaum (1997), Armstrong and Sloan (1989), Clogg and Shihadeh (1994), Liu (2009, 2016a, 2016b), Liu et al. (2018), Fullerton and Xu (2016), Long (1997), Long and Freese (2014), McCullagh (1980), McCullagh and Nelder (1989), Menard (2010), O'Connell (2000, 2006), Powers and Xie (2008), Smithson and Merkle (2014), and Tutz (2012).

4.1.1 Odds and Odds Ratios in PO Models

In binary logistic regression, the values of the outcome variable are either 1 or 0, and we model the odds of success or of having an event when the outcome variable takes the value of 1 ($Y = 1$). The odds of success are the probability of success (p) divided by the probability of failure ($1 - p$).

In proportional odds models, the outcome variable is ordered with multiple levels, and we estimate the odds of being at or below a particular category ($Y \leq j$). Similar to the odds in binary logistic regression, the odds of being at or below a category in ordinal logistic regression equals the probability of being at or below a category divided by the probability of being above that category:

$$\text{Odds}(Y \leq j) = \frac{P(Y \leq j)}{P(Y > j)}$$

where $P(Y \leq j)$ is the cumulative probability of being at or below a category j or the cumulative probability of the ordinal response variable Y less than or equal to a category j. Since the probability of being at or below a category and the probability of being above that category is complementary, $P(Y \leq j) + P(Y > j) = 1$, this equation can be rewritten as:

$$\text{Odds}(Y \leq j) = \frac{P(Y \leq j)}{1 - P(Y \leq j)}.$$

It reads as follows: The odds of being at or below a category j in ordinal logistic regression equal the probability of being at or below a category divided by its complimentary probability, 1 minus the probability of being at or below that category.

The probability of being at or below a category $P(Y \leq j)$ is the cumulative probability since it equals the sum of the probabilities of all categories at or below that category:

$$P(Y \leq j) = P(Y = 1) + P(Y = 2) + \ldots + P(Y = j) \;\; \text{When } j = 1, 2, \ldots, J$$

For example, an outcome variable, health status, is ordinal with four levels from 1 to 4, where 1 = poor, 2 = fair, 3 = good, and 4 = excellent:

$$\begin{aligned}
P(Y \leq 4) &= P(Y = 1) + P(Y = 2) + P(Y = 3) + P(Y = 4) = 1 \\
P(Y \leq 3) &= P(Y = 1) + P(Y = 2) + P(Y = 3) \\
P(Y \leq 2) &= P(Y = 1) + P(Y = 2) \\
P(Y \leq 1) &= P(Y = 1)
\end{aligned}$$

The probability of being at a category $P(Y = j)$ is equal to the difference between the cumulative probability $P(Y \leq j)$ and the cumulative probability $P(Y \leq j - 1)$. It is written as: $P(Y = j) = P(Y \leq j) - P(Y \leq j - 1)$. Therefore, the probability of being at each category in the above example can be computed as follows:

$$\begin{aligned}
P(Y = 4) &= P(Y \leq 4) - P(Y \leq 3) \\
P(Y = 3) &= P(Y \leq 3) - P(Y \leq 2) \\
P(Y = 2) &= P(Y \leq 2) - P(Y \leq 1) \\
P(Y = 1) &= P(Y \leq 1)
\end{aligned}$$

Since this outcome variable has four categories, we can estimate the following cumulative odds: the odds of being at or below category 1, the odds of being at or below category 2, and the odds of being at or below category 3. The odds of being at or below a category in ordinal logistic regression are also called the cumulative odds.

Odds ($Y \leq 1$) equal the ratio of probability of being at or below category 1 to the probability of being above this category. The probability, $P(Y > 1) = P(Y = 2) + P(Y = 3) + P(Y = 4)$, which is the sum of the probabilities when $Y = 2$, 3, and 4. We define $P(Y = j)$ to be $P(j)$, so the equation can be written as:

$$\text{Odds}(Y \leq 1) = \frac{P(Y \leq j)}{1 - P(Y \leq j)} = \frac{P(Y = 1)}{P(Y = 2) + P(Y = 3) + P(Y = 4)}$$
$$= \frac{P(1)}{P(2) + P(3) + P(4)}$$

Odds ($Y \leq 2$) equal the ratio of probability of being at or below category 2 to the probability of being above this category. Since $P(Y \leq 2) = P(1) + P(2)$, and $P(Y > 2) = P(3) + P(4)$, the odds of being at or below category 2, can be expressed as follows:

$$\text{Odds}(Y \leq 2) = \frac{P(Y \leq 2)}{1 - P(Y \leq 2)} = \frac{P(1) + P(2)}{P(3) + P(4)}$$

Odds ($Y \leq 3$) equal the ratio of probability of being at or below category 3 to the probability of being above this category. Using the same method, we get the following equation:

$$\text{Odds}(Y \leq 3) = \frac{P(1) + P(2) + P(3)}{P(4)}$$

The odds of being at or below category 1 are the probability comparisons between category 1 and categories 2, 3, and 4; the odds of being at or below category 2 compare the probabilities of categories 1 and 2 with the probabilities of categories 3 and 4; and the odds of being at or below category 3 compare the probabilities of categories 1, 2, and 3 with the probability of category 4. Therefore, the cumulative odds in ordinal logistic regression are basically comparisons between two complimentary probabilities [i.e., $P(Y \leq j)$ and $P(Y > j)$]. Table 4.1 presents the logits, odds, and category comparisons for the PO model for the health status with four levels.

TABLE 4.1 ● Category Comparisons for the Proportional Odds Model With Four Levels of Health Status (j = 1, 2, 3, 4)

Category	Logit $P(Y \leq j)$	Odds	Probability Comparisons
Level 1	logit $P(Y \leq 1)$	$\frac{P(Y \leq 1)}{P(Y > 1)}$	Category 1 vs. categories 2–4
Level 2	logit $P(Y \leq 2)$	$\frac{P(Y \leq 2)}{P(Y > 2)}$	Categories 1 and 2 vs. categories 3 and 4
Level 3	logit $P(Y \leq 3)$	$\frac{P(Y \leq 3)}{P(Y > 3)}$	Categories 1, 2, and 3 vs. category 4

Odds Ratios in PO Models

In binary logistic regression, the odds ratio is the ratio of two odds, the odds of success when the value of a predictor is $(x + 1)$ relative to the odds when the predictor has a value of x. In other words, it is the change in the odds for a one-unit increase in the predictor variable. Similar to binary logistic regression, the odds ratio in PO models is the change in the odds (i.e., the odds of being above a particular category versus being at or below that category) for a one-unit increase from any value of x to the value of $(x + 1)$, and it is the exponentiated logit coefficient, $\exp(\beta)$. In contrast, the odds ratio of being at or below a particular category is the multiplicative inverse or reciprocal of the odds of being above that category. It is the exponentiated logit coefficient with a negative sign before that [i.e., $\exp(-\beta)$].

4.1.2 The PO Assumption

PO Assumption

In the proportional odds models, we assume that each predictor has the same effects across the categories of the ordinal outcome variable. In other words, the logit regression coefficients for each predictor are the same across the ordinal categories. For example, if we predict the ordinal outcome variable, health status, from the predictor, marital status, we estimate the odds of being at or below a category of health status relative to above that category, given that predictor variable. The estimated logits and the corresponding odds ratios of being at or below category 1, category 2, and category 3 for the predictor, marital status, are assumed to be the same. Although we assume that they are equal, how can we know whether the assumption holds?

Likelihood Ratio Test

To test whether the PO assumption is met, we can use the likelihood ratio test to look at the logit coefficients of a series of underlying binary logistic regression models for the dichotomized ordinal outcome variable, comparing outcomes at or below a category versus above that category.

The likelihood ratio test of the PO assumption can be examined using the `nominal_test()` function in the `ordinal` package. It provides the likelihood ratio test result for each predictor. We can also use the `lrtest()` function in the `VGAM` package to test the PO assumption, which provides the omnibus test for the overall model.

4.1.3 Goodness-of-Fit Statistics

Since ordinal logistic regression is an extension of binary logistic regression, all measures-of-fit statistics in binary logistic regression models, such as pseudo R^2 statistics, the deviance, the likelihood ratio test, and Akaike's information criterion (AIC) and Bayesian information criterion (BIC), can also be applied to proportional odds models.

4.1.4 Interpretation of Model Parameter Estimates

The odds ratio in ordinal logistic regression can be interpreted in a similar way as that of the binary logistic regression. In the binary logistic regression, we estimate the odds of success when the outcome takes the value of 1 (i.e., $Y = 1$), whereas in ordinal logistic regression, the odds are the ones when the outcomes are at or below a particular category (i.e., $Y \leq j$).

Recall that the signs before the logit coefficients in the equation of the ordinal logistic regression (Equation 4.1) are negative. To get the odds ratio (OR) of being at or below a category, we need to exponentiate the logit coefficient with a negative sign before that. This odds ratio can be interpreted as the change in the predicted logit or the log odds of being at or below a particular category for a one-unit increase in the predictor variable. By removing the negative sign and then exponentiating the logit coefficient, we get the OR of being above a category. In contrast, taking the multiplicative inverse of the odds of being at or below a particular category also gives us the odds of being above that category. The odds ratio of being a particular category can be interpreted as the change in the predicted logit or the log odds of being above that particular category for a one-unit increase in the predictor variable.

When the logit coefficient itself is positive, it indicates the relationship between the predictor variable and the logit function of the probability is positive. In other words, a positive coefficient increases the probability of being above a category. By exponentiating the logit coefficient, you get the OR, which is greater than 1. This means that the odds of being above a particular category increases for a one-unit increase in the predictor variable.

When the logit coefficient itself is negative, it indicates that the relationship between the predictor variable and the logit function is negative. A negative coefficient decreases the probability of being above a category. The exponentiated coefficient, the OR, is less than 1. This means that the odds of being above a particular category decreases for a one-unit increase in the predictor variable.

When the logit coefficient equals 0, the OR equals 1. This indicates that there is no relationship between the predictor and the odds, so there is no change in the odds when the values of the predictor variable change.

4.2 RESEARCH EXAMPLE AND DESCRIPTION OF THE DATA AND SAMPLE

Research Problem and Questions: In this chapter, the purpose of the research example is to investigate the relationships between the ordinal response variable, health status, and the three predictor variables: marital status, the highest education completed, and gender. The research question is as follows: Do the three predictor variables predict the ordinal response variable, health status? Specifically, do the three predictor variables predict the cumulative odds and then the cumulative probabilities of being at or below a particular level of health status, or the cumulative odds and then the cumulative probabilities of being above that health status level?

Description of the Data and Sample: The data for the following analyses were from the General Social Survey 2016 (GSS 2016). The following are the variables used for data analysis in this chapter:

- `healthre`: the recoded variable of health (health status) with four ordinal categories (1 = poor health, 2 = fair health, 3 = good health, and 4 = excellent health)
- `maritals`: the recoded variable of marital (marital status) with 1 = currently married and 0 = not currently married
- `educ`: the highest education completed
- `female`: recoded variable of sex with 1 = female and 0 = male

4.3 FITTING A ONE-PREDICTOR PO MODEL USING THE `clm()` FUNCTION

4.3.1 Packages and Functions for Proportional Odds Models in R

Several packages in R can be used for fitting PO models. This chapter focuses on the `ordinal` package (Christensen, 2019) and the `VGAM` package (Yee, 2010, 2015, 2021). The `clm()` function in `ordinal` and the `vglm()` function in `VGAM` are both used. The `clm()` function is introduced first.

4.3.2 The `clm()` Function in the `Ordinal` Package

The `clm()` function in the `ordinal` package is used for the ordinal logistic regression analysis, where `clm` stands for the cumulative link model with the logit link as the default. Since `ordinal` is a user-written package, you need to install it first by typing `install.packages("ordinal")` and then load the package by typing `library(ordinal)`. The basic syntax is `clm()` with the model formula which is specified within the parentheses after the function name `clm`. Writing the model formula for ordinal logistic regression in `clm()` is similar to that for the linear regression in `lm()`. The ordinal response variable and the independent variable(s) in the model are separated by the tilde (~). For example, the command `clm(y ~ x)` tells R to run a simple ordinal logistic regression analysis predicting the ordinal dependent variable *y* with an independent variable *x*. When there are more than multiple predictor variables in the formula, they are connected by plus (+) signs. For example, the model formula in `clm(y ~ x1 + x2)` includes two predictor variables, *x1* and *x2*. The default link function is the logit function, which can be omitted. To fit an ordinal probit model, we can use the `link = "probit"` argument. For more details on how to use this command, type `help(clm)` in the command prompt after loading the `ordinal` package.

4.3.3 The PO Model: One-Predictor Model With the `clm()` Function

The command `PO.1 <- clm(healthre ~ educ, data = chp4.po)` tells R to conduct the ordinal logistic regression to estimate the ordinal outcome variable `healthre` using the predictor variable `educ` with the `clm()` function. In the function, the outcome variable `healthre` is estimated by the predictor variable `educ` with a tilde (~). The `data = chp4.po` argument specifies the data frame. The output of the fitted model is defined as an object named `PO.1`. The `summary(PO.1)` command prints out the output, which is displayed as follows.

```
> library(foreign)
> chp4.po <- read.dta("C:/CDA/gss2016.dta")
> chp4.po$healthre <- factor(chp4.po$healthre, ordered=TRUE)
> chp4.po$educ <- as.numeric(chp4.po$educ)
> chp4.po$wrkfull <- as.numeric(chp4.po$wrkfull)
> chp4.po$maritals <- as.numeric(chp4.po$maritals)
> attach(chp4.po)

> # One-predictor model with the clm() function in ordinal
> library(ordinal)
> PO.1 <- clm(healthre ~ educ, data = chp4.po)
> summary(PO.1)
formula: healthre ~ educ
data:    chp4.po

 link   threshold  nobs  logLik    AIC      niter  max.grad  cond.H
 logit  flexible   1873  -2166.16  4340.32  5(0)   5.85e-08  1.4e+04

Coefficients:
        Estimate   Std. Error   z value   Pr(>|z|)
educ      0.1790       0.0152     11.78    <2e-16 ***

---
Signif. codes: 0 '***' 0.001 '**' 0.01 '*' 0.05 '.' 0.1 ' ' 1

Threshold coefficients:
       Estimate  Std. Error  z value
1|2     -0.3554      0.2162   -1.644
2|3      1.5258      0.2092    7.294
3|4      3.8056      0.2262   16.823
```

4.3.4 Interpreting R Output

The output at the beginning displays the model formula and the data. It then shows the link function, threshold type, number of observations, log likelihood, AIC statistic, number of Newton-Raphson iterations, maximum absolute gradient of the log-likelihood function, and condition number of the Hessian in sequence.

The link function of the PO model is logit. The estimated thresholds or intercepts are unconstrained. The number of observations for the analysis is 1,873. The maximum log likelihood value is −2,166.16, and the AIC statistic is 4,340.32.

The next part shows the coefficients table (labeled `Coefficients:`). It includes the parameter estimates for the predictor variables, their standard errors, the Wald z statistics, and the associated p values. The null hypothesis for the Wald test is that the coefficient of the predictor variable is 0, and the alternative hypothesis is that the coefficient of the predictor variable is significantly different from 0. The logit regression coefficient of the predictor variable `educ` $\beta = .179$ and its Wald $z = 11.78$. The associated p value `Pr(>|z|) < .001`, so we rejected the null hypothesis. Therefore, the predictor variable `educ` is a significant predictor of the ordinal outcome variable, health status.

The final part of the output displays the intercepts or the threshold coefficients table (labeled `Threshold Coefficients:`). It includes the parameter estimates for the intercepts or the thresholds, their standard errors, and the Wald z values.

4.3.5 Interpreting the Coefficients and the Intercepts/Thresholds

The logit coefficients can also be obtained using `coef(PO.1)` and their confident intervals can be obtained with `confint(PO.1)`.

```
> coef(PO.1)
        1|2          2|3          3|4         educ
-0.3553797    1.5258035    3.8055757    0.1790195
> confint(PO.1)
             2.5 %       97.5 %
educ    0.1493574    0.2089509
```

$\beta = .179$. It can be interpreted as follows: for a one-unit increase in the years of education completed, the change in the logit or log odds of being above a category of health status (i.e., better health status) is .179. The 95% confidence interval of the regression coefficient is [`.149, .209`]. It does not contain 0, which indicates the coefficient is significantly different from 0.

The threshold coefficients table reports the three intercepts or thresholds: `1|2`, `2|3`, and `3|4`. These are the estimated thresholds on the latent variable Y^* used to differentiate the adjacent levels of health status. When the response category is 1, the latent variable falls at or below the first cut point α_1. When the response category is 2, the latent variable falls between the first cut point α_1 and the second cut point α_2; when the response category reaches 3, the latent variable falls between the second α_2 and the third cut point α_3; and when the response category reaches 4, the latent variable is at or above the third cut point α_3. These thresholds are also called intercepts or cut points.

They can be thought of as the intercepts for three underlying binary logistic regression models if we dichotomize the ordinal outcome variable.

4.3.6 Odds Ratios

We use the `exp(coef(PO.1))` command to get the odds ratios and the `exp(confint(PO.1))` command to produce the corresponding confidence intervals. The output is shown as follows.

```
> exp(coef(PO.1))
      1|2           2|3           3|4          educ
0.7009073     4.5988372    44.9511225     1.1960440
> exp(confint(PO.1))
           2.5 %       97.5 %
educ    1.161088     1.232384
```

The odds ratio for the predictor variable `educ` is `1.196`. It equals the exponentiated regression coefficient exp(.179). The 95% confidence interval of the odds ratio is `[1.161, 1.232]`.

4.3.7 Interpreting the Odds Ratio of Being at or Below a Particular Category

The estimated logit regression coefficient β = .533, z = 5.11, $p < .001$, which indicates that education is a significant predictor of the ordinal response variable, health status. By substituting the value of the coefficient into Equation 4.2, logit $[P(Y \leq j \mid$ `educ`$)] = \alpha_j + (-\beta_1 X_1)$, we calculated logit $[P(Y \leq j \mid$ `educ`$)] = \alpha_j - .179$ (`educ`). OR = $e^{(-.179)}$ = .836, which indicates that for a one-unit increase in the years of education the odds of being at or below any category of health status (i.e., less healthy) decrease by .836.

To estimate the cumulative odds being at or below a certain category j for `educ`, let us take a look at the logit form of proportional odds model, logit $[P(Y \leq j \mid$ `educ`$)] = \alpha_j$ –.179 (`educ`). For example, when $Y \leq 1$, α_1, −.355 is the first cut point for the model. By substituting it into Equation 4.2, we get logit $[P(Y \leq j \mid$ `educ`$)]$ = −.355 −.179(`educ`). For `educ` (x = 1), logit $[P(Y \leq 1 \mid$ `educ`$)]$ = −.534. By exponentiating the logit, we calculate the odds of being at or below the category 1 (poor health) when `educ` = 1, $e^{(-.534)}$ = .586. For `educ` (x = 2), logit $[P(Y \leq 1 \mid$ `educ`$)]$ = −.355 −.179× 2 = −.713, so the odds of being at or below the category 1 (poor health) when `educ` = 2, $e^{-.713}$ =.490. The odds ratio of `educ` (x = 2) relative to `educ` (x = 1) = .490/.586 = .836.

4.3.8 Interpreting the Odds Ratio of Being Above a Particular Category

The proportional odds model can also estimate the ln(odds) of being above a category j. Again, these ln(odds) can be transformed into the cumulative odds and cumulative probabilities. For example, we can estimate the cumulative probability of health status above category 3, $P(Y > 3)$; above category 2, $P(Y > 2)$; and above category 1, $P(Y > 1)$. The cumulative logit form can be expressed as logit $[P(Y > j \mid \text{educ})] = -\alpha_j + (\beta_1 X_1)$. When estimating the odds of being above category j, the sign of the cut points needs to be reversed and their magnitude remains unchanged since we estimate the cut points from the right to the left of the latent variable Y^*, that is, from the direction when $Y = 4$ approaches $Y = 1$. Therefore, three cut points from right to left turn to –3.806, –1.526, and .355.

When the predictor is dichotomous, a positive sign of the logit coefficient indicates that it is more likely for the group ($x = 1$) to be above a particular category than for the relative group ($x = 0$). When the predictor is continuous, a positive coefficient indicates that when the value of the predictor variable increases, the odds of being above a particular category increase.

The `exp(coef(PO.1))` command provides the odds ratios of being above a particular category: OR = 1.196. It can be interpreted that the odds of being above a particular category of health status (better health status) increase by a factor of 1.196 for each unit increase in years of education. In other words, the odds of being above a particular category of health status increase by 19.6% for each unit increase in education.

4.3.9 Model Fit Statistics

Testing the Overall Model Using the Likelihood Ratio Test

To test if the overall model is significant, we fit a null model with the intercept only and compare the one-predictor PO model with the null model using the `anova()` function. The command `PO.0 <- clm(healthre ~ 1, data = chp4.po)` is used to fit the null model. The output is displayed below by the `summary(PO.0)` command.

```
> # Null model with the intercept only
> PO.0 <- clm(healthre ~ 1, data = chp4.po)
> summary(PO.0)
formula: healthre ~ 1
data:    chp4.po

 link   threshold  nobs  logLik    AIC      niter  max.grad  cond.H
 logit  flexible   1873  -2238.22  4482.44  5(0)   4.56e-09  5.9e+00


Threshold coefficients:
       Estimate   Std. Error   z value
1|2   -2.69954     0.09510    -28.39
2|3   -0.89064     0.05087    -17.51
3|4    1.25964     0.05569     22.62
```

The anova(PO.0, PO.1) command compares the log-likelihood statistics of the fitted model PO.1 and the null model PO.0 using the likelihood ratio test.

```
> # Testing the overall model using the likelihood ratio test
> anova(PO.0, PO.1)
Likelihood ratio tests of cumulative link models:


      formula:              link: threshold:
PO.0  healthre ~ 1          logit flexible
PO.1  healthre ~ educ       logit flexible


      no.par    AIC     logLik   LR.stat   df   Pr(>Chisq)
PO.0       3  4482.4   -2238.2
PO.1       4  4340.3   -2166.2    144.11    1    < 2.2e-16 ***

---
Signif. codes: 0 '***' 0.001 '**' 0.01 '*' 0.05 '.' 0.1 ' ' 1
```

The null hypothesis of the test for the overall model is that the predictor variable does not contribute to the model, and the alternative hypothesis is that the one-predictor PO model is better than the null model with no independent variables. The likelihood ratio test statistic $LR\ \chi^2_{(1)} = 144.11, p < .001$, which indicated that the overall model with one predictor was significantly different from zero. Therefore, the one-predictor PO model provides a better fit than the null model with no independent variables.

Pseudo R^2

The nagelkerke(PO.1) function in the rcompanion package (Mangiafico, 2021) produces the three types of pseudo R^2 statistics and the likelihood ratio test statistic for the PO model. You need to install rcompanion first by typing install.packages("rcompanion") and then load it by typing library(rcompanion).

```
> # Pseudo R2
> library(rcompanion)
> nagelkerke(PO.1)
$`Models`

Model: "clm, healthre ~ educ, chp4.po"
Null:  "clm, healthre ~ 1, chp4.po"

$Pseudo.R.squared.for.model.vs.null
                              Pseudo.R.squared
McFadden                             0.0321935
Cox and Snell (ML)                   0.0740563
Nagelkerke (Cragg and Uhler)         0.0815267
```

```
$Likelihood.ratio.test
 Df.diff    LogLik.diff    Chisq      p.value
      -1        -72.056   144.11    3.358e-33

$Number.of.observations

Model:    1873
Null:     1873

$Messages
[1] "Note: For models fit with REML, these statistics are based on refitting with ML"

$Warnings
[1] "None"
```

The McFadden R^2 is .032, the Cox and Snell R^2 is .074, and the Nagelkerke R^2 is .082. The same results can be computed using the equations introduced in the previous section. In the R command below, `LLM` is the log-likelihood value for the single-predictor model and `LL0` is the log-likelihood value for the null model. In addition, `McFadden` is the object name for the McFadden R^2, `CS` for the Cox and Snell R^2, and `NG` for the Nagelkerke R^2.

```
> LLM <- logLik(PO.1)
> LL0 <- logLik(PO.0)
> McFadden <- 1-(LLM/LL0)
'log Lik.' 0.03219349 (df=4)
> CS <- 1-exp(2*(LL0-LLM)/1873)
> CS
'log Lik.' 0.07405631 (df=3)
> NG <- CS/(1-exp(2*LL0/1873))
> NG
'log Lik.' 0.08152671 (df=3)
```

4.3.10 Using the Likelihood Ratio Test to Test the PO Assumption

The `nominal_test()` function in the `ordinal` package is used to test the PO assumption. It provides the likelihood ratio test result for each predictor. A nonsignificant test indicates that the proportional odds assumption is not violated for that predictor. The results of the `nominal_test(PO.1)` function are shown as follows.

```
> # PO assumption test
> nominal_test(PO.1)
Tests of nominal effects
```

```
formula: healthre ~ educ
          Df    logLik      AIC        LRT    Pr(>Chi)
<none>          -2166.2    4340.3
educ       2    -2165.8    4343.5    0.82398    0.6623
```

The likelihood ratio test yields $\chi^2_{(2)} = .824$, $p = .662$, which indicates that the proportional odds assumptions for the model is upheld, suggesting that the effect of the explanatory variable `educ` is constant across the underlying binary models.

4.4 FITTING A MULTIPLE-PREDICTOR PO MODEL USING THE `clm()` FUNCTION

4.4.1 The PO Model: Multiple-Predictor Model With the `clm()` Function

The command `PO.2 <- clm(healthre ~ maritals + educ + female, data = chp4.po)` tells R to predict the ordinal response variable `healthre` from the three predictor variables `maritals`, `educ`, and `female` using ordinal logistic regression. The predictor variables are connected by plus signs in the model formula. The output is shown as follows after typing the `summary(PO.2)` command.

```
> # Multiple-predictor model with the clm() function
> PO.2 <- clm(healthre ~ maritals + educ + female, data = chp4.po)
> summary(PO.2)
formula: healthre ~ maritals + educ + female
data:    chp4.po

 link   threshold  nobs  logLik    AIC      niter  max.grad  cond.H
 logit  flexible   1873  -2160.51  4333.01  5(0)   7.25e-08  1.5e+04


Coefficients:
           Estimate  Std. Error  z value  Pr(>|z|)
maritals    0.29157     0.08819    3.306  0.000946 ***
educ        0.17502     0.01523   11.490   < 2e-16 ***
female      0.06702     0.08760    0.765  0.444232
---
Signif. codes: 0 '***' 0.001 '**' 0.01 '*' 0.05 '.' 0.1 ' ' 1

Threshold coefficients:
      Estimate  Std. Error  z value
1|2    -0.2543      0.2222   -1.145
2|3     1.6344      0.2158    7.574
3|4     3.9243      0.2331   16.834
```

4.4.2 Interpreting R Output

The R output for the multiple-predictor PO model includes the model formula and the data. It then displays the link function, the threshold option, the number of observations, the log likelihood value, AIC, the number of Newton-Raphson iterations, the maximum absolute gradient of the log-likelihood function, and the condition number of the Hessian. The fourth and fifth parts show the coefficients table and the threshold coefficients table, respectively.

The number of observations for the analysis is 1,873. The maximum log likelihood value is −2,160.51, and the AIC statistic is 4,333.01.

The coefficients table (labeled `Coefficients:`) displays the parameter estimates for the three predictor variables, their standard errors, the Wald z statistics, and the associated p values.

For the predictor variable `maritals`, Wald $z = 3.306$. The associated p value, `Pr(>|z|) < .001`, so we rejected the null hypothesis.

For the predictor variable `educ`, the Wald $z = 11.490$. The associated p value, `Pr(>|z|) < .001`, so we also reject the null hypothesis. Therefore, `maritals` and `educ` are significant predictors of the outcome variable.

For the predictor variable `female`, the Wald $z = .765$. The associated p value `Pr(>|z|) = .444`, so we fail to reject the null hypothesis and conclude that there is no significant effect of being female on the outcome variable.

The threshold coefficients table (labeled `Threshold Coefficients:`) includes the parameter estimates for the intercepts or the thresholds, their standard errors, and the Wald z values.

4.4.3 Interpreting the Coefficients and the Intercepts/Thresholds

The logit coefficients can also be obtained using the `coef(PO.2)` command and their confident intervals can be obtained with `confint(PO.2)`.

```
> coef(PO.2)
         1|2          2|3          3|4     maritals         educ       female
 -0.25430764   1.63443046   3.92426852   0.29157322   0.17502373   0.06701637
> confint(PO.2)
                 2.5 %      97.5 %
maritals     0.1189005   0.4646685
educ         0.1452924   0.2050181
female      -0.1046664   0.2387539
```

The logit coefficient for `maritals` $\beta = .292$. This means that the logit or log odds of being above a category of health status for the married is .292 points higher than that for the unmarried.

The logit coefficient for `educ` $\beta = .175$. This can be interpreted as the increase in the logit or log odds of being above a category of health status (i.e., better health status) is .179 for a one-unit increase in the years of education.

The logit coefficient for `female` $\beta = .067$. Since it is not significant ($p = .444$), being a female does not impact the logit or log odds of being above a category of health status.

The threshold coefficients table reports the three thresholds: `1|2`, `2|3`, and `3|4`. These are the estimated thresholds or intercepts to differentiate the adjacent categories of health status. The first intercept α_1 is $-.254$; the second intercept α_2 is 1.634; and the third intercept α_3 is 3.924.

4.4.4 Interpreting the Odds Ratios of Being Above a Particular Category

The `exp(coef(PO.2))` command provides the odds ratios of being above a category and the `exp(confint(PO.2))` command produces the corresponding confidence intervals. The following output is displayed.

```
> exp(coef(PO.2))
       1|2          2|3          3|4     maritals         educ       female
 0.7754532    5.1265374   50.6160401    1.3385316    1.1912745    1.0693130
> exp(confint(PO.2))
              2.5 %     97.5 %
maritals  1.1262579   1.591487
educ      1.1563777   1.227547
female    0.9006249   1.269666
```

For `maritals`, $\beta = .292$, which is positive; OR = 1.339, which is greater than 1. This indicates that the odds of being above a particular category of health status (better health status) for the married are 1.339 times the odds for the unmarried when holding all the other predictors constant.

For `educ`, $\beta = .175$, which is positive; OR = 1.191, which is greater than 1. This indicates that the odds of being above a particular category of health status (better health status) increase by a factor of 1.191 for a one-unit increase in the predictor, education, when holding all the other predictors constant. In other words, for a one-unit increase in education, the odds of being healthier increase by 19.1%.

For `female`, $\beta = .067$, $p = .444$, which is not significantly different from 0; OR = 1.069, which almost equals 1. This indicates that there is no relationship between being a female and the cumulative odds of being in better health status. In other words, there is no significant difference between the males and females in better health status.

4.4.5 Interpreting the Odds Ratios of Being at or Below a Particular Category

In the preceding section, we interpreted the odds ratio of being above a category. We can also interpret how these predictor variables contribute to the odds of being at or below a particular category if we reverse the sign before the estimated logit coefficients and then compute the corresponding odds ratios.

The `exp(-coef(PO.2)` command tells R to reverse the odds of being above a category versus being at or below that category to the odds of being at or below a category versus above that category. The following is the output produced by the command.

```
> exp(-coef(PO.2))
       1|2          2|3          3|4     maritals         educ       female
1.28956847   0.19506344   0.01975658   0.74708731   0.83943710   0.93517989
> exp(-confint(PO.2))
                2.5 %      97.5 %
maritals    0.8878961   0.6283434
educ        0.8647694   0.8146326
female      1.1103402   0.7876087
```

By substituting the values of the four logit coefficients into Equation 4.2, we get logit $[P(Y \leq j)] = \alpha_j + (-.292 \times$ `maritals` $-.175 \times$ `educ` $-.067 \times$ `female`). The exponentiated logit coefficients are the odds ratios of being at or below a particular category.

For the predictor `maritals`, OR = .747, which is less than 1. This indicates that the odds of being at or below a particular category of health status (worse health status) for the married are .747 times the odds for the unmarried when holding all the other predictors constant.

For the predictor `educ`, OR = .839, which is less than 1. This indicates that the odds of being at or below a particular category of health status (poorer health status) decrease by a factor of .839 for a one-unit increase in education when holding all the other predictors constant.

For the predictor `female`, OR = .935 (p = .444), which is close to 1. This indicates that there is no relationship between being a female and the cumulative odds of being in poorer health status.

4.4.6 Computing the Predicted Probabilities With the `ggpredict()` Function in the `ggeffects` Package

Since the `margins` package (Leeper, 2021) has not been fully developed for the ordinal regression models, introduction of the marginal effects is omitted. We use the `ggpredict()` function in the `ggeffects` package (Lüdecke, 2018b) compute the predicted probabilities of being in a particular category of the ordinal response variable at specified values of predictor variables. The command is as follows: `margins.e <- ggpredict(PO.2, terms = "educ[12, 14, 16]")`. In the `ggpredict()` function, `PO.2` is the fitted model; and the `terms = "educ[12, 14, 16]"` option specifies the predictor variable `educ` at the values of 12, 14, and 16 when holding the other predictor variables at their means. The `terms` option can specify up to four variables, including the second to fourth grouping variables. The output is assigned to an object named `margins.e`. To request the standard errors of the predicted probabilities, we can use either the `as.data.frame()` or the `sqrt(diag(vcov()))` function.

```
> # Predicted probabilities with ggpredict() in ggeffects
> library(ggeffects)
> margins.e <- ggpredict(PO.2, terms = "educ[12, 14, 16]")
> margins.e

# Predicted probabilities of healthre

# Response Level = 1


educ    |   Predicted    |          95% CI
------------------------------------------
  12    |        0.07    |    [0.06, 0.09]
  14    |        0.05    |    [0.04, 0.07]
  16    |        0.04    |    [0.03, 0.05]

# Response Level = 2

educ    |   Predicted    |          95% CI
------------------------------------------
  12    |        0.27    |    [0.25, 0.30]
  14    |        0.22    |    [0.20, 0.24]
  16    |        0.17    |    [0.15, 0.19]

# Response Level = 3

educ    |   Predicted    |          95% CI
------------------------------------------
  12    |        0.49    |    [0.47, 0.52]
  14    |        0.51    |    [0.49, 0.54]
  16    |        0.51    |    [0.49, 0.54]
```

```
# Response Level = 4

educ     |   Predicted    |            95% CI
-----------------------------------------------
  12     |        0.16    |    [0.14, 0.18]
  14     |        0.21    |    [0.19, 0.23]
  16     |        0.28    |    [0.25, 0.31]

Adjusted for:
* maritals = 0.44
*    female = 0.56

> plot(margins.e)
```

The predicted probabilities for each response level are listed in sequence. For each response level, the first column in the table of the output lists `educ` at the values of 12, 14, and 16. The remaining columns list the predicted probabilities and the lower and upper confidence intervals. The predicted probabilities of being in poor health (i.e., response level = 1) are .07, .05, and .04, respectively. The predicted probabilities of being in the other three categories are also listed in the output. The last section titled "`Adjusted for`" lists the means of the other variables.

The predicted probabilities for all four response levels are plotted using the `plot (margins.e)` command. Figure 4.1 shows the predicted probabilities of being in each category (i.e., $Y = 1, 2, 3$, and 4) for `educ` at 12, 14, and 16.

The graph shows that with the increase in the years of education, the probabilities of being in poor and fair health condition (categories 1 and 2) decrease. In other words, people with higher levels of education are less likely to be associated with poor and fair health conditions. In addition, with the increase in the years of education, the probabilities of being in good and excellent health conditions (categories 3 and 4) increase. In other words, people with a higher level of education are more likely to be in good and excellent health conditions.

4.4.7 Model Fit Statistics

Testing the Overall Model Using the Likelihood Ratio Test

To test if the overall model is significant, we fit a null model with the intercept only and compare it with the multiple-predictor PO model by using the `anova()` function. Since the null model is fitted in the previous section, the output is omitted here. The `anova(PO.0, PO.2)` command compares the log-likelihood statistics of the fitted model `PO.2` and the null model `PO.0` using the likelihood ratio test.

FIGURE 4.1 ● Predicted Probabilities of Being in Categories 1, 2, 3, and 4 for educ

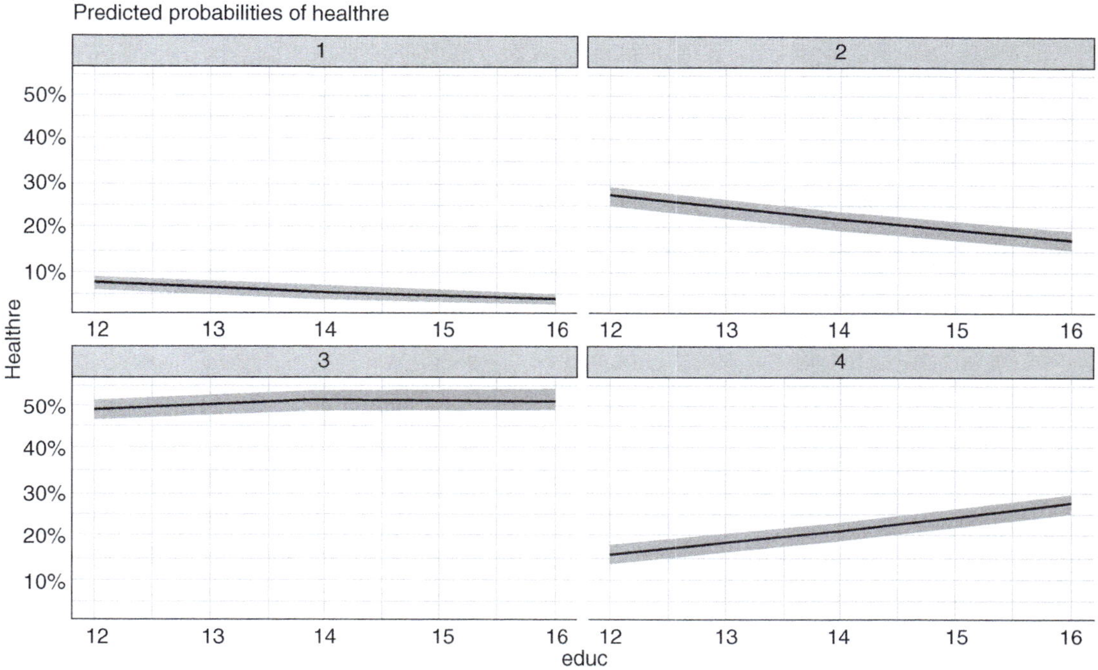


```
> # Testing the overall model using the likelihood ratio test
> anova(PO.0, PO.2)
Likelihood ratio tests of cumulative link models:

       formula:                                    link: threshold:
PO.0 healthre ~ 1                                  logit flexible
PO.2 healthre ~ maritals + educ + female           logit flexible

         no.par     AIC    logLik   LR.stat   df   Pr(>Chisq)
PO.0          3  4482.4   -2238.2
PO.2          6  4333.0   -2160.5    155.42    3    < 2.2e-16 ***

---
Signif. codes: 0 '***' 0.001 '**' 0.01 '*' 0.05 '.' 0.1 ' ' 1
```

The null hypothesis of the test for the overall model is that the predictor variables do not contribute to the model, and the alternative hypothesis is that the multiple-predictor PO model is better than the null model with no independent variables. The likelihood ratio test statistic $LR\ \chi^2_{(3)} = 155.42$, $p < .001$, which indicated that the overall model with the three predictors was significantly different from 0. Therefore, the multiple-predictor PO model provides a better fit than the null model with no independent variables.

Pseudo R^2

The `nagelkerke(PO.2)` command produces the three types of pseudo R^2 statistics and the likelihood ratio test statistic for the PO model.

```
> #PseudoR2
> nagelkerke(PO.2)
$`Models`

Model: "clm, healthre ~ maritals + educ + female, chp4.po"
Null:  "clm, healthre ~ 1, chp4.po"

$Pseudo.R.squared.for.model.vs.null
                             Pseudo.R.squared
McFadden                            0.0347206
Cox and Snell (ML)                  0.0796320
Nagelkerke (Cragg and Uhler)        0.0876648

 $Likelihood.ratio.test
Df.diff    LogLik.diff    Chisq       p.value
     -3        -77.712   155.42    1.7801e-33

$Number.of.observations

Model:     1873
Null:      1873

$Messages
[1] "Note: For models fit with REML, these statistics are based on refitting with ML"

$Warnings
[1] "None"
```

McFadden's R^2 is .035, Cox and Snell's R^2 is .080, and Nagelkerke's R^2 is .088.

4.4.8 Using the Likelihood Ratio Test to Test the PO Assumption

We use the `nominal_test()` function in the `ordinal` package to test the PO assumption. It provides the likelihood ratio test result for each predictor. A nonsignificant test indicates that the proportional odds assumption is upheld for that predictor. The results of the `nominal_test(PO.2)` command are shown as follows.

```
> # PO assumption test
> nominal_test(PO.2)
Tests of nominal effects

formula: healthre ~ maritals + educ + female
          Df    logLik     AIC      LRT     Pr(>Chi)
<none>          -2160.5   4333.0
maritals   2    -2156.3   4328.5   8.4986   0.01427  *
educ       2    -2160.1   4336.1   0.8962   0.63884
female     2    -2160.3   4336.6   0.3922   0.82194
---
Signif. codes: 0 '***' 0.001 '**' 0.01 '*' 0.05 '.' 0.1 ' ' 1
```

The proportional odds assumption is upheld for `educ` and `female`, whereas it is violated for `maritals`. For `maritals`, the likelihood ratio test $\chi^2_{(2)} = 8.499$, $p = .014$, which is significant.

4.4.9 Model Comparison Using the Likelihood Ratio Test

The likelihood ratio test or the deviance difference test is used to compare the full model and the one-predictor model. Recall that this test compares the reduced model, which contains less parameters, and the full model, which contains all parameters. The difference in deviance is often expressed as G = Deviance for the reduced model − Deviance for the full model or as $D_{\text{Reduced}} - D_{\text{Full}}$. The difference in deviance between nested models has a chi-square distribution with the degrees of freedom equal to the difference in the number of parameters between these two models.

The `anova()` function is used for the likelihood ratio test or the deviance difference test. Next, we compare the simple-predictor PO model and the multiple-predictor PO model with the `anova(PO.1, PO.2)` command.

```
> # Model comparison using the likelihood ratio test
> anova(PO.1, PO.2)
Likelihood ratio tests of cumulative link models:

       formula:                                link: threshold:
PO.1   healthre ~ educ                         logit flexible
PO.2   healthre ~ maritals + educ + female     logit flexible
```

```
      no.par     AIC    logLik   LR.stat   df   Pr(>Chisq)
PO.1       4  4340.3   -2166.2
PO.2       6  4333.0   -2160.5    11.313    2     0.003496 **
---
Signif. codes: 0 '***' 0.001 '**' 0.01 '*' 0.05 '.' 0.1 ' ' 1
```

The likelihood ratio chi-square test $\chi^2_{(2)} = 11.313, p < .001$. This result indicates that the full model has a better fit than the one-predictor model. The same result can be obtained if we compute it using the following equation:

$$G = D_{\text{Reduced}} - D_{\text{Full}} = -2 \times [-2166.2 - (-2160.5)] \\ = 11.4, \text{ df} = 6 - 4 = 2$$

4.5 FITTING A SINGLE-PREDICTOR PO MODEL USING THE `vglm()` FUNCTION

4.5.1 The `vglm()` Function in the VGAM Package

The `vglm()` function in the VGAM package can also be used for the ordinal logistic regression analysis, where `vglm` stands for vector generalized linear models. You need to install the VGAM package first by typing `install.packages("VGAM")` since it is a user-written package. After installation, load the package by typing `library (VGAM)`. The basic model formula command for `vglm()` is similar to that for either introduced in this chapter or `glm()` introduced in Chapter 2. In addition to the model formula, the `family` argument is needed for different types of models. For example, the command `vglm(y ~ x, family = cumulative(parallel = TRUE), data = data1)` tells R to fit a simple cumulative odds model predicting the ordinal dependent variable *y* with an independent variable *x*. The ordinal response variable and the independent variable in the model are separated by the tilde (~). The argument `family = cumulative(parallel = TRUE)` specifies the VGAM family function. It tells R to fit a cumulative odds model with the proportional odds assumption being specified. The `data` argument specifies the data frame used for the analysis. For more details on how to use this command, type `help(vglm)`, `help(propodds)`, and `help(cumulative)` in the command prompt after loading the VGAM package.

4.5.2 Using the `vglm()` Function to Fit a Single-Predictor PO Model

To fit the same single-predictor PO model introduced in the earlier section, we use the command `model1 <- vglm(healthre ~ educ, cumulative(parallel = TRUE, reverse = FALSE), data = chp4.po)`. Following the model equation, `healthre ~ educ`, the argument `cumulative(parallel = TRUE, reverse = FALSE)` tells R to fit a cumulative odds model with the parallel odds or proportional odds assumption and nonreversed ordinal categories. The `data = chp4.po` argument specifies the data frame. The `summary(model1)` command produces the following output.

```
> # One-predictor model with the vglm() function in VGAM
> library(VGAM)
> model1 <- vglm(healthre ~ educ, cumulative(parallel = TRUE, reverse = FALSE),
data = chp4.po)
> summary(model1)

Call:
vglm(formula = healthre ~ educ, family = cumulative(parallel = TRUE,
     reverse = FALSE), data = chp4.po)

Pearson residuals:
                        Min        1Q    Median        3Q     Max
logitlink(P[Y<=1])  -0.9413   -0.2425   -0.1716   -0.1312   7.102
logitlink(P[Y<=2])  -2.2568   -0.7070   -0.3115    0.5197   3.031
logitlink(P[Y<=3])  -6.6746    0.1568    0.3433    0.6484   1.013

Coefficients:
                Estimate   Std. Error    z value    Pr(>|z|)
(Intercept):1   -0.35539      0.21332     -1.666      0.0957 .
(Intercept):2    1.52580      0.20608      7.404    1.32e-13 ***
(Intercept):3    3.80557      0.22311     17.057     < 2e-16 ***
educ            -0.17902      0.01497    -11.955     < 2e-16 ***
---
Signif. codes: 0 '***' 0.001 '**' 0.01 '*' 0.05 '.' 0.1 ' ' 1

Names of linear predictors: logitlink(P[Y<=1]), logitlink(P[Y<=2]),
logitlink(P[Y<=3])

Residual deviance: 4332.323 on 5615 degrees of freedom

Log-likelihood: -2166.162 on 5615 degrees of freedom

Number of Fisher scoring iterations: 4

No Hauck-Donner effect found in any of the estimates

Exponentiated coefficients:
     educ
0.8360901
```

4.5.3 Interpreting R Output

The R output includes the call, the Pearson residuals, the coefficients, the number and names of the three linear predictors, the residual deviance, the log-likelihood value, the number of iterations, and the exponentiated coefficients.

The intercepts in the coefficients table are the same as those in the threshold coefficients table from the `clm ()` function. The logit coefficient of the `educ` predictor has the same magnitude as that in the coefficients table from the `clm ()` function, but has a negative sign. This is due to different parameterizations between the `clm ()` function and the `vglm ()` function. In the PO model equation for the `vglm ()` function, the signs before the coefficients are positive as follows.

$$\text{logit}\left[P\left(Y \leq j|x_1, x_2, \ldots, x_p\right)\right] = \alpha_j + \beta_1 X_1 + \beta_2 X_2 + \ldots + \beta_p X_p \quad (4.3)$$

In PO models, we estimate the logit or the log odds of being at or below a particular category ($Y \leq j$). The link functions for the three linear predictors in the model are `logit (P[Y<=1])`, `logit(P[Y<=2])`, and `logit(P[Y<=3])`. The log odds of being at or below category 1, `logit(P[Y<=1])`, compares the probability of category 1 to the probabilities of categories 2, 3, and 4; the log odds of being at or below category 2, `logit(P[Y<=2])`, compares the probabilities of categories1 and 2 to the probabilities of categories 3 and 4; and the log odds of being at or below category 3, `logit(P[Y<=3])`, compares the probabilities of categories 1, 2, and 3 to the probability of category 4.

4.5.4 Odds Ratios

The `exp(coef(model1, matrix = TRUE))` command provides the odds ratios of being at or below a category and the `exp(confint(model1, matrix = TRUE))` command produces the corresponding confidence intervals. We use the `cbind (exp(coef(model1)), exp(confint(model1)))` command to combine the odds ratios and the confidence intervals. The following output is displayed.

```
> exp(coef(model1, matrix = TRUE))
               logit(P[Y<=1])   logit(P[Y<=2])   logit(P[Y<=3])
(Intercept)         0.7009020        4.5988016       44.9507590
educ                0.8360901        0.8360901        0.8360901
> exp(confint(model1, matrix = TRUE))
                     2.5 %      97.5 %
(Intercept):1    0.4614005   1.0647226
(Intercept):2    3.0706560   6.8874455
(Intercept):3   29.0288233  69.6056713
educ             0.8119083   0.8609921
> cbind(exp(coef(model1)), exp(confint(model1)))
                                 2.5 %      97.5 %
(Intercept):1    0.7009020   0.4614005   1.0647226
(Intercept):2    4.5988016   3.0706560   6.8874455
(Intercept):3   44.9507590  29.0288233  69.6056713
educ             0.8360901   0.8119083   0.8609921
```

4.5.5 AIC Statistic

We can get the AIC statistic using `AIC(model1)`.

```
> AIC(model1)
[1] 4340.323
```

4.5.6 Logit Coefficients of Being at or Above a Category

With the `reverse = TRUE` option, we can estimate the logit coefficients of being at or above a particular category of an ordinal outcome variable. The `summary(model1b)` command produces the following output.

```
> # Logit coefficients of being at or above a category
> model1b <- vglm(healthre ~ educ, cumulative(parallel = TRUE, reverse = TRUE),
data = chp4.po)
> summary(model1b)

Call:
vglm(formula = healthre ~ educ, family = cumulative(parallel = TRUE,
     reverse = TRUE), data = chp4.po)

Pearson residuals:
                         Min       1Q    Median       3Q    Max
logitlink(P[Y>=2])    -7.102   0.1312    0.1716   0.2425 0.9413
logitlink(P[Y>=3])    -3.031  -0.5197    0.3115   0.7070 2.2568
logitlink(P[Y>=4])    -1.013  -0.6484   -0.3433  -0.1568 6.6746

Coefficients:
               Estimate  Std. Error   z value   Pr(>|z|)
(Intercept):1   0.35539     0.21332     1.666   0.0957 .
(Intercept):2  -1.52580     0.20608    -7.404   1.32e-13 ***
(Intercept):3  -3.80557     0.22311   -17.057   < 2e-16 ***
educ            0.17902     0.01497    11.955   < 2e-16 ***
---
Signif. codes: 0 '***' 0.001 '**' 0.01 '*' 0.05 '.' 0.1 ' ' 1

Names of linear predictors: logitlink(P[Y>=2]), logitlink(P[Y>=3]),
logitlink(P[Y>=4])

Residual deviance: 4332.323 on 5615 degrees of freedom

Log-likelihood: -2166.162 on 5615 degrees of freedom

Number of Fisher scoring iterations: 4

No Hauck-Donner effect found in any of the estimates

Exponentiated coefficients:
    educ
1.196043
```

The intercepts and the coefficient in the output by the `summary(model1b)` command and in that by the `summary(model1)` command have opposite signs since the former model estimates of the log odds of being at or below a particular category ($Y \leq j$), whereas the latter model estimates the log odds of being at or above a particular category ($Y \geq j+1$). Please note that the log odds ($Y \geq j+1$) equal the log odds ($Y > j$).

4.5.7 Odds Ratios of Being at or Above a Category

We again use the `exp(coef(model1b, matrix = TRUE))` command to obtain the odds ratios of being at or above a category and use the `exp(confint(model1b, matrix = TRUE))` command to produce the corresponding confidence intervals. The results are combined using `cbind(exp(coef(model1b)), exp(confint(model1b)))`. The following output is created.

```
> exp(coef(model1b, matrix = TRUE))
              logit(P[Y>=2])   logit(P[Y>=3])   logit(P[Y>=4])
(Intercept)         1.426733         0.217448       0.02224657
Educ                1.196043         1.196043       1.19604334
> exp(confint(model1b, matrix = TRUE))
                      2.5 %       97.5 %
(Intercept):1    0.93921178   2.16731441
(Intercept):2    0.14519171   0.32566331
(Intercept):3    0.01436665   0.03444852
 educ            1.16145082   1.23166615
> cbind(exp(coef(model1b)), exp(confint(model1b)))
                                  2.5 %       97.5 %
(Intercept):1    1.42673306   0.93921178   2.16731441
(Intercept):2    0.21744795   0.14519171   0.32566331
(Intercept):3    0.02224657   0.01436665   0.03444852
 educ            1.19604334   1.16145082   1.23166615
```

4.6 FITTING A MULTIPLE-PREDICTOR PO MODEL USING THE `vglm()` FUNCTION

4.6.1 Using the `vglm()` Function to Fit a Multiple-Predictor PO Model

To fit the same multiple-predictor PO model in the preceding section, we use the following command: `model2 <- vglm(healthre ~ educ + maritals + female, cumulative(parallel = TRUE, reverse = FALSE), data = chp4.po)`. The resulting output is displayed as follows.

```
> # Multiple-predictor model with the vglm() function in VGAM
> model2 <- vglm(healthre ~ educ + maritals + female, cumulative(parallel = TRUE,
reverse = FALSE), data = chp4.po)
> summary(model2)

Call:
vglm(formula = healthre ~ educ + maritals + female, family = cumulative(parallel =
TRUE,
  reverse = FALSE), data = chp4.po)

Pearson residuals:
                       Min       1Q    Median      3Q    Max
logitlink(P[Y<=1])  -0.8733  -0.2335  -0.1697  -0.1288  7.507
logitlink(P[Y<=2])  -2.3796  -0.7354  -0.3226   0.5107  3.293
logitlink(P[Y<=3])  -5.9188   0.1555   0.3464   0.6210  1.086

Coefficients:
               Estimate  Std. Error   z value   Pr(>|z|)
(Intercept):1  -0.25432     0.21950    -1.159   0.246613
(Intercept):2   1.63442     0.21261     7.687    1.5e-14 ***
(Intercept):3   3.92426     0.23005    17.058    < 2e-16 ***
educ           -0.17502     0.01503   -11.646    < 2e-16 ***
maritals       -0.29157     0.08842    -3.297   0.000976 ***
female         -0.06702     0.08754    -0.766   0.443957
---
Signif. codes:  0 '***' 0.001 '**' 0.01 '*' 0.05 '.' 0.1 ' ' 1

Names of linear predictors: logitlink(P[Y<=1]), logitlink(P[Y<=2]),
logitlink(P[Y<=3])

Residual deviance: 4321.01 on 5613 degrees of freedom

Log-likelihood: -2160.505 on 5613 degrees of freedom

Number of Fisher scoring iterations: 4

No Hauck-Donner effect found in any of the estimates



Exponentiated coefficients:
     educ    maritals      female
0.8394376   0.7470883   0.9351809
```

The `coef(model2, matrix = TRUE)` command produces the coefficients table. Again, we can get the odds ratios and the corresponding confidence intervals using `exp(coef(model2, matrix = TRUE))` and `exp(confint(model2,` matrix = TRUE)), respectively. The cbind(exp(coef(model2)), exp(confint(model2))) command combines the results.

```
> coef(model2, matrix = TRUE)
            logit(P[Y<=1]) logit(P[Y<=2]) logit(P[Y<=3])
(Intercept)     -0.2543170      1.6344197      3.9242577
educ            -0.1750231     -0.1750231     -0.1750231
maritals        -0.2915719     -0.2915719     -0.2915719
female          -0.0670153     -0.0670153     -0.0670153

> exp(coef(model2, matrix = TRUE))
            logit(P[Y<=1]) logit(P[Y<=2]) logit(P[Y<=3])
(Intercept)      0.7754460      5.1264821     50.6154915
educ             0.8394376      0.8394376      0.8394376
maritals         0.7470883      0.7470883      0.7470883
female           0.9351809      0.9351809      0.9351809

> exp(confint(model2, matrix = TRUE))
                    2.5 %     97.5 %
(Intercept):1   0.5043268  1.1923151
(Intercept):2   3.3794230  7.7767178
(Intercept):3  32.2449950 79.4519578
educ            0.8150731  0.8645305
maritals        0.6282100  0.8884625
female          0.7877356  1.1102245

> cbind(exp(coef(model2)), exp(confint(model2)))
                                2.5 %     97.5 %
(Intercept):1   0.7754460   0.5043268  1.1923151
(Intercept):2   5.1264821   3.3794230  7.7767178
(Intercept):3  50.6154915  32.2449950 79.4519578
educ            0.8394376   0.8150731  0.8645305
maritals        0.7470883   0.6282100  0.8884625
female          0.9351809   0.7877356  1.1102245
```

The AIC statistic of the fitted model can be obtained with AIC(model2).

```
> AIC(model2)
[1] 4333.01
```

We can also use the nagelkerke() function in the rcompanion package to obtain the three types of pseudo R^2 statistics for model2. The syntax is nagelkerke(model2). The results are omitted here.

4.6.2 Logit Coefficients of Being at or Above a Category in the Multiple-Predictor PO Model

We add the `reverse = TRUE` option to the multiple-predictor PO model so we can estimate the logit coefficients of being at or above a particular category of the ordinal outcome variable. The `summary(model2b)` command produces the following output.

```
> # Logit coefficients of being at or above a category with reverse = TRUE
> model2b <- vglm(healthre ~ educ + maritals + female, cumulative(parallel = TRUE,
reverse = TRUE), data = chp4.po)
> summary(model2b)

Call:
vglm(formula = healthre ~ educ + maritals + female, family = cumulative(parallel =
TRUE,
  reverse = TRUE), data = chp4.po)

Pearson residuals:
                         Min        1Q     Median        3Q      Max
logitlink(P[Y>=2])    -7.507    0.1288     0.1697    0.2335   0.8733
logitlink(P[Y>=3])    -3.293   -0.5107     0.3226    0.7354   2.3796
logitlink(P[Y>=4])    -1.086   -0.6210    -0.3464   -0.1555   5.9188

Coefficients:
               Estimate   Std. Error   z value   Pr(>|z|)
(Intercept):1   0.25432      0.21950     1.159   0.246613
(Intercept):2  -1.63442      0.21261    -7.687    1.5e-14 ***
(Intercept):3  -3.92426      0.23005   -17.058    < 2e-16 ***
educ            0.17502      0.01503    11.646    < 2e-16 ***
maritals        0.29157      0.08842     3.297   0.000976 ***
female          0.06702      0.08754     0.766   0.443957

---
Signif. codes:  0 '***' 0.001 '**' 0.01 '*' 0.05 '.' 0.1 ' ' 1

Names of linear predictors: logitlink(P[Y>=2]), logitlink(P[Y>=3]),
logitlink(P[Y>=4])

Residual deviance: 4321.01 on 5613 degrees of freedom


Log-likelihood: -2160.505 on 5613 degrees of freedom


Number of Fisher scoring iterations: 4


No Hauck-Donner effect found in any of the estimates



Exponentiated coefficients:
    educ   maritals     female
1.191274   1.338530   1.069312
```

We can obtain the coefficients table, the odds ratios, and the corresponding confidence intervals with the `coef()`, `exp(coef())`, `exp(confint())`, and `cbind()` functions. The output is omitted here.

4.6.3 Computing the Predicted Probabilities With the `predict()` Function

We can use the `predict()` function to compute the predicted probabilities of being in a particular category of the ordinal response variable. For example, we would like to compute the predicted probabilities for `educ` at the specified values of 12, 14, and 16 when holding the other predictor variables at their means. We first create a data frame with the `data.frame()` function and then apply the `predict()` function. In the `data.frame()` function, `educ = c(12, 14, 16)` specifies the values of `educ`; `maritals = rep(mean(maritals), 3)` repeats the mean of `maritals` three times; and `female = rep(mean(female), 3)` repeats the mean of `female` three times. The created data frame is assigned to an object named `new1`.

```
> new1 <- data.frame(educ = c(12, 14, 16),
+                    maritals = rep(mean(maritals), 3),
+                    female = rep(mean(female), 3))
> new1
  educ  maritals    female
1   12 0.4372664 0.5563267
2   14 0.4372664 0.5563267
3   16 0.4372664 0.5563267
```

In the `predict()` function, we first specify the model object `model2` and then the `newdata = new1` argument, followed by the `type = "response"` argument for the predicted probabilities. The predicted probabilities labeled from `pred.prob.1` to `pred.prob.4` are provided in the data frame named `new1`.

```
> new1[ , c('pred.prob')] <- predict(model2, newdata = new1, type = "response")
> new1
  educ  maritals    female pred.prob.1 pred.prob.2 pred.prob.3 pred.prob.4
1   12 0.4372664 0.5563267  0.07451130  0.27285538  0.49276437  0.15986894
2   14 0.4372664 0.5563267  0.05368624  0.21907049  0.51461537  0.21262789
3   16 0.4372664 0.5563267  0.03843981  0.17059935  0.51390465  0.27705620
```

4.6.4 Computing the Predicted Probabilities With the `ggpredict()` Function in the `ggeffects` Package

We can also use the `ggpredict()` function in the `ggeffects` package (Lüdecke, 2018b) to compute the predicted probabilities of being in a particular category of the ordinal response variable at specified values of the predictor variables. The command is as follows: `margins.e2.ciNA <- ggpredict(model2, terms = "educ[12, 14, 16]", ci = NA)`. In the `ggpredict()` function, `model2` is the fitted model; the `terms = "educ[12, 14, 16]"` option specifies the predictor variable `educ` at the values of 12, 14, and 16 when holding the other predictor variables at their means; and `ci = NA` specifies no confidence intervals. The `terms` option can specify up to four variables, including the second to fourth grouping variables. The `ci = NA` option is needed there since the confidence intervals are not available for the predicted probabilities of a particular category in the models estimated by the `vglm()` function. Currently the confidence intervals can only be obtained for the cumulative probabilities. The output is assigned to an object named `margins.e2.ciNA`.

```
> margins.e2.ciNA <- ggpredict(model2, terms = "educ[12, 14, 16]", ci = NA)
> margins.e2.ciNA

# Predicted probabilities of healthre

# Response Level = 1

educ    |   Predicted
---------------------
  12    |        0.07
  14    |        0.05
  16    |        0.04

# Response Level = 2

educ    |   Predicted
---------------------
  12    |        0.27
  14    |        0.22
  16    |        0.17

# Response Level = 3

educ    |   Predicted
---------------------
  12    |        0.49
  14    |        0.51
  16    |        0.51

# Response Level = 4

educ    |   Predicted
---------------------
  12    |        0.16
  14    |        0.21
  16    |        0.28

Adjusted for:
* maritals = 0.44
*   female = 0.56
```

FIGURE 4.2 ● Estimated Probabilities of Being in Categories 1, 2, 3, and 4 for educ

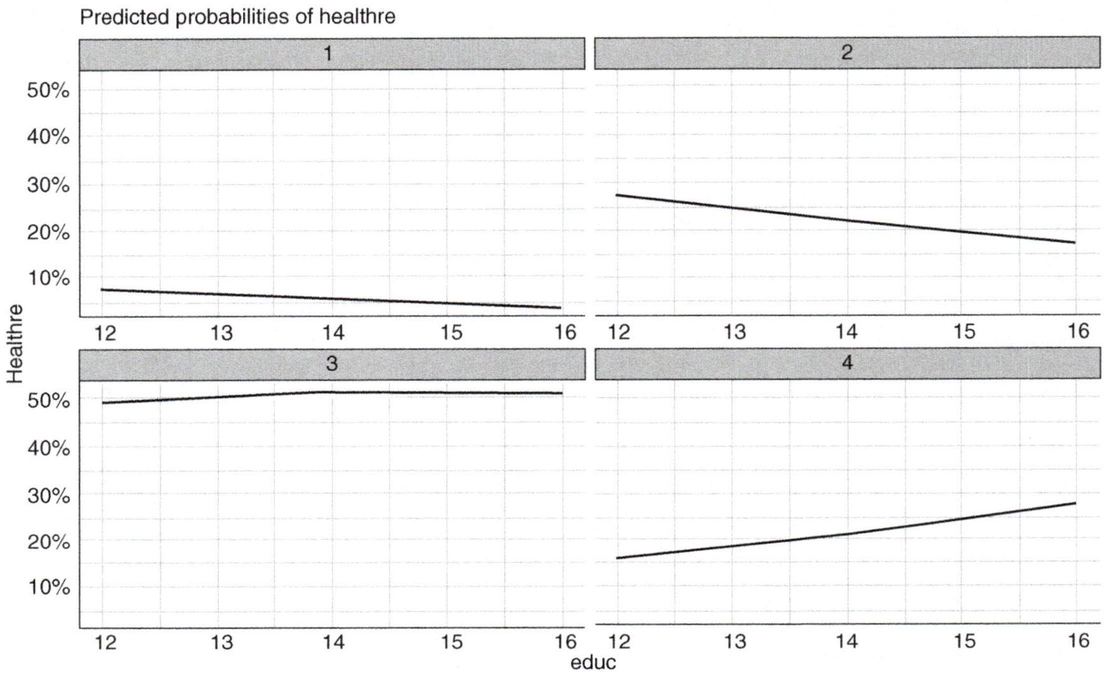

When educ equals 12, 14, and 16, and other predictor variables are held at their means, the predicted probabilities of being in each category (i.e., $Y = 1, 2, 3$, and 4) are displayed in the output. The last section titled "Adjusted for" lists the means of the other two variables.

The predicted probabilities for all four response levels are plotted using the plot (margins.e2.ciNA) function. Figure 4.2 shows the estimated probabilities of being in each category (i.e., $Y = 1, 2, 3$, and 4) for educ at 12, 14, and 16.

The graph shows that people with higher levels of education are less likely associated with poor and fair health conditions (categories 1 and 2). In addition, with the increase in the years of education, the probabilities of being in good and excellent health conditions (categories 3 and 4) increase.

4.6.5 Computing the Cumulative Probabilities With the ggpredict() Function

We can also compute the cumulative probabilities of being at or above a particular category of the ordinal response variable at specified values of the predictor variables.

The command `margins.e2 <- ggpredict(model2, terms = "educ [12, 14, 16]")` tells R to compute the cumulative probabilities of being at or above a category of the ordinal response variable using the `ggpredict()` function by removing the `ci = NA` option. The output is assigned an objected named `margins.e2`. The `as.data.frame()` function is used to request the standard errors.

```
> margins.e2 <- ggpredict(model2, terms = "educ[12, 14, 16]")
> margins.e2

# Predicted probabilities of healthre

# Response Level = P[Y >= 2]
educ     |   Predicted    |          95% CI
--------------------------------------------
  12     |        0.93    |    [0.94, 0.91]
  14     |        0.95    |    [0.96, 0.94]
  16     |        0.96    |    [0.97, 0.95]

# Response Level = P[Y >= 3]
educ     |   Predicted    |          95% CI
--------------------------------------------
  12     |        0.65    |    [0.68, 0.63]
  14     |        0.73    |    [0.75, 0.71]
  16     |        0.79    |    [0.81, 0.77]

# Response Level = P[Y >= 4]
educ     |   Predicted    |          95% CI
--------------------------------------------
  12     |        0.16    |    [0.18, 0.14]
  14     |        0.21    |    [0.23, 0.19]
  16     |        0.28    |    [0.30, 0.25]

Adjusted for:
* maritals = 0.44
*   female = 0.56

> as.data.frame(margins.e2)
     x   predicted   std.error    conf.low   conf.high   response.level   group
1   12   0.9254887  0.09662647   0.9375413   0.9113303        P[Y >= 2]       1
2   12   0.6526333  0.05581345   0.6770017   0.6274382        P[Y >= 3]       1
3   12   0.1598689  0.06749109   0.1784443   0.1428909        P[Y >= 4]       1
4   14   0.9463138  0.09773966   0.9552544   0.9357070        P[Y >= 2]       1
5   14   0.7272433  0.05309915   0.7473924   0.7061189        P[Y >= 3]       1
6   14   0.2126279  0.05739799   0.2320700   0.1944022        P[Y >= 4]       1
7   16   0.9615602  0.10759248   0.9686394   0.9529605        P[Y >= 2]       1
8   16   0.7909608  0.06580753   0.8114879   0.7688353        P[Y >= 3]       1
9   16   0.2770562  0.06197374   0.3020310   0.2533968        P[Y >= 4]       1

> plot(margins.e2)
```

FIGURE 4.3 ● Cumulative Probabilities of Being at or Above Categories 2, 3, and 4 for educ

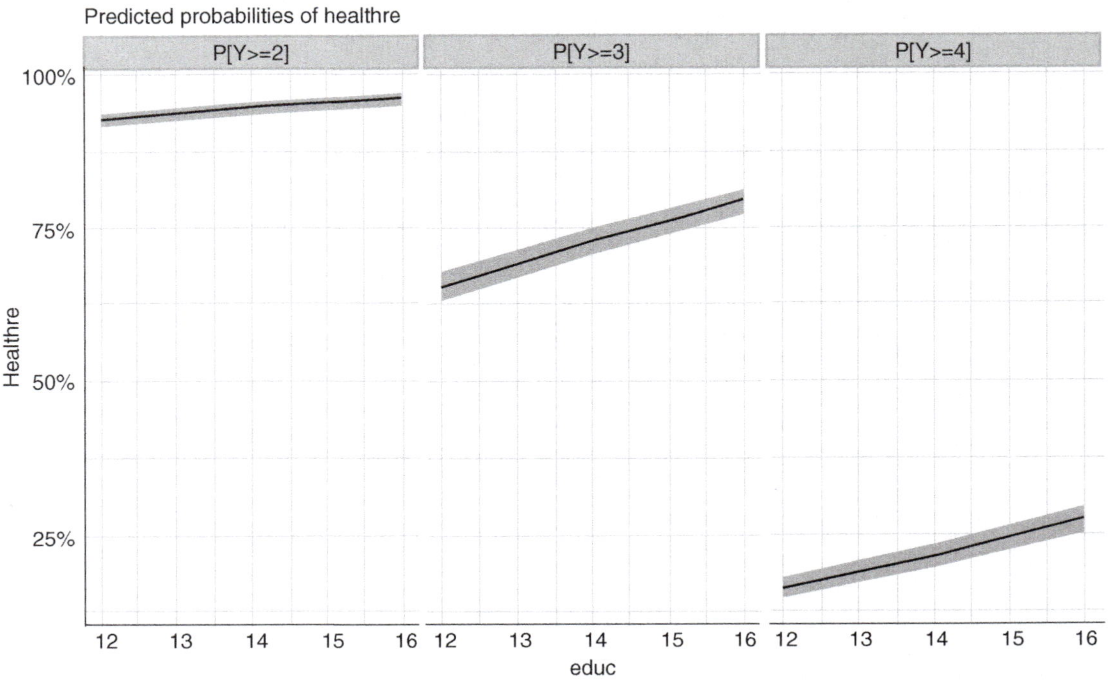

The output provides the three cumulative probabilities with the confidence intervals for `educ` at 12, 14, and 16, while other predictor variables are held at their means. Please note that the standard errors are on the logit-link scale and are not transformed back to the probabilities. The results are plotted by using the `plot(margins.e2)` function. Figure 4.3 shows the cumulative probabilities of being at or above categories 2, 3, and 4 for `educ`.

With the increase in the years of education, people are more likely to be in better health conditions.

4.6.6 Using the `lrtest()` Function to Test the PO Assumption

The `lrtest()` function is used to test the PO assumption. We fit a cumulative odds model, `model2c`, with the `parallel = FALSE` option and then compare it with the PO model, `model2`, with the `parallel = TRUE` option. A nonsignificant test indicates that the proportional odds assumption is upheld for the PO model. The results of the `lrtest(model2, model2c)` command are shown as follows.

```
> # PO assumption test
> model2c <- vglm(healthre ~ educ + maritals + female, cumulative(parallel = FALSE,
reverse = FALSE),data = chp4.po)
> lrtest(model2, model2c)
Likelihood ratio test


Model 1: healthre ~ educ + maritals + female
Model 2: healthre ~ educ + maritals + female
     #Df     LogLik    Df     Chisq    Pr(>Chisq)
1    5613    -2160.5
2    5607    -2155.7   -6    9.6479        0.1403
```

The likelihood ratio test yields $\chi^2_{(6)} = 9.648$, $p = .140$, which indicates that the proportional odds assumption for the overall model is met.

4.6.7 Model Comparison Using the Likelihood Ratio Test With the `lrtest()` Function

Since the `anova()` function for model comparisons does not work with the `vglm()` function, the `lrtest()` function is used. The `lrtest(model1, model2)` command compares the simple-predictor PO model and the multiple-predictor PO model using the likelihood ratio test. The resulting output is as follows.

```
> lrtest(model1, model2)
Likelihood ratio test


Model 1: healthre ~ educ
Model 2: healthre ~ educ + maritals + female
     #Df     LogLik    Df     Chisq    Pr(>Chisq)
1    5615    -2166.2
2    5613    -2160.5   -2    11.313      0.003496   **
---
Signif. codes: 0 '***' 0.001 '**' 0.01 '*' 0.05 '.' 0.1 ' ' 1
```

The difference in deviance, $G = D_{\text{Reduced}} - D_{\text{Full}} = 2 \times (2{,}166.2 - 2{,}160.5) = 11.4$. The likelihood ratio test $\chi^2_{(2)} = 11.313$, $p < .001$. This result indicates that the full model has a better fit than the one-predictor model.

4.7 MAKING PUBLICATION-QUALITY TABLES

4.7.1 Presenting the Results of the `clm` Models Using the `stargazer` Package

We can use the `stargazer` package (Hlavac, 2018) to make a table containing the results of the fitted models with the `clm()` function. Since the package has been installed in earlier chapters, we only need to load the package by typing `library(stargazer)`. After fitting the single-predictor model `PO.1` and the multiple-predictor model `PO.2`, we load the `stargazer` package and then use the command as follows: `stargazer(PO.1, PO.2, type = "text", align = TRUE, out = "po2mod.txt")`. In the `stargazer()` function, we first specify the two model objects to be presented and then the type of table. The option `type = "text"` specifies the table type and the `align = TRUE` option aligns the results of the two models. The `out = "po2mod.txt"` argument saves the output named `po2mod.txt`.

```
> library(stargazer)
> stargazer(PO.1, PO.2, type = "text", align = TRUE, out = "po2mod.txt")

============================================
                     Dependent variable:
                ----------------------------
                          healthre
                      (1)             (2)
--------------------------------------------
maritals                            0.292***
                                    (0.088)

educ                0.179***        0.175***
                    (0.015)         (0.015)

female                               0.067
                                    (0.088)

--------------------------------------------
Observations          1,873           1,873
Log Likelihood     -2,166.161      -2,160.505
============================================
Note:           *p<0.1;  **p<0.05;  ***p<0.01
```

We can also create the table in the HTML format and copy it into Microsoft Word. The command is as follows: `stargazer(PO.1, PO.2, type = "html", align = TRUE, out = "po2mod.htm")`. The resulting table is omitted here.

4.7.2 Presenting the Results of the `vglm` Models Using the `texreg` Package

The `stargazer()` function currently cannot directly produce the results table from the `vglm` models, so we use the `screenreg()` and `htmlreg()` functions from the `texreg` package (Leifeld, 2013). Since `texreg` is a user-written package, you need to install it first by typing `install.packages("texreg")` and then load the package by typing `library(texreg)`.

After we use the `vglm()` function to fit the single-predictor model `model1` and the multiple-predictor model `model2`, we create a table containing the results of both models with the following command: `screenreg(list(model1, model2))`. In the `screenreg()` function, we specify the two model objects to be presented with the `list()` function. The output is a plain text table.

```
> # Presenting the results of the vglm Models using the texreg package
> library(texreg)
Version:    1.37.5
Date:       2020-06-17
Author:     Philip Leifeld (University of Essex)

Consider submitting praise using the praise or praise_interactive functions.
Please cite the JSS article in your publications -- see citation("texreg").
> screenreg(list(model1, model2))


=========================================================
                      Model 1             Model 2
---------------------------------------------------------
(Intercept):1          -0.36                 -0.25
                       (0.21)                (0.22)
(Intercept):2           1.53      ***         1.63    ***
                       (0.21)                (0.21)
(Intercept):3           3.81      ***         3.92    ***
                       (0.22)                (0.23)
educ                   -0.18      ***        -0.18    ***
                       (0.01)                (0.02)
maritals                                     -0.29    ***
                                             (0.09)
female                                       -0.07
                                             (0.09)
---------------------------------------------------------
Log Likelihood      -2166.16              -2160.51
DF                   5615                  5613
Num. obs.            5619                  5619
=========================================================
*** p < 0.001;    ** p < 0.01;    * p < 0.05

> htmlreg(list(model1, model2), file="chap4po.doc", doctype=TRUE, html.tag=TRUE,
head.tag=TRUE)
The table was written to the file 'chap4po.doc'.
```

TABLE 4.2 ● Results of the Proportional Odds Models: Single-Predictor Model and Multiple-Predictor PO Model (Shown in Original Format Generated by R)

	Model 1	Model 2
(Intercept):1	−0.36	−0.25
	(0.21)	(0.22)
(Intercept):2	1.53***	1.63***
	(0.21)	(0.21)
(Intercept):3	3.81***	3.92***
	(0.22)	(0.23)
Educ	−0.18***	−0.18***
	(0.01)	(0.02)
maritals		−0.29***
		(0.09)
female		−0.07
		(0.09)
Log Likelihood	−2,166.16	−2,160.51
DF	5,615	5,613
Num. obs.	5,619	5,619

***$p < 0.001$
**$p < 0.01$
*$p < 0.05$

We can also use the `htmlreg()` function to create a regression table for the estimated results and save it to a Microsoft Word file named `chap4po.doc` with the following command: `htmlreg(list(model1, model2), file = "chap4po.doc", doctype = TRUE, html.tag = TRUE, head.tag = TRUE)`. It automatically produces Table 4.2, as shown here in its original format, presenting the results of both the single-predictor and the multiple-predictor PO models.

4.8 REPORTING THE RESULTS

Writing the results of ordinal logistic regression models is similar to that of binary logistic regression models.

First, describe the statistical method you used for data analysis, the dependent variable and the independent variables in the models, and your research hypothesis, or the purpose of your study.

Second, report the model fit statistics, including but not limited to the likelihood ratio statistic and the associated p value, and the pseudo R^2, followed by a concise statement of interpretation on whether the fitted model is better than the null model. If more fit statistics, such as various pseudo R^2 values, deviance statistic, and AIC and BIC statistics are computed, then include them in a table.

Third, report the parameter estimates for the predictor variables, their standard errors, the associated p values, and odds ratios either in a table or in the text. A table is preferable for models with multiple predictors. The odds ratios for each predictor should be interpreted.

If more than one model is fitted, then the results of all the competing models from the simple model to the full model should be presented in a table. The following is an example of summarizing the results from the ordinal logistic regression model.

> The proportional odds model was fitted to estimate the ordinal outcome variable, health status, from a set of predictor variables, such as marital status, years of education, and gender. A single-predictor model with marital status as the predictor was fitted first, and then the full model with all the predictors was fitted. The likelihood ratio test is used to compare the two models, $\chi^2_{(2)} = 11.313$, $p < .001$. The result indicated that the full model fitted data better than the single-predictor model.
>
> For the `maritals` predictor, OR = 1.339, which was greater than 1. This indicated that the odds of being above a particular category of health status (better health status) for the married were 1.339 times the odds for the unmarried when holding all the other predictors constant.
>
> For the `educ` predictor, OR = 1.191, which was greater than 1. This indicated that the odds of being above a particular category of health status (better health status) increased by a factor of 1.191 for a one-unit increase in the predictor, education, when holding other predictors constant. In other words, for a one-unit increase in education, the odds of being healthier increased by 19.1%.
>
> For `female`, $\beta = .067$, $p = .444$, which was not significantly different from 0; OR = 1.069, which almost equaled 1. This indicated that there was no relationship between being a female and the cumulative odds of being in better health status.

4.9 SUMMARY OF R COMMANDS IN THIS CHAPTER

```
# Chap 4 R Script

# Remove all objects
rm(list = ls(all = TRUE))

# The following user-written packages need to be installed first by using
install.packages(" ") and then by loading it with library()
# library(ordinal)
# library(rcompanion)              # It is already installed for Chapter 3
# library(ggeffects)               # It is already installed for Chapter 2
# library(stargazer)               # It is already installed for Chapter 2
# library(VGAM)
# library(texreg)

# Import GSS 2016 Stata data file
chp4.po <- read.dta("C:/CDA/gss2016.dta")
chp4.po$healthre <- factor(chp4.po$healthre, ordered=TRUE)
chp4.po$educ <- as.numeric(chp4.po$educ)
chp4.po$wrkfull <- as.numeric(chp4.po$wrkfull)
chp4.po$maritals <- as.numeric(chp4.po$maritals)
attach(chp4.po)
str(healthre)

# One-predictor model with the clm() function in ordinal
library(ordinal)
PO.1 <- clm(healthre ~ educ, data = chp4.po)
summary(PO.1)
coef(PO.1)
confint(PO.1)
exp(coef(PO.1))
exp(confint(PO.1))

# Null model with the intercept only
PO.0 <- clm(healthre ~ 1, data = chp4.po)
summary(PO.0)

# Testing the overall model using the likelihood ratio test
anova(PO.0, PO.1)

# Pseudo R2
library(rcompanion)
nagelkerke(PO.1)
LLM <- logLik(PO.1)
LL0 <- logLik(PO.0)
McFadden <- 1-(LLM/LL0)
```

```
McFadden
CS <- 1-exp(2*(LL0-LLM)/1873)
CS
NG <- CS/(1-exp(2*LL0/1873))
NG

# PO assumption test
nominal_test(PO.1)

# Multiple-predictor model with the clm() function
PO.2 <- clm(healthre ~ maritals + educ + female, data = chp4.po)
summary(PO.2)
coef(PO.2)
confint(PO.2)
exp(coef(PO.2))
exp(confint(PO.2))

exp(-coef(PO.2))
exp(-confint(PO.2))

# Predicted probabilities with ggpredict() in ggeffects
library(ggeffects)
margins.e <- ggpredict(PO.2, terms = "educ[12, 14, 16]")
margins.e
plot(margins.e)

# Predicted probabilities with predict(): Omitted in the chapter
New <- data.frame(educ=c(12,14,16),
                  maritals=rep(mean(maritals), 3),
                  female=rep(mean(female), 3))
new
new[,c('pred.prob')] <- predict(PO.2, newdata=new, type="prob", se.fit=TRUE,
interval=TRUE)
new

# Testing the overall model using the likelihood ratio test
anova(PO.0, PO.2)

# Pseudo R2
nagelkerke(PO.2)

# PO assumption test
nominal_test(PO.2)

# Model comparison using the likelihood ratio test
anova(PO.1, PO.2)

# Presenting the results of the clm models using the stargazer package
library(stargazer)
stargazer(PO.1, PO.2, type="text", align=TRUE, out="po2mod.txt")
stargazer(PO.1, PO.2, type="html", align=TRUE, out="po2mod.htm")

# One-predictor model with the vglm() function in VGAM
library(VGAM)
```

```
model1 <- vglm(healthre ~ educ, cumulative(parallel = TRUE, reverse = FALSE),
data = chp4.po)
summary(model1)
exp(coef(model1, matrix = TRUE))
exp(confint(model1, matrix = TRUE))
cbind(exp(coef(model1)), exp(confint(model1)))
nagelkerke(model1)
AIC(model1)


# Logit coefficients of being at or above a category
model1b <- vglm(healthre ~ educ, cumulative(parallel = TRUE, reverse = TRUE),
data = chp4.po)
summary(model1b)
exp(coef(model1b, matrix = TRUE))
exp(confint(model1b, matrix = TRUE))
cbind(exp(coef(model1b)), exp(confint(model1b)))


# Multiple-predictor model with the vglm() function in VGAM
model2 <- vglm(healthre ~ educ + maritals + female, cumulative(parallel = TRUE,
reverse = FALSE),data = chp4.po)
summary(model2)
coef(model2, matrix = TRUE)
confint(model2, matrix = TRUE)
exp(coef(model2, matrix = TRUE))
exp(confint(model2, matrix = TRUE))
cbind(exp(coef(model2)), exp(confint(model2)))
AIC(model2)
# nagelkerke(model2)


# Predicted probabilities with predict()
new1 <- data.frame(educ=c(12,14,16),
                   maritals=rep(mean(maritals), 3),
                   female=rep(mean(female), 3))
new1
new1[,c('pred.prob')] <- predict(model2, newdata=new1, type="response")
new1


# Predicted probabilities with ggpredict() in ggeffects
library(ggeffects)
margins.e2.ciNA <- ggpredict(model2, terms="educ[12, 14, 16]", ci=NA)
margins.e2.ciNA
plot(margins.e2.ciNA)


margins.e2 <- ggpredict(model2, terms="educ[12, 14, 16]")
margins.e2
as.data.frame(margins.e2)
plot(margins.e2)


# Logit coefficients of being at or above a category with reverse = TRUE
model2b <- vglm(healthre ~ educ + maritals + female, cumulative(parallel = TRUE,
reverse = TRUE), data = chp4.po)
summary(model2b)
coef(model2b, matrix = TRUE)
confint(model2b, matrix = TRUE)
exp(coef(model2b, matrix = TRUE))
exp(confint(model2b, matrix = TRUE))
cbind(exp(coef(model2b)), exp(confint(model2b)))
# AIC(model2b)
```

```
# Testing the Overall Model Using the Likelihood Ratio Test: Omitted in the chapter
model0 <- vglm(healthre ~ 1, cumulative(parallel = TRUE, reverse = FALSE), data =
chp4.po)
summary(model0)
lrtest(model0, model1)
lrtest(model0, model2)

# PO assumption test
model2c <- vglm(healthre ~ educ + maritals + female, cumulative(parallel = FALSE,
reverse = FALSE),data = chp4.po)
lrtest(model2, model2c)

# Model comparison with the likelihood ratio test
lrtest(model1, model2)

# Presenting the results of the vglm Models using the texreg package
library(texreg)
screenreg(list(model1, model2))
htmlreg(list(model1, model2), file="chap4po.doc", doctype=TRUE, html.tag=TRUE,
head.tag=TRUE)

detach(chp4.po)
```

Glossary

An ordinal probit regression model is a regression model for an ordinal response variable with the probit link.

Ordinal logistic regression models are regression models for ordinal response variables with the logistic function or the logit link.

The cumulative probability of being at or below a category $P(Y \le j)$ equals the sum of the probabilities of all categories at or below that category.

The odds of being at or below a category in ordinal logistic regression equals the probability of being at or below a category divided by the probability of being above that category.

The proportional odds (PO) model is one of the most commonly used models for the analysis of ordinal response variables. The odds ratio of any predictor is assumed to be constant across all categories, so it is referred to as the proportional odds assumption or the parallel lines assumption.

Exercises

Use the GSS 2016 data available at **https://edge.sagepub.com/liu1e** for the following problems.

1. Conduct an analysis for a proportional odds model to estimate the ordinal response variable `fechld` from the three predictor variables `sex`, `educ`, and `age`.
2. Identify the likelihood ratio test of the model and interpret it.
3. Compute the deviance statistic for the model.
4. List three measures of pseudo R^2 and the AIC statistic.
5. Identify the logit coefficient, the Wald z test, and the 95% confidence interval for the predictor variables `sex` and `educ`. Are they statistically significant?
6. Compute the odds ratios for `sex` and `educ`.
7. Test the proportional odds assumption and interpret the results.
8. What are the important criteria you may use for model comparisons?
9. Make a publication-quality table containing the estimated logit coefficients.
10. Write a report to summarize the results from the output.

5

GENERALIZED ORDINAL LOGISTIC REGRESSION MODELS AND PARTIAL PROPORTIONAL ODDS MODELS

OBJECTIVES OF THIS CHAPTER

This chapter presents partial proportional odds models and generalized ordinal logistic regression models when the proportional odds assumption is untenable. It begins with an introduction to the models, odds and odds ratio, model fit statistics, and interpretations of parameter estimates. After a description of the data, two examples on how to fit these two models using R are illustrated. The `vglm()` function in the `VGAM` package is used to fit the models. R functions are explained, and output is interpreted in detail. This chapter focuses on fitting partial proportional odds models and generalized ordinal logistic regression models using R and on interpreting and presenting the results. After reading this chapter, you should be able to:

- Identify when a partial proportional odds model or a generalized ordinal logistic regression model is used.
- Conduct analysis for both models using R.
- Interpret the output.
- Interpret the model in terms of odds ratios.
- Compute the estimated probabilities.
- Compare models using the likelihood ratio test and other fit statistics.
- Present results in publication-quality tables using R.
- Write the results for publication.

5.1 PARTIAL PROPORTIONAL ODDS MODELS AND GENERALIZED ORDINAL LOGISTIC REGRESSION MODELS: AN INTRODUCTION

In Chapter 4, we discussed the proportional odds (PO) model, which is used to estimate the cumulative probability of being at or below a particular level of the ordinal response variable or the complementary probability of being above that particular level. This model follows the assumption that the coefficient of each predictor is the same across the categories of the ordinal response variable. In other words, for each predictor, its effect on the log odds of being at or below any category remains the same. This restriction is referred to as the proportional odds assumption or the parallel lines assumption.

The PO assumption is strict and is often violated in real data analysis. An alternative solution is to fit the partial proportional odds (PPO) model or the generalized ordinal logit model (Fu, 1998; Fullerton & Xu, 2016; Liu, 2016a; Liu & Koirala, 2012; Peterson & Harrell, 1990; Williams, 2006). In the PPO model, not all predictor variables violate the PO assumption, so the effects of those predictors violating the assumption are allowed to vary across categories. The generalized ordinal logit model can be considered an extreme case of the PPO model, and it allows the effect of each explanatory variable to vary. The original PPO model proposed by Peterson and Harrell (1990) specifies an interaction between a predictor variable that violates the PO assumption and different categories of the ordinal outcome variable and requires data restructuring. Fu (1998) and William (2006) developed Stata add-on programs which made the analysis of generalized ordinal logit models and PPO models easier without data restructuring.

The generalized ordinal logistic regression model is an extension of the PO model by relaxing the PO assumption for all predictor variables. This model in the `vglm()` function from the `VGAM` package is expressed as follows:

$$\ln\left(\frac{\pi_j(x)}{1-\pi_j(x)}\right) = \alpha_j + \left(\beta_{1j}X_1 + \beta_{2j}X_2 + \cdots + \beta_{pj}X_p\right) \tag{5.1}$$

To estimate the odds of being at or below a certain category relative to being above that category, Equation 5.1 also can be expressed as follows:

$$\begin{aligned}\text{logit}\left[P\left(Y \le j \middle| x_1, x_2, \ldots, x_p\right)\right] &= \ln\left(\frac{P\left(Y \le j \middle| x_1, x_2, \ldots, x_p\right)}{P\left(Y > j \middle| x_1, x_2, \ldots, x_p\right)}\right) \\ &= \alpha_j + \left(\beta_{1j}X_1 + \beta_{2j}X_2 + \cdots + \beta_{pj}X_p\right)\end{aligned} \tag{5.2}$$

where in both equations α_j are the intercepts or cut points and $\beta_{1j}, \beta_{2j}, \ldots, \beta_{pj}$ are the logit coefficients. This model estimates the odds of being at or below a certain category relative to being above that category. A positive logit coefficient normally indicates that an individual is more likely to be in a lower category rather than in a higher category of the outcome variable. To estimate the odds of being above a particular category, however, the signs before both the intercepts and logit coefficients in Equation 5.2 need to be reversed as follows:

$$\begin{aligned} \text{logit}[P(Y>j|x_1, x_2, \ldots, x_p)] &= \ln\left(\frac{P(Y>j|x_1, x_2, \ldots, x_p)}{P(Y \le j|x_1, x_2, \ldots, x_p)}\right) \\ &= -\alpha_j - \left(\beta_{1j}X_1 + \beta_{2j}X_2 + \cdots + \beta_{pj}X_p\right) \end{aligned} \tag{5.3}$$

The model in Equation 5.3 estimates the odds of being above a certain category relative to being at or below that category. A positive logit coefficient normally indicates that an individual is more likely to be in a higher category rather than in a lower category of the outcome variable.

Some software uses a modified form estimating the cumulative probability of being at or above a category as follows:

$$\begin{aligned} \text{logit}[P(Y \ge j|x_1, x_2, \ldots, x_p)] &= \ln\left(\frac{P(Y \ge j|x_1, x_2, \ldots, x_p)}{P(Y<j|x_1, x_2, \ldots, x_p)}\right) \\ &= -\alpha_j - \left(\beta_{1j}X_1 + \beta_{2j}X_2 + \cdots + \beta_{pj}X_p\right) \end{aligned} \tag{5.4}$$

Equations 5.3 and 5.4 are actually equivalent since Equation 5.3 estimates the probabilities of $J - 1$ categories from $j = 1$ to $J - 1$, whereas Equation 5.4 estimates the probabilities of $J - 1$ categories from $j = 2$ to J. Therefore, $P(Y > 1)$ in Equation 5.3 is equal to in $P(Y \ge 2)$ in Equation 5.4.

Compared to the generalized ordinal logistic regression model, the PPO model is more parsimonious since it only allows predictor variables violating the PO assumption to vary. Although the PPO model or the generalized ordinal logistic regression model provides a better fit than a PO model does, the former two models are not a panacea to deal with the violation of the PO assumption. We need to look at the underlying binary logistic models and examine the discrepancy of the coefficients. When the discrepancy is minor, the PO model may still be useful. Williams (2016) explained five possible reasons why the PO assumption was violated and suggested that researchers justify the use of the PPO model or the generalized ordinal logistic regression model.

5.1.1 Odds and Odds Ratios

As with the PO model, the partial proportional odds model and the generalized ordinal logit model estimate the odds of being at or below a certain category versus being above that category. The cumulative odds in partial proportional odds models and generalized ordinal logit models are comparing $P(Y \le j)$ and $P(Y > j)$. In other words, the odds of being at or below a category are the probability of being at or below a category divided by the probability of being above that category:

$$\text{Odds}(Y \le j) = \frac{P(Y \le j)}{P(Y>j)}$$

Odds ($Y \leq 1$) equal the ratio of probability of being at or below category 1 to the probability of being above this category.

$$\text{Odds}(Y \leq 1) = \frac{P(Y \leq 1)}{1 - P(Y \leq 1)} = \frac{P(Y = 1)}{P(Y = 2) + P(Y = 3) + P(Y = 4)}$$
$$= \frac{P(1)}{P(2) + P(3) + P(4)}$$

Similarly, the odds of being at or below category 2 compare the probabilities of categories 1 and 2 with the probabilities of categories 3 and 4.

$$\text{Odds}(Y \leq 2) = \frac{P(Y \leq 2)}{1 - P(Y \leq 2)} = \frac{P(1) + P(2)}{P(3) + P(4)}$$

In addition, the odds of being at or below category 3 compare the probabilities of categories 1, 2, and 3 with the probability of category 4.

$$\text{Odds}(Y \leq 3) = \frac{P(1) + P(2) + P(3)}{P(4)}$$

The category comparisons for the odds of being at or below a certain category versus being above that category are also discussed in the previous chapter, so the table is omitted here.

The model in Equation 5.4 also estimates the odds of being at or above a certain category relative to being below that category. The `vglm()` function with the `reverse = TRUE` argument in the VGAM package (Yee, 2010) follows this model equation. In this example, since $P(Y \geq 1) = P(1) + P(2) + P(3) + P(4) = 1$, we estimate the odds ($Y \geq 2$), ($Y \geq 3$), and ($Y \geq 4$), respectively.

$$\text{Odds}(Y \geq j) = \frac{P(Y \geq j)}{P(Y < j)}$$

Odds ($Y \geq 2$) equal the ratio of the probability of being at or above category 2 [i.e., $P(Y \geq 2)$] to the probability of being below this category [i.e., $P(Y < 2)$], which also equals the ratio of the probability of being above category 1 [i.e., $P(Y > 1)$] to the probability of being at or below this category [i.e., $P(Y \leq 1)$]. The probability $P(Y \geq 2) = P(2) + P(3) + P(4)$ and the probability $P(Y < 2) = P(Y = 1) = P(1)$:

$$\text{Odds}(Y > 1) = \frac{P(Y > 1)}{1 - P(Y > 1)} = \text{Odds}(Y \geq 2) = \frac{P(Y \geq 2)}{1 - P(Y \geq 2)}$$
$$= \frac{P(2) + P(3) + P(4)}{P(1)}$$

Similarly, odds ($Y \geq 3$) = odds ($Y > 2$), which equal the ratio of probability of being at or above category 3 to the probability of being below this category. Since

$P(Y \geq 3) = P(Y > 2) = P(3) + P(4)$ and $P(Y < 3) = P(Y \leq 2) = P(1) + P(2)$, the odds of being at or above category 3, odds $(Y \geq 3)$, can be expressed as follows:

$$\text{Odds}(Y \geq 3) = \text{Odds}(Y > 2) = \frac{P(Y > 2)}{1 - P(Y > 2)} = \frac{P(3) + P(4)}{P(1) + P(2)}$$

Finally, odds $(Y \geq 4)$ or odds $(Y > 3)$ equal the ratio of probability of being above category 3 to the probability of being at or below this category. The equation is as follows:

$$\text{Odds}(Y \geq 4) = \text{Odds}(Y > 3) = \frac{P(Y > 3)}{1 - P(Y > 3)} = \frac{P(4)}{P(1) + P(2) + P(3)}$$

Table 5.1 presents the logits, odds, and category comparisons for the PPO model/generalized ordinal logistic model for the health status with four levels.

Odds Ratios in PPO Models/Generalized Ordinal Logistic Regression Models

Just like the odds ratios in the PO model, the odds ratios of being above a particular category in the PPO model and generalized ordinal logistic regression model are the exponentiated logit coefficients. An odds ratio is the change in the odds of being above a particular category for a one-unit increase from any value of x to the value of $(x + 1)$.

5.1.2 Goodness of Fit

Since partial proportional odds models and generalized ordinal logistic regression models are extensions of proportional odds models, measures of fit statistics for the latter, such as the deviance, likelihood ratio test, and pseudo R^2 measures, can be applied to the former models.

TABLE 5.1 ● Category Comparisons for the Partial Proportional Odds Model/Generalized Ordinal Logistic Model with Four Levels of Health Status ($j = 1, 2, 3, 4$)

Category	Logit $P(Y \geq j)$	Logit $P(Y > j)$	Odds	Probability Comparisons
Level 2	logit $P(Y \geq 2)$	logit $P(Y > 1)$	$\frac{P(Y>1)}{P(Y\leq 1)}$	Categories 2 through 4 vs. Category 1
Level 3	logit $P(Y \geq 3)$	logit $P(Y > 2)$	$\frac{P(Y>2)}{P(Y\leq 2)}$	Categories 3 and 4 vs. Categories 1 and 2
Level 4	logit $P(Y \geq 4)$	logit $P(Y > 3)$	$\frac{P(Y>3)}{P(Y\leq 3)}$	Category 4 vs. Categories 1 through 3

5.1.3 Interpretation of Model Parameter Estimates

The odds ratio in partial proportional odds models and generalized ordinal logistic regression models can be interpreted in the same way as that in the proportional odds regression. It can be interpreted as the change in the predicted logit or the log odds of being above a particular category relative to being at or below that category for a one-unit increase in the predictor variable.

To estimate the odds of being at or below a particular category, however, the signs before both the intercepts and logit coefficients in Equation 5.2 need to be reversed. Taking the multiplicative inverse of the odds of being above a particular category gives us the odds of being at or below that category.

5.2 RESEARCH EXAMPLE AND DESCRIPTION OF THE DATA AND SAMPLE

Research Problem and Questions: This chapter focuses on the same research problem as that in Chapter 4. We will still investigate the relationships between the ordinal response variable, health status, and the four predictor variables, including the highest education, marital status, gender, and working status. Unlike Chapter 3, however, here the research interest focuses on using the PPO model and the generalized ordinal logistic regression model when the PO assumption is violated. The research question is as follows: Do the four predictor variables predict the ordinal response variable, health status? Specifically, do the four predictor variables predict the cumulative odds and then the cumulative probabilities of being above a particular level of health status when the proportional odds assumption is violated?

Description of the Data and Sample: The data for the following analyses were the General Social Survey 2016 (GSS 2016). The following are the variables used for data analysis in this chapter:

- `healthre`: the recoded variable of health (health status) with four ordinal categories (1 = poor health, 2 = fair health, 3 = good health, and 4 = excellent health)
- `educ`: the highest education completed
- `maritals`: the recoded variable of marital (marital status) with 1 = currently married and 0 = not currently married
- `female`: recoded variable of sex with 1 = female and 0 = male
- `wrkfull`: working full time or not

5.3 GENERALIZED ORDINAL LOGISTIC REGRESSION MODELS WITH R

5.3.1 The `vglm()` Function in the `VGAM` Package

Both the `ordinal` package (Christensen, 2019) and the `VGAM` package (Yee, 2010, 2015, 2021) were introduced to fit proportional odds models in the last chapter. They can also be used to fit both the partial proportional odds models and the generalized ordinal logistic models. We focus on the use of the `VGAM` package only in this chapter. You need to install the `VGAM` package first by typing `install.packages("VGAM")` if it is not already installed. Since the package has been installed in the preceding chapter, we only need to load the package by typing `library(VGAM)`. The syntax for partial proportional odds models and generalized ordinal logistic models is the same as that for proportional odds models using `vglm()` except that the different `parallel=` arguments are specified. We specify `parallel = TRUE` for proportional odds models and `parallel = FALSE` for nonproportional odds models or generalized ordinal logistic models. For example, the command `vglm(y ~ x, family = cumulative(parallel = FALSE), data = data1)` tells R to fit a generalized ordinal logistic model predicting the ordinal dependent variable *y* with an independent variable *x*. The `parallel = FALSE` argument tells us the predictor variable does not meet the proportional odds assumption and is allowed to vary across the ordinal categories. To fit a partial proportional odds model, we need to specify variables which violate the proportional odds assumption. For example, the command `vglm(y ~ x1 + x2, family = cumulative(parallel = FALSE ~ x2), data = data1)` tells R to fit a partial proportional odds model predicting the ordinal dependent variable *y* with two independent variables. The `parallel = FALSE ~ x2` argument tells R that the second predictor variable `x2` in the model violates the proportional odds assumption. When there are multiple predictor variables violate the assumption, they are connected by plus (+) symbols following the tilde (~). For more details on how to use this function, type `help(vglm)` and `help(cumulative)` in the command prompt after loading the `VGAM` package.

5.3.2 The Multiple-Predictor PO Model

We first fit a PO model including the four predictor variables with the following command: `mod.po <- vglm(healthre ~ educ + maritals + female + wrkfull, cumulative(parallel = TRUE, reverse = FALSE), data = chp5.gpo)`. The resulting output is displayed as follows.

```
> # Import the GSS 2016 data
> library(foreign)
> chp5.gpo <- read.dta("C:/CDA/gss2016.dta")
> chp5.gpo$healthre <- factor(chp5.gpo$healthre, ordered=TRUE)
> chp5.gpo$educ <- as.numeric(chp5.gpo$educ)
> chp5.gpo$wrkfull <- as.numeric(chp5.gpo$wrkfull)
> chp5.gpo$maritals <- as.numeric(chp5.gpo$maritals)
> attach(chp5.gpo)

> # PO model with vglm() in VGAM
> library(VGAM)
> mod.po <- vglm(healthre ~ educ + maritals + female + wrkfull, cumulative(parallel =
 TRUE, reverse = FALSE), data = chp5.gpo)
> summary(mod.po)

Call:
vglm(formula = healthre ~ educ + maritals + female + wrkfull,
  family = cumulative(parallel = TRUE, reverse = FALSE), data = chp5.gpo)

Pearson residuals:
                        Min       1Q     Median       3Q     Max
logitlink(P[Y<=1])  -0.7759  -0.2336   -0.1641  -0.1267   7.038
logitlink(P[Y<=2])  -2.0262  -0.7009   -0.3364   0.4918   3.589
logitlink(P[Y<=3])  -6.1518   0.1530    0.3416   0.6290   1.163


Coefficients:
               Estimate  Std. Error  z value   Pr(>|z|)
(Intercept):1  -0.18286     0.22020   -0.830     0.4063
(Intercept):2   1.72231     0.21374    8.058   7.75e-16 ***
(Intercept):3   4.03153     0.23183   17.390    < 2e-16 ***
educ           -0.16610     0.01515  -10.966    < 2e-16 ***
maritals       -0.26248     0.08876   -2.957     0.0031 **
female         -0.12648     0.08854   -1.429     0.1531
wrkfull        -0.41262     0.09002   -4.584   4.57e-06 ***
---
Signif. codes:  0 '***' 0.001 '**' 0.01 '*' 0.05 '.' 0.1 ' ' 1

Names of linear predictors: logitlink(P[Y<=1]), logitlink(P[Y<=2]),
logitlink(P[Y<=3])

Residual deviance: 4299.823 on 5612 degrees of freedom

Log-likelihood: -2149.912 on 5612 degrees of freedom

Number of Fisher scoring iterations: 4

No Hauck-Donner effect found in any of the estimates

Exponentiated coefficients:
     educ     maritals       female      wrkfull
0.8469646    0.7691443    0.8811878    0.6619166
```

5.3.3 Using the `lrtest()` Function to Test the PO Assumption

We then fit a generalized proportional odds model, or a nonproportional odds model, `mod.gpo`, with the `parallel = FALSE` option and test the PO assumption by using the `lrtest()` function. A significant test indicates that the proportional odds assumption is violated. The output of the `lrtest(mod.po, mod.gpo)` command is shown as follows.

```
> # GPO model with vglm() in VGAM
>mod.gpo<-vglm(healthre~educ+maritals+female+wrkfull, cumulative(parallel=
FALSE, reverse = FALSE), data = chp5.gpo)
> # Using the lrtest() function to test the PO assumption
> lrtest(mod.po, mod.gpo)
Likelihood ratio test

Model 1: healthre ~ educ + maritals + female + wrkfull
Model 2: healthre ~ educ + maritals + female + wrkfull
      #Df    LogLik    Df     Chisq     Pr(>Chisq)
1    5612    -2149.9
2    5604    -2136.1    -8    27.617     0.0005528 ***

---
Signif. codes: 0 '***' 0.001 '**' 0.01 '*' 0.05 '.' 0.1 ' ' 1
```

The likelihood ratio test yields $\chi^2_{(8)} = 27.617$, $p < .001$, which indicates that the proportional odds assumption for the overall model is violated.

5.3.4 The Multiple-Predictor Generalized Ordinal Logistic Regression Model

The generalized ordinal logistic regression model relaxes the proportionality assumption by allowing the logits coefficients of all four predictor variables to vary across the ordinal response variable. Similar to a series of underlying binary logistic regression models, where the data are dichotomized across different categories, the effects of the predictor variables estimated by the generalized ordinal logistic regression model can vary freely. After fitting the generalized ordinal logistic regression model named `mod.gpo` above, we can request the results by using the `summary(mod.gpo)` command.

```
> summary(mod.gpo)
Call:
vglm(formula = healthre ~ educ + maritals + female + wrkfull,
     family = cumulative(parallel = FALSE, reverse = FALSE), data = chp5.gpo)

Pearson residuals:
                          Min         1Q      Median         3Q       Max
logitlink(P[Y<=1])    -0.8896    -0.2524    -0.1560    -0.1116     9.009
logitlink(P[Y<=2])    -2.0847    -0.6751    -0.3579     0.4991     3.981
logitlink(P[Y<=3])    -6.0010     0.1532     0.3391     0.6317     1.000
```

```
Coefficients:
               Estimate   Std. Error   z value   Pr(>|z|)
(Intercept):1   0.19647     0.41350     0.475    0.634683
(Intercept):2   1.86004     0.25942     7.170    7.50e-13 ***
(Intercept):3   3.73087     0.30303    12.312     < 2e-16 ***
educ:1         -0.18037     0.03137    -5.749    8.98e-09 ***
educ:2         -0.16361     0.01857    -8.810     < 2e-16 ***
educ:3         -0.16597     0.02013    -8.246     < 2e-16 ***
maritals:1     -0.59747     0.21385    -2.794    0.005208 **
maritals:2     -0.38257     0.10878    -3.517    0.000437 ***
maritals:3     -0.08731     0.11480    -0.761    0.446910
female:1       -0.11250     0.19631    -0.573    0.566580
female:2       -0.20045     0.10708    -1.872    0.061211 .
female:3       -0.05112     0.11537    -0.443    0.657686
wrkfull:1      -0.75615     0.22072    -3.426    0.000613 ***
wrkfull:2      -0.65609     0.11068    -5.928    3.07e-09 ***
wrkfull:3      -0.10049     0.11562    -0.869    0.384772

---
Signif. codes: 0 '***' 0.001 '**' 0.01 '*' 0.05 '.' 0.1 ' ' 1

Names of linear predictors: logitlink(P[Y<=1]), logitlink(P[Y<=2]),
logitlink(P[Y<=3])

Residual deviance: 4272.206 on 5604 degrees of freedom

Log-likelihood: -2136.103 on 5604 degrees of freedom

Number of Fisher scoring iterations: 5

Warning: Hauck-Donner effect detected in the following estimate(s):
'(Intercept):3'

Exponentiated coefficients:
   educ:1      educ:2      educ:3   maritals:1   maritals:2   maritals:3
0.8349644   0.8490712   0.8470711    0.5502042    0.6821031    0.9163904
 female:1    female:2    female:3    wrkfull:1    wrkfull:2    wrkfull:3
0.8935952   0.8183624   0.9501620    0.4694709    0.5188783    0.9043933
```

5.3.5 Interpreting R Output

As with the output for the PO model, the R output for the generalized ordinal logistic regression model also includes the call, the Pearson residuals, the coefficients, the number and names of the three linear predictors, the residual deviance, the log-likelihood value, the number of iterations, and the exponentiated coefficients. Just like the PO model, the coefficients section displays the parameter estimates, standard errors, z statistics, and associated p values. Unlike the coefficients section in the output for the PO model, the output produced by the `vglm()` function displays the parameter estimates for the three underlying binary logistic models whose outcome variables are dichotomized from the ordinal response variable.

This generalized ordinal logistic regression model estimates the logit or the log odds of being at or below a particular category ($Y \leq j$). Since the ordinal outcome variable in the model has four levels, three underlying binary logistic regression models are fitted, including the models for `logit(P[Y<=1])`, `logit(P[Y<=2])`, and `logit(P[Y<=3])`. The estimated intercepts and logit coefficients for these three submodels are numbered 1, 2, and 3 in the output. As explained in Chapter 3, `logit(P[Y<=1])` is the log odds of being at or below category 1, which compares the probability of category 1 with the probabilities of categories 2, 3, and 4; `logit(P[Y<=2])` compares the probabilities of categories 1 and 2 with those of categories 3 and 4; and `logit(P[Y<=3])` compares the probabilities of categories 1, 2, and 3 with the probability of category 4.

5.3.6 Logit Coefficients of Being at or Below a Category

The `coef(mod.gpo, matrix = TRUE)` command produces the coefficients table for the three binary logistic regression models estimating `logit(P[Y<=1])`, `logit(P[Y<=2])`, and `logit(P[Y<=3])`.

```
> coef(mod.gpo, matrix = TRUE)
            logit(P[Y<=1])  logit(P[Y<=2])  logit(P[Y<=3])
(Intercept)      0.1964733       1.8600429      3.73086647
educ            -0.1803662      -0.1636122     -0.16597062
maritals        -0.5974657      -0.3825745     -0.08731283
female          -0.1125023      -0.2004501     -0.05112275
wrkfull         -0.7561489      -0.6560860     -0.10049090
```

The logit coefficients of all four variables are different across the three equations/models. For example, the regression coefficients for the first predictor variable, `educ`, are −.180, −.164, and −.166, respectively. A total of 12 logit coefficients are estimated, so it would be tedious to interpret each coefficient.

5.3.7 Odds Ratios of Being at or Below a Category

With `exp(coef(mod.gpo, matrix = TRUE))` and `exp(confint(mod.gpo, matrix = TRUE))`, we can get the odds ratios and the corresponding confidence intervals, respectively. The `cbind(exp(coef(mod.gpo)), exp(confint(mod.gpo)))` command combines the results.

```
> exp(coef(mod.gpo, matrix = TRUE))
            logit(P[Y<=1])  logit(P[Y<=2])  logit(P[Y<=3])
(Intercept)      1.2171028       6.4240125      41.7152375
educ             0.8349644       0.8490712       0.8470711
maritals         0.5502042       0.6821031       0.9163904
female           0.8935952       0.8183624       0.9501620
wrkfull          0.4694709       0.5188783       0.9043933
```

```
> exp(confint(mod.gpo, matrix = TRUE))
                   2.5 %     97.5 %
(Intercept):1  0.5411958  2.7371596
(Intercept):2  3.8635822 10.6812627
(Intercept):3 23.0331168 75.5503937
educ:1         0.7851678  0.8879192
educ:2         0.8187227  0.8805446
educ:3         0.8143041  0.8811566
maritals:1     0.3618213  0.8366691
maritals:2     0.5511299  0.8442013
maritals:3     0.7317522  1.1476172
female:1       0.6081972  1.3129169
female:2       0.6634361  1.0094671
female:3       0.7578662  1.1912498
wrkfull:1      0.3046025  0.7235757
wrkfull:2      0.4176918  0.6445774
wrkfull:3      0.7210082  1.1344217

> cbind(exp(coef(mod.gpo)), exp(confint(mod.gpo)))
                               2.5 %     97.5 %
(Intercept):1  1.2171028  0.5411958  2.7371596
(Intercept):2  6.4240125  3.8635822 10.6812627
(Intercept):3 41.7152375 23.0331168 75.5503937
educ:1         0.8349644  0.7851678  0.8879192
educ:2         0.8490712  0.8187227  0.8805446
educ:3         0.8470711  0.8143041  0.8811566
maritals:1     0.5502042  0.3618213  0.8366691
maritals:2     0.6821031  0.5511299  0.8442013
maritals:3     0.9163904  0.7317522  1.1476172
female:1       0.8935952  0.6081972  1.3129169
female:2       0.8183624  0.6634361  1.0094671
female:3       0.9501620  0.7578662  1.1912498
wrkfull:1      0.4694709  0.3046025  0.7235757
wrkfull:2      0.5188783  0.4176918  0.6445774
wrkfull:3      0.9043933  0.7210082  1.1344217
```

Just like the logit coefficients, the corresponding odds ratios of all four predictor variables vary across the three binary models. The odds ratios can still be interpreted as the change in the odds of being at or below a category versus beyond that category for a one-unit change in the predictor variable when holding all other predictors constant. For example, the odds ratios for `wrkfull` across the three equations are .469, .519, and .904, respectively.

5.3.8 Model Fit Statistics

Testing the Overall Model Using the Likelihood Ratio Test

To test if the overall model is significant, we fit a null model with the intercept only and compare the generalized ordinal logistic regression model with the null model using the `lrtest()` function. The command `gpomodel0 <- vglm(healthre ~ 1,` `cumulative(parallel = TRUE, reverse = FALSE))` is used to fit the null model. The output is displayed below by the `summary(gpomodel0)` command.

```
> # Testing the overall model using the likelihood ratio test
> gpomodel0 <- vglm(healthre ~ 1, cumulative(parallel = TRUE, reverse = FALSE))
> summary(gpomodel0)

Call:
vglm(formula = healthre ~ 1, family = cumulative(parallel = TRUE,
    reverse = FALSE))

Pearson residuals:
                        Min       1Q      Median       3Q      Max
logitlink(P[Y<=1])   -0.5249   -0.1844   -0.1844   -0.1386   3.8138
logitlink(P[Y<=2])   -0.7477   -0.7477   -0.2980    0.5477   1.7380
logitlink(P[Y<=3])   -1.8483    0.1667    0.3002    0.6754   0.6754

Coefficients:
                 Estimate    Std. Error    z value    Pr(>|z|)
(Intercept):1    -2.69954      0.09510     -28.39      <2e-16 ***
(Intercept):2    -0.89064      0.05087     -17.51      <2e-16 ***
(Intercept):3     1.25964      0.05569      22.62      <2e-16 ***

---
Signif. codes: 0 '***' 0.001 '**' 0.01 '*' 0.05 '.' 0.1 ' ' 1

Names of linear predictors: logitlink(P[Y<=1]), logitlink(P[Y<=2]),
logitlink(P[Y<=3])

Residual deviance: 4476.435 on 5616 degrees of freedom

Log-likelihood: -2238.218 on 5616 degrees of freedom

Number of Fisher scoring iterations: 1

Warning: Hauck-Donner effect detected in the following estimate(s):
'(Intercept):1'
```

The `lrtest(gpomodel0, mod.gpo)` command compares the log-likelihood statistics of the fitted model `mod.gpo` and the null model `gpomodel0` using the likelihood ratio test.

```
> lrtest(gpomodel0, mod.gpo)
Likelihood ratio test

Model 1: healthre ~ 1
Model 2: healthre ~ educ + maritals + female + wrkfull
```

```
    #Df    LogLik    Df    Chisq    Pr(>Chisq)
1   5616   -2238.2
2   5604   -2136.1   -12   204.23    < 2.2e-16 ***

---
Signif. codes: 0 '***' 0.001 '**' 0.01 '*' 0.05 '.' 0.1 ' ' 1
```

The null hypothesis of the test for the overall model is that the predictor variables do not contribute to the model, and the alternative hypothesis is that the generalized ordinal logistic regression model is better than the null model with no independent variables. The likelihood ratio test statistic $LR\ \chi^2_{(12)} = 204.23$, $p < .001$, which indicated that the overall model with the four predictors was significantly different from zero. Therefore, the generalized ordinal logistic regression model provides a better fit than the null model with no independent variables.

Pseudo R^2

The `nagelkerke(mod.gpo)` command produces the three types of pseudo R^2 statistics and the likelihood ratio test statistic for the generalized ordinal logistic regression model. You need to load `rcompanion` first (Mangiafico, 2021) by typing `library(rcompanion)`.

```
> # PseudoR2
> library(rcompanion)
> nagelkerke(mod.gpo)
$`Models`

Model: "vglm, healthre ~ educ + maritals + female + wrkfull, cumulative(parallel =
FALSE, reverse = FALSE), chp5.gpo"
Null: "vglm, healthre ~ 1, cumulative(parallel = FALSE, reverse = FALSE), chp5.gpo"

$Pseudo.R.squared.for.model.vs.null
                                    Pseudo.R.squared
McFadden                                   0.0456232
Cox and Snell (ML)                         0.1033040
Nagelkerke (Cragg and Uhler)               0.1137250

$Likelihood.ratio.test
Df.diff    LogLik.diff    Chisq      p.value
     12        -102.11   204.23   4.3651e-37

$Number.of.observations

Model: 1873
Null:  1873
```

```
$Messages
[1] "Note: For models fit with REML, these statistics are based on refitting with ML"

$Warnings
[1] "None"
```

McFadden's R^2 is .046, Cox and Snell's R^2 is .103, and Nagelkerke's R^2 is .114.

AIC and BIC Statistics

The AIC and BIC statistics of the fitted model can be obtained using `AIC(mod.gpo)` and `BIC(mod.gpo)`, respectively.

```
> AIC(mod.gpo)
[1] 4302.206
> BIC(mod.gpo)
[1] 4385.235
```

5.3.9 Logit Coefficients of Being at or Above a Category

With the `reverse = TRUE` option, we can estimate the logit coefficients of being at or above a particular category of the ordinal outcome variable. The `summary(mod.gpo2)` command produces the following output.

```
> # Logit coefficients of being at or above a category with reverse = TRUE
> mod.gpo2 <- vglm(healthre ~ educ + maritals + female + wrkfull, cumulative(parallel =
FALSE, reverse = TRUE), data = chp5.gpo)
> summary(mod.gpo2)

Call:
vglm(formula = healthre ~ educ + maritals + female + wrkfull,
  family = cumulative(parallel = FALSE, reverse = TRUE), data = chp5.gpo)

Pearson residuals:
                        Min       1Q    Median       3Q     Max
logitlink(P[Y>=2])   -9.009   0.1116    0.1560   0.2524  0.8896
logitlink(P[Y>=3])   -3.981  -0.4991    0.3579   0.6751  2.0847
logitlink(P[Y>=4])   -1.000  -0.6317   -0.3391  -0.1532  6.0010


Coefficients:
               Estimate   Std. Error   z value   Pr(>|z|)
(Intercept):1  -0.19647      0.41350    -0.475   0.634683
(Intercept):2  -1.86004      0.25942    -7.170   7.50e-13 ***
(Intercept):3  -3.73087      0.30303   -12.312    < 2e-16 ***
educ:1          0.18037      0.03137     5.749   8.98e-09 ***
educ:2          0.16361      0.01857     8.810    < 2e-16 ***
```

```
educ:3          0.16597    0.02013    8.246    < 2e-16 ***
maritals:1      0.59747    0.21385    2.794    0.005208 **
maritals:2      0.38257    0.10878    3.517    0.000437 ***
maritals:3      0.08731    0.11480    0.761    0.446910
female:1        0.11250    0.19631    0.573    0.566580
female:2        0.20045    0.10708    1.872    0.061211 .
female:3        0.05112    0.11537    0.443    0.657686
wrkfull:1       0.75615    0.22072    3.426    0.000613 ***
wrkfull:2       0.65609    0.11068    5.928    3.07e-09 ***
wrkfull:3       0.10049    0.11562    0.869    0.384772

---
Signif. codes: 0 '***' 0.001 '**' 0.01 '*' 0.05 '.' 0.1 ' ' 1

Names of linear predictors: logitlink(P[Y>=2]), logitlink(P[Y>=3]),
logitlink(P[Y>=4])

Residual deviance: 4272.206 on 5604 degrees of freedom

Log-likelihood: -2136.103 on 5604 degrees of freedom

Number of Fisher scoring iterations: 5

Warning: Hauck-Donner effect detected in the following estimate(s):
'(Intercept):3'

Exponentiated coefficients:
   educ:1     educ:2     educ:3   maritals:1   maritals:2   maritals:3
 1.197656   1.177758   1.180538     1.817507     1.466054     1.091238
 female:1   female:2   female:3    wrkfull:1    wrkfull:2    wrkfull:3
 1.119075   1.221953   1.052452     2.130057     1.927234     1.105714
```

In the output, the three binary logistic regression models estimate `logit(P[Y>=2])`, `logit(P[Y>=3])`, and `logit(P[Y>=4])` and are numbered 1, 2, and 3, respectively. The logit coefficients can also be displayed as a matrix with `coef(mod.gpo2, matrix = TRUE)`. The output is omitted here.

5.3.10 Odds Ratios of Being at or Above a Category

The odds ratios of being at or above a category and the corresponding confidence intervals can be obtained by using the following functions.

```
> exp(coef(mod.gpo2, matrix = TRUE))
             logit(P[Y>=2])   logit(P[Y>=3])   logit(P[Y>=4])
(Intercept)       0.8216233        0.1556659       0.02397206
educ              1.1976558        1.1777575       1.18053842
maritals          1.8175069        1.4660541       1.09123800
female            1.1190749        1.2219526       1.05245207
wrkfull           2.1300574        1.9272343       1.10571358

> exp(confint(mod.gpo2, matrix = TRUE))
                    2.5 %       97.5 %
(Intercept):1  0.36534223   1.84776002
(Intercept):2  0.09362189   0.25882716
(Intercept):3  0.01323620   0.04341575
educ:1         1.12622857   1.27361311
educ:2         1.13566076   1.22141471
educ:3         1.13487202   1.22804239
maritals:1     1.19521569   2.76379514
maritals:2     1.18455160   1.81445415
maritals:3     0.87137073   1.36658293
female:1       0.76166281   1.64420343
female:2       0.99062171   1.50730406
female:3       0.83945453   1.31949418
wrkfull:1      1.38202537   3.28296763
wrkfull:2      1.55140405   2.39410997
wrkfull:3      0.88150640   1.38694685

> cbind(exp(coef(mod.gpo2)), exp(confint(mod.gpo2)))
                                 2.5 %       97.5 %
(Intercept):1  0.82162326   0.36534223   1.84776002
(Intercept):2  0.15566595   0.09362189   0.25882716
(Intercept):3  0.02397206   0.01323620   0.04341575
educ:1         1.19765582   1.12622857   1.27361311
educ:2         1.17775751   1.13566076   1.22141471
educ:3         1.18053842   1.13487202   1.22804239
maritals:1     1.81750690   1.19521569   2.76379514
maritals:2     1.46605408   1.18455160   1.81445415
maritals:3     1.09123800   0.87137073   1.36658293
female:1       1.11907489   0.76166281   1.64420343
female:2       1.22195259   0.99062171   1.50730406
female:3       1.05245207   0.83945453   1.31949418
wrkfull:1      2.13005741   1.38202537   3.28296763
wrkfull:2      1.92723427   1.55140405   2.39410997
wrkfull:3      1.10571358   0.88150640   1.38694685
```

In the output, the odd ratios of being at or above a category versus being below that category for all four predictor variables are different across the three binary models. They can be interpreted as the change in the odds of being at or above a particular category for a one-unit increase in the predictor variable when holding all other predictors constant. For example, the odds ratios for `educ` across three equations are 1.198, 1.178, and 1.181, respectively. They can be interpreted as the odds of being at or above a category increase by 19.8%, 17.8%, and 18.1% across three comparisons, respectively, for a one-unit increase in `educ`.

5.3.11 Computing the Predicted Probabilities With the `ggpredict()` Function in the `ggeffects` Package

We use the `ggpredict()` function in the `ggeffects` package (Lüdecke, 2018b) to compute the predicted probabilities of being in a particular category of the ordinal response variable at specified values of predictor variables. Since the package has been installed in earlier chapters, we only need to load the package by typing `library(ggeffects)`. The command `prob.e <- ggpredict(mod.gpo, terms = "educ[12, 14, 16]", ci = NA)` tells R to compute the predicted probabilities for each category of the ordinal response variable using the `ggpredict()` function. In the function, `mod.gpo` is the fitted model; the `terms = "educ[12, 14, 16]"` option specifies the predictor variable `educ` at the values of 12, 14, and 16 when holding the other predictor variables at their means; and the `ci = NA` option specifies no confidence intervals. The `terms` option can specify up to four variables, including the second to fourth grouping variables. Please also note that the confidence intervals can only be obtained for the cumulative probabilities, so the `ci = NA` option is needed there. The output is assigned to the object named `prob.e`.

```
> # Predicted probabilities with ggpredict() in ggeffects
> library(ggeffects)
> prob.e <- ggpredict(mod.gpo, terms = "educ[12, 14, 16]", ci = NA)
> prob.e
# Predicted probabilities of healthre


# Response Level = 1

educ | Predicted
----------------
  12 |      0.07
  14 |      0.05
  16 |      0.03


# Response Level = 2

educ | Predicted
----------------
  12 |      0.27
  14 |      0.22
  16 |      0.17


# Response Level = 3

educ | Predicted
----------------
  12 |      0.50
  14 |      0.52
  16 |      0.52
```

```
# Response Level = 4

educ | Predicted
----------------
  12 |      0.16
  14 |      0.22
  16 |      0.28


Adjusted for:
* maritals = 0.44
*   female = 0.56
*  wrkfull = 0.47

> plot(prob.e)
```

When educ equals 12, 14, and 16, and other predictor variables are held at their means, the estimated probabilities of being in each category (i.e., $Y = 1, 2, 3$, and 4) are displayed in the output. The last section under the title "Adjusted for" lists the means of the other three variables.

The predicted probabilities for all four response levels are plotted using plot (prob.e). Figure 5.1 shows the estimated probabilities of being in each category (i.e., $Y = 1, 2, 3$, and 4) for educ at 12, 14, and 16.

FIGURE 5.1 ● Predicted Probabilities of Being in Categories 1, 2, 3, and 4 for educ

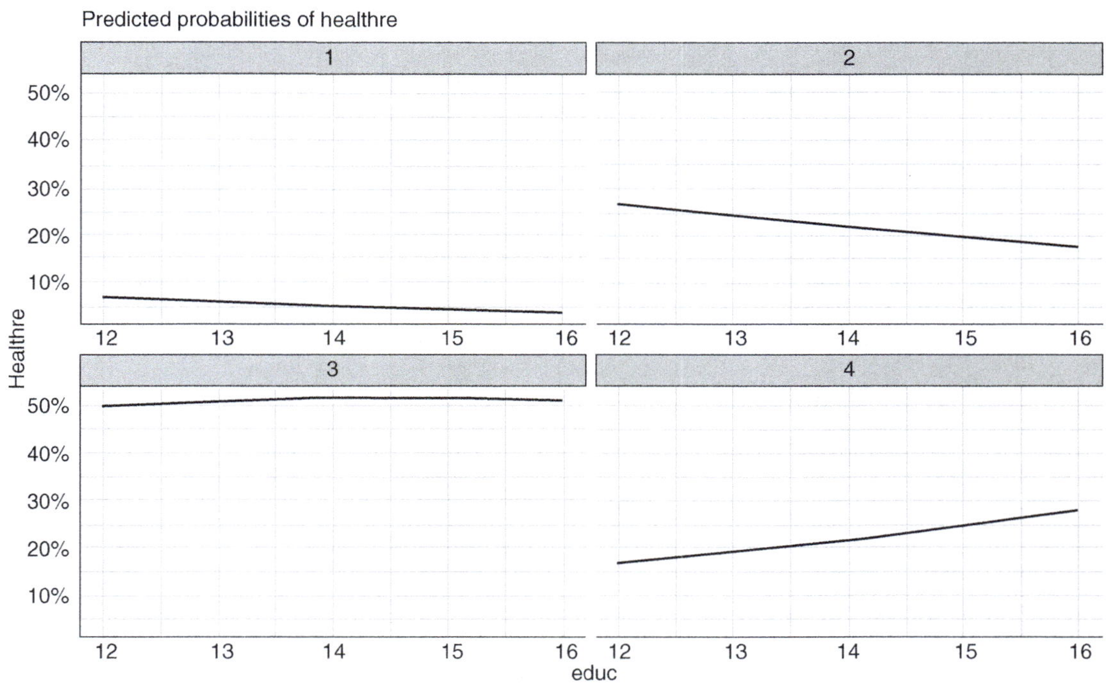

The graph shows that with an increase in the years of education, the probabilities of being in poor and fair health conditions (categories 1 and 2) decrease. In other words, people with higher levels of education are less likely to be associated with poor and fair health conditions. In addition, with an increase in the years of education, the probabilities of being in good and excellent health conditions (categories 3 and 4) increase. In other words, people with a higher level of education are more likely to be in good and excellent health conditions.

5.3.12 Computing the Predicted Cumulative Probabilities With the `ggpredict()` Function

We can also compute the predicted cumulative probabilities of being at or above a particular category of the ordinal response variable at specified values of predictor variables. The command `cumprob.e <- ggpredict(mod.gpo, terms = "educ [12, 14, 16]")` tells R to compute the cumulative probabilities of being at or above a category of the ordinal response variable with the `ggpredict()` function by removing the `ci = NA` option. The output is assigned to the object named `cumprob.e`. The `as.data.frame()` function is used to request the standard errors.

```
> # Predicted cumulative probabilities with ggpredict() in ggeffects
> cumprob.e <- ggpredict(mod.gpo, terms = "educ[12, 14, 16]")
> cumprob.e
# Predicted probabilities of healthre

# Response Level = P[Y >= 2]

educ | Predicted |          95% CI
-------------------------------
  12 |      0.93 | [0.95, 0.92]
  14 |      0.95 | [0.96, 0.94]
  16 |      0.97 | [0.98, 0.96]


# Response Level = P[Y >= 3]

educ | Predicted |          95% CI
-------------------------------
  12 |      0.67 | [0.69, 0.64]
  14 |      0.73 | [0.75, 0.71]
  16 |      0.79 | [0.82, 0.77]

# Response Level = P[Y >= 4]

educ | Predicted |          95% CI
-------------------------------
  12 |      0.16 | [0.19, 0.14]
  14 |      0.22 | [0.24, 0.20]
  16 |      0.28 | [0.30, 0.25]
```

```
Adjusted for:
* maritals = 0.44
*   female = 0.56
*  wrkfull = 0.47

> as.data.frame(cumprob.e)
      x    predicted    std.error    conf.low    conf.high    response.level    group
1    12    0.9337026    0.10979500   0.9458436   0.9190726          P[Y >= 2]        1
2    12    0.6656569    0.05886990   0.6908273   0.6395067          P[Y >= 3]        1
3    12    0.1644267    0.07658484   0.1861005   0.1448279          P[Y >= 4]        1
4    14    0.9528327    0.12049652   0.9623812   0.9410091          P[Y >= 2]        1
5    14    0.7341598    0.05587161   0.7549774   0.7122475          P[Y >= 3]        1
6    14    0.2152252    0.05808976   0.2350777   0.1966183          P[Y >= 4]        1
7    16    0.9666401    0.15766559   0.9752891   0.9551032          P[Y >= 2]        1
8    16    0.7929920    0.07440783   0.8159120   0.7680297          P[Y >= 3]        1
9    16    0.2765238    0.06422452   0.3024005   0.2520613          P[Y >= 4]        1

# Standard errors are on link-scale (untransformed).
> plot(cumprob.e)
```

The output provides the three cumulative probabilities with the confidence intervals for `educ` at 12, 14, and 16, and other predictor variables are held at their means. Please note that the standard errors are on the logit-link scale and are not transformed back to the probabilities. The results are plotted with `plot(cumprob.e)`. Figure 5.2 shows the predicted cumulative probabilities of being at or above categories 2, 3, and 4 for `educ`.

With an increase in the years of education, people are more likely to be in better health conditions.

5.4 PARTIAL PROPORTIONAL ODDS MODELS WITH R

5.4.1 The Partial Proportional Odds (PPO) Model With the `vglm()` Function

In a PPO model, the coefficients of the predictor variables violating the assumption are allowed to vary across categories, therefore, if they are identified, we can specify them in the `vglm()` function. In the following example, `wrkfull` violates the PO assumption. The `parallel = FALSE ~ wrkfull` argument tells R that `wrkfull` in the model violates the proportional odds assumption. We run the command for the PPO model, `mod.ppo <- vglm(healthre ~ educ + maritals + female + wrkfull, cumulative(parallel = FALSE ~ wrkfull, reverse = FALSE), data = chp5.gpo)` and the results are shown as follows.

FIGURE 5.2 ● Predicted Cumulative Probabilities of Being at or Above Categories 2, 3, and 4 for educ

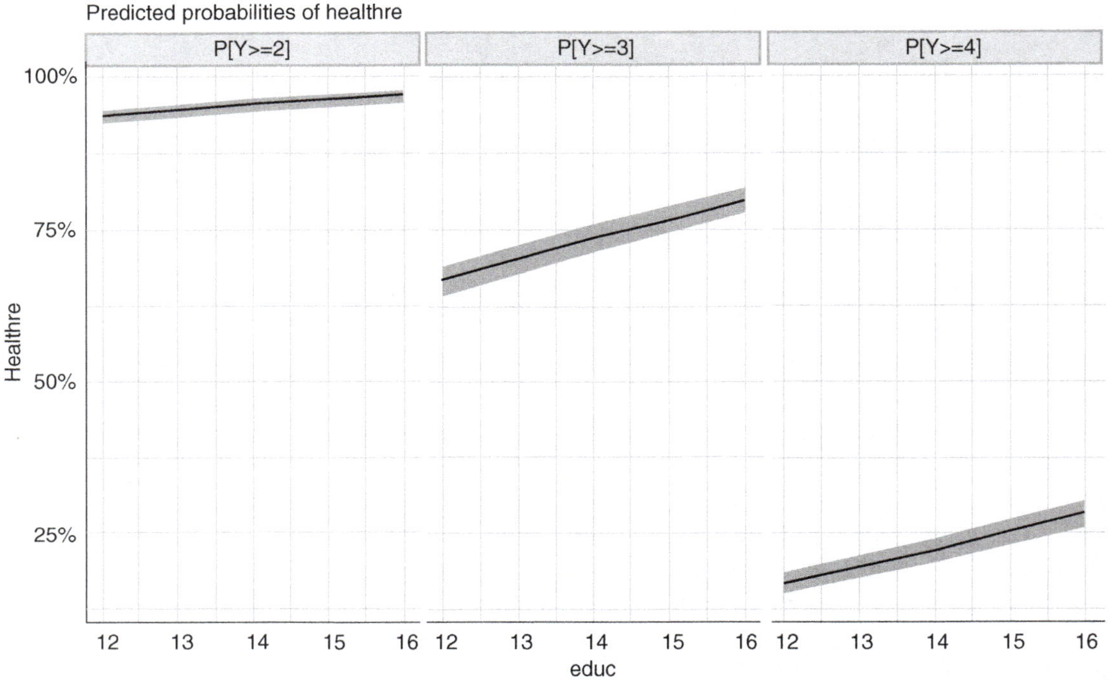


```
> # PPO model with vglm() in VGAM
>mod.ppo<-vglm(healthre~educ+maritals+female+wrkfull, cumulative(parallel=
FALSE~wrkfull, reverse=FALSE), data=chp5.gpo)
> summary(mod.ppo)

Call:
vglm(formula = healthre ~ educ + maritals + female + wrkfull,
  family = cumulative(parallel = FALSE ~ wrkfull, reverse = FALSE),
  data = chp5.gpo)

Pearson residuals:
                        Min       1Q   Median       3Q    Max
logitlink(P[Y<=1])  -0.7817  -0.2403  -0.1562  -0.1189  8.028
logitlink(P[Y<=2])  -2.0978  -0.6728  -0.3658   0.4940  3.839
logitlink(P[Y<=3])  -5.5984   0.1583   0.3406   0.6397  1.077
```

```
Coefficients:
                 Estimate    Std. Error    z value    Pr(>|z|)
(Intercept):1    -0.07075      0.22603     -0.313    0.754271
(Intercept):2     1.80233      0.21568      8.357     < 2e-16 ***
(Intercept):3     3.85156      0.23357     16.490     < 2e-16 ***
educ             -0.16541      0.01516    -10.908     < 2e-16 ***
maritals         -0.26358      0.08901     -2.961    0.003064 **
female           -0.12968      0.08881     -1.460    0.144237
wrkfull:1        -0.79925      0.21638     -3.694    0.000221 ***
wrkfull:2        -0.65348      0.10972     -5.956    2.59e-09 ***
wrkfull:3        -0.09649      0.11521     -0.837    0.402327

---
Signif. codes: 0 '***' 0.001 '**' 0.01 '*' 0.05 '.' 0.1 ' ' 1

Names of linear predictors: logitlink(P[Y<=1]), logitlink(P[Y<=2]),
logitlink(P[Y<=3])

Residual deviance: 4280.54 on 5610 degrees of freedom

Log-likelihood: -2140.27 on 5610 degrees of freedom

Number of Fisher scoring iterations: 4

No Hauck-Donner effect found in any of the estimates

Exponentiated coefficients:
     educ      maritals       female     wrkfull:1    wrkfull:2    wrkfull:3
0.8475485     0.7682963    0.8783731    0.4496664    0.5202321    0.9080234
```

5.4.2 Interpreting R Output

As with the generalized ordinal logistic regression model, the coefficients section displays the parameter estimates of the predictor variables for the three underlying binary logistic models estimating `logit(P[Y<=1])`, `logit(P[Y<=2])`, and `logit(P[Y<=3])`. Again, model 1 compares category 1 with categories 2, 3 and 4, model 2 compares categories 1 and 2 with categories 3 and 4, and model 3 compares categories 1 through 3 with category 4, respectively.

Let us look at the estimates for the predictor variables that meet the PO assumption and those violating the assumption separately. The coefficients of the former variables are constrained but the coefficients of those latter variables are free to vary across the three binary logistic models. In the output, `educ`, `maritals`, and `female` meet the PO assumption, so the equal-slope or the proportional odds constraints are placed on those variables. Each of these three variables has the same logit coefficient across all three binary models. For `educ`, $\beta = -.165$; for `maritals`, $\beta = -.264$; and for `female`, $\beta = -.130$.

The `wrkfull` predictor variable is the only one that violates the PO assumption, so its coefficients are allowed to vary across the three binary models. The estimated logit coefficients are $-.799$, $-.653$, and $-.096$ for each respective model. In model 1, the Wald z test for `wrkfull` $= -3.694$, $p < .001$; in model 2, Wald $z = -5.956$, $p < .001$; and in model 3, Wald $z = -.837$, $p > .05$. The results of the Wald z tests indicate that logit coefficients for `wrkfull` are significant across the first two models but not the third model.

The `coef(mod.ppo, matrix = TRUE)` command produces the coefficients table in the matrix form. The output is omitted.

Odds Ratios of Being at or Below a Category

With the `exp(coef(mod.ppo, matrix = TRUE))` and the `exp(confint (mod.ppo, matrix = TRUE))` commands, we can get the odds ratios and the corresponding confidence intervals, respectively. The results are combined using `cbind(exp(coef(mod.ppo)), exp(confint(mod.ppo)))`. The following output displays the odds ratio and the corresponding confidence intervals.

```
>> exp(coef(mod.ppo, matrix = TRUE))
              logit(P[Y<=1])  logit(P[Y<=2])  logit(P[Y<=3])
(Intercept)       0.9316959       6.0637586      47.0665040
educ              0.8475485       0.8475485       0.8475485
maritals          0.7682963       0.7682963       0.7682963
female            0.8783731       0.8783731       0.8783731
wrkfull           0.4496664       0.5202321       0.9080234

> exp(confint(mod.ppo, matrix = TRUE))
                    2.5 %      97.5 %
(Intercept):1   0.5982474   1.4510005
(Intercept):2   3.9733536   9.2539381
(Intercept):3  29.7782630  74.3917064
educ            0.8227302   0.8731154
maritals        0.6453035   0.9147311
female          0.7380428   1.0453857
wrkfull:1       0.2942466   0.6871782
wrkfull:2       0.4195704   0.6450442
wrkfull:3       0.7244863   1.1380567

> cbind(exp(coef(mod.ppo)), exp(confint(mod.ppo)))
                                2.5 %      97.5 %
(Intercept):1   0.9316959   0.5982474   1.4510005
(Intercept):2   6.0637586   3.9733536   9.2539381
(Intercept):3  47.0665040  29.7782630  74.3917064
educ            0.8475485   0.8227302   0.8731154
maritals        0.7682963   0.6453035   0.9147311
female          0.8783731   0.7380428   1.0453857
wrkfull:1       0.4496664   0.2942466   0.6871782
wrkfull:2       0.5202321   0.4195704   0.6450442
wrkfull:3       0.9080234   0.7244863   1.1380567
```

5.4.3 Interpreting the Odds Ratios of Being at or Below a Particular Category

With the `reverse = FALSE` option, the `vglm()` function estimates the odds ratios of being at or below a particular category versus being above that category in the PPO model and the generalized ordinal logistic model. The odds ratios in both models can be interpreted in a similar way as that in the PO model. It can be interpreted as the change in the odds of being at or below a particular category for a one-unit increase in the predictor variable when holding all the other predictors constant.

Unlike the PO model, the partial PO model allows the effects of some of the predictor variables to vary, so we need to interpret the odds ratios for the predictor variables that meet the PO assumption and those violating the assumption separately.

Three predictor variables, `educ`, `maritals`, and `female`, meet the PO assumption in the model. Let us interpret their odds ratios first. For the `educ` predictor variable, $\beta = -.165$, which is negative, so there is a negative relationship between age and the log odds of being at or below a category of health status; OR = .848, which is less than 1. This indicates that the odds of being at or below a particular category of health status (poorer health status) decrease by a factor of .848 for a one-unit increase in the `educ` predictor when holding other variables constant. In other words, a one-unit increase in the number of years of education is associated with a decrease by 15.2% in the odds of being less healthy.

For the `maritals` predictor variable, $\beta = -.264$, and its corresponding OR = .768. This indicates that the odds of being at or below a particular category of health status (poorer health status) versus being beyond that category (better health status) for the married are .768 times the odds for the unmarried when holding the other predictors constant.

For the `female` predictor variable, $\beta = -.130$, $p = .114$, which is not significantly different from 0; OR = .878, which is close to 1. This indicates that there is no relationship between being female and the cumulative odds of being less healthy.

Next, let us interpret the predictor variables that violate the PO assumption. Since only one predictor variable `wrkfull` violates the assumption, in the output its odds ratios are different across the three binary models. The three odds ratios are .450, .520, and .908, respectively. Overall, working full time decreases the odds of being at or below a particular category of health status. In other words, working full time is associated with the odds of being in better health status.

5.4.4 Model Fit Statistics

Testing the Overall Model Using the Likelihood Ratio Test

To test if the overall PPO model is significant, we compare the PPO model with the null model using the `lrtest()` function. The `lrtest(gpomodel0, mod.ppo)` command compares the log-likelihood statistics of the fitted model `mod.ppo` and the null model `gpomodel0` using the likelihood ratio test.

```
> # Testing the overall model using the likelihood ratio test
> lrtest(gpomodel0, mod.ppo)
Likelihood ratio test


Model 1: healthre ~ 1
Model 2: healthre ~ educ + maritals + female + wrkfull
      #Df     LogLik    Df     Chisq     Pr(>Chisq)
1     5616    -2238.2
2     5610    -2140.3   -6     195.9      < 2.2e-16 ***

---
Signif. codes: 0 '***' 0.001 '**' 0.01 '*' 0.05 '.' 0.1 ' ' 1
```

The likelihood ratio test statistic $LR\ \chi^2_{(6)} = 195.9$, $p < .001$, which indicated that the overall PPO model with the four predictors was significantly different from zero. Therefore, the PPO model provides a better fit than the null model with no independent variables.

Pseudo R^2

The `nagelkerke(mod.ppo)` command produces the three types of pseudo R^2 statistics and the likelihood ratio test statistic for the PPO model. The McFadden R^2 is .044, the Cox and Snell R^2 is .099, and the Nagelkerke R^2 is .109. The output is omitted here.

AIC and BIC Statistics

The AIC and BIC statistics of the fitted model can be obtained using `AIC(mod.ppo)` and `BIC(mod.ppo)`, respectively.

```
> AIC(mod.ppo)
[1] 4298.54
> BIC(mod.ppo)
[1] 4348.358
```

5.4.5 Logit Coefficients of Being at or Above a Category

Just like the generalized ordinal logistic regression model, the PPO model can also estimate the logit coefficients of being at or above a particular category of the ordinal outcome variable with the `reverse = TRUE` option. We fit the model named `mod.ppo2`. The `summary(mod.ppo2)` command produces the following output.

```
> # Logit coefficients of being at or above a category with reverse = TRUE
>mod.ppo2<-vglm(healthre~educ+maritals+female+wrkfull, cumulative(parallel =
FALSE ~ wrkfull, reverse = TRUE), data = chp5.gpo)
> summary(mod.ppo2)
```

```
Call:
vglm(formula = healthre ~ educ + maritals + female + wrkfull,
  family = cumulative(parallel = FALSE ~ wrkfull, reverse = TRUE),
  data = chp5.gpo)

Pearson residuals:
                      Min       1Q     Median       3Q      Max
logitlink(P[Y>=2])  -8.028    0.1189   0.1562    0.2403   0.7817
logitlink(P[Y>=3])  -3.839   -0.4940   0.3658    0.6728   2.0978
logitlink(P[Y>=4])  -1.077   -0.6397  -0.3406   -0.1583   5.5984


Coefficients:
               Estimate  Std. Error   z value   Pr(>|z|)
(Intercept):1   0.07075    0.22603     0.313   0.754271
(Intercept):2  -1.80233    0.21568    -8.357    < 2e-16 ***
(Intercept):3  -3.85156    0.23357   -16.490    < 2e-16 ***
educ            0.16541    0.01516    10.908    < 2e-16 ***
maritals        0.26358    0.08901     2.961   0.003064 **
female          0.12968    0.08881     1.460   0.144237
wrkfull:1       0.79925    0.21638     3.694   0.000221 ***
wrkfull:2       0.65348    0.10972     5.956   2.59e-09 ***
wrkfull:3       0.09649    0.11521     0.837   0.402327

---
Signif. codes:  0 '***' 0.001 '**' 0.01 '*' 0.05 '.' 0.1 ' ' 1

Names of linear predictors: logitlink(P[Y>=2]), logitlink(P[Y>=3]),
logitlink(P[Y>=4])

Residual deviance: 4280.54 on 5610 degrees of freedom

Log-likelihood: -2140.27 on 5610 degrees of freedom

Number of Fisher scoring iterations: 4

No Hauck-Donner effect found in any of the estimates


Exponentiated coefficients:
    educ    maritals      female   wrkfull:1   wrkfull:2   wrkfull:3
1.179873   1.301581    1.138468    2.223871    1.922219    1.101293
```

5.4.6 Interpreting R Output

The coefficients section displays the parameter estimates of predictor variables for the three underlying binary logistic models estimating `logit(P[Y>=2])`, `logit(P[Y>=3])`, and `logit(P[Y>=4])`. Specifically, model 1 compares categories

2, 3, and 4 with category 1, model 2 compares categories 3 and 4 with categories 1 and 2, and model 3 compares category 4 with categories 1 through 3, respectively.

In the coefficients section, the logit coefficients in the `mod.ppo2` model are the same as those in the `mod.ppo` model in magnitude but are reversed in sign since the former model estimates the log odds of being at or above a category, logit[$P(Y \geq j + 1)$], whereas the latter model estimates the log odds of being at or below a category, logit[$P(Y \leq j)$].

The three predictor variables, `educ`, `maritals`, and `female` meet the PO assumption, so they have the same logit coefficients across all three binary models. For `educ`, $\beta = .165$; for `maritals`, $\beta = .264$; and for `female`, $\beta = .130$.

The `wrkfull` predictor variable violates the PO assumption, so its coefficients are allowed to vary across the three binary models. The estimated logit coefficients are .799, .653, and .096 for each respective model. In model 1, the Wald z test for `wrkfull` = 3.694, $p < .001$; in model 2, Wald $z = 5.956$, $p < .001$; and in model 3, Wald $z = .837$, $p > .05$.

We again use `coef(mod.ppo, matrix = TRUE)` to obtain the coefficients table in the matrix form and use the `exp(coef(mod.ppo2, matrix = TRUE))` and `exp(confint(mod.ppo2, matrix = TRUE))` commands to request the odds ratios and the corresponding confidence intervals. The results are combined at the end.

```
> coef(mod.ppo2, matrix = TRUE)
              logit(P[Y>=2])    logit(P[Y>=3])    logit(P[Y>=4])
(Intercept)       0.07074881        -1.8023298       -3.85156158
educ              0.16540723         0.1654072        0.16540723
maritals          0.26357980         0.2635798        0.26357980
female            0.12968378         0.1296838        0.12968378
wrkfull           0.79924931         0.6534802        0.09648514

> exp(coef(mod.ppo2, matrix = TRUE))
              logit(P[Y>=2])    logit(P[Y>=3])    logit(P[Y>=4])
(Intercept)         1.073312         0.1649142        0.02124653
educ                1.179873         1.1798735        1.17987350
maritals            1.301581         1.3015812        1.30158116
female              1.138468         1.1384683        1.13846832
wrkfull             2.223871         1.9222190        1.10129321

> exp(confint(mod.ppo2, matrix = TRUE))
                     2.5 %       97.5 %
(Intercept):1   0.68917966   1.67154930
(Intercept):2   0.10806210   0.25167657
(Intercept):3   0.01344236   0.03358154
educ            1.14532398   1.21546523
maritals        1.09321745   1.54965832
female          0.95658471   1.35493503
wrkfull:1       1.45522669   3.39850942
wrkfull:2       1.55028134   2.38339049
wrkfull:3       0.87869084   1.38028835

> cbind(exp(coef(mod.ppo2)), exp(confint(mod.ppo2)))
                                  2.5 %       97.5 %
(Intercept):1   1.07331159   0.68917966   1.67154930
(Intercept):2   0.16491422   0.10806210   0.25167657
```

```
(Intercept):3    0.02124653    0.01344236    0.03358154
educ             1.17987350    1.14532398    1.21546523
maritals         1.30158116    1.09321745    1.54965832
female           1.13846832    0.95658471    1.35493503
wrkfull:1        2.22387087    1.45522669    3.39850942
wrkfull:2        1.92221898    1.55028134    2.38339049
wrkfull:3        1.10129321    0.87869084    1.38028835
```

5.4.7 Interpreting the Odds Ratios of Being at or Above a Particular Category

We are interested in the odds ratios reported in the coefficients table in the second section in the output. First, let us take a look at the odds ratios for the predictor variables that meet the PO assumption. They are 1.180, 1.302, and 1.138, for `educ`, `maritals`, and `female`, respectively.

Second, we look at the predictor variables that violate the PO assumption. Only one predictor variable `wrkfull` violates the assumption. Its odds ratios are different across the three binary models. They are 2.224, 1.922, and 1.101, respectively.

The odds of being at or above a particular category versus being below that category are the multiplicative inverse of the odds of being below that category since $\exp(-\beta) = 1/\exp(\beta)$. The odds ratios in the partial PO model can also be interpreted as the change in the odds of being at or above a particular category for a one-unit increase in the predictor variable when holding all other predictors constant.

Three predictor variables `maritals`, `age`, and `male` meet the PO assumption in the model. For the `educ` predictor, OR = 1.180, which is greater than 1. This indicates that a one-unit increase in education is associated with an increase of 18% in the odds of being healthier.

For `maritals`, OR = 1.302. This indicates that the odds of being at or above a particular category of health status (better health status) versus being below that category (poorer health status) for the married are 1.302 times the odds for the unmarried when holding the other predictors constant.

For `female`, $\beta = .130$, $p = .114$, which is not significantly different from 0; OR = 1.138, which is close to 1. This indicates that there is no change in the odds for being female.

Next, let us interpret the predictor variables that violate the PO assumption. The odds ratios for `wrkfull` are different across the three binary models. The three odds ratios are 2.224, 1.922, and 1.101, respectively. Overall, working full time increases the odds of being at or above a particular category of health status. The odds of being at or above a particular category of health status for working full time are 122.4%, 92.2%, and 10.1% higher than the odds for not working full time in each of the binary logistic models, respectively, when holding other variables constant. The largest odds ratio is identified in the first binary model comparing categories 2, 3, and 4 with category 1, and the smallest odds ratio is found in the third binary model comparing category 4 with categories 1 through 3.

5.4.8 Computing the Predicted Probabilities With the `ggpredict()` Function for the PPO Model

To compute the predicted probabilities of being in a particular category of the ordinal response variable at specified values of predictor variables in the PPO model, we again use the `ggpredict()` function in the `ggeffects` package (Lüdecke, 2018b). The command `prob.e2 <- ggpredict(mod.ppo, terms = "educ[12, 14, 16]", ci = NA)` tells R to compute the predicted probabilities for each category of the ordinal response variable using the `ggpredict()` function. In the function, `mod.ppo` is the fitted model; the `terms = "educ[12, 14, 16]"` option specifies the predictor variable `educ` at the values of 12, 14, and 16 when holding the other predictor variables at their means; and the `ci = NA` option specifies no confidence intervals. The output is assigned to the object named `prob.e2`.

```
>> library(ggeffects)
> prob.e2 <- ggpredict(mod.ppo, terms = "educ[12, 14, 16]", ci = NA)
> prob.e2

# Predicted probabilities of healthre

# Response Level = 1

educ | Predicted
----------------
  12 |      0.07
  14 |      0.05
  16 |      0.04

# Response Level = 2

educ | Predicted
----------------
  12 |      0.27
  14 |      0.22
  16 |      0.17

# Response Level = 3

educ | Predicted
----------------
  12 |      0.50
  14 |      0.52
  16 |      0.52

# Response Level = 4

educ | Predicted
----------------
  12 |      0.16
  14 |      0.21
  16 |      0.27
```

```
Adjusted for:
* maritals = 0.44
*   female = 0.56
*  wrkfull = 0.47

> plot(prob.e2)
```

When educ equals 12, 14, and 16, and other predictor variables are held at their means, the predicted probabilities of being in each category (i.e., $Y = 1, 2, 3$, and 4) are displayed in the output. The last section under the title "Adjusted for" lists the means of the other three variables.

The predicted probabilities for all four response levels are plotted using plot (prob.e2). Figure 5.3 shows the predicted probabilities of being in each category (i.e., $Y = 1, 2, 3$, and 4) for educ at 12, 14, and 16.

The graph for the PPO model looks similar to the one for the generalized ordinal logistic regression model introduced in the previous section.

FIGURE 5.3 ● Predicted Probabilities of Being in Categories 1, 2, 3, and 4 for educ

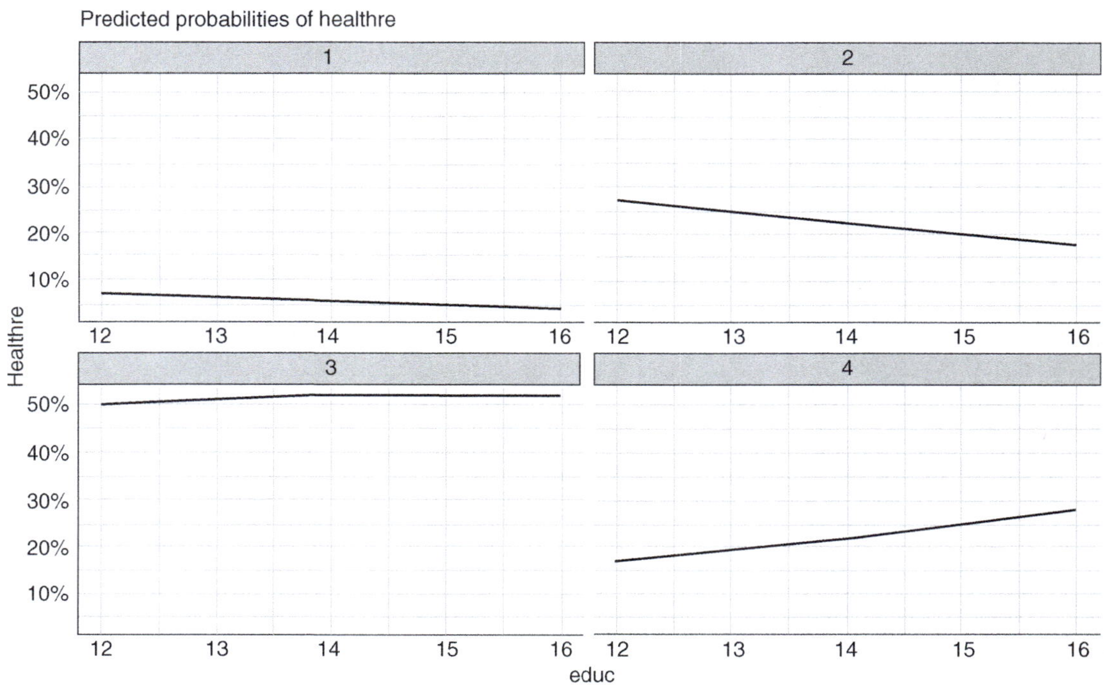

5.4.9 Computing the Predicted Cumulative Probabilities With the ggpredict() Function

We can also compute the cumulative probabilities of being at or above a particular category of the ordinal response variable at specified values of predictor variables. By removing the `ci = NA` option, we use the command `cumprob.e2 <- ggpredict(mod.ppo, terms = "educ[12, 14, 16]")` to compute the predicted cumulative probabilities of being at or above a category. The output is assigned to the object named `cumprob.e2`. The `as.data.frame()` function can be used to request the standard errors. The output is omitted here.

5.5 MAKING PUBLICATION-QUALITY TABLES

Since the `stargazer()` function (Hlavac, 2018) currently cannot directly produce the results table from the `vglm` models, we use the `screenreg()` and `htmlreg()` functions from the `texreg` package (Leifeld, 2013). You need to install `texreg` first by typing `install.packages("texreg")` if you have not done so and then load the package by typing `library(texreg)`. This package needs to be installed only once.

After we use the `vglm()` function to fit the PPO model `mod.ppo`, we create a table containing the results of the model with the following command: `screenreg(mod.ppo)`. In the `screenreg()` function, we specify the model object to be presented. The output is a plain text table.

```
> # Presenting the results of the vglm Models using the texreg package
> library(texreg)
Version:  1.37.5
Date:     2020-06-17
Author:   Philip Leifeld (University of Essex)

Consider submitting praise using the praise or praise_interactive functions.
Please cite the JSS article in your publications - see citation("texreg").
> screenreg(mod.ppo)


=============================
                  Model 1
-----------------------------
(Intercept):1      -0.07
                   (0.23)
(Intercept):2       1.80 ***
                   (0.22)
(Intercept):3       3.85 ***
                   (0.23)
educ               -0.17 ***
                   (0.02)
maritals           -0.26 **
                   (0.09)
```

```
female                  -0.13
                        (0.09)
wrkfull:1               -0.80 ***
                        (0.22)
wrkfull:2               -0.65 ***
                        (0.11)
wrkfull:3               -0.10
                        (0.12)
---------------------------
Log Likelihood        -2140.27
DF                     5610
Num. obs.              5619
===========================
*** p < 0.001; ** p < 0.01; * p < 0.05

> htmlreg(list(mod.ppo), file="chap5ppo.doc", doctype=TRUE, html.tag=TRUE,
head.tag=TRUE)
The table was written to the file 'chap5ppo.doc'.
```

We can also use the `htmlreg()` function to create a regression table for the estimated results and save it to a Microsoft Word file named `chap5ppo.doc` with the following command: `htmlreg(list(mod.ppo), file = "chap5ppo.doc", doctype = TRUE, html.tag = TRUE, head.tag = TRUE)`. It automatically produces Table 5.2, as shown here in its original format, presenting the results of the PPO model.

5.6 REPORTING THE RESULTS

Reporting the results of the PPO model and the generalized ordinal logistic regression model is similar to that of the PO models. You need to test the PO assumption and justify why the PPO model or the generalized ordinal logistic regression model needs to be used to address your research question(s). Report the results of the test for the PO assumption.

Unlike the PO models, you need to report the logit coefficients and corresponding odds ratios for the predictor variables for the underlying binary logistic models. In the PPO model, the logit coefficients of some predictor variables may vary across the binary models, whereas the logit coefficients of all predictors vary freely in the generalized ordinal logistic regression model. For each binary model, label the category comparisons, such as above category j versus at or below category j.

Have a table containing all the parameter estimates, their standard errors, and associated p values. The fit statistics, such as the likelihood ratio chi-square test statistic and the likelihood ratio R^2, may also be reported in the table. If more than one model is fitted, the results of all the competing models may be presented in the table.

In the body of the text, briefly interpret the model fit statistics. In addition, interpret the odds ratios of the estimates across binary comparisons or splits. In the interpretation, explain the comparisons between categories, for example, whether the odds of

TABLE 5.2 ● Results of the Partial Proportional Odds Model (Shown in Original Format Generated by R)

	Model 1
(Intercept):1	−0.07
	(0.23)
(Intercept):2	1.80***
	(0.22)
(Intercept):3	3.85***
	(0.23)
Educ	−0.17***
	(0.02)
maritals	−0.26**
	(0.09)
female	−0.13
	(0.09)
wrkfull:1	−0.80***
	(0.22)
wrkfull:2	−0.65***
	(0.11)
wrkfull:3	−0.10
	(0.12)
Log Likelihood	−2140.27
DF	5610
Num. obs.	5619

***$p < 0.001$
**$p < 0.01$
*$p < 0.05$.

being above a category versus at or below that category, or the reciprocal odds of being at or below a category. The following is an example of summarizing the results for the partial PO model.

The partial proportional odds model was fitted to estimate the ordinal outcome variable, health status, from a set of predictor variables, such as years of education, marital status, gender, and working status. This model was used as it allows the effects of some predictor variables to vary when the proportional odds assumption (PO) does not hold.

The likelihood ratio statistic $LR\ \chi^2_{(6)} = 195.9$, $p < .001$, which indicated that the overall PPO model with the four predictors provided a better fit than the null model with no independent variables in predicting the ordinal response variable.

Table 5.2 presents the coefficients and standard errors of the predictor variables. Three predictor variables, `educ`, `maritals`, and `female`, meeting the PO assumption had the same logit regression coefficients across all three binary models. For `educ`, $\beta = -.165$, $p < .001$; for `maritals`, $\beta = -.264$, $p < .001$; and for `female`, $\beta = -.130$, $p > .05$.

Three predictor variables, `edu`, `maritals`, and `female`, meet the PO assumption in the model. For the `educ` predictor, OR = 1.180, which is greater than 1. This indicates that the odds of being at or above a particular category of health status (better health status) increase by a factor of 1.180 for a one-unit increase in the number of years of education when holding other variables constant. In other words, a one-unit increase in education is associated with an increase of 18% in the odds of being healthier.

For `maritals`, OR = 1.302. This indicates that the odds of being at or above a particular category of health status (better health status) versus being below that category (poorer health status) for the married are 1.302 times the odds for the unmarried when holding other predictors constant.

For `female`, $\beta = .130$, $p = .114$, which is not significantly different from 0; OR = 1.138, which is close to 1. This indicates that there is no change in the odds for being female.

The odds ratios for `wrkfull` are different across the three binary models since this predictor violates the PO assumption. The three odds ratios are 2.224, 1.922, and 1.101, respectively. Overall, working full time increases the odds of being at or above a particular category of health status. The odds of being at or above a particular category of health status for working full time are 122.4%, 92.2%, and 10.1% higher than the odds for not working full time in each of the binary logistic models, respectively, when holding other variables constant. The largest odds ratio is identified in the first binary model comparing categories 2, 3, and 4 with category 1, and the smallest odds ratio is found in the third binary model comparing category 4 with categories 1 through 3.

5.7 SUMMARY OF R COMMANDS IN THIS CHAPTER

```
# Chap 5 R Script
# Remove all objects
rm(list = ls(all = TRUE))

# The following user-written packages need to be installed first by using
install.packages(" ") and then by loading it with library()
# library(VGAM)              # It is already installed for Chapter 4
# library(rcompanion)        # It is already installed for Chapter 3
# library(ggeffects)         # It is already installed for Chapter 2
# library(texreg)            # It is already installed for Chapter 4

# Import the GSS 2016 data
library(foreign)
chp5.gpo <- read.dta("C:/CDA/gss2016.dta")
chp5.gpo$healthre <- factor(chp5.gpo$healthre, ordered=TRUE)
chp5.gpo$educ <- as.numeric(chp5.gpo$educ)
chp5.gpo$wrkfull <- as.numeric(chp5.gpo$wrkfull)
chp5.gpo$maritals <- as.numeric(chp5.gpo$maritals)
attach(chp5.gpo)

# PO model with vglm() in VGAM
library(VGAM)
mod.po <- vglm(healthre ~ educ + maritals + female + wrkfull, cumulative(parallel =
TRUE, reverse = FALSE), data = chp5.gpo)
summary(mod.po)

# GPO model with vglm() in VGAM
mod.gpo <- vglm(healthre ~ educ + maritals + female + wrkfull, cumulative(parallel =
FALSE, reverse = FALSE), data = chp5.gpo)

# Using the lrtest() function to test the PO assumption
lrtest(mod.po, mod.gpo)

summary(mod.gpo)
coef(mod.gpo, matrix = TRUE)
confint(mod.gpo, matrix = TRUE)
exp(coef(mod.gpo, matrix = TRUE))
exp(confint(mod.gpo, matrix = TRUE))
cbind(exp(coef(mod.gpo)), exp(confint(mod.gpo)))

# Testing the overall model using the likelihood ratio test
gpomodel0 <- vglm(healthre ~ 1, cumulative(parallel = TRUE, reverse = FALSE))
summary(gpomodel0)
lrtest(gpomodel0, mod.gpo)
```

```
# Pseudo R2
library(rcompanion)
nagelkerke(mod.gpo)
AIC(mod.gpo)
BIC(mod.gpo)

# Predicted probabilities with ggpredict() in ggeffects
library(ggeffects)
prob.e <- ggpredict(mod.gpo, terms = "educ[12, 14, 16]", ci = NA)
prob.e
plot(prob.e)

# Predicted cumulative probabilities with ggpredict() in ggeffects
cumprob.e <- ggpredict(mod.gpo, terms = "educ[12, 14, 16]")
cumprob.e
as.data.frame(cumprob.e)
plot(cumprob.e)

# Logit coefficients of being at or above a category with reverse = TRUE
mod.gpo2 <- vglm(healthre ~ educ + maritals + female + wrkfull, cumulative(parallel =
FALSE, reverse = TRUE), data = chp5.gpo)
summary(mod.gpo2)
coef(mod.gpo2, matrix = TRUE)
confint(mod.gpo2, matrix = TRUE)
exp(coef(mod.gpo2, matrix = TRUE))
exp(confint(mod.gpo2, matrix = TRUE))
cbind(exp(coef(mod.gpo2)), exp(confint(mod.gpo2)))
AIC(mod.gpo2)
BIC(mod.gpo2)
nagelkerke(mod.gpo2)

# PPO model with vglm() in VGAM
mod.ppo <- vglm(healthre ~ educ + maritals + female + wrkfull, cumulative(parallel =
FALSE ~ wrkfull, reverse = FALSE), data = chp5.gpo)
summary(mod.ppo)
coef(mod.ppo, matrix = TRUE)
confint(mod.ppo, matrix = TRUE)
exp(coef(mod.ppo, matrix = TRUE))
exp(confint(mod.ppo, matrix = TRUE))
cbind(exp(coef(mod.ppo)), exp(confint(mod.ppo)))

# Testing the overall model using the likelihood ratio test
lrtest(gpomodel0, mod.ppo)

# Pseudo R2
nagelkerke(mod.ppo)
AIC(mod.ppo)
BIC(mod.ppo)

# Predicted probabilities with ggpredict() in ggeffects
prob.e2 <- ggpredict(mod.ppo, terms = "educ[12, 14, 16]", ci = NA)
prob.e2
plot(prob.e2)

# Predicted cumulative probabilities with ggpredict() in ggeffects
cumprob.e2 <- ggpredict(mod.ppo, terms = "educ[12, 14, 16]")
cumprob.e2
as.data.frame(cumprob.e2)
plot(cumprob.e2)
```

```
# Logit coefficients of being at or above a category with reverse = TRUE
mod.ppo2<-vglm(healthre~educ+maritals+female+wrkfull, cumulative(parallel=
FALSE~wrkfull, reverse = TRUE), data = chp5.gpo)
summary(mod.ppo2)
coef(mod.ppo2, matrix = TRUE)
exp(coef(mod.ppo2, matrix = TRUE))
exp(confint(mod.ppo2, matrix = TRUE))
cbind(exp(coef(mod.ppo2)), exp(confint(mod.ppo2)))
AIC(mod.ppo2)
BIC(mod.ppo)

# Presenting the results of the vglm Models using the texreg package
library(texreg)
screenreg(mod.ppo)
htmlreg(list(mod.ppo), file="chap5ppo.doc", doctype=TRUE, html.tag=TRUE,
head.tag=TRUE)

detach(chp5.gpo)
```

Glossary

The cumulative probability of being at or below a category $P(Y \leq j)$ equals the sum of the probabilities of all categories at or below that category.

The generalized ordinal logit model is an extreme case of the PPO model. It relaxes the PO assumption for all predictor variables and allows the effect of each explanatory variable to vary.

The odds of being at or above a certain category relative to being below that category equals the probability of being at or above a category divided by the probability of being below that category.

The odds of being at or below a category equals the probability of being at or below a category divided by the probability of being above that category.

The partial proportional odds (PPO) model allows the effects of those predictors violating the PO assumption to vary across categories.

Exercises

Use the GSS 2016 data available at **https://edge.sagepub.com/liu1e** for the following problems.

1. Conduct an analysis for a proportional odds model to estimate the ordinal response variable `happy` from the three predictor variables `educ`, `age`, and `satfin`.
2. With the same variables, conduct an analysis for a generalized ordinal logistic regression model.
3. List the logit coefficients, the Wald z tests, and the 95% confidence intervals for the predictor variables violating the assumption.
4. Compute the odds ratios for `educ` and `satfin`.
5. Fit a partial proportional odds (PPO) model using the same variables as those in the preceding PPO model in Exercise 2. Which model is more parsimonious?

6

OTHER ORDINAL LOGISTIC REGRESSION MODELS

OBJECTIVES OF THIS CHAPTER

This chapter discusses continuation ratio models, adjacent categories models, and stereotype logistic regression models. The first section starts with an introduction to the continuation ratio models followed by a discussion of the conditional probabilities, odds, and odds ratios and interpretations of parameter estimates. This chapter shows how to conduct the analysis of the continuation ratio models with and without the parallel assumption using R with step-by-step instructions. The `vglm()` function in the `VGAM` package is used to fit the models. The R syntax is explained and the output is interpreted. The second and third section shows how to fit the adjacent categories model and the stereotype logistic regression model with R as well as on interpreting the results. After reading this chapter, you should be able to:

- Identify when a continuation ratio model, adjacent categories model, and a stereotype logistic regression model are used.
- Fit the model using R.
- Interpret the output.
- Interpret the model in terms of odds ratios.
- Compute and plot the estimated probabilities.
- Compare models using the likelihood ratio test and other fit statistics.
- Present results in publication-quality tables using R.
- Write the results for publication.

6.1 CONTINUATION RATIO MODELS

6.1.1 Continuation Ratio Models: An Introduction

In the previous two chapters, we introduced the proportional odds (PO) model, the partial proportional odds (PPO) model, and the generalized ordinal logistic regression model. These models estimate either the cumulative odds of being at or below a particular category of the ordinal response variable or the odds of being above a particular level, which is the complementary direction. Sometimes the ordinal categories represent successive stages or proficiency levels. Individuals need to reach a lower stage first before moving up to a higher stage. Once a higher stage is reached, they cannot go back to an earlier stage. We can use the continuation ratio (CR) model to predict this type of ordinal response variable. The CR model estimates the odds of being in a certain category versus being above that category. In terms of probability, this model estimates the conditional probability of being in a category given that an individual has been in that category or above. In addition, it estimates the conditional probability of being above a category given that a person has attained that particular category since these two conditional probabilities are complementary.

The CR model has different names. It is also called the sequential model or sequential logit model (Buis, 2007; Liao, 1994; Tutz, 1991, 2012), the stage approach (Fullerton, 2009; Fullerton & Xu, 2016), a special case of the sequential logit model (Long & Freese, 2014), or the model for nested dichotomies (Fox, 2016).

When estimating the conditional probability of being above a category given that an individual has attained that particular category, that is, $P(Y > j \mid Y \geq j)$, the CR model can be expressed in this form (Allison, 2012; O'Connell, 2006; Tutz, 2012):

$$\ln\left(\frac{P(Y > j|x_1, x_2, \ldots, x_p)}{P(Y = j|x_1, x_2, \ldots, x_p)}\right) = \alpha_j + \beta_1 X_1 + \beta_2 X_2 + \ldots + \beta_p X_p \tag{6.1}$$

where $P(Y > j \mid x_1, x_2, \ldots, x_p)$ is the conditional probability of being above a category j, conditional on being in that category, given a set of predictors, that is, $P(Y > j \mid Y \geq j)$; $j = 1, 2, \ldots, J - 1$; α_j are the cut points; and $\beta_1, \beta_2, \ldots, \beta_p$ are the logit coefficients.

The CR model also estimates the odds of being in a particular category j relative to being above that category.

$$\ln\left(\frac{P(Y = j|x_1, x_2, \ldots, x_p)}{P(Y > j|x_1, x_2, \ldots, x_p)}\right) = \alpha_j + \beta_1 X_1 + \beta_2 X_2 + \ldots + \beta_p X_p \tag{6.2}$$

where $P(Y = j \mid x_1, x_2, \ldots, x_p)$ is the conditional probability of being in category j conditional on being in that category or beyond given a set of predictors, that is, $P(Y = j \mid Y \geq j)$; $j = 1, 2, \ldots, J - 1$. α_j are the cut points; and $\beta_1, \beta_2, \ldots, \beta_p$ are the logit coefficients. Just like the PO models, the CR model here also assumes that the logit coefficients are parallel across ordinal categories. This model is also called the constrained CR model (Cole & Ananth, 2001).

The `vglm()` function in the `VGAM` package (Yee, 2010) can be used to estimate the odds of being in a particular category *j* relative to being above that category when the `sratio` family is specified, or to estimate the reciprocal odds, the odds of being above a category versus at a category when the `cratio` family is requested. In this section, the `vglm()` function with the `sratio` family in the `VGAM` package uses Equation 6.2.

The CR model can be formulated in a different form with negative signs preceding logit coefficients (Ananth & Kleinbaum, 1997; Armstrong & Sloan, 1989; Fienberg, 1980; Liu, 2016a; Liu et al., 2011; Long & Freese, 2014):

$$\ln\left(\frac{P(Y = j|x_1, x_2, \ldots, x_p)}{P(Y > j|x_1, x_2, \ldots, x_p)}\right) = \alpha_j + \left(-\beta_1 X_1 - \beta_2 X_2 - \ldots - \beta_p X_p\right) \quad (6.3)$$

The CR model may be parameterized differently to estimate the odds of being in a particular category *j* relative to being below that category (Hosmer et al., 2013), which is not equivalent to the models for Equations 6.1 and 6.2. This type of the CR model is referred to as the backward CR model (Liu & Bai, 2019). The introduction of this model is beyond the scope of this chapter.

6.1.2 Conditional Probabilities, Odds, and Odds Ratios

The PO model estimates $J - 1$ cumulative odds and then the cumulative probabilities of being at or below a category of the ordinal response variable with *j* levels. When the ordinal response variable, health status, has four levels from 1 to 4, the proportional odds model estimates three cumulative probabilities, which include $P(Y \leq 1)$, $P(Y \leq 2)$, and $P(Y \leq 3)$. The cumulative probabilities of being above a category can also be estimated since they are the complementary probabilities of being at or below a particular category.

Unlike the cumulative probabilities in the PO model, the conditional probabilities are estimated in the CR logit model. The CR model estimates conditional probabilities of being in category *j*, conditional on being at or above that category, that is, $P(Y = j \mid Y \geq j)$. This CR model can also estimate the conditional probability of being above a category given that the individual has achieved that particular category since $P(Y > j \mid Y \geq j)$ is the complementary form of $P(Y = j \mid Y \geq j)$.

The logit or log odds of being in a category relative to being above that category $\ln[P(Y = j)/P(Y > j)]$ is the conditional logit. The odds in the CR model are the probability of being in a category divided by the probability of being above that category. In other words, they compare the probability when $P(Y = j)$ and $P(Y > j)$.

Since the probability of being in a category in the CR model is a conditional probability, the odds are the conditional odds:

$$\text{Odds} = \frac{P(Y = j)}{P(Y > j)}$$

The odds ($Y = 1 \mid Y \geq 1$ vs. $Y > 1 \mid Y \geq 1$), abbreviated as the odds ($Y = 1 \mid Y \geq 1$), equal the ratio of the conditional probability of being in the category 1 to the probability of being above this category. Since the probability $P(Y > 1) = P(2) + P(3) + P(4)$, the conditional odds of being in the category 1, given an individual has reached this category, can be expressed in the following form:

$$\text{Odds}(Y = 1|Y \geq 1) = \frac{P(Y = 1)}{P(Y > 1)} = \frac{P(1)}{P(2) + P(3) + P(4)}$$

Similarly, the odds ($Y = 2 \mid Y \geq 2$) equal the ratio of conditional probability of being in the category 2 to the probability of being above this category. Since $P(Y > 2) = P(3) + P(4)$, the conditional odds of being in the category 2, given an individual has reached this category, can be expressed as follows:

$$\text{Odds}(Y = 2|Y \geq 2) = \frac{P(Y = 2)}{P(Y > 2)} = \frac{P(2)}{P(3) + P(4)}$$

The odds ($Y = 3 \mid Y \geq 3$) equal the ratio of conditional probability of being in the category 3 to the probability of being above this category. Using the same method, we get the following equation:

$$\text{Odds}(Y = 3|Y \geq 3) = \frac{P(Y = 3)}{P(Y > 3)} = \frac{P(3)}{P(4)}$$

Table 6.1 presents the logits, odds, and category comparisons for the continuation ratio model for the ordinal response variable with four levels.

Similar to the odds ratios in other logistic regression models, the odds ratio in the CR model is the change in the odds for a one-unit increase from any value of x to the value of $(x + 1)$. The odds ratios in the CR model can be either the odds of being in a category relative to being above that category, or their multiplicative inverse or reciprocal, the odds of being above a category versus in that category. The former is the exponentiated logit coefficient $\exp(\beta)$ in Equation 6.2, and the latter is the exponentiated negative logit coefficient $\exp(-\beta)$.

TABLE 6.1 ● Category Comparisons for the Continuation Ratio Model With Four Levels of Health Status (j = 1, 2, 3, and 4)

Category	Logit $P(Y = j \mid Y \geq j)$	Odds	Probability Comparisons
Level 1	logit $P(Y = 1 \mid Y \geq 1)$	$\frac{P(Y=1)}{P(Y>1)}$	Category 1 vs. Categories 2, 3, and 4
Level 2	logit $P(Y = 2 \mid Y \geq 2)$	$\frac{P(Y=2)}{P(Y>2)}$	Category 2 vs. Categories 3 and 4
Level 3	logit $P(Y = 3 \mid Y \geq 3)$	$\frac{P(Y=3)}{P(Y>3)}$	Category 3 vs. Category 4

6.1.3 Interpretation of Model Parameter Estimates

As with other logistic regression models, the regression coefficient in the CR model (Equation 6.2) is also the logit coefficient. Since the odd ratios in the CR model are different from other models, so is the interpretation of the logit coefficients. A logit coefficient can be interpreted as the change in the predicted logit or the log odds of being in a particular category versus being above that category for a one-unit increase in the predictor variable.

Positive logit coefficients indicate positive relationships between the predictor variables and the logit or log odds of being in a category versus being above that category. A positive coefficient corresponds to an OR greater than 1, so the odds of being in a particular category increase for a one-unit increase in the predictor variable.

A negative logit coefficient corresponds to the exponentiated coefficient, the OR, less than 1. This means that the odds of being in a particular category decrease for a one-unit increase in the predictor variable.

A logit coefficient of 0 corresponds to the OR equal to 1, which indicates that there is no relationship between the predictor and the log odds and, thus, that there is no change in the odds with an increase in the predictor variable.

To estimate the OR of being above a particular category versus being in that category, however, the signs before both the intercepts and logit coefficients in Equation 6.2 need to be reversed. Taking the multiplicative inverse or the reciprocal of the conditional odds of being in a particular category gives us the conditional odds of being above that category.

6.1.4 Research Example and Description of the Data and Sample

Research Problem and Questions: In this chapter, the purpose of the research example is still to investigate the relationships between the ordinal response variable, health status, and the four predictor variables, marital status, the highest education, gender, and working status. Unlike the previous two chapters, however, here the research interest will focus on the odds of being in a particular level of health status versus being above that level using the CR model. The research question is as follows: Do the four predictor variables predict the ordinal response variable, health status? Specifically, do the four predictor variables predict the conditional probabilities of being in a particular level of health status given that higher levels of health status have been achieved?

Description of the Data and Sample: The data for the following analyses are from the General Social Survey 2016 (GSS 2016). For illustration purposes, the same variables from the last chapter are used for data analysis in this chapter.

6.1.5 The `vglm()` Function With `sratio` or `cratio` in the `VGAM` Package

The `vglm()` function in the `VGAM` package can also be used for the analysis of continuation ration models. You need to install the user-written `VGAM` package first by typing `install.packages("VGAM")` before fitting the model. Since the package has been installed, we only need to load the package by typing `library(VGAM)`.

The syntax for continuation ratio models is the same as that for proportional odds models using `vglm()` except that the different `VGAM` family functions are specified. Either the `sratio` or `cratio` family function can be specified, which estimates the stopping ratios of being a particular category, $P(Y = j)/P(Y > j)$, or the continuation ratios of being above a category in the CR model, $P(Y > j)/P(Y = j)$, respectively. For example, the command `vglm(y ~ x, family = sratio(parallel = TRUE), data = data1)` tells R to fit a CR model predicting the ordinal dependent variable y with an independent variable x. The argument `family = sratio()` tells R that the `VGAM` family function is `sratio` and the model is a CR model estimating the stopping ratios or a stopping ratio model. The `parallel = TRUE` argument tells us the predictor variable meets the proportional odds assumption. We can also use the `parallel = FALSE` argument to estimate a CR model with non-proportional odds. When there are multiple predictor variables in the formula, they are connected by plus (+) symbols. For more details on how to use this command, type `help(sratio)` in the command prompt after loading the `VGAM` package.

6.1.6 The CR Model: Multiple-Predictor Model With the `vglm()` Function

The command `crmodel2<-vglm(healthre ~ educ + maritals + female + wrkfull, sratio(parallel = TRUE, reverse = FALSE))` fits the CR model with the `sratio` family using the four predictor variables `educ`, `maritals`, `female`, and `wrkfull`. The fitted model is referred to as the full model. The results of the full model are shown as follows with `summary(crmodel2)`.

```
> # CR model with vglm() in VGAM
> library(VGAM)
> crmodel2 <- vglm(healthre ~ educ + maritals + female + wrkfull, sratio(parallel = TRUE, reverse =
FALSE))
> summary(crmodel2)

Call:
vglm(formula = healthre ~ educ + maritals + female + wrkfull,
    family = sratio(parallel = TRUE, reverse = FALSE))

Pearson residuals:
                              Min          1Q       Median          3Q     Max
logitlink(P[Y=1|Y>=1])    -0.6594  -2.792e-01   -2.406e-01  -0.1967603   6.470
logitlink(P[Y=2|Y>=2])    -1.7925  -6.159e-01   -4.754e-01  -0.0000139   3.337
logitlink(P[Y=3|Y>=3])    -9.1957  -2.453e-05   -2.920e-06   0.7367910   1.223
```

```
Coefficients:

               Estimate  Std. Error   z value   Pr(>|z|)
(Intercept):1  -0.53561    0.19783    -2.707    0.00678 **
(Intercept):2   1.10382    0.19076     5.786   7.20e-09 ***
(Intercept):3   3.19896    0.21070    15.183    < 2e-16 ***
educ           -0.14402    0.01337   -10.772    < 2e-16 ***
maritals       -0.20459    0.07776    -2.631    0.00851 **
female         -0.09285    0.07758    -1.197    0.23141
wrkfull        -0.31860    0.07878    -4.044   5.25e-05 ***

---
Signif. codes: 0 '***' 0.001 '**' 0.01 '*' 0.05 '.' 0.1 ' ' 1

Names of linear predictors: logitlink(P[Y=1|Y>=1]),
logitlink(P[Y=2|Y>=2]), logitlink(P[Y=3|Y>=3])

Residual deviance: 4312.173 on 5612 degrees of freedom

Log-likelihood: -2156.087 on 5612 degrees of freedom

Number of Fisher scoring iterations: 5

No Hauck-Donner effect found in any of the estimates
```

6.1.7 Interpreting R Output

Similar to the output for the PO model, the R output for the CR model also includes the call, the Pearson residuals, the coefficients, the number and names of the three linear predictors, the residual deviance, the log-likelihood value, and the number of iterations. As with the PO model, the coefficients section displays the parameter estimates for the predictor variable `educ` and the three intercepts, their standard errors, the z statistics, and the associated p values.

In the CR model, we estimate the logit or the log odds of being in a particular category relative to being above that category. The link functions for the three linear predictors in the model are `logit(P[Y=1|Y>=1])`, `logit(P[Y=2|Y>=2])`, and `logit (P[Y=3|Y>=3])`. The log odds of being in category 1 given an individual has reached this category, `logit(P[Y=1|Y>=1])`, compares the probability of category 1 with the probabilities of categories 2, 3, and 4; the log odds of being in category 2 given an individual has reached this category, `logit(P[Y=2|Y>=2])`, compares the probability of category 2 with the probabilities of categories 3 and 4; and the log odds of being in category 3 given an individual has reached this category, `logit(P[Y=3|Y>=3])`, compares the probability of category 3 with that of category 4.

The coefficients section (labeled `Coefficients:`) displays the parameter estimates for the four predictor variables and the three intercepts. Of them, the coefficients of the three variables `educ`, `maritals`, and `wrkfull` are significant.

For `educ`, $\beta = -.144$, Wald $z = -10.772$, $p < .001$, which indicates that the logit coefficient of years of education are significant. The logit or log odds of staying in a particular category versus being above that category decreases by a factor of .144 for a one-unit increase in years of education (`educ`) when holding other variables constant.

For `maritals`, $\beta = -.205$, Wald $z = -2.631$, $p < .001$, which is significant. The logit or log odds of staying in a particular category versus being above that category for the married is $-.205$ times the logit for the unmarried when holding other variables constant.

The coefficient of `wrkfull` is also negative, $\beta = -.319$, Wald $z = -4.044$, $p < .001$, which indicates the relationship between the predictor variable, working status, and the logit of the conditional probability of staying in in a particular category is significantly negative. The logit or log odds of staying in a particular category versus being above that category for working full time is $-.319$ times the logit for not working full time when holding other variables constant.

One coefficient is not significant. For `female`, $\beta = -.093$, Wald $z = -1.197$, $p = .231$, which is not significant, indicating there is no significant relationship between being female and the logit of the conditional probability.

We can also extract the coefficients table with `coef(crmodel2, matrix = TRUE)`. The output is omitted here.

6.1.8 Interpreting the Odds Ratio of Stopping in a Particular Category

Odds Ratios of Being in a Category

The odds ratios and the corresponding confidence intervals are reported in the following with `exp(coef(crmodel2, matrix = TRUE))` and `exp(confint(crmodel2, matrix = TRUE))`. The results are combined with the `cbind()` function.

```
> exp(coef(crmodel2, matrix = TRUE))
              logit(P[Y=1|Y>=1])  logit(P[Y=2|Y>=2])  logit(P[Y=3|Y>=3])
(Intercept)            0.5853125           3.0156615          24.5069110
educ                   0.8658700           0.8658700           0.8658700
maritals               0.8149824           0.8149824           0.8149824
female                 0.9113327           0.9113327           0.9113327
wrkfull                0.7271688           0.7271688           0.7271688

> exp(confint(crmodel2, matrix = TRUE))
                    2.5 %      97.5 %
(Intercept):1   0.3971874   0.8625418
(Intercept):2   2.0749303   4.3829009
(Intercept):3  16.2158872  37.0370540
educ            0.8434754   0.8888591
maritals        0.6997790   0.9491516
female          0.7827766   1.0610018
wrkfull         0.6231343   0.8485722
```

```
> cbind(exp(coef(crmodel2)), exp(confint(crmodel2)))
                                   2.5 %      97.5 %
(Intercept):1     0.5853125    0.3971874   0.8625418
(Intercept):2     3.0156615    2.0749303   4.3829009
(Intercept):3    24.5069110   16.2158872  37.0370540
educ              0.8658700    0.8434754   0.8888591
maritals          0.8149824    0.6997790   0.9491516
female            0.9113327    0.7827766   1.0610018
wrkfull           0.7271688    0.6231343   0.8485722
```

Interpretation

For `educ`, OR =.866, indicating that for a one-unit increase in years of education the odds of being in a category compared with being in higher categories decreases by a factor of .866, while holding the other predictors constant.

For `maritals`, OR =.815, which indicates that the odds for the married of being in a category relative to being in better health status are .815 times the odds for the unmarried when holding the other predictors constant in the model.

For `wrkfull`, OR =.727, which indicates that the odds of being in a particular health status level relative to being in better health status for working full time are .727 times the odds for not working full time after controlling for the effects of other predictors in the model.

Finally, the odds ratio for the predictor `female` is .911, $p = .231$, which is not significant, indicating being female does not impact the odds of stopping in a particular category.

6.1.9 Interpreting Odds Ratios of Being Above a Particular Category

Odds Ratios of Being Above a Category

To obtained the inversed odds, the odds of being above a category versus being in that category, we can exponentiate ($-\beta$) with the `exp(-coef(crmodel2, matrix = TRUE))` command. The corresponding confidence intervals can also be computed. The output is displayed below.

```
> exp(-coef(crmodel2, matrix = TRUE))
             logit(P[Y=1|Y>=1])  logit(P[Y=2|Y>=2])  logit(P[Y=3|Y>=3])
(Intercept)            1.708489           0.3316022          0.04080482
educ                   1.154908           1.1549078          1.15490783
maritals               1.227020           1.2270204          1.22702036
female                 1.097294           1.0972941          1.09729405
wrkfull                1.375197           1.3751966          1.37519658
```

```
> exp(-confint(crmodel2, matrix = TRUE))
                    2.5 %       97.5 %
(Intercept):1   2.51770318   1.15936409
(Intercept):2   0.48194390   0.22815939
(Intercept):3   0.06166792   0.02699999
educ            1.18557103   1.12503769
maritals        1.42902262   1.05357252
female          1.27750369   0.94250548
wrkfull         1.60479056   1.17845012

> cbind(exp(-coef(crmodel2)), exp(-confint(crmodel2)))
                                  2.5 %       97.5 %
(Intercept):1   1.70848900   2.51770318   1.15936409
(Intercept):2   0.33160221   0.48194390   0.22815939
(Intercept):3   0.04080482   0.06166792   0.02699999
educ            1.15490783   1.18557103   1.12503769
maritals        1.22702036   1.42902262   1.05357252
female          1.09729405   1.27750369   0.94250548
wrkfull         1.37519658   1.60479056   1.17845012
```

We can also use the `vglm()` function with the `cratio` family to get the same results. The R code is omitted here.

Interpretation

All the four predictor variables have the odds ratios greater than 1, so they are positively associated with the odds of being beyond a particular health status level rather than staying in that category. However, the predictor variable being female (`female`) is not significantly associated with the odds.

For `educ`, OR = 1.155. It indicates that the odds of being above a particular category versus being in that category increase by a factor of 1.155 or 15.5% for a one-unit increase in years of education (`educ`) when holding the other variables constant.

For `maritals`, OR = 1.227, which indicates that the odds for the married of being in a better health status level relative to being in that level are 1.227 times the odds for the unmarried after controlling for the effects of other predictors in the model.

Similarly, for `wrkfull`, OR = 1.375, which indicates that the odds of being in a better health status level relative to being in that level for working full time are 1.375 times the odds for not working full time after controlling for the effects of other predictors in the model.

Finally, the odds ratio for the predictor `female` is 1.097, p = .231, which is not significant, indicating being female does not impact the odds of being in a higher category.

6.1.10 Model Fit Statistics

Testing the Overall Model Using the Likelihood Ratio Test

To test if the overall full model is significant, we fit a null model with the intercept only and compare the multiple-predictor model with the null model using the `lrtest()` function. The command `crmodel0 <- vglm(healthre ~ 1, sratio(parallel = TRUE, reverse = FALSE))` is used to fit the null model. The output is displayed below by `summary(crmodel0)`.

```
> # Testing the overall model using the likelihood ratio test
> crmodel0 <- vglm(healthre ~ 1, sratio(parallel = TRUE, reverse = FALSE))
> summary(crmodel0)

Call:
vglm(formula = healthre ~ 1, family = sratio(parallel = TRUE,
    reverse = FALSE))

Pearson residuals:
                              Min          1Q      Median          3Q    Max
logitlink(P[Y=1|Y>=1])    -0.2593  -2.593e-01  -2.593e-01  -2.593e-01  3.8565
logitlink(P[Y=2|Y>=2])    -0.5858  -5.858e-01  -5.858e-01   5.990e-09  1.8219
logitlink(P[Y=3|Y>=3])    -1.7646  -2.121e-08  -2.121e-08   7.993e-01  0.7993

Coefficients:
               Estimate   Std. Error   z value   Pr(>|z|)
(Intercept):1  -2.69954      0.09510    -28.39    <2e-16 ***
(Intercept):2  -1.13465      0.05563    -20.39    <2e-16 ***
(Intercept):3   0.79196      0.05924     13.37    <2e-16 ***

---
Signif. codes: 0 '***' 0.001 '**' 0.01 '*' 0.05 '.' 0.1 ' ' 1

Names of linear predictors: logitlink(P[Y=1|Y>=1]),
logitlink(P[Y=2|Y>=2]), logitlink(P[Y=3|Y>=3])

Residual deviance: 4476.435 on 5616 degrees of freedom

Log-likelihood: -2238.218 on 5616 degrees of freedom

Number of Fisher scoring iterations: 5
```

We use the `lrtest(crmodel0, crmodel2)` command to compare the log-likelihood statistics of the fitted model `crmodel2` and the null model `crmodel0`. The output is displayed below.

```
> lrtest(crmodel0, crmodel2)
Likelihood ratio test

Model 1: healthre ~ 1
Model 2: healthre ~ educ + maritals + female + wrkfull
     #Df    LogLik    Df     Chisq       Pr(>Chisq)
1    5616   -2238.2
2    5612   -2156.1   -4    164.26     < 2.2e-16 ***

---
Signif. codes: 0 '***' 0.001 '**' 0.01 '*' 0.05 '.' 0.1 ' ' 1
```

The null hypothesis of the test for the overall model is that the predictor variable does not contribute to the model, and the alternative hypothesis is that the multiple-predictor model is better than the null model with no independent variables. The likelihood ratio test $LR\,\chi^2_{(4)} = 164.26$, $p < .001$, indicates that the full model with the four predictor variables provides a better fit than the null model with no independent variables.

Pseudo R^2

The `nagelkerke(crmodel2)` command in the `rcompanion` package (Mangiafico, 2021) produces the three types of pseudo R^2 statistics and the likelihood ratio test statistic for the multiple-predictor model. McFadden's R^2 is .037, Cox and Snell's R^2 is .084, and Nagelkerke's R^2 is .092. The output is omitted here.

AIC and BIC Statistics

The AIC and BIC statistics of the fitted model can be obtained using `AIC(crmodel2)` and `BIC(crmodel2)`, respectively.

```
> AIC(crmodel2)
[1] 4326.173
> BIC(crmodel2)
[1] 4364.92
```

6.1.11 Computing the Predicted Probabilities With the `ggpredict()` Function in the `ggeffects` Package

We use the `ggpredict()` function in the `ggeffects` package (Lüdecke, 2018b) to compute the predicted probabilities of being in a particular category of the ordinal response variable at specified values of predictor variables. The command `pr.cr2m <- ggpredict(crmodel2, terms = "educ[12, 14, 16]", ci = NA)` tells R to compute the estimated probabilities for each category of the ordinal response variable using the `ggpredict()` function. In the function, `crmodel2` is the estimated model; the `terms = "educ[12, 14, 16]"` option specifies the predictor variable `educ` at the values of 12, 14, and 16 when holding the other predictor variables at their means; and the `ci = NA` option specifies no confidence intervals. The `terms` option can specify up to four variables, including the second to fourth grouping variables. Please also note that the confidence intervals can only be obtained for the cumulative probabilities, so the `ci = NA` option is needed there. The output is assigned to the object named `pr.cr2m`.

```
> library(ggeffects)
> pr.cr2m <- ggpredict(crmodel2, terms = "educ[12, 14, 16]", ci = NA)
> pr.cr2m

# Predicted probabilities of healthre

# Response Level = 1

educ    |   Predicted
---------------------
12      |        0.07
14      |        0.06
16      |        0.04


# Response Level = 2

educ    |   Predicted
---------------------
12      |        0.27
14      |        0.22
16      |        0.18


# Response Level = 3

educ    |   Predicted
---------------------
12      |        0.51
14      |        0.52
16      |        0.51


# Response Level = 4

educ    |   Predicted
---------------------
12      |        0.16
14      |        0.21
16      |        0.28
```

```
Adjusted for:
* maritals = 0.44
*   female = 0.56
*  wrkfull = 0.47

> plot(pr.cr2m)
```

When educ equals 12, 14, and 16, and other predictor variables are held at their means, the predicted probabilities of being in each category (i.e., $Y = 1, 2, 3$, and 4) are shown in the output. The last section under the title "Adjusted for" lists the means of the other three variables.

The predicted probabilities for all the four response levels are plotted with plot(pr.cr2m). Figure 6.1 shows the predicted probabilities of being in each category (i.e., $Y = 1, 2, 3$, and 4) for educ at 12, 14, and 16.

The graph shows that with the increase of the years of education, the probabilities of being in poor and fair health condition (categories 1 and 2) decrease. In other words, people with higher levels of education are less likely associated with poor and fair health

FIGURE 6.1 ● Predicted Probabilities of Being in Categories 1, 2, 3, and 4 for educ

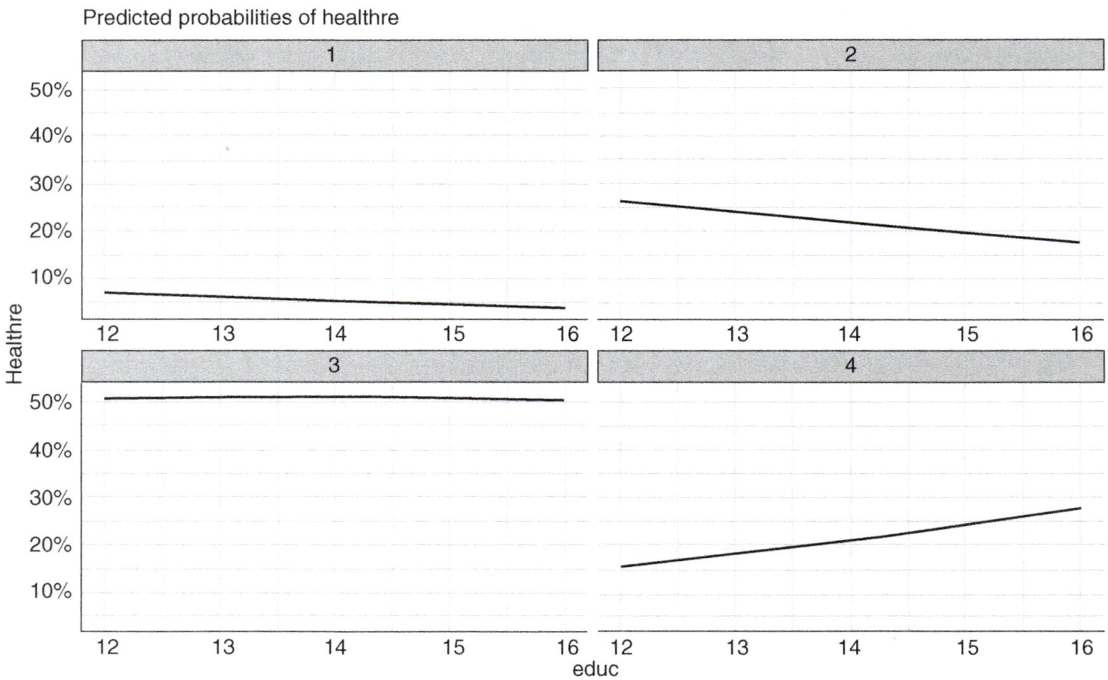

conditions. In addition, with the increase of the years of education, the probabilities of being in excellent health condition (category 4) increase. In other words, people with a higher level of education are more likely to be in excellent health condition.

With the `terms = c("educ[12, 14, 16]", "maritals")` argument, the predicted probabilities for `educ` at 12, 14, and 16 can be grouped by `maritals`. We name the object `pr.cr2`, and the output is omitted here.

```
> pr.cr2 <- ggpredict(crmodel2, terms = c("educ[12, 14, 16]", "maritals"), ci = NA)
> pr.cr2

# The output is omitted.

> plot(pr.cr2)
```

The results are plotted by using `plot(pr.cr2)`. The graph is also omitted here.

6.1.12 The CR Model With Non-Proportional Odds

We can fit a CR model with non-proportional odds by relaxing the proportionality assumption to allow the logits coefficients of each predictor variable to vary across the ordinal response variable. The following is an example using the `vglm()` function with the `parallel = FALSE` option. The ordinal response variable and the four predictor variables `educ`, `maritals`, `female`, and `wrkfull` are the same as those in the CR model introduced in the last section.

The command `crmodel3 <- vglm(healthre ~ educ + maritals + female + wrkfull, sratio(parallel = FALSE, reverse = FALSE))` fits the CR model with non-proportional odds. The fitted model is named `crmodel3`. The `summary(crmodel3)` command displays the results of the model.

```
> # CR model with non-proportional odds
> crmodel3 <- vglm(healthre ~ educ + maritals + female + wrkfull, sratio(parallel = FALSE, reverse =
FALSE))
> summary(crmodel3)

Call:
vglm(formula = healthre ~ educ + maritals + female + wrkfull,
    family = sratio(parallel = FALSE, reverse = FALSE))

Pearson residuals:
                              Min         1Q      Median          3Q    Max
logitlink(P[Y=1|Y>=1])     -0.865  -2.858e-01  -2.137e-01  -1.585e-01  9.021
logitlink(P[Y=2|Y>=2])     -2.026  -6.152e-01  -4.496e-01   1.601e-10  3.808
logitlink(P[Y=3|Y>=3])     -8.140  -1.754e-07   2.641e-07   7.735e-01  1.109
```

```
Coefficients:
                Estimate   Std. Error   z value   Pr(>|z|)
(Intercept):1    0.17635      0.41922     0.421   0.673996
(Intercept):2    1.33051      0.28428     4.680   2.86e-06 ***
(Intercept):3    2.48505      0.32643     7.613   2.68e-14 ***
educ:1          -0.17876      0.03178    -5.626   1.85e-08 ***
educ:2          -0.14581      0.02029    -7.186   6.66e-13 ***
educ:3          -0.12364      0.02161    -5.722   1.05e-08 ***
maritals:1      -0.59761      0.21474    -2.783   0.005387 **
maritals:2      -0.30490      0.11778    -2.589   0.009631 **
maritals:3       0.03576      0.12115     0.295   0.767857
female:1        -0.10896      0.19766    -0.551   0.581460
female:2        -0.18578      0.11669    -1.592   0.111347
female:3         0.01124      0.12227     0.092   0.926774
wrkfull:1       -0.76014      0.22160    -3.430   0.000603 ***
wrkfull:2       -0.59004      0.11974    -4.928   8.33e-07 ***
wrkfull:3        0.12821      0.12226     1.049   0.294308

---
Signif. codes: 0 '***' 0.001 '**' 0.01 '*' 0.05 '.' 0.1 ' ' 1

Names of linear predictors: logitlink(P[Y=1|Y>=1]),
logitlink(P[Y=2|Y>=2]), logitlink(P[Y=3|Y>=3])

Residual deviance: 4274.367 on 5604 degrees of freedom

Log-likelihood: -2137.183 on 5604 degrees of freedom

Number of Fisher scoring iterations: 6

No Hauck-Donner effect found in any of the estimates
```

Interpreting R Output

The link functions for the three linear predictors in the model are `logit(P[Y=1|Y>=1])`, `logit(P[Y=2|Y>=2])`, and `logit(P[Y=3|Y>=3])`. The first binary model estimates the logit or log odds of being in category 1 versus categories 2, 3, and 4; the second model estimates the logit of being in category 2 versus categories 3 and 4; and the third model estimates the logit of being in category 3 versus category 4. The estimated intercepts and logit coefficients for these three sub-models are numbered 1, 2, and 3.

The logit coefficient table here looks similar to that of the generalized ordinal logistic regression model introduced in Chapter 4 since the logit coefficients of all predictor variables are allowed to vary freely across the underlying binary models. For example, the estimated logit coefficients for `educ` are −.179, −.146, and −.124 for each binary model. Therefore, this model is also called the unconstrained continuation ratio model. If an ordinal response variable has J levels in an unconstrained continuation ratio model, we need to estimate $J - 1$ logit coefficients for each predictor variable. A more parsimonious model is the constrained continuation ratio model where the logit coefficients for each predictor variable are constrained to be equal across the underlying binary comparisons. The CR model introduced in the preceding sections is the constrained model. Similar to the PO model, the logit coefficients for each predictor variables are assumed to be equal in the constrained CR model.

We can also extract the coefficients table with `coef(crmodel3, matrix = TRUE)`. The output is omitted here.

Odds Ratios of Being in a Category

The odds ratios and the corresponding confidence intervals are reported in the following with the `exp(coef(crmodel3, matrix = TRUE))` and the `exp(confint(crmodel3, matrix = TRUE))` commands. The results are combined with the `cbind()` function.

```
> exp(coef(crmodel3, matrix = TRUE))
             logit(P[Y=1|Y>=1])  logit(P[Y=2|Y>=2])  logit(P[Y=3|Y>=3])
(Intercept)           1.1928610           3.7829870          12.0017485
educ                  0.8363025           0.8643232           0.8836989
maritals              0.5501272           0.7372006           1.0364081
female                0.8967646           0.8304521           1.0113005
wrkfull               0.4676031           0.5543058           1.1367952

> exp(confint(crmodel3, matrix = TRUE))
                   2.5 %       97.5 %
(Intercept):1   0.5245010    2.7128972
(Intercept):2   2.1669791    6.6041201
(Intercept):3   6.3297690   22.7562752
educ:1          0.7858062    0.8900437
educ:2          0.8306256    0.8993878
educ:3          0.8470573    0.9219256
maritals:1      0.3611405    0.8380115
maritals:2      0.5852418    0.9286156
maritals:3      0.8173491    1.3141776
female:1        0.6087352    1.3210782
female:2        0.6606791    1.0438513
female:3        0.7957990    1.2851596
wrkfull:1       0.3028668    0.7219434
wrkfull:2       0.4383534    0.7009297
wrkfull:3       0.8945752    1.4446000

> cbind(exp(coef(crmodel3)), exp(confint(crmodel3)))
                                 2.5 %       97.5 %
(Intercept):1    1.1928610   0.5245010    2.7128972
(Intercept):2    3.7829870   2.1669791    6.6041201
(Intercept):3   12.0017485   6.3297690   22.7562752
educ:1           0.8363025   0.7858062    0.8900437
educ:2           0.8643232   0.8306256    0.8993878
educ:3           0.8836989   0.8470573    0.9219256
maritals:1       0.5501272   0.3611405    0.8380115
maritals:2       0.7372006   0.5852418    0.9286156
```

```
maritals:3       1.0364081    0.8173491    1.3141776
female:1         0.8967646    0.6087352    1.3210782
female:2         0.8304521    0.6606791    1.0438513
female:3         1.0113005    0.7957990    1.2851596
wrkfull:1        0.4676031    0.3028668    0.7219434
wrkfull:2        0.5543058    0.4383534    0.7009297
wrkfull:3        1.1367952    0.8945752    1.4446000
```

Odds Ratios of Being Above a Category

To obtained the inversed odds, the odds of being above a category versus being in that category, we can exponentiate $(-\beta)$ with the `exp(-coef(crmodel3, matrix = TRUE))` command. The corresponding confidence intervals can also be computed. The output is omitted here.

AIC and BIC Statistic

The AIC and BIC statistics of the fitted model can be obtained using `AIC(crmodel3)` and `BIC(crmodel3)`, respectively.

```
> AIC(crmodel3)
[1] 4304.367
> BIC(crmodel3)
[1] 4387.396
```

Model Comparison Using the Likelihood Ratio Test

The likelihood ratio test is used to compare the multiple-predictor CR model and the CR model with non-proportional odds with `lrtest(crmodel2, crmodel3)`.

```
> # Model comparison with the likelihood ratio test
> lrtest(crmodel2, crmodel3)
Likelihood ratio test

Model 1: healthre ~ educ + maritals + female + wrkfull
Model 2: healthre ~ educ + maritals + female + wrkfull
     #Df    LogLik    Df    Chisq     Pr(>Chisq)
1    5612   -2156.1
2    5604   -2137.2   -8    37.806    8.174e-06 ***

---
Signif. codes: 0 '***' 0.001 '**' 0.01 '*' 0.05 '.' 0.1 ' ' 1
```

The likelihood ratio chi-square test $\chi^2_{(8)} = 37.806, p < .001$. This result indicates that the CR model with non-proportional odds has a better fit than the multiple-predictor CR model with proportional odds.

6.2 ADJACENT CATEGORIES MODELS

6.2.1 Adjacent Categories Models: An Introduction

The adjacent categories model compares the pairs of adjacent categories for ordinal response variables. Specifically, the adjacent categories model (Agresti, 2010; Clogg & Shihadeh, 1994; Fullerton & Xu, 2016; Goodman, 1983; Hosmer et al., 2013; Liu, 2016; O'Connell, 2006) estimates the odds of being in a category $j + 1$ versus being in the next lower category j of an ordinal response variable, or the odds comparing a category j with the next higher category $j + 1$. If an ordinal response variable has J levels, then there are $J - 1$ pairs of adjacent categories, which is the number of comparisons between these pairs of adjacent categories.

The adjacent categories model can be written as:

$$\begin{aligned} \operatorname{logit}\left[P\left(Y = j+1 \middle| x_1, x_2, \ldots, x_p\right)\right] &= \ln\left(\frac{P\left(Y = j+1 \middle| x_1, x_2, \ldots, x_p\right)}{P\left(Y = j \middle| x_1, x_2, \ldots, x_p\right)}\right) \\ &= \alpha_j + \beta_1 X_1 + \beta_2 X_2 + \ldots + \beta_p X_p \end{aligned} \tag{6.4}$$

where j is the number of the categories of the ordinal response variable from 1 to $J - 1$. α_j are the intercepts, and $\beta_1, \beta_2, \ldots, \beta_j$ are the logit coefficients for each predictor variable. The adjacent categories model is a special case of the multinomial logistic regression with constrained regression coefficients.

6.2.2 Odds and Odds Ratios in AC Models

Let us now look at the difference in the odds among the PO model, the CR model, and the AC model.

In PO models, we estimate the odds of being at or below a particular category ($Y \leq j$) versus being above that category ($Y > j$). The odds of being at or below a category in ordinal logistic regression are the probability of being at or below a category divided by the probability of being above that category:

$$\text{Odds}(Y \leq j) = \frac{P(Y \leq j)}{P(Y > j)} \tag{6.5}$$

The CR model estimates the conditional probabilities of being in category j, conditional on being at or above that category; that is, $P(Y = j \mid Y \geq j)$. The odds in the CR model are the ratio of the probability of being in a category to the probability of being above that category:

$$\text{Odds}(Y = j|Y \geq j) = \frac{P(Y = j)}{P(Y > j)} \tag{6.6}$$

In comparison, in adjacent categories models, we estimate the odds of adjacent categories, which are the ratio of the probabilities of one category to the next lower category. The odds can be expressed as:

$$\text{Odds}(Y = j+1 \text{ vs. } Y = j) = \frac{P(Y = j+1)}{P(Y = j)} \tag{6.7}$$

where j can be any categories from 0 to J categories.

Or:

$$\text{Odds}(Y = j \text{ vs. } Y = j-1) = \frac{P(Y = j)}{P(Y = j-1)} \tag{6.8}$$

where j is greater than 0.

For example, an outcome variable, health status, has four ordinal levels from 1 to 4, with 1 = poor, 2 = fair, 3 = good, and 4 = excellent. We estimate the following three odds: the odds of being in the category 2 versus 1, the odds of being in the category 3 versus 2, and the odds of being in the category 4 versus 3.

Odds ($Y = 2$ vs. $Y = 1$) equal the ratio of probability of being in category 2 to the probability of being in category 1:

$$\text{Odds}(Y = 2 \text{ vs. } 1) = \frac{P(Y = 2)}{P(Y = 1)} = \frac{P(2)}{P(1)}$$

Odds ($Y = 3$ vs. $Y = 2$) equal the ratio of probability of being in category 3 to the probability of being in category 2. It can be expressed as follows:

$$\text{Odds}(Y = 3 \text{ vs. } 2) = \frac{P(Y = 3)}{P(Y = 2)} = \frac{P(3)}{P(2)}$$

Odds ($Y = 4$ vs. $Y = 3$) equal the probability of being in category 4 divided by the probability of being in category 3. It can be expressed as follows:

$$\text{Odds}(Y = 4 \text{ vs. } 3) = \frac{P(Y = 4)}{P(Y = 3)} = \frac{P(4)}{P(3)}$$

In sum, the odds in the AC model compare two adjacent probabilities; that is, $P(Y = j + 1)$ and $P(Y = j)$. Table 6.2 presents the logits, odds, and category comparisons for the adjacent categories model for the ordinal response variable with four levels.

TABLE 6.2 ● Category Comparisons for the Adjacent Categories Model With Four Levels of Health Status (j = 1, 2, 3, 4)

Category	Logit $P(Y = j + 1)$	Odds	Probability Comparisons
Level 2	logit $P(Y = 2)$	$\frac{P(Y=2)}{P(Y=1)}$	Category 2 vs. category 1
Level 3	logit $P(Y = 3)$	$\frac{P(Y=3)}{P(Y=2)}$	Category 3 vs. category 2
Level 4	logit $P(Y = 4)$	$\frac{P(Y=4)}{P(Y=3)}$	Category 4 vs. category 3

Odds Ratios in Adjacent Categories Logistic Regression

Just like the odds ratios in the PO model and the CR model, the odds ratios in the adjacent categories model are also the exponentiated logistic coefficients, exp(β). Since the odds in the AC model are the probability comparisons between the higher category and the lower category for two adjacent categories, the odds ratio is the ratio of the two odds when the value of a predictor is (x + 1) relative to the odds when the predictor has a value of x. It is the change in the odds of being in a higher category relative to the next lower category for a one-unit increase in a predictor variable when holding others constant.

6.2.3 Interpretation of Model Parameter Estimates

The estimated regression coefficients in the AC model are the logit coefficients, which are the log odds of being in a higher category relative to the next lower category for any pairs of adjacent categories. It is the change in the predicted logit or the log odds for a one-unit increase in a predictor variable when holding others constant. Exponentiating the logit coefficients gives us the odds ratios.

What about the odds of being in a lower category for any pair of the adjacent categories? They are just the reciprocal of the odds of being in a higher category for a pair of the adjacent categories. Whereas the latter compares a higher category j + 1 with the next lower category j, the odds of being in a lower category compares the category j with the next higher category j + 1.

To get the OR of being in a lower category, we need to exponentiate the logit coefficient with a negative sign before that, that is, exp($-\beta$), or take the reciprocal of the odds of being in a higher category for a pair of the adjacent categories.

6.2.4 Research Example and Description of the Data and Sample

The research interest in this section focuses on the odds comparing adjacent categories by using the AC model. The research question is as follows: Do the four predictor variables predict the odds comparing adjacent levels of health status?

The data for the following analyses were taken from the General Social Survey 2016 (GSS 2016). The same variables in Section I are used for data analysis in this section.

The ordinal response variable is `healthre`, the recoded variable of health (health status) with four ordinal categories; the four predictor variables are `educ`, `maritals`, `female`, and `wrkfull`.

6.2.5 The `vglm()` Function With the `acat` Family in the `VGAM` Package

The `vglm()` function in the `VGAM` package can also be used for the analysis of adjacent categories (AC) models. The syntax for AC models is the same as that for proportional odds models and continuation ratio models using `vglm()` except that the different `family` arguments are specified. The `acat` family needs to be specified for the AC model. For example, the command `vglm(y ~ x, family = acat(parallel = TRUE), data = data1)` tells R to fit a AC model predicting the ordinal dependent variable *y* with an independent variable *x*. The argument `family = acat()` tells R that a AC model is fitted. The `parallel = TRUE` argument tells us the predictor variable meets the proportional odds assumption. When there is more than one predictor variable in the formula, they are connected by plus (+) symbols. For more details on how to use this command, type `help(acat)` in the command prompt after loading the `VGAM` package.

6.2.6 The AC Model: Multiple-Predictor Model

The multiple-predictor model is fitted after including the four predictor variables `educ`, `maritals`, `female`, and `wrkfull` in the model equation. The syntax for the model is `acmodel2 <- vglm(healthre ~ educ + maritals + female + wrkfull, acat(parallel = TRUE, reverse = FALSE))`. The `acat` family with the proportional odds option `parallel = TRUE` is also used. We specify the `reverse = FALSE` option since the ordinal response variable is not reversed. The results of the model are shown as follows with `summary(acmodel2)`.

```
> # Section II
> # AC model with vglm() in VGAM
> acmodel2 <- vglm(healthre ~ educ + maritals + female + wrkfull, acat(parallel =
TRUE, reverse = FALSE))
> summary(acmodel2)


Call:
vglm(formula = healthre ~ educ + maritals + female + wrkfull,
     family = acat(parallel = TRUE, reverse = FALSE))


Pearson residuals:
                            Min        1Q      Median       3Q      Max
loglink(P[Y=2]/P[Y=1])   -8.929   0.04851    0.08429   0.1679   0.8604
loglink(P[Y=3]/P[Y=2])   -3.384  -0.90710    0.49112   0.7162   1.8580
loglink(P[Y=4]/P[Y=3])   -1.015  -0.65718   -0.43109  -0.1139   7.1919
```

```
Coefficients:
                Estimate   Std. Error   z value    Pr(>|z|)
(Intercept):1   -0.27542    0.16180      -1.702    0.08872 .
(Intercept):2   -0.94752    0.15520      -6.105   1.03e-09 ***
(Intercept):3   -2.64539    0.17399     -15.204    < 2e-16 ***
educ             0.10889    0.01052      10.348    < 2e-16 ***
maritals         0.18676    0.06019       3.103    0.00192 **
female           0.07770    0.05971       1.301    0.19315
wrkfull          0.28940    0.06118       4.730   2.24e-06 ***

---
Signif. codes: 0 '***' 0.001 '**' 0.01 '*' 0.05 '.' 0.1 ' ' 1

Names of linear predictors: loglink(P[Y=2]/P[Y=1]),
loglink(P[Y=3]/P[Y=2]), loglink(P[Y=4]/P[Y=3])

Residual deviance: 4298.71 on 5612 degrees of freedom

Log-likelihood: -2149.355 on 5612 degrees of freedom

Number of Fisher scoring iterations: 5

No Hauck-Donner effect found in any of the estimates
```

6.2.7 Interpreting R Output

The R output for the AC model also includes the call, Pearson residuals, coefficients, names of the three linear predictors, residual deviance, log-likelihood value, and the number of iterations.

In the AC model, we estimate the logit or the log odds of being in a particular category relative to the next lower category. The link functions for the three linear predictors in the model are `loglink(P[Y=2]|P[Y=1])`, `loglink(P[Y=3]|P[Y=2])`, and `loglink(P[Y=4]|P[Y=3])`. The log odds of being in category 2 versus category 1, `loglink(P[Y=2]|P[Y=1])`, compares the probabilities of being in categories 2 and 1; `loglink(P[Y=3]|P[Y=2])` compares the probabilities of being in categories 3 and 2; and `loglink(P[Y=4]|P[Y=3])` compares the probabilities of being in categories 4 and 3.

The coefficients section (labeled `Coefficients:`) displays the parameter estimates for the three intercepts and four predictor variables. The coefficients of the three variables `educ`, `maritals`, and `wrkfull` are significantly different from zero.

For `educ`, $\beta = .109$, Wald $z = 10.348$, $p < .001$, which indicates that the logit coefficient of years of education are significant. The logit or log odds of being in a particular category versus being in the next lower category increases by a factor of .109

for a one-unit increase in years of education (educ) when holding the other variables constant.

For maritals, $\beta = .187$, Wald $z = 3.103$, $p < .001$, which is significant. The logit or log odds of being in a particular category versus being in the next lower category for the married is .187 times the logit for the unmarried when holding other variables constant.

The coefficient of wrkfull is also positive, $\beta = .289$, Wald $z = 4.730$, $p < .001$. The logit or log odds of being in a particular category versus being in the next lower category for working full time is .289 times the logit for not working full time when holding the other variables constant.

The coefficient for female is not significant. $\beta = .078$, Wald $z = 1.301$, $p = .193$, indicating there is no significant relationship between being female and the logit comparing pairs adjacent categories.

The coef(crmodel2, matrix = TRUE) command can be used to extract the coefficients table for the three underlying binary models. The output is omitted here.

We request the odds ratios of being in a higher category versus the next lower category for a pair of adjacent categories and the corresponding confidence intervals using exp(coef(acmodel2, matrix = TRUE)) and exp(confint(acmodel2, matrix = TRUE)), respectively. The results are combined with the cbind() function.

```
> exp(coef(acmodel2, matrix = TRUE))
             loglink(P[Y=2]/P[Y=1])  loglink(P[Y=3]/P[Y=2])
(Intercept)               0.7592517               0.3877015
educ                      1.1150417               1.1150417
maritals                  1.2053402               1.2053402
female                    1.0807936               1.0807936
wrkfull                   1.3356240               1.3356240

             loglink(P[Y=4]/P[Y=3])
(Intercept)              0.07097765
educ                     1.11504172
maritals                 1.20534020
female                   1.08079364
wrkfull                  1.33562400

> exp(confint(acmodel2, matrix = TRUE))
                    2.5 %      97.5 %
(Intercept):1  0.55291503  1.04258915
(Intercept):2  0.28601549  0.52553958
(Intercept):3  0.05046859  0.09982104
educ           1.09227964  1.13827814
maritals       1.07122018  1.35625247
female         0.96143776  1.21496674
wrkfull        1.18469223  1.50578474
```

```
> cbind(exp(coef(acmodel2)), exp(confint(acmodel2)))
                                     2.5 %      97.5 %
(Intercept):1    0.75925174    0.55291503    1.04258915
(Intercept):2    0.38770151    0.28601549    0.52553958
(Intercept):3    0.07097765    0.05046859    0.09982104
educ             1.11504172    1.09227964    1.13827814
maritals         1.20534020    1.07122018    1.35625247
female           1.08079364    0.96143776    1.21496674
wrkfull          1.33562400    1.18469223    1.50578474
```

6.2.8 Interpreting the Odds Ratios of Being in a Higher Category (j + 1) Versus the Next Lower Category j

The odds ratios can be interpreted just like the other logistic regression models except that the comparisons between categories are different. A positive logit regression coefficient corresponds to an odds ratio being a higher category versus the next lower category greater than 1, whereas a negative logit coefficient corresponds to an odds ratio less than 1.

For the `educ` predictor variable, β = .109, which is positive; OR = 1.115, which is greater than 1. This indicates that the odds of being in a higher category of health status versus the next lower category increase by a factor of 1.115 for a one-unit increase in years of education when holding the other predictors constant. In other words, for a one-unit increase in education, the odds of being healthier increase by 11.5%.

For the `maritals` predictor variable, β = .187, which is positive; OR = 1.205, which is greater than 1. This indicates that the odds of being in a higher category of health status versus the next lower category for the married are 1.205 times the odds for the unmarried when holding the other predictors constant.

For the `wrkfull` predictor variable, β = .289, which is positive; OR = 1.336, which is greater than 1. This indicates that the odds of being in a higher category of health status versus the next lower category for working full time are 1.336 times the odds for not working full time with all the other predictors remaining constant.

For the `female` predictor variable, β = .078, p = .193, which is not significantly different from 0; OR = 1.081, which almost equals 1. This indicates that there is no relationship between being a female and the odds of being in a higher category of health status versus the next lower category. In other words, there is no significant difference between the female and the male in the odds of being in better health status versus the next lower category (i.e., categories 2 vs. 1, categories 3 vs. 2, and categories 4 vs. 3).

Odds Ratios of Being in a Category Versus Being in the Next Higher Category

To obtained the inversed odds, the odds of being in a category versus being in the next higher category, we can exponentiate ($-\beta$) with `exp(-coef(acmodel2,` matrix = TRUE)). The output is displayed below. The same results can be obtained if we use the reverse = TRUE option in the model command.

```
> exp(-coef(acmodel2, matrix = TRUE))
              loglink(P[Y=2]/P[Y=1])    loglink(P[Y=3]/P[Y=2])
(Intercept)                1.3170862                 2.5793039
educ                       0.8968274                 0.8968274
maritals                   0.8296413                 0.8296413
female                     0.9252460                 0.9252460
wrkfull                    0.7487137                 0.7487137

              loglink(P[Y=4]/P[Y=3])
(Intercept)               14.0889413
educ                       0.8968274
maritals                   0.8296413
female                     0.9252460
wrkfull                    0.7487137

> exp(-confint(acmodel2, matrix = TRUE))
                    2.5 %       97.5 %
(Intercept):1   1.8085962    0.9591506
(Intercept):2   3.4963142    1.9028063
(Intercept):3  19.8143031   10.0179283
educ            0.9155165    0.8785199
maritals        0.9335149    0.7373258
female          1.0401089    0.8230678
wrkfull         0.8441011    0.6641055

> cbind(exp(-coef(acmodel2)), exp(-confint(acmodel2)))
                                2.5 %       97.5 %
(Intercept):1   1.3170862   1.8085962    0.9591506
(Intercept):2   2.5793039   3.4963142    1.9028063
(Intercept):3  14.0889413  19.8143031   10.0179283
educ            0.8968274   0.9155165    0.8785199
maritals        0.8296413   0.9335149    0.7373258
female          0.9252460   1.0401089    0.8230678
wrkfull         0.7487137   0.8441011    0.6641055
```

6.2.9 Interpreting the Odds Ratios of Being in a Lower Category for the AC Model

Just as the odds in other ordinal regression models have directions, they also have directions in the AC model. In other words, we can reverse the order of categories that the odds compare. In AC models, we can either estimate the odds of being a higher category versus the next lower category or the odds of being in a lower category versus the next higher category since they are reciprocal of each other.

For the `educ` predictor variable, OR = .899, which is less than 1. This indicates that the odds of being in a lower category of the pair decrease by 10.1% for a one-unit increase in education when the effects of other predictors are held constant in the model.

For the `maritals` predictor variable, OR = .830, which is less than 1. This indicates that the odds of being in a lower category out of the adjacent categories of health status for the married are .830 times the odds for the unmarried when holding other predictors constant.

For the `wrkfull` predictor variable, OR =.749, which is less than 1. This indicates that the odds of being in a lower category of health status for working full time are .749 times the odds for not working full time with other predictors remaining constant.

For the `female` predictor variable, OR = .925 (p =.193), which is close to 1. This indicates that there is no relationship between being a female and the odds of being in a lower category of health status.

6.2.10 Model Fit Statistics

Testing the Overall Model Using the Likelihood Ratio Test

To test if the overall full model is significant, we fit a null model with the intercept only and compare the multiple-predictor model with the null model using the `lrtest()` function. The command `acmodel0 <- vglm(healthre ~ 1, acat(parallel = TRUE, reverse = FALSE))` is used to fit the null model. The output is omitted here.

```
> # Testing the overall model using the likelihood ratio test
> acmodel0 <- vglm(healthre ~ 1, acat(parallel = TRUE, reverse = FALSE))
> summary(acmodel0)
# The output is omitted.
```

We use the `lrtest(acmodel0, acmodel2)` command to compare the log-likelihood statistics of the fitted model `acmodel2` and the null model `acmodel0`. The output is displayed below.

```
> lrtest(acmodel0, acmodel2)
Likelihood ratio test

Model 1: healthre ~ 1
Model 2: healthre ~ educ + maritals + female + wrkfull
     #Df    LogLik   Df     Chisq   Pr(>Chisq)
1   5616   -2238.2
2   5612   -2149.3   -4    177.73    < 2.2e-16 ***

---
Signif. codes: 0 '***' 0.001 '**' 0.01 '*' 0.05 '.' 0.1 ' ' 1
```

The likelihood ratio test $LR\,\chi^2_{(4)} = 177.73, p < .001$, indicates that the full model with the four predictor variables provides a better fit than the null model with no independent variables.

Pseudo R^2

The `nagelkerke(acmodel2)` command can be used to produce the three types of pseudo R^2 statistics and the likelihood ratio test statistic for the multiple-predictor model. The output is omitted here.

AIC and BIC Statistics

The `AIC(acmodel2)` and `BIC(acmodel2)` commands compute the AIC and BIC statistics, respectively.

```
> AIC(acmodel2)
[1] 4312.71
> BIC(acmodel2)
[1] 4351.457
```

6.3 STEREOTYPE LOGISTIC REGRESSION MODELS

6.3.1 Stereotype Logistic Regression Models: An Introduction

The stereotype logistic (SL) regression model is an alternative model for ordinal response variables when the parallel lines or the proportional odds (PO) assumption does not hold. It was first developed by Anderson (1984) and later introduced by Greenland (1994), Hilbe (2009), Kuss (2006), Liu (2014, 2016a), Long and Freese (2014), and Menard (2010). As discussed in Chapter 4, a proportional odds model assumes that the underlying binary models have the same coefficients. When the PO assumption is violated, one option is to fit the partial proportional odds (PPO) model or the generalized ordinal logit model (Fu, 1998; Fullerton & Xu, 2016; Liu, 2016a; Liu & Koirala, 2012; Peterson & Harrell, 1990; Williams, 2006), which has been discussed in Chapter 5. Another alternative is the SL model, which is the focus of this section.

The Stereotype Logistic Model

Anderson's SL model (1984) can be written in the following form:

$$\begin{aligned}\text{logit}[\pi(j, J)] &= \ln\left(\frac{P(Y = j|x_1, x_2, \ldots, x_p)}{P(Y = J|x_1, x_2, \ldots, x_p)}\right) \\ &= \alpha_j + \phi_j\left(\beta_1 X_1 + \beta_2 X_2 + \ldots + \beta_p X_p\right)\end{aligned} \tag{6.9}$$

where $j = 1, 2, \ldots, J - 1$; J is the baseline or reference category, which is the last category here, but it can be the first category or any of the other categories decided by the researcher; Y is the ordinal response variable with categories from j to J; α_j are the intercepts; $\beta_1, \beta_2, \ldots$, and β_p are the logit coefficients for the predictors, $X_1, X_2, \ldots, X_p$, respectively; and ϕ_j are the constraints or scale parameters that are used to ensure the outcome variable is ordinal if the following condition is satisfied:

$$1 = \phi_1 > \phi_2 > \phi_3 > \ldots \phi_{J-1} > \phi_J = 0 \tag{6.10}$$

These constraints are ordered from 1 to 0 with each constraint getting smaller. The first constraint ϕ_1 is set to be 1, and the last constraint ϕ_J is equal to 0 so that the estimated SL model can be identified. For example, when the ordinal response variable has four categories, only two constraints (i.e., ϕ_2 and ϕ_3) need to be estimated and they meet the following condition: $1 = \phi_1 > \phi_2 > \phi_3 > \phi_4 = 0$. Lunt (2001) considered the SL model to be the constrained multinomial logistic model.

When the constraints (ϕ_j) are multiplied by a logit coefficient (β) of a predictor variable, we obtain $\phi_1\beta$, $\phi_2\beta$, $\phi_3\beta$, and $\phi_4\beta$ and $\phi_1\beta > \phi_2\beta > \phi_3\beta > \phi_4\beta$. Since the constraints are ordered with the first constraint set to 1 and the last one set to 0, the function of the constraints or scale parameters here is to ensure the ordinality of the model.

To calculate the odds of being in a category j versus a category m, we just need to take the exponential of $[(\alpha_j - \alpha_m) + (\phi_j - \phi_m)\beta]$. When the category m becomes the baseline category J, we need to substitute it into the equation. Since $\phi_J = 0$, we get $[(\alpha_j - 0) + (\phi_j - 0)\beta] = \alpha_j + \phi_j\beta$. By exponentiating ($\phi_j\beta$), we get the odds of being in a category j versus the baseline category J for a unit change in a predictor variable.

The SL model can also be expressed in a slightly different form with a negative sign before the logit coefficients so that $\phi_j\beta$ can be ordered in an opposite direction: $-\phi_1\beta < -\phi_2\beta < -\phi_3\beta < -\phi_4\beta$. The `rrvglm()` function in the `VGAM` package in R follows Equation 6.9.

6.3.2 Odds and Odds Ratios in Stereotype Logistic Regression Models

The SL model estimates the logit or log odds of being in a particular category relative to the baseline category. The odds in the SL model can be defined as the ratio of the probability of being in a particular category to the probability of being in the base category. This is expressed as:

$$\text{Odds}(Y = j \text{ vs. } Y = J) = \frac{P(Y = j)}{P(Y = J)} \tag{6.11}$$

where j can be any categories from 1 to $J - 1$ categories.

In the following example, an ordinal response variable, health status, has four ordinal levels from 1 to 4, with 1 = poor, 2 = fair, 3 = good, and 4 = excellent. We estimate three odds with category 4 as the base or reference category: the odds of being in category 1 versus category 4, the odds of being in category 2 versus category 4, and the odds of in category 3 versus category 4.

Odds ($Y = 1$ vs. $Y = 4$) equal the ratio of probability of being in category 1 to the probability of being in category 4:

$$\text{Odds}(Y = 1 \quad \text{vs.} \quad 4) = \frac{P(Y = 1)}{P(Y = 4)} = \frac{P(1)}{P(4)}$$

Odds ($Y = 2$ vs. $Y = 4$) and odds ($Y = 3$ vs. $Y = 4$) can be expressed in a similar way:

$$\text{Odds}(Y = 2 \quad \text{vs.} \quad 4) = \frac{P(Y = 2)}{P(Y = 4)} = \frac{P(2)}{P(4)}$$

$$\text{Odds}(Y = 3 \quad \text{vs.} \quad 4) = \frac{P(Y = 3)}{P(Y = 4)} = \frac{P(3)}{P(4)}$$

In sum, the odds in the SL model compare the probability of being in a particular category $P(Y = j)$ with the probability of being in the base category $P(Y = J)$. Please note that the logit or log odds of being a particular category versus the base category in the SL model need to be multiplied by the scale parameters ϕ_j. Table 6.3 presents the scale parameters, logits, odds, and category comparisons for the SL model for the ordinal response variable with four levels.

Odds Ratios in Stereotype Logistic Regression

The odds ratio of being in a category j versus the baseline category J is obtained by taking the exponential of $[(\alpha_j - \alpha_J) + (\phi_j - \phi_J)\beta]$. Since the estimated α_J and ϕ_J equal 0, the equation could be simplified to be $(\alpha_j + \phi_j\beta)$. Exponentiating $(\phi_j\beta)$ gives us the odds ratio of being a category j versus the baseline J for a unit change in a predictor variable when holding other predictor variables constant.

TABLE 6.3 ● Category Comparisons for the Stereotype Logistic Regression Model With Four Levels of Health Status (j = 1, 2, 3, 4)

Equation	Scale Parameters (ϕ_j)	Logit $P(Y = j$ vs. $J)$	Odds	Probability Comparisons
1	$\phi_1 = 1$	logit $P(Y = 1$ vs. $4)$	$\frac{P(Y=1)}{P(Y=4)}$	Category 1 vs. category 4
2	ϕ_2	logit $P(Y = 2$ vs. $4)$	$\frac{P(Y=2)}{P(Y=4)}$	Category 2 vs. Category 4
3	ϕ_3	logit $P(Y = 3$ vs. $4)$	$\frac{P(Y=3)}{P(Y=4)}$	Category 3 vs. Category 4

6.3.3 Interpretation of Model Parameter Estimates

A logit coefficient in the SL model is the log odds of being a particular category relative to the base category. The logit coefficients need to be multiplied by the constraints or the scale parameters ϕ_j. After multiplication, they can be interpreted as the change in the predicted logit or the log odds for a one-unit increase in a predictor variable when holding other predictor variables constant. Exponentiating the product of the logit coefficients and the scale parameters gives us the odds ratios of being a category j versus the baseline J. The interpretation of odds ratios in the SL model is similar to that of other logistic regression models. The odds ratios are the change in the predicted odds of being in a particular category compared with the base category for a one-unit increase in the predictor variable when holding other predictor variables constant.

The odds of being in the base category compared with a particular category can also be estimated since they are just the reciprocal of the odds of being in a particular category versus the base category. These two odds are different in the order when comparing categories. The odds of being in a particular category versus the base category compares category j and the base category J, whereas the odds of being in the base category compared with a particular category compares categories in the reversed order, that is, the base category J versus a particular category j.

To get the odds ratio of being in the base category relative to a particular category, we can either exponentiate $(-\phi_j\beta)$ or take the reciprocal of the odds of being in a particular category versus the base category.

6.3.4 Research Example and Description of the Data and Sample

In this section, the purpose of the research example is still to investigate the relationships between the ordinal response variable, health status, and four predictor variables. Different from previous chapters, however, the SL model is used for data analysis since it relaxes the proportional odds assumption.

The data for the following analyses is from the General Social Survey 2016 (GSS 2016). The same variables from previous sections are used for data analysis in this section.

6.3.5 The `rrvglm()` Function With the `multinomial` Family in the `VGAM` Package

The `rrvglm()` function in the `VGAM` package can be used for the analysis of stereotype logistic (SL) regression models. Since the package has been installed, we only need to load the package by typing `library(VGAM)`. The syntax for SL models is similar to that for other models using `vglm()` except that the different `family` arguments are specified. The `multinomial` family needs to be specified for the SL model. For example, the command `rrvglm(y ~ x, family = multinomial, rank = 1, data = data1)` tells R to fit a SL model predicting the ordinal dependent variable y with an independent variable x. The argument `family = multinomial, rank = 1` tells R that a one-dimension SL model is fitted. When there are more than one predictor variable in the formula, they are connected by plus (+) symbols. For more details on how to use this command, type `help(rrvglm)` in the command prompt after loading the VGAM package.

6.3.6 The SL Model: Multiple-Predictor Model With R

The command `slmodel2 <- rrvglm(healthre ~ educ + maritals + female + wrkfull, multinomial, rank = 1)` tells R to predict the ordinal response variable `healthre` from the four predictor variables `educ`, `maritals`, `female` and `wrkfull` using the stereotype logistic regression. The output is shown by the `summary(slmodel2)` command.

```
> # Section III
> # SL model with vglm() in VGAM
> slmodel2 <- rrvglm(healthre ~ educ + maritals + female + wrkfull, multinomial,
rank=1)
> summary(slmodel2)

Call:
rrvglm(formula = healthre ~ educ + maritals + female + wrkfull,
       family = multinomial, rank = 1)

Pearson residuals:
                         Min        1Q     Median          3Q      Max
log(mu[,1]/mu[,4])   -3.6720  -0.23104   -0.11682   -0.060837   9.1500
log(mu[,2]/mu[,4])   -3.7610  -0.56993   -0.27016   -0.123319   3.7973
log(mu[,3]/mu[,4])   -3.2473  -0.75050   -0.40877    0.937293   2.0443

Coefficients:
                   Estimate   Std. Error   z value    Pr(>|z|)
I(latvar.mat):1    0.683932     0.074513    9.1787    < 2.2e-16 ***
I(latvar.mat):2    0.232269     0.053012    4.3815    5.894e-06 ***
(Intercept):1      3.520793     0.500196    7.0388    9.693e-13 ***
(Intercept):2      3.427660     0.380129    9.0171    < 2.2e-16 ***
(Intercept):3      2.008639     0.321816    6.2416    2.166e-10 ***
educ              -0.300684     0.035828   -8.3925    < 2.2e-16 ***
maritals          -0.584325     0.179114   -3.2623    0.0005526 ***
female            -0.252006     0.169530   -1.4865    0.0685738 .
wrkfull           -0.935306     0.187253   -4.9949    2.944e-07 ***

---
Signif. codes: 0 '***' 0.001 '**' 0.01 '*' 0.05 '.' 0.1 ' ' 1

Number of linear predictors: 3

Names of linear predictors:
log(mu[,1]/mu[,4]), log(mu[,2]/mu[,4]), log(mu[,3]/mu[,4])

Residual deviance: 4292.94 on 5608 degrees of freedom
```

```
Log-likelihood: -2146.47 on 5608 degrees of freedom

Number of iterations: 5

Reference group is level  of the response
```

6.3.7 Interpreting the Output

The coefficients section displays the parameter estimates for the four predictor variables, the constraints, and the intercepts, their standard errors, the Wald z statistics, and the associated p values. Among four predictor variables, the logit coefficients for `educ`, `maritals`, and `wrkfull` are statistically significant, whereas the coefficient for `female` is not significant (p = .069).

The two constraints are .684 and .232, respectively. Recall that the first constraint is set to 1 and the fourth constraint is set to 0. These constraints are useful when we compute the logit coefficients comparing a particular category to the base category with the formula ($\phi_j\beta$).

We can also extract the coefficients table with `coef(slmodel2, matrix = TRUE)`.

```
> coef(slmodel2, matrix = TRUE)
            log(mu[,1]/mu[,4])  log(mu[,2]/mu[,4])  log(mu[,3]/mu[,4])
(Intercept)          3.5207927           3.4276595          2.00863942
educ                -0.3006839          -0.2056473         -0.06983959
maritals            -0.5843250          -0.3996385         -0.13572064
female              -0.2520056          -0.1723547         -0.05853311
wrkfull             -0.9353057          -0.6396854         -0.21724261
```

Odds Ratios of Being in a Particular Category versus the Base Category

The odds ratios are reported in the following with `exp(coef(slmodel2, matrix = TRUE))`.

```
> exp(coef(slmodel2, matrix = TRUE))
            log(mu[,1]/mu[,4])  log(mu[,2]/mu[,4])  log(mu[,3]/mu[,4])
(Intercept)         33.8112196          30.8044617           7.4531698
educ                 0.7403117           0.8141201           0.9325434
maritals             0.5574820           0.6705624           0.8730865
female               0.7772404           0.8416806           0.9431470
wrkfull              0.3924659           0.5274583           0.8047347
```

The output displays the odds ratios for the four predictor variables comparing a particular category (1, 2, or 3) with the base category 4. For example, the column labeled as `log(mu[,1]/mu[,4])` on the top lists the odds ratios comparing the categories 1 and 4. ORs for `educ`, `maritals`, `wrkfull`, and `female` are .740, .557, .777, and .392, respectively. The same results can be computed, and the process is explained as follows.

The odds ratios comparing a particular category with the base category are the exponentiated ($\phi_j\beta$), whereas the odds ratios comparing the base category 4 with any other category are the exponentiated ($-\phi_j\beta$). For each of the four predictor variables, we can obtain the odds of being in any other category versus the baseline category *J*. For example, for the `educ` predictor, the odds ratio of being in category 1 versus the base category 4, OR (1, 4) = $e^{(1 \times -.301)}$ = $e^{(-.301)}$ = .740; the odds ratio of being in category 2 versus the base category 4, OR(2, 4) = $e^{(.684 \times (-.301))}$ = $e^{(-.206)}$ = .814; and the odds ratio of being in category 3 versus category 4, OR(3, 4) = $e^{(.232 \times (-.301))}$ = $e^{(-.070)}$ = .932. The odds ratios for the other three predictors can be calculated in a similar way.

6.3.8 Interpreting the Odds Ratios of Being in a Particular Category Versus the Base Category

For `educ`, the odds ratios of being in category 1 versus category 4, category 2 versus category 4, and category 3 versus category 4 are .740, .814, and .933, respectively. Therefore, for each one-unit increase in `educ`, the odds across the three comparisons decrease by .740, .814, and .933, respectively. Regarding `maritals`, the odds for the married are .557, .671, and .873 times the odds for the unmarried across the three comparisons, respectively; with regard to `wrkfull`, the odds for working full time are .392, .527, and .805 times the odds for not working full time across the three comparisons, respectively. Similarly, the odds for females are .777, .842, and .943 times the odds for males across the three comparisons, respectively.

6.3.9 Interpreting the Odds Ratios of Being in the Base Category Versus a Particular Category

The odds ratios comparing the base category with a particular category can be obtained by exponentiating ($-\phi_j\beta$), which are the inverse of the odds ratios above comparing any other category with the base category 4. The output by the `exp(-coef(slmodel2, matrix = TRUE))` command provides odds ratios of being in the base category versus a particular category across three comparisons.

```
> exp(-coef(slmodel2, matrix = TRUE))
              log(mu[,1]/mu[,4])  log(mu[,2]/mu[,4])  log(mu[,3]/mu[,4])
(Intercept)           0.02957598          0.03246283           0.1341711
educ                  1.35078234          1.22831994           1.0723362
maritals              1.79377973          1.49128548           1.1453619
female                1.28660323          1.18809913           1.0602801
wrkfull               2.54799222          1.89588431           1.2426455
```

The same results can be computed as follows. For example, for the `educ` predictor, the odds ratio of being in base category 4 compared with category 1, OR (4, 1) = $e^{(1 \times .301)}$ = 1.351; the odds ratio of being in base category 4 compared with category 2, OR(4, 2) = $e^{(.684 \times .301)}$ = $e^{(.206)}$ = 1.228; and the odds ratio of being in base category 4 compared with category 3, OR(4, 3) = $e^{(.232 \times .301)}$ = $e^{(.070)}$ = 1.072. The odds ratios for the other two predictors can be calculated in a similar way.

The odds ratio can be interpreted as the change in the odds of being in the baseline category versus any other category for a one-unit increase in the predictor variable when holding other predictors constant. For each predictor variable, there are $J - 1$ comparisons, so there are $J - 1$ odds ratios.

In terms of odds ratio, the odds of being in the base category 4 versus category 1 increase by a factor of 1.351 with a one-unit increase in `educ`. For the predictor `maritals`, OR = 1.794, which indicates that the odds of being in the base category 4 versus category 1 for the married are 1.794 times the odds for the unmarried when holding all the other predictors constant. With regard to `wrkfull`, the odds for working full time are 2.548 times the odds for not working full time. In addition, for `female`, OR = 1.287, indicating that the odds for females are 1.287 times the odds for males.

The odds ratio for each predictor also needs to be interpreted across three comparisons. For example, regarding `educ`, the odds ratios of being in category 4 versus category 1, category 4 versus category 2, and category 4 versus category 3 are 1.351, 1.228, and 1.072, respectively. Therefore, for each one-unit increase in `educ`, the increases in the odds across three comparisons are 1.351, 1.228, and 1.072, respectively. The odds ratios for the other three variables can be interpreted in the similar way.

6.4 MAKING PUBLICATION-QUALITY TABLES

6.4.1 Presenting the Results of the `vglm` Models Using the `texreg` Package

The `stargazer()` function currently cannot directly produce the results table from the `vglm` models, so we use the `screenreg()` and `htmlreg()` functions from the `texreg` package (Leifeld, 2013). Since the package has been installed in earlier chapters, we only need to load the package by typing `library(texreg)`.

In this example, we use the `screenreg()` and `htmlreg()` functions from the `texreg` package to display the results of the CR models. The results of the AC and SL models are omitted here. We first use the `vglm()` function to fit the single-predictor model `crmodel1` and the multiple-predictor model `crmodel2`. We then create a table containing the results of the both model with the following command: `screenreg(list(crmodel1, crmodel2))`. In the `screenreg()`

function, we specify the two model objects to be presented with the `list()` function. The output is a plain text table.

```
> # Presenting the results of the CR Models using the texreg package
> library(texreg)
Version:     1.37.5
Date:        2020-06-17
Author:      Philip Leifeld (University of Essex)
> screenreg(list(crmodel1, crmodel2))

==============================================
                    Model 1       Model 2
----------------------------------------------
(Intercept):1         -0.69 ***     -0.54 **
                      (0.19)        (0.20)
(Intercept):2          0.94 ***      1.10 ***
                      (0.18)        (0.19)
(Intercept):3          3.00 ***      3.20 ***
                      (0.20)        (0.21)
educ                  -0.15 ***     -0.14 ***
                      (0.01)        (0.01)
maritals                            -0.20 **
                                    (0.08)
female                              -0.09
                                    (0.08)
wrkfull                             -0.32 ***
                                    (0.08)
----------------------------------------------
Log Likelihood     -2168.90      -2156.09
DF                  5615          5612
Num. obs.           5619          5619
==============================================
*** p < 0.001; ** p < 0.01; * p < 0.05

> htmlreg(list(crmodel1, crmodel2), file = "chap6cr.doc", doctype = TRUE,
html.tag = TRUE, head.tag = TRUE)
The table was written to the file 'chap6cr.doc'.
```

We can also use the `htmlreg()` function to create a regression table for the estimated and save it to a Microsoft Word file named `chap6cr.doc` with the following command: `htmlreg(list(crmodel1, crmodel2), file = "chap6cr.doc", doctype = TRUE, html.tag = TRUE, head.tag = TRUE)`. It automatically produces Table 6.4, as shown here in its original format, presenting the results of both the single-predictor and the multiple-predictor CR models.

6.5 REPORTING THE RESULTS

Writing the results for the continuation ratio models, adjacent categories models, and stereotype logistic regression models is similar to that for the proportional odds models.

TABLE 6.4 ● Results of the Continuation Ratio Models: Single-Predictor Model and Multiple-Predictor CR Models (Shown in Original Format Generated by R)

	Model 1	Model 2
(Intercept):1	−0.69***	−0.54**
	(0.19)	(0.20)
(Intercept):2	0.94***	1.10***
	(0.18)	(0.19)
(Intercept):3	3.00***	3.20***
	(0.20)	(0.21)
educ	−0.15***	−0.14***
	(0.01)	(0.01)
maritals		−0.20**
		(0.08)
female		−0.09
		(0.08)
wrkfull		−0.32***
		(0.08)
Log Likelihood	−2168.90	−2156.09
DF	5615	5612
Num. obs.	5619	5619

***$p < 0.001$
**$p < 0.01$
*$p < 0.05$

First, describe the statistical method you use for data analysis and justify why it is needed for your research. Also describe the dependent variable and the independent variables in the models, and your research hypothesis or the purpose of your study. Furthermore, report and interpret the available model fit statistics, including the likelihood ratio test statistic and the pseudo R^2.

Second, have a summary table containing all the parameter estimates, their standard errors, p values or whether the estimates are significant, and odds ratios.

Third, in the body of the text, concisely interpret the parameter estimates for the predictor variables. When interpreting the odds ratios of the estimates, pay attention to the directions of the comparisons. Explain clearly the odds comparing the probabilities of being in a higher category versus being in a particular category, or vice versa.

Fourth, when more than one model is fitted, results of all the competing models may be presented in the table. The following is an example of summarizing results for the continuation ratio model with all four predictors (full model). Examples for the other two models are omitted here.

> The continuation ratio model was fitted to estimate the ordinal outcome variable, health status, from a set of predictor variables, including marital status, years of education, gender, and working status. This model was used to address the category comparisons between outcomes being in a particular category and those being above that category.
>
> The likelihood ratio test statistic $LR\ \chi^2_{(4)} = 164.26$, $p < .001$, indicated that the full model with the four predictor variables provided a better fit than the null model with no independent variables in predicting conditional probability for health status. The likelihood ratio $R^2_L = .037$, which was small.
>
> Table 6.4 displays the parameter estimates for the four predictor variables. Of them, the coefficients of the three variables, `educ` ($\beta = -.144$, Wald $z = -10.772$, $p < .001$), `maritals`, ($\beta = -.205$, Wald $z = -2.631$, $p <$ .001), and `wrkfull` ($\beta = -.319$, Wald $z = -4.044$, $p < .001$) were negatively significant; and `female` ($\beta = -.093$, Wald $z = -1.197$, $p = .231$) was not significant.
>
> In terms of odds ratios, the odds of being in a better health status level relative to being in that level for the married were 1.227 times the odds for the unmarried (OR = 1.227) after controlling for other predictors in the model. In addition, the odds of being in a better health status for working full time were 1.375 times the odds for not working full time. Further, the odds of being above a particular category versus being in that category increased by a factor of 1.155 or by 15.5% for a one-unit increase in years of education (`educ`) when holding other variables constant. Finally, the odds ratio for the predictor `female` is 1.097, $p = .231$, which is not significant, indicating being female does not impact the odds of being in a higher category.

6.6 SUMMARY OF R COMMANDS IN THIS CHAPTER

```
# Chap 6 R Script
# Remove all objects
rm(list = ls(all = TRUE))

# The following user-written packages need to be installed first by using
install.packages(" ") and then by loading it with library()
# library(VGAM)            # It is already installed for Chapter 4
# library(rcompanion)      # It is already installed for Chapter 3
# library(ggeffects)       # It is already installed for Chapter 2
# library(texreg)          # It is already installed for Chapter 4

# Section I
# Import GSS 2016 data
library(foreign)
chp6.cr <- read.dta("C:/CDA/gss2016.dta")
chp6.cr$healthre <- factor(chp6.cr$healthre, ordered=TRUE)
chp6.cr1$educ <- as.numeric(chp6.cr$educ)
chp6.cr$wrkfull <- as.numeric(chp6.cr$wrkfull)
chp6.cr$maritals <- as.numeric(chp6.cr$maritals)
attach(chp6.cr)

# CR model with vglm() in VGAM
library(VGAM)
crmodel2 <- vglm(healthre ~ educ + maritals + female + wrkfull, sratio(parallel =
TRUE, reverse = FALSE))
summary(crmodel2)
coef(crmodel2, matrix = TRUE)
confint(crmodel2, matrix = TRUE)
exp(coef(crmodel2, matrix = TRUE))
exp(confint(crmodel2, matrix = TRUE))
cbind(exp(coef(crmodel2)), exp(confint(crmodel2)))

exp(-coef(crmodel2, matrix = TRUE))
exp(-confint(crmodel2, matrix = TRUE))
cbind(exp(-coef(crmodel2)), exp(-confint(crmodel2)))

# Testing the overall model using the likelihood ratio test
crmodel0 <- vglm(healthre ~ 1, sratio(parallel = TRUE, reverse = FALSE))
summary(crmodel0)
lrtest(crmodel0, crmodel2)

# Pseudo R2
library(rcompanion)
nagelkerke(crmodel2)

# AIC and BIC Statistics
AIC(crmodel2)
BIC(crmodel2)
```

```
# Predicted probabilities with ggpredict() in ggeffects
library(ggeffects)
pr.cr2m <- ggpredict(crmodel2, terms = "educ[12, 14, 16]", ci = NA)
pr.cr2m
plot(pr.cr2m)

pr.cr2 <- ggpredict(crmodel2, terms=c("educ[12, 14, 16]","maritals"), ci=NA)
pr.cr2
plot(pr.cr2)

# CR model with non-proportional odds
crmodel3 <- vglm(healthre ~ educ + maritals + female + wrkfull, sratio(parallel =
FALSE, reverse = FALSE))
summary(crmodel3)
coef(crmodel3, matrix = TRUE)
confint(crmodel3, matrix = TRUE)
exp(coef(crmodel3, matrix = TRUE))
exp(confint(crmodel3, matrix = TRUE))
cbind(exp(coef(crmodel3)), exp(confint(crmodel3)))

exp(-coef(crmodel3, matrix = TRUE))
exp(-confint(crmodel3, matrix = TRUE))
cbind(exp(-coef(crmodel3)), exp(-confint(crmodel3)))

lrtest(crmodel0, crmodel3)
AIC(crmodel3)
BIC(crmodel3)

# Model comparison with the likelihood ratio test
lrtest(crmodel2, crmodel3)

# Single-predictor CR model
crmodel1 <- vglm(healthre ~ educ, sratio(parallel = TRUE, reverse = FALSE))

# Presenting the results of the CR Models using the texreg package
library(texreg)
screenreg(list(crmodel1, crmodel2))
htmlreg(list(crmodel1, crmodel2), file="chap6cr.doc", doctype=TRUE, html.tag=
TRUE, head.tag=TRUE)

# Section II
# AC model with vglm() in VGAM
acmodel2 <- vglm(healthre ~ educ + maritals + female + wrkfull, acat(parallel =
TRUE, reverse = FALSE))
summary(acmodel2)
coef(acmodel2, matrix = TRUE)
confint(acmodel2, matrix = TRUE)
exp(coef(acmodel2, matrix = TRUE))
exp(confint(acmodel2, matrix = TRUE))
cbind(exp(coef(acmodel2)), exp(confint(acmodel2)))

exp(-coef(acmodel2, matrix = TRUE))
exp(-confint(acmodel2, matrix = TRUE))
cbind(exp(-coef(acmodel2)), exp(-confint(acmodel2)))
```

```
# Testing the overall model using the likelihood ratio test
acmodel0 <- vglm(healthre ~ 1, acat(parallel = TRUE, reverse = FALSE))
summary(acmodel0)
lrtest(acmodel0, acmodel2)

# Pseudo R2
nagelkerke(acmodel2)

# AIC and BIC Statistics
AIC(acmodel2)
BIC(acmodel2)

# Section III
# SL model with vglm() in VGAM
slmodel2 <- rrvglm(healthre ~ educ + maritals + female + wrkfull, multinomial,
rank=1)
summary(slmodel2)
coef(slmodel2, matrix = TRUE)
exp(coef(slmodel2, matrix = TRUE))
exp(-coef(slmodel2, matrix = TRUE))

detach(chp6.cr)
```

Glossary

A logit coefficient in the SL model is the log odds of being a particular category relative to the base category. The logit coefficients need to be multiplied by the constraints or the scale parameters ϕ_j.

The adjacent categories (AC) model focuses on the pairs of adjacent categories for ordinal response variables.

The continuation ratio (CR) model estimates the conditional probability of being in a category of an ordinal response variable given that an individual has been in that category or above.

The odds in the CR model are the probability of being in a category divided by the probability of being above that category.

The odds of adjacent categories are the ratio of the probabilities of one category to the next lower category.

The stereotype logistic (SL) regression model is an alternative model for ordinal response variables when the parallel lines or the proportional odds (PO) assumption does not hold. In the model, the constraints or scale parameters are used to ensure the outcome variable is ordinal.

Exercises

Use the GSS 2016 data available at **https://edge.sagepub.com/liu1e** for the following problems.

1. Conduct an analysis for a continuation ratio model using the `vglm()` function to examine the relationship between the outcome variable `degree` and three predictor variables `sex`, `class`, and `sibs`.
2. Use the likelihood ratio test to compare the model with the null model and interpret it.
3. Identify the logit coefficients, the Wald *z* tests, and the 95% confidence intervals for `sex` and `sibs`. Are they statistically significant?
4. Compute the odds ratios for `class` and `sibs`.
5. Make a publication-quality table containing the estimated logit coefficients.
6. Write a report to summarize the results from the output.

7

MULTINOMIAL LOGISTIC REGRESSION MODELS

OBJECTIVES OF THIS CHAPTER

This chapter introduces multinomial logistic regression models. It first starts with an introduction to the multinomial logistic regression model followed by a discussion of the odds and odds ratios or relative risk ratios in the model, goodness-of-fit statistics, and how to interpret parameter estimates. After a description of the research example, the data, and the sample, the multinomial logistic regression models are illustrated with the `VGAM`, `nnet`, and `mlogit` packages. R commands and output are explained in detail. This chapter focuses on fitting the multinomial logistic regression models with R, as well as on interpreting and presenting the results. After reading this chapter, you should be able to:

- Identify when multinomial logistic regression models are used.
- Fit a multinomial logistic regression model using R.
- Interpret the output.
- Compute, plot and interpret the predicted probabilities.
- Compare models using the likelihood ratio test.
- Present results in publication-quality tables using R.
- Write the results for publication.

7.1 MULTINOMIAL LOGISTIC REGRESSION MODELS: AN INTRODUCTION

The multinomial logistic regression model is used to estimate nominal response variables that have multiple unordered categories. For example, the nominal response variable in Hoffmann (2016) and Kaufman (2019) was the three-category political views including liberal, moderate, and conservative. Another example in Menard (2010) was the four-category political party affiliation including Democrat, Independent, Republican, and other. This model is a generalization of binary logistic regression when there are more than two nominal categories in a response variable. It can also be used for an ordinal response variable when the proportional odds assumption does not hold. It estimates the odds of being in a category versus the base category of a nominal variable. Although the proportional odds model compares the cumulative probabilities of being at or below a particular category and the probabilities of being above that category, the multinomial logistic regression model compares a particular category with the base category. If a nominal response variable has J levels, there are $J - 1$ comparisons between any other categories and the base category. For example, if we disregard the ordinal nature of the ordinal response variable, health status, and treat it as a nominal response variable with four categories, then we compare category 2 and category 1, category 3 and category 1, and category 4 and category 1 in the multinomial logistic model where the base category is set to be one.

The multinomial logistic model can be expressed as follows:

$$\ln\left(\frac{P(Y = j|x_1, x_2, \ldots, x_p)}{P(Y = J|x_1, x_2, \ldots, x_p)}\right) = \alpha_j + \beta_{j1}X_1 + \beta_{j2}X_2 + \ldots + \beta_{jp}X_p \tag{7.1}$$

where $j = 1, 2, \ldots, J - 1$; J is the base category, which can be any category but is generally the highest one; α_j are the intercepts; and $\beta_{j1}, \beta_{j2}, \ldots, \beta_{jp}$ are the logit coefficients for each comparison. The model estimates $J - 1$ logit coefficients for each predictor. If we set the base category to be category 1, the lowest category, then the model can be rewritten as follows:

$$\ln\left(\frac{P(Y = j|x_1, x_2, \ldots, x_p)}{P(Y = 1|x_1, x_2, \ldots, x_p)}\right) = \alpha_j + \beta_{j1}X_1 + \beta_{j2}X_2 + \ldots + \beta_{jp}X_p \tag{7.2}$$

where $j = 2, 3, \ldots, J$; and category 1 is the base category.

It can be treated as a simultaneous estimation of a series of binary logistic regression models comparing a particular category and the base category. In each binary model, being in a particular category is coded as the binary outcome of 1 and being in the base category is coded as 0. For example, when the base category is 1, the dichotomized outcome in the first binary model compares category 2 with category 1.

7.1.1 The Multinomial Distribution

The multinomial distribution is an extension of the binomial distribution when the discrete random variable is a nominal variable with more than two categories. Recall that the binomial distribution is expressed.

$$P(Y = k) = \binom{n}{k} p^k (1 - p)^{n-k} \tag{7.3}$$

where $\binom{n}{k}$ is the binomial coefficient, k is the number of successes, n is the number of trials, and p is the success probability when the binary outcome is 1.

$$\binom{n}{k} = \frac{n!}{(n-k)!k!}$$

where $n!$ is n factorial or the factorial of n. $n! = n*(n-1) \ldots 2*1$.

When the outcome of a nominal variable has more than two categories, the probability function or the PDF is expressed as:

$$P(n_1, n_2, \ldots n_j) = \frac{n!}{n_1!n_2!\ldots n_j!} p_1^{n_1} p_2^{n_2} \ldots p_j^{n_j} \tag{7.4}$$

where j is the number of categories in the nominal variable, n_j is the number of observations for a particular category, and p_j is the probability of choosing each category. The total number of observations across all the categories, $n = n_1 + n_2 + \ldots + n_j$. In addition, the total probability across all the categories is 1. $p_1 + p_2 + \ldots + p_j = 1$. When the nominal variable has only two categories (i.e., $j = 2$), the multinomial distribution becomes the binomial distribution.

The log likelihood function for the multinomial distribution is expressed as:

$$l(p, n) = \sum_{i=1}^{j} n_i \ln p_i + \ln \frac{n!}{n_1!n_2!\ldots n_j!} \tag{7.5}$$

where $\ln \frac{n!}{n_1!n_2!\ldots n_j!}$ is a constant term and does not involve the parameter p. $\sum_{i=1}^{j} n_i \ln p_i$ is the summation term which adds the product of the number of observations for a particular category (n_i) and the log of the probability of choosing that category ($\ln p_i$). For example, if the nominal outcome has four categories, $\sum_{i=1}^{j} n_i \ln p_i = n_1 \ln p_1 + n_2 \ln p_2 + n_3 \ln p_3 + n_4 \ln p_4$.

7.1.2 Odds in Multinomial Logistic Models

The multinomial logistic model estimates the logit or log odds of being in a particular category relative to the baseline category. The odds in the multinomial logistic model can be defined as the ratio of the probability of being in a particular category to the probability of being in the base category. It is expressed as:

$$\text{Odds}(Y = j \text{ vs. } Y = J) = \frac{P(Y = j)}{P(Y = J)} \tag{7.6}$$

where j can be any categories from 1 to $J - 1$ categories.

For example, if we treat the ordinal response variable, health status, as nominal with four categories from 1 to 4, with 1 = poor, 2 = fair, 3 = good, and 4 = excellent, then we estimate three odds with category 1 as the base category: The odds of being in category 2 versus category 1, the odds of being in category 3 versus category 1, and the odds of being in category 4 versus category 1.

Specifically, odds ($Y = 2$ vs. $Y = 1$) equal the ratio of the probability of being in category 2 to the probability of being in category 1:

$$\text{Odds}(Y = 2 \text{ vs. } 1) = \frac{P(Y = 2)}{P(Y = 1)} = \frac{P(2)}{P(1)}$$

The other two odds, odds ($Y = 3$ vs. $Y = 1$) and odds ($Y = 4$ vs. $Y = 1$), are expressed as follows:

$$\text{Odds}(Y = 3 \text{ vs. } 1) = \frac{P(Y = 3)}{P(Y = 1)} = \frac{P(3)}{P(1)}$$

$$\text{Odds}(Y = 4 \text{ vs. } 1) = \frac{P(Y = 4)}{P(Y = 1)} = \frac{P(4)}{P(1)}$$

Table 7.1 presents the logits, odds, and category comparisons for the multinomial logistic regression model for the nominal response variable with four levels.

7.1.3 Odds Ratios or Relative Risk Ratios in Multinomial Logistic Regression Models

Since the multinomial logistic model can be treated as a series of binary logistic regression models estimated simultaneously with the comparison of any other categories to the base category, the logit coefficients can be interpreted in a similar way as that for the binary logistic regression. The odds ratio of being in a category j versus the baseline category J is obtained by taking the exponential of the logit coefficient β. Although the relative risk can be defined differently, the odds ratio in the multinomial logistic

TABLE 7.1 ● Category Comparisons for the Multinomial Logistic Regression Model With Four Levels of Health Status (j = 1, 2, 3, 4)

Equation	Logit $P(Y = j \text{ vs. } J)$	Odds	Probability Comparisons
1	logit $P(Y = 2 \text{ vs. } 1)$	$\frac{P(Y=2)}{P(Y=1)}$	Category 2 vs. category 1
2	logit $P(Y = 3 \text{ vs. } 1)$	$\frac{P(Y=3)}{P(Y=1)}$	Category 3 vs. category 1
3	logit $P(Y = 4 \text{ vs. } 1)$	$\frac{P(Y=4)}{P(Y=1)}$	Category 4 vs. category 1

regression is also called the relative risk ratio. Hilbe (2009) preferred the use of the relative risk ratio rather than the odds ratio since the categories of the nominal response variable are independent of each other. The odds ratio or relative risk ratio in multinomial logistic regression can be interpreted as the change in the odds or the relative risk for a one-unit change in a predictor variable when holding other predictor variables constant. To obtain the multiplicative inverse or reciprocal of the odds, the odds of being in the base category versus a particular category, we exponentiate the logit coefficient with a negative sign $\exp(-\beta)$.

7.1.4 Model Fit Statistics

Same as those discussed in the previous chapters, model fit statistics, such as the log likelihood statistic, the residual deviance, the model chi-square statistic, the AIC and BIC statistics, and the pseudo R^2 statistics, can be computed for the multinomial logistic regression model. The likelihood ratio test and the AIC and BIC statistics can also be used for model comparisons.

7.1.5 Interpretation of Model Parameter Estimates

A logit coefficient in the multinomial logistic regression model is the log odds of being a particular category relative to the base category. Exponentiating the product of the logit coefficients gives us the odds ratios of being a category j versus the baseline J. The interpretation of odds ratios is similar to that of other logistic regression models. The odds ratios are the change in the predicted odds of being in a particular category compared with the base category for a one-unit increase in the predictor variable when holding other predictor variables constant.

When an OR is larger than 1, the odds of being in a particular category versus the base category increase for a one-unit increase in the predictor variable.

When an OR is less than 1, the odds of being in a particular category versus the base category decrease for a one-unit increase in the predictor variable.

An OR of 1 indicates that there is no relationship between the predictor variable and the estimated odds.

The odds of being in the base category compared with a particular category can also be estimated since they are just the reciprocal of the odds of being in a particular category versus the base category. These two odds are different in the order when comparing categories. The odds of being in a particular category versus the base category compares category j and the base category J, whereas the odds of being in the base category compared with a particular category compares categories in the reversed order, that is, the base category J versus a particular category j.

7.2 RESEARCH EXAMPLE AND DESCRIPTION OF THE DATA AND SAMPLE

We investigate the relationships between the nominal response variable, health status, and four predictor variables. Unlike other chapters, however, here the research interest focuses on using multinomial logistic regression to predict the nominal response variable. The GSS 2016 data are used for the following analyses. The following are the variables used for data analysis in this chapter:

- `healthre`: the recoded variable of health (health status) with four categories (1 = poor health, 2 = fair health, 3 = good health, and 4 = excellent health)
- `maritals`: the recoded variable of marital (marital status) with 1 = currently married and 0 = not currently married
- `educ`: the highest education completed
- `female`: recoded variable of sex with 1 = female and 0 = male
- `wrkfull`: working full time or not

7.3 FITTING A ONE-PREDICTOR MULTINOMIAL LOGISTIC REGRESSION MODEL WITH R

7.3.1 Packages and Functions for Multinomial Logistic Regression Models in R

Several packages in R can be used for fitting multinomial logistic regression models. This chapter introduces the `VGAM`, `nnet`, and `mlogit` packages with the first package as the main focus. The `vglm()` function in `VGAM` (Yee, 2010), the `multimon()` function in `nnet` (Venables & Ripley, 2002), and the `mlogit()` function in `mlogit` (Croissant, 2020) are introduced in sequence. Since the `nnet` package is part of the R base distribution, we do not need to install it. You just need to install the other two packages first by using the `install.packages()` function and then load them with the `library()` function.

7.3.2 The `vglm()` Function With the `multinomial` Family in the `VGAM` Package

The `vglm()` function in the `VGAM` package can be used for the analysis of multinomial logistic regression models. If the user-written `VGAM` package is not installed, you need to install it first by typing `install.packages("VGAM")` before fitting the model. Since the package has been installed in earlier chapters, we only need to load the package by typing `library(VGAM)`.

The syntax for multinomial logistic regression models is similar to that for other models using the `vglm()` function. The `multinomial` family needs to be specified for the `family` argument. For example, the command `vglm(y ~ x, family = multinomial(refLevel = 1), data = data1)` tells R to fit a multinomial logistic regression model predicting the dependent variable *y* with an independent variable *x*. The argument `family = multinomial(refLevel = 1)` tells R that it is the `multinomial` family and the reference level is the first level. If not specified, the default reference level is the highest level of the nominal response variable. For more details on how to use this command, type `help(multinomial)` in the command prompt after loading the `VGAM` package.

7.3.3 The Multinomial Logistic Regression Model: One-Predictor Model

The command `mulmodel1 <- vglm(healthre ~ educ, multinomial(refLevel = 1), data = chp7.mul)` tells R to fit the multinomial logistic regression model for the nominal response variable `healthre` with the predictor variable `educ`. In the `vglm()` function, the `multinomial(refLevel = 1)` argument tells us that the `multinomial` family is used to fit the model and the reference level is the first level of `healthre`. The `summary(mulmodel)` command displays the output of the fitted model.

```
> # One-predictor multinomial logistic regression model with vglm() in VGAM
> library(VGAM)
> mulmodel1 <- vglm(healthre ~ educ, multinomial(refLevel = 1), data=chp7.mul)
> summary(mulmodel1)

Call:
vglm(formula = healthre ~ educ, family = multinomial(refLevel = 1),
    data = chp7.mul)

Pearson residuals:
                       Min       1Q    Median       3Q     Max
log(mu[,2]/mu[,1])  -4.378  -0.3625   -0.3029  -0.2267   2.704
log(mu[,3]/mu[,1])  -5.234  -0.6881   -0.4458   0.8864   2.145
log(mu[,4]/mu[,1])  -5.159  -0.3677   -0.2720  -0.1541   8.317

Coefficients:
               Estimate  Std. Error   z value   Pr(>|z|)
(Intercept):1   0.17577     0.42732     0.411    0.68082
(Intercept):2  -0.78176     0.42076    -1.858    0.06318 .
(Intercept):3  -3.27536     0.48121    -6.807   1.00e-11 ***
educ:1          0.08907     0.03382     2.633    0.00845 **
educ:2          0.21724     0.03305     6.573   4.93e-11 ***
educ:3          0.33528     0.03640     9.212    < 2e-16 ***

---
Signif. codes: 0 '***' 0.001 '**' 0.01 '*' 0.05 '.' 0.1 ' ' 1

Names of linear predictors: log(mu[,2]/mu[,1]), log(mu[,3]/mu[,1]),
log(mu[,4]/mu[,1])

Residual deviance: 4333.297 on 5613 degrees of freedom

Log-likelihood: -2166.649 on 5613 degrees of freedom

Number of Fisher scoring iterations: 5

No Hauck-Donner effect found in any of the estimates

Reference group is level 1 of the response
```

7.3.4 Interpreting the Output

The output for the multinomial logistic regression model looks similar to that for binary logistic regression models except that there are multiple binary comparisons to the reference group. As with the other logistic regression models, the R output for the multinomial logistic regression model also includes the call, the Pearson residuals, the coefficients, the number and names of the three linear predictors, the residual deviance, the log-likelihood value, and the number of iterations. In addition, it lists the reference group number.

The coefficients section displays the parameter estimates for the intercepts and the predictor variable, their standard errors, the Wald z statistics, and the associated p values.

The null hypothesis for the Wald z test is that the coefficient of the predictor variable is 0, and the alternative hypothesis is that the coefficient of the predictor variable is significantly different from 0.

Three coefficients for the predictor variable `educ` are displayed as `educ:1`, `educ:2`, and `educ: 3` since they are the parameter estimates for the three binary logistic models comparing each category with the base category. Only three binary models are estimated since the base outcome is category 1. These three equations, labeled `log(mu[,2]/mu[,1])`, `log(mu[,3]/mu[,1])`, and `log(mu[,4]/mu[,1])`, compare categories 2 with 1, categories 3 with 1, and categories 4 with 1, respectively. The estimated intercepts and logit coefficients for these three sub-models are numbered 1, 2, and 3 in the output.

Based on the parameter estimates in the output, the three equations can be expressed as:

$$\ln\left(\frac{P(Y=2)}{P(Y=1)}\right) = .176 + .089\text{educ}$$

$$\ln\left(\frac{P(Y=3)}{P(Y=1)}\right) = -.782 + .217\text{educ}$$

$$\ln\left(\frac{P(Y=4)}{P(Y=1)}\right) = -3.275 + .335\text{educ}$$

The first equation compares category 2 and category 1. The coefficient for `educ`, displayed as `educ:1`, $\beta = .089$, Wald $z = 2.633$. The associated p value, `Pr(>|z|) < .01`, so we reject the null hypothesis and conclude that `educ` is significant in predicting the log odds of being in category 2 versus category 1.

The second equation compares category 3 and category 1. The predictor variable `educ`, displayed as `educ:2`, is also significant. For `educ:2`, $\beta = .217$, Wald $z = 6.573$, $p < .001$.

The third equation compares category 4 and category 1. The predictor variable educ, displayed as educ:3, is also significant. For educ:3, $\beta = .335$, Wald $z = 9.212$, $p < .001$.

The Coefficients section also reports the intercepts (labeled as (Intercept):1, (Intercept):2, and (Intercept):3). They are the intercepts for each equation comparing a particular category with the reference group or base category. The reference group in this example is level 1 (i.e., healthre = 1). If not specified, the default is the highest outcome, but you can specify any category as the reference group.

We can extract the coefficients with coef(mulmodel1, matrix = TRUE) and obtain the confidence intervals with confint(mulmodel1, matrix = TRUE) as follows.

```
> coef(mulmodel1, matrix = TRUE)
            log(mu[,2]/mu[,1])  log(mu[,3]/mu[,1])  log(mu[,4]/mu[,1])
(Intercept)         0.17577310          -0.7817562          -3.2753578
educ                0.08906957           0.2172438           0.3352757

> confint(mulmodel1, matrix = TRUE)
                     2.5 %      97.5 %
(Intercept):1  -0.66175738  1.01330357
(Intercept):2  -1.60643404  0.04292172
(Intercept):3  -4.21851259 -2.33220303
educ:1          0.02277699  0.15536215
educ:2          0.15246679  0.28202081
educ:3          0.26393850  0.40661298
```

7.3.5 Interpreting the Odds Ratios of Being in a Particular Category Versus the Base Category for the Multinomial Logistic Regression Model

The multinomial logistic regression model estimates the logit odds of being a category relative to the baseline category. Recall that the exponentiated (β_j) is the odds ratio of being in a category j versus the baseline J for a one-unit change in a predictor variable. In this model, we define the odds ratio of being in category 2 compared with the base category 1 as OR (2, 1). Since $\beta = .089$ for educ:1, OR(2, 1) = $e^{(.089)} = 1.093$, which indicates that for a one-unit increase in education the odds of being in category 2 of health condition versus the base category 1 increase by a factor of 1.093. In other words, people who work full time have greater odds of being in the highest health condition (category 4) rather than being in category 1.

The odds ratio of being in category 3 versus category 1, OR(3, 1) = $e^{(.217)} = 1.242$, which indicates that for a one-unit increase in education the odds of being in category 2 of health condition versus the base category 1 increase by 24.2%. Similarly, the odds ratio of being in category 4 versus category 1, OR(4, 1) = $e^{(.335)} = 1.398$, which indicates that for a one-unit increase in education the odds of being in category 4 of health condition versus the base category 1 increase by 39.8%.

The above results can be obtained using the `exp(coef(slmodel1, matrix = TRUE))` command. We also use the `exp(confint(mulmodel1, matrix = TRUE))` command to obtain the corresponding confidence intervals. Both results are combined with the `cbind(exp(coef(mulmodel1)), exp(confint(mulmodel1)))` command.

```
> exp(coef(mulmodel1, matrix = TRUE))
            log(mu[,2]/mu[,1])  log(mu[,3]/mu[,1])  log(mu[,4]/mu[,1])
(Intercept)           1.192168           0.4576017          0.03780334
educ                  1.093157           1.2426470          1.39832590

> exp(confint(mulmodel1, matrix = TRUE))
                   2.5 %      97.5 %
(Intercept):1  0.51594383  2.75468631
(Intercept):2  0.20060168  1.04385618
(Intercept):3  0.01472052  0.09708164
educ:1         1.02303836  1.16808091
educ:2         1.16470378  1.32580631
educ:3         1.30204812  1.50172279

> cbind(exp(coef(mulmodel1)), exp(confint(mulmodel1)))
                               2.5 %      97.5 %
(Intercept):1  1.19216752  0.51594383  2.75468631
(Intercept):2  0.45760169  0.20060168  1.04385618
(Intercept):3  0.03780334  0.01472052  0.09708164
educ:1         1.09315670  1.02303836  1.16808091
educ:2         1.24264702  1.16470378  1.32580631
educ:3         1.39832590  1.30204812  1.50172279
```

7.3.6 Model Fit Statistics

Testing the Overall Model Using the Likelihood Ratio Test

To test if the overall model is significant, we fit a null model with the intercept only and compare the single-predictor model with the null model using the `lrtest()` function. The null model is fitted using the `vglm()` function with the predictor as 1 for the intercept in the model equation. The command and the output are displayed below.

```
> # Testing the overall model using the likelihood ratio test
> mulmodel0 <- vglm(healthre ~ 1, multinomial(refLevel = 1), data=chp7.mul)
> summary(mulmodel0)

Call:
vglm(formula = healthre ~ 1, family = multinomial(refLevel = 1),
    data = chp7.mul)

Pearson residuals:
                      Min       1Q    Median       3Q     Max
log(mu[,2]/mu[,1])  -2.107  -0.4351  -0.3087  -0.3087  1.6649
log(mu[,3]/mu[,1])  -2.459  -0.6678  -0.6608   0.9287  0.9287
log(mu[,4]/mu[,1])  -2.094  -0.4218  -0.3025  -0.3025  1.6997
```

```
Coefficients:
                 Estimate   Std. Error   z value   Pr(>|z|)
(Intercept):1    1.28610     0.10400      12.37     <2e-16 ***
(Intercept):2    2.04715     0.09782      20.93     <2e-16 ***
(Intercept):3    1.25518     0.10436      12.03     <2e-16 ***
---
Signif. codes: 0 '***' 0.001 '**' 0.01 '*' 0.05 '.' 0.1 ' ' 1

Names of linear predictors: log(mu[,2]/mu[,1]), log(mu[,3]/mu[,1]),
log(mu[,4]/mu[,1])

Residual deviance: 4476.435 on 5616 degrees of freedom

Log-likelihood: -2238.218 on 5616 degrees of freedom

Number of Fisher scoring iterations: 5

No Hauck-Donner effect found in any of the estimates

Reference group is level 1 of the response
```

The `lrtest(mulmodel0, mulmodel1)` command compares the log-likelihood statistics of the fitted model `mulmodel1` and the null model `mulmodel0` using the likelihood ratio test.

```
> lrtest(mulmodel0, mulmodel1)
Likelihood ratio test


Model 1: healthre ~ 1
Model 2: healthre ~ educ
   #Df    LogLik   Df    Chisq    Pr(>Chisq)
1  5616   -2238.2
2  5613   -2166.7  -3   143.14    < 2.2e-16 ***
---
Signif. codes: 0 '***' 0.001 '**' 0.01 '*' 0.05 '.' 0.1 ' ' 1
```

The null hypothesis of the test for the overall model is that the predictor variable does not contribute to the model, and the alternative hypothesis is that the one-predictor model is better than the null model with no independent variables. The likelihood ratio chi-square test statistic $LR\ \chi^2_{(3)} = 143.14$, $p < .001$, which indicates that the overall model with one predictor `educ` is significantly different from zero. Therefore, the one-predictor model provides a better fit than the null model in predicting the logit or log odds of being in a particular category relative to the base category.

Pseudo R^2

We use the `nagelkerke()` function in the `rcompanion` package (Mangiafico, 2021) to compute the pseudo R^2 statistics for the single-predictor model. We load the package first with `library(rcompanion)` and then use `nagelkerke(mulmodel1)`.

```
> # Pseudo R2 with nagelkerke()
> library(rcompanion)
> nagelkerke(mulmodel1)
$`Models`

Model: "vglm, healthre ~ educ, multinomial(refLevel = 1), chp7.mul"
Null:  "vglm, healthre ~ 1, multinomial(refLevel = 1), chp7.mul"

$Pseudo.R.squared.for.model.vs.null
                             Pseudo.R.squared
McFadden                            0.0319758
Cox and Snell (ML)                  0.0735744
Nagelkerke (Cragg and Uhler)        0.0809962

$Likelihood.ratio.test
Df.diff   LogLik.diff    Chisq      p.value
      3       -71.569   143.14   7.9595e-31

$Number.of.observations

Model: 1873
Null:  1873

$Messages
[1] "Note: For models fit with REML, these statistics are based on refitting with ML"

$Warnings
[1] "None"
```

The McFadden R^2 is .032, the Cox and Snell R^2 is .074, and the Nagelkerke R^2 is .081. The same results can be computed using the equations for these three pseudo R^2 statistics. In the R commands below, `LLM1` is the log-likelihood value for the single-predictor model and `LL0` is the log-likelihood value for the null model. In addition, `McFadden1` is the object name for McFadden's R^2, `CS1` for Cox and Snell's R^2, and `NG1` for Nagelkerke's R^2.

```
> LLM1 <- logLik(mulmodel1)
> LL0 <- logLik(mulmodel0)
> McFadden1 <- 1-(LLM1/LL0)
> McFadden1
[1] 0.0319758
> CS1 <- 1-exp(2*(LL0-LLM1)/1873)
> CS1
[1] 0.07357443
> NG1 <- CS1/(1-exp(2*LL0/1873))
> NG1
[1] 0.08099622
```

AIC and BIC Statistics

The AIC and BIC statistics can also be computed from the AIC() and BIC() functions. The output is shown as follows.

```
> AIC(mulmodel1)
[1] 4345.297
> BIC(mulmodel1)
[1] 4378.509
```

7.4 FITTING A MULTIPLE-PREDICTOR MULTINOMIAL LOGISTIC REGRESSION MODEL WITH R

7.4.1 The Multinomial Logistic Regression Model: Multiple-Predictor Model

The command mulmodel2 <- vglm(healthre ~ educ + maritals + female + wrkfull, multinomial(refLevel = 1), data = chp7.mul) tells R to predict the nominal response variable healthre from the four predictor variables educ, maritals, female and wrkfull with multinomial logistic regression. In the vglm() function, the model equation is specified as healthre ~ educ + maritals + female + wrkfull; the multinomial(refLevel = 1) argument specifies that the multinomial family is used to fit the model and the reference level is the first level of healthre; and the data argument specifies data = chp7.mul. The output is shown by the summary(mulmodel2) command.

```
> # Multiple-predictor multinomial logistic regression model with vglm() in VGAM
> mulmodel2 <- vglm(healthre ~ educ + maritals + female + wrkfull, multinomial(refLevel = 1),
data=chp7.mul)
> summary(mulmodel2)

Call:
vglm(formula = healthre ~ educ + maritals + female + wrkfull,
    family = multinomial(refLevel = 1), data = chp7.mul)


Pearson residuals:
                       Min        1Q    Median        3Q    Max
log(mu[,2]/mu[,1])  -5.057   -0.3768   -0.2919   -0.2047  3.343
log(mu[,3]/mu[,1])  -5.982   -0.6924   -0.3757    0.8763  2.003
log(mu[,4]/mu[,1])  -5.734   -0.3810   -0.2745   -0.1603  7.289


Coefficients:
                Estimate   Std. Error   z value   Pr(>|z|)
(Intercept):1    0.02270      0.45925     0.049   0.960576
(Intercept):2   -1.19945      0.45236    -2.652   0.008013 **
(Intercept):3   -3.63716      0.51091    -7.119   1.09e-12 ***
```

```
educ:1          0.08421     0.03483     2.417    0.015632 *
educ:2          0.19416     0.03403     5.706    1.16e-08 ***
educ:3          0.31509     0.03729     8.450     < 2e-16 ***
maritals:1      0.38625     0.22994     1.680    0.093000 .
maritals:2      0.68849     0.21996     3.130    0.001747 **
maritals:3      0.64602     0.23401     2.761    0.005769 **
female:1       -0.02156     0.21283    -0.101    0.919298
female:2        0.16979     0.20413     0.832    0.405546
female:3        0.15659     0.21921     0.714    0.475009
wrkfull:1       0.32714     0.23802     1.374    0.169321
wrkfull:2       0.95782     0.22646     4.230    2.34e-05 ***
wrkfull:3       0.82250     0.24018     3.424    0.000616 ***

---
Signif. codes:  0 '***' 0.001 '**' 0.01 '*' 0.05 '.' 0.1 ' ' 1

Names of linear predictors: log(mu[,2]/mu[,1]), log(mu[,3]/mu[,1]),
log(mu[,4]/mu[,1])

Residual deviance: 4275.242 on 5604 degrees of freedom

Log-likelihood: -2137.621 on 5604 degrees of freedom

Number of Fisher scoring iterations: 6

No Hauck-Donner effect found in any of the estimates

Reference group is level  1  of the response
```

7.4.2 Interpreting R Output

The coefficients section (labeled `Coefficients:`) displays the parameter estimates for the three intercepts and the four predictor variables. Since the multinomial logistic regression model includes a series of binary logistic regression models, the table displays the parameter estimates for the three binary logistic models comparing each category versus the base category. These three equations, labeled `log(mu[,2]/mu[,1])`, `log(mu[,3]/mu[,1])`, and `log(mu[,4]/mu[,1])`, compare categories 2 with 1, categories 3 with 1, and categories 4 with 1, respectively. The estimated intercepts and logit coefficients for these three sub-models are numbered 1, 2, and 3 in the output.

Based on the parameter estimates in the output, the three equations for the model can be expressed as:

$$\ln\left(\frac{P(Y=2)}{P(Y=1)}\right) = .023 + .084\texttt{educ} + .386\texttt{maritals} - .022\texttt{female} + .327\texttt{wrkfull}$$

$$\ln\left(\frac{P(Y=3)}{P(Y=1)}\right) = -1.199 + .194\texttt{educ} + .688\texttt{maritals} + .170\texttt{female} + .958\texttt{wrkfull}$$

$$\ln\left(\frac{P(Y=4)}{P(Y=1)}\right) = -3.637 + .315\texttt{educ} + .646\texttt{maritals} + .157\texttt{female} + .823\texttt{wrkfull}$$

The first equation compares category 2 and category 1. Among the four predictor variables, only `educ` is significant, whereas the other three predictor variables `maritals`, `female`, and `wrkfull` are not significant in predicting the log odds of being in category 2 versus category 1. The coefficient for `educ`, displayed as `educ:1`, $\beta = .084$, Wald $z = 2.417$. The associated p value, `Pr(>|z|) < .05`, so we reject the null hypothesis and conclude that `educ` is significant in predicting the log odds of being in category 2 versus category1. The coefficient for `maritals`, displayed as `maritals:1`, $\beta = .386$, Wald $z = 1.680$, $p > .05$, which is not significant; the coefficient for `female`, displayed as `female:1`, $\beta = -.022$, Wald $z = -.101$, $p > .05$, which is not significant, either; the coefficient for `wrkfull`, displayed as `wrkfull:1`, $\beta = .327$, Wald $z = 1.374$, $p > .05$, which is not significant, either.

The second equation compares category 3 and category 1. The three predictor variables `educ`, `maritals`, and `wrkfull` are significant, whereas `female` is not significant. The predictor variable `educ`, displayed as `educ:2`, $\beta = .194$, Wald $z = 5.706$, $p < .001$. For the predictor variable `maritals`, displayed as `maritals:2`, $\beta = .688$, Wald $z = 3.130$. The associated p value, `P>|z| < .01`, so we reject the null hypothesis; for `wrkfull`, displayed as `wrkfull:2`, $\beta = .958$, Wald $z = 4.230$, $p < .001$; however, for `female`, displayed as `female:2`, $\beta = .170$, Wald $z = .830$, $p = .406$, so the coefficient for `female` is not significantly different from 0.

The third equation compares category 4 and category 1. The three predictor variables `educ`, `maritals`, and `wrkfull` are significant, whereas `female` is not significant. The predictor variable `educ`, displayed as `educ:3`, is also significant. For `educ:3`, $\beta = .315$, Wald $z = 8.450$, $p < .001$. For the predictor variable `maritals`, displayed as `maritals:3`, $\beta = .646$, Wald $z = 2.761$, $p < .01$, so we reject the null hypothesis; for `wrkfull`, displayed as `wrkfull:3`, $\beta = .823$, Wald $z = 3.424$, $p < .001$; however, for `female`, displayed as `female:3`, $\beta = .157$, Wald $z = .714$, $p > .05$, so the coefficient for `female` is not significantly different from 0.

We use the `coef(mulmodel2, matrix = TRUE)` command to extract the coefficients table for the three underlying binary models which compare categories 2 with 1, categories 3 with 1, and categories 4 with 1, respectively. We also use the `confint(mulmodel2, matrix = TRUE)` command to compute the corresponding confidence intervals. The output is omitted here.

We request the odds ratios of being in a particular category versus the base category and the corresponding confidence intervals using the `exp(coef(mulmodel2, matrix = TRUE))` and the `exp(confint(mulmodel2, matrix = TRUE))` commands, respectively. The results are combined with the `cbind()` function.

```
> exp(coef(mulmodel2, matrix = TRUE))
             log(mu[,2]/mu[,1])  log(mu[,3]/mu[,1])  log(mu[,4]/mu[,1])
(Intercept)           1.0229608           0.3013613          0.02632712
educ                  1.0878549           1.2142874          1.37038037
maritals              1.4714529           1.9907077          1.90794028
female                0.9786676           1.1850530          1.16951660
wrkfull               1.3869905           2.6060030          2.27617493

> exp(confint(mulmodel2, matrix = TRUE))
                    2.5 %      97.5 %
(Intercept):1 0.415858123  2.51636007
(Intercept):2 0.124176341  0.73136819
(Intercept):3 0.009672075  0.07166171
educ:1        1.016061645  1.16472089
educ:2        1.135942132  1.29803614
educ:3        1.273795085  1.47428923
maritals:1    0.937605488  2.30925872
maritals:2    1.293540526  3.06362029
maritals:3    1.206069994  3.01826272
female:1      0.644877457  1.48522842
female:2      0.794293786  1.76804926
female:3      0.761058336  1.79719349
wrkfull:1     0.869896953  2.21146033
wrkfull:2     1.671903678  4.06198751
wrkfull:3     1.421546528  3.64460269

> cbind(exp(coef(mulmodel2)), exp(confint(mulmodel2)))
                                2.5 %      97.5 %
(Intercept):1 1.02296079  0.415858123  2.51636007
(Intercept):2 0.30136129  0.124176341  0.73136819
(Intercept):3 0.02632712  0.009672075  0.07166171
educ:1        1.08785487  1.016061645  1.16472089
educ:2        1.21428742  1.135942132  1.29803614
educ:3        1.37038037  1.273795085  1.47428923
maritals:1    1.47145290  0.937605488  2.30925872
maritals:2    1.99070766  1.293540526  3.06362029
maritals:3    1.90794028  1.206069994  3.01826272
female:1      0.97866763  0.644877457  1.48522842
female:2      1.18505297  0.794293786  1.76804926
female:3      1.16951660  0.761058336  1.79719349
wrkfull:1     1.38699048  0.869896953  2.21146033
wrkfull:2     2.60600304  1.671903678  4.06198751
wrkfull:3     2.27617493  1.421546528  3.64460269
```

The results of the odds ratios across the three binary comparisons are summarized in Table 7.2.

7.4.3 Interpreting the Odds Ratios of Being in a Category *j* Versus the Base Category 1

The odds ratio of being in a particular category compared with the base category can be interpreted as the change in the odds of being in that category versus the base category for a one-unit increase in the predictor variable when holding other predictors constant. Recall there are $J - 1$ binary comparisons for a nominal response variable with J categories.

TABLE 7.2 ● Odds Ratios for All Four Predictor Variables Across Three Comparisons ($Y = j$ vs. $Y = 1$)

Category Comparisons	$Y = 2$ vs. $Y = 1$	$Y = 3$ vs. $Y = 1$	$Y = 4$ vs. $Y = 1$
Variables	OR	OR	OR
Educ	1.088*	1.214**	1.370**
Maritals	1.471	1.991**	1.908**
Female	.979	1.185	1.170
Wrkfull	1.387	2.606**	2.276**

*$p < 0.05$
**$p < 0.01$

The odds ratio for each predictor needs to be interpreted across three comparisons. For `educ`, the odds ratios of being in category 2 versus category 1, category 3 versus category 1, and category 4 versus category 1 are 1.088, 1.214, and 1.370, respectively. The results indicate that the odds of being in category 2 versus the base category, category 3 versus the base category, and category 4 versus the base category increase by a factor of 1.088, 1.214, and 1.370, respectively, for a one-unit increase in the `educ` predictor when holding all other predictors constant.

For `maritals`, the odds ratios for the three binary comparisons (i.e., categories 2 vs. 1, 3 vs. 1, and 4 vs. 1) are 1.471, 1.991, and 1.908, respectively. OR(2,1) = 1.471, $p > .05$, which indicates that there is no relationship between `maritals` and the odds of being in category 2 versus the base category 1. The odds ratios for the other two comparisons are significant, OR(3,1) = 1.991, OR(4,1) = 1.908, which indicates that the odds of being in category 3 versus the base category and category 4 versus the base category for the married are 1.991 and 1.908 times as large as the odds for the unmarried, respectively, when holding other predictors constant.

The odds ratios for `wrkfull` can be interpreted in the similar way. The odds ratios for three binary comparisons (i.e., categories 2 vs. 1, 3 vs. 1, and 4 vs. 1) are 1.387, 2.606, and 2.276, respectively. OR(2,1) = 1.387, $p > .05$, which indicates that there is no relationship between `wrkfull` and the odds of being in category 2 versus the base category 1. The odds ratios for the other two comparisons are significant. The odds of being in category 3 versus the base category and category 4 versus the base category for those working full time are 2.606 and 2.267 times as large as the odds for those not working full time, respectively, when holding other predictors constant.

With regard to `female`, none of the odds ratios for the binary comparisons are significant. It indicates that being female does not impact the odds of being in any particular category versus the base category when holding other predictors constant.

7.4.4 Model Fit Statistics

Testing the Overall Model Using the Likelihood Ratio Test

To test if the overall model is significant, we compare the multiple-predictor model with the null model using the `lrtest()` function. The `lrtest(mulmodel0, mulmodel2)` command compares the log-likelihood statistics of the fitted model `mulmodel2` and the null model `mulmodel0` using the likelihood ratio test. The resulting output is displayed below.

```
> # Testing the overall model using the likelihood ratio test
> lrtest(mulmodel0, mulmodel2)
Likelihood ratio test

Model 1: healthre ~ 1
Model 2: healthre ~ educ + maritals + female + wrkfull
    #Df    LogLik    Df    Chisq    Pr(>Chisq)
1   5616   -2238.2
2   5604   -2137.6   -12   201.19    < 2.2e-16 ***

---
Signif. codes: 0 '***' 0.001 '**' 0.01 '*' 0.05 '.' 0.1 ' '
```

The likelihood ratio test $\chi^2_{(12)} = 201.19$, $p < .001$, indicates that the full model with the four predictors provides a better fit than the null model with no independent variables in predicting the nominal response variable.

Pseudo R^2

The `nagelkerke(mulmodel2)` command computes the three types of pseudo R^2 statistics and the likelihood ratio test statistic for the overall multiple-predictor model. The output is omitted here. To obtain the same results, we can also compute the three types of pseudo R^2 statistics with their equations for the multiple-predictor model as follows.

```
> LLM2 <- logLik(mulmodel2)
> McFadden2 <- 1-(LLM2/LL0)
> McFadden2
[1] 0.04494503
> CS2 <- 1-exp(2*(LL0-LLM2)/1873)
> CS2
[1] 0.1018496
> NG2 <- CS2/(1-exp(2*LL0/1873))
> NG2
[1] 0.1121237
```

In the output, `LLM2` is the log-likelihood value for the multiple-predictor model and `LL0` is the log-likelihood value for the null model. The McFadden R^2 is .045, the Cox and Snell R^2 is .102, and the Nagelkerke R^2 is .112.

AIC and BIC Statistics

The `AIC(mulmodel2)` and `BIC(mulmodel2)` commands produce the AIC and BIC statistics.

```
> AIC(mulmodel2)
[1] 4305.241
> BIC(mulmodel2)
[1] 4388.271
```

The AIC and BIC statistics are 4,305.241 and 4,388.271, respectively. Recall the AIC and BIC in the single predictor model are 4,345.297 and 4,378.509, respectively. Compared with the single-variable model, both AIC and BIC indicate that the multiple-predictor model fits the data better.

7.4.5 Interpreting the Predicted Probabilities With the `ggpredict()` Function in the `ggeffects` Package

By using the `ggpredict()` function in the `ggeffects` package (Lüdecke, 2018b), we can compute the predicted probabilities for each category of the nominal response variable at specified values of the predictor variables. The command `pr.mul2m <- ggpredict(mulmodel2, terms = "educ[12, 14, 16]", ci = NA)` tells R to compute the predicted probabilities for each category of the nominal response variable using the `ggpredict()` function. The argument inside the function includes the estimated model, `mulmodel2`, the `terms = "educ[12, 14, 16]"` option, which specifies the predictor variable `educ` at the values of 12, 14, and 16 when holding the other predictor variables at their means, and the `ci = NA` option for not specifying the confidence intervals. The `terms` option can specify up to four variables, including the second to fourth grouping variables. Please also note that the confidence intervals can only be obtained for the cumulative probabilities, so the `ci = NA` option is needed there. The output is assigned to the object named `pr.mul2m`.

```
> library(ggeffects)
> pr.mul2m <- ggpredict(mulmodel2, terms="educ[12, 14, 16]", ci=NA)
> pr.mul2m

# Predicted probabilities of healthre

# Response Level = 1

educ     |    Predicted
--------------------
  12     |         0.07
  14     |         0.05
  16     |         0.03
```

```
# Response Level = 2

educ      |     Predicted
-------------------
  12      |          0.26
  14      |          0.22
  16      |          0.17

# Response Level = 3

educ      |     Predicted
-------------------
  12      |          0.50
  14      |          0.51
  16      |          0.51

# Response Level = 4

educ      |     Predicted
-------------------
  12      |          0.17
  14      |          0.22
  16      |          0.28


Adjusted for:
* maritals = 0.44
*   female = 0.56
*  wrkfull = 0.47

> plot(pr.mul2m)
```

When `educ` = 12, and the other predictor variables are held at their means (`maritals` = .44, `female` = .56, and `wrkfull` = .47), the predicted probability for the response level 1 (i.e., $Y = 1$) is .07.

When `educ` = 14, and the other three predictor variables are held at their means, the predicted probability for $Y = 1$ is .05.

When `educ` = 16, and the other predictor variables are held at their means, the predicted probability for $Y = 1$ is .03.

The predicted probabilities for all the four response levels are plotted using `plot(pr.mul2m)`. Figure 7.1 shows the predicted probabilities of being in each category (i.e., $Y = 1$, 2, 3, and 4).

The graph shows that with the increase of the years of education, the probabilities of being in poor and fair health condition (categories 1 and 2) decrease. In other words, people with higher levels of education are less likely associated with poor and fair health conditions. In addition, with the increase of the years of education, the probabilities of being in good and excellent health condition (categories 3 and 4) increase. In other words, people with a higher level of education are more likely to be in good and excellent health condition.

FIGURE 7.1 ⬢ Predicted Probabilities for **educ** at **12, 14,** and **16**

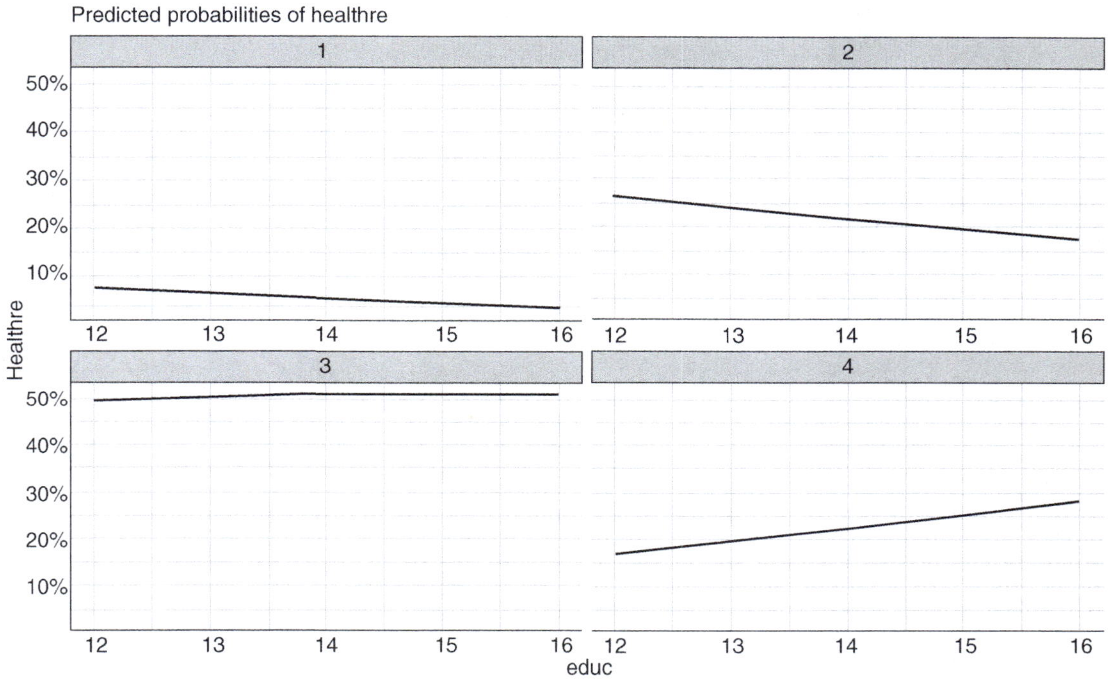

With the `terms = c("educ[12, 14, 16]", "maritals")` argument, the predicted probabilities for `educ` at 12, 14, and 16 can be grouped by `maritals`. The output is assigned to the object named `pr.mul2`.

```
> pr.mul2 <- ggpredict(mulmodel2, terms=c("educ[12, 14, 16]", "maritals"), ci=NA)
> pr.mul2

# Predicted probabilities of healthre

# Response Level = 1
# maritals = 0

educ    |    Predicted
--------------------
  12    |         0.09
  14    |         0.06
  16    |         0.04
```

```
# Response Level = 2
# maritals = 0

educ     |    Predicted
-------------------
  12     |         0.28
  14     |         0.24
  16     |         0.19

# Response Level = 3
# maritals = 0


educ     |    Predicted
-------------------
  12     |         0.47
  14     |         0.49
  16     |         0.49

# Response Level = 4
# maritals = 0


educ     |    Predicted
-------------------
  12     |         0.16
  14     |         0.22
  16     |         0.28

# Response Level = 1
# maritals = 1


educ     |    Predicted
-------------------
  12     |         0.05
  14     |         0.03
  16     |         0.02

# Response Level = 2
# maritals = 1


educ     |    Predicted
-------------------
  12     |         0.24
  14     |         0.19
  16     |         0.15

# Response Level = 3
# maritals = 1


educ     |    Predicted
-------------------
  12     |         0.53
  14     |         0.54
  16     |         0.53
```

```
# Response Level = 4
# maritals = 1

educ    |   Predicted
--------------------
  12    |        0.18
  14    |        0.23
  16    |        0.29

Adjusted for:
*  female = 0.56
* wrkfull = 0.47

> plot(pr.mul2)
```

The results are plotted with `plot(pr.mul2)`. Figure 7.2 shows the predicted probabilities of being in Categories 1, 2, 3, and 4 for `educ` by the grouping variable `maritals`.

FIGURE 7.2 ● Predicted Probabilities of Being in Categories 1, 2, 3, and 4 for educ by the Grouping Variable `maritals`

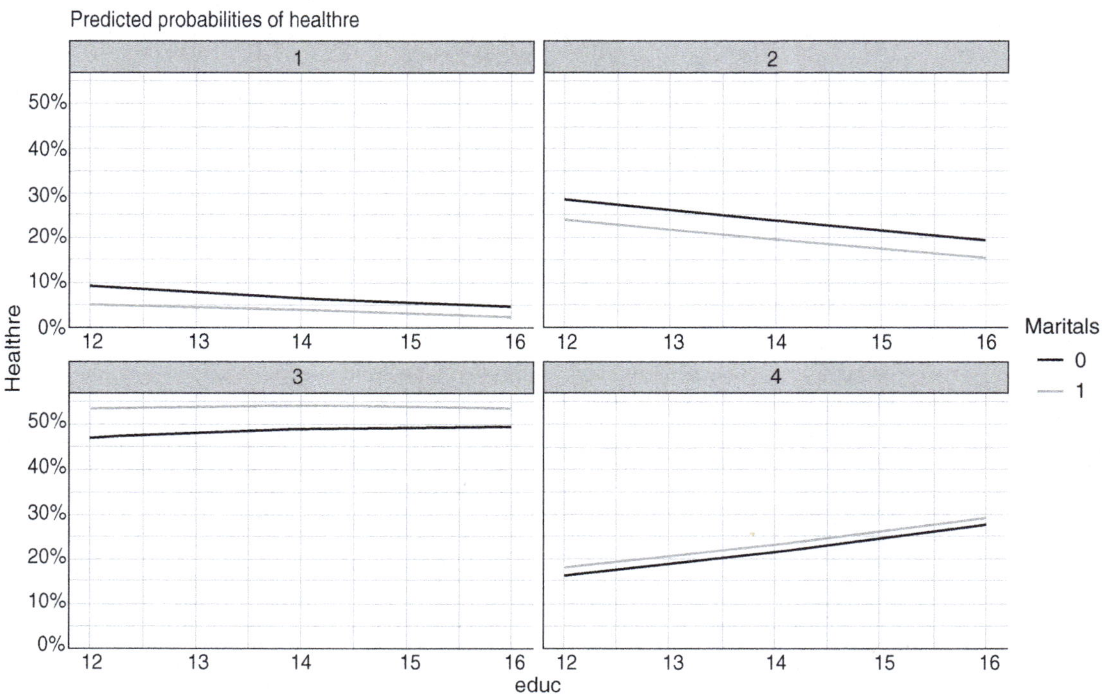

7.4.6 Model Comparisons Using the Likelihood Ratio Test

The likelihood ratio test, or the deviance difference test, is used to compare the multiple-predictor model and the one-predictor model. In the `lrtest(mulmodel1, mulmodel2)` syntax, `mulmodel1` and `mulmodel2` are the two models being compared. The following output is displayed.

```
> # Model comparison with the likelihood ratio test
> lrtest(mulmodel1, mulmodel2)
Likelihood ratio test

Model 1: healthre ~ educ
Model 2: healthre ~ educ + maritals + female + wrkfull
    #Df    LogLik   Df     Chisq    Pr(>Chisq)
1   5613   -2166.7
2   5604   -2137.6  -9    58.056    3.171e-09 ***

---
Signif. codes: 0 '***' 0.001 '**' 0.01 '*' 0.05 '.' 0.1 ' ' 1
```

The likelihood ratio test, $\chi^2_{(9)} = 58.056$, $p < .001$, which indicates that the full model with the four predictor variables fits the data better than the single-predictor model.

7.5 MULTINOMIAL LOGISTIC REGRESSION WITH THE `multinom()` FUNCTION IN THE `nnet` PACKAGE

We can also use the `multinom()` function in the `nnet` package to fit multinomial logistic regression models. Since `nnet` is a user-written package, you need to install it first by typing `install.packages("nnet")` and then load the package by typing `library(nnet)`.

In the following example, the `mulmodel2b <- multinom(healthre ~ educ + maritals + female + wrkfull, data = chp7.mul)` command tells R to predict the nominal response variable `healthre` from the four independent variables. In the model formula for the `multinom()` function, the dependent variable `healthre` and the four predictor variables are separated by the tilde (~). The four predictor variables include, `educ`, `maritals`, `female`, and `wrkfull` which are connected by plus (+) symbols. We also specify the data arguments `data = chp7.mul` in the function. The fitted model is named `mulmodel2b`. The following output is shown by the `summary(mulmodel2b)` command. Please note that the default reference category or base category in the function is the lowest category. If you would like to change the reference category, you need to use the `relevel()` function to specify it before model fitting. For example, `healthre <- relevel(healthre, ref = 4)` defines category 4 to be the base category for the nominal response variable.

```
> # Using multinom() in nnet
> library(nnet)
> mulmodel2b <- multinom(healthre ~ educ + maritals + female + wrkfull, data=chp7.mul)
# weights: 24 (15 variable)
initial  value 2596.529338
iter  10 value 2150.935582
iter  20 value 2137.622895
final  value 2137.620746
converged
> summary(mulmodel2b)
Call:
multinom(formula = healthre ~ educ + maritals + female + wrkfull,
    data = chp7.mul)

Coefficients:
   (Intercept)       educ    maritals      female    wrkfull
2   0.02279756 0.08420084 0.3861814 -0.02155419 0.3271689
3  -1.19940172 0.19415272 0.6884447  0.16981115 0.9578518
4  -3.63712240 0.31508366 0.6459960  0.15661149 0.8225368

Std. Errors:
   (Intercept)       educ  maritals    female   wrkfull
2    0.4592498 0.03483414 0.2299392 0.2128258 0.2380238
3    0.4523597 0.03402868 0.2199547 0.2041307 0.2264614
4    0.5109066 0.03729027 0.2340112 0.2192057 0.2401842

Residual Deviance: 4275.241
AIC: 4305.241
```

The R output includes the call of the model command, the coefficients, the standard errors, the residual deviance, and the AIC statistic. The call shows the R command for the model. The second section shows the coefficients table including the parameter estimates for the intercept and the four predictor variables. The third section shows the standard errors. Finally, the deviance residuals and AIC are displayed at the end.

In the coefficients section (labeled `Coefficients:`), the first column lists 2, 3, and 4 which are the categories of the nominal response variable. They are the outcomes of these three binary logistic regression models which compare categories 2 with 1, categories 3 with 1, and categories 4 with 1, respectively. For example, when the nominal response variable is 2, the binary model compares categories 2 with 1. The first row lists the intercept and the four predictor variables. Each predictor has three coefficients for the underlying binary models since the reference level is category 1. In addition, the three intercepts are also estimated for the binary models.

The standard errors section (labeled `Std. Errors:`) provides the corresponding standard errors of the coefficients above. Please note that the Wald z statistics and the associated p values are not reported in the output.

We can compute the Wald *z* statistics and the associated *p* values with the following command `z <- summary(mulmodel2b)$coefficients/summary(mulmodel2b)$standard.errors`. We first compute the Wald *z* statistics by dividing the coefficients to their standard errors and name the object `z`. We then compute the associated *p* values with the command `p <-(1-pnorm(abs(z),0,1))*2`.

```
> z <- summary(mulmodel2b)$coefficients/summary(mulmodel2b)$standard.errors
> z
   (Intercept)      educ   maritals     female  wrkfull
2   0.04964088  2.417193   1.679493 -0.1012762 1.374522
3  -2.65143354  5.705561   3.129938  0.8318747 4.229647
4  -7.11895783  8.449488   2.760535  0.7144498 3.424608

> p <- (1-pnorm(abs(z),0,1))*2
> p
   (Intercept)         educ    maritals    female      wrkfull
2 9.604086e-01 1.564071e-02 0.093055936 0.9193312 1.692796e-01
3 8.015089e-03 1.159605e-08 0.001748430 0.4054797 2.340586e-05
4 1.087352e-12 0.000000e+00 0.005770681 0.4749491 6.156865e-04
```

The logit coefficients can be extracted with `coef(mulmodel2b)`.

```
> coef(mulmodel2b)
   (Intercept)       educ  maritals      female   wrkfull
2   0.02279756 0.08420084 0.3861814 -0.02155419 0.3271689
3  -1.19940172 0.19415272 0.6884447  0.16981115 0.9578518
4  -3.63712240 0.31508366 0.6459960  0.15661149 0.8225368
```

The odds ratios can be obtained by using `exp(coef(mulmodel2b))`. We also use the `exp(confint(mulmodel2b))` command to obtain the corresponding confidence intervals.

```
> exp(coef(mulmodel2b))
  (Intercept)     educ maritals    female  wrkfull
2   1.0230594 1.087847 1.471352 0.9786764 1.387036
3   0.3013745 1.214282 1.990617 1.1850810 2.606092
4   0.0263280 1.370374 1.907886 1.1695411 2.276267

> exp(confint(mulmodel2b))
, , 2

                2.5 %   97.5 %
(Intercept) 0.4158979 2.516604
educ        1.0160547 1.164713
```

```
maritals       0.9375444     2.309091
female         0.6448838     1.485241
wrkfull        0.8699243     2.211535

, , 3

                   2.5 %        97.5 %
(Intercept)    0.1241816     0.7314013
educ           1.1359369     1.2980300
maritals       1.2934866     3.0634692
female         0.7943130     1.7680901
wrkfull        1.6719583     4.0621322

, , 4

                   2.5 %         97.5 %
(Intercept)    0.00967238     0.0716642
educ           1.27378919     1.4742822
maritals       1.20604029     3.0181662
female         0.76107465     1.7972304
wrkfull        1.42160204     3.6447554
```

We can use the `ggpredict()` function in the `ggeffects` package to compute the predicted probabilities for each category of the nominal response variable at the specified values of the predictor variables. The same command introduced in the previous section on the `vglm()` function in this chapter can be used. The output is omitted here.

7.6 MULTINOMIAL LOGISTIC REGRESSION WITH THE `mlogit()` FUNCTION IN THE `mlogit` PACKAGE

We can also use the `mlogit()` function in the `mlogit` package to fit multinomial logistic regression models. Since `mlogit` is a user-written package, you need to install it first by typing `install.packages("mlogit")` and then load the package by typing `library(mlogit)`.

Different from the `vglm()` and `multinom()` functions, we need to follow two steps to use the `mlogit()` function. First, before fitting the model, we need to create the dataset in the wide format by using the `mlogit.data()` function so that the `mlogit()` function can load the data. In the command `chp7 <- mlogit.data (chp7.mul, choice = "healthre", shape = "wide")`, we specify the data `chp7.mul`, followed by the `choice = "healthre"` argument for the nominal response variable and the `shape = "wide"` argument. The new data are named `chp7`. Second, we run the model command. The model equation for the `mlogit()` function is slightly different from those for the `vglm()` and `multinom()` functions. In the following example, the `mulmodel2c <- mlogit` (healthre ~ 1 | educ + maritals + female + wrkfull, reflevel = 1, data = chp7) command tells R to predict the nominal response variable healthre from the intercept and four independent variables. In the model formula for the mlogit() function, the dependent variable healthre and the intercept are separated by the tilde (~); 1 is the intercept, which is separated from the predictor variables by the vertical line (|); the four predictor variables include, educ, maritals, female, and wrkfull which are connected by plus (+) symbols. The reflevel = 1 argument defines the reference level. We also specify the data arguments data = chp7 in the function. The fitted model is named mulmodel2c. The following output is shown by the summary(mulmodel2c) command.

```
> # Using mlogit() in mlogit
> library(mlogit)
> chp7 <- mlogit.data(chp7.mul, choice="healthre", shape="wide")
> mulmodel2c <- mlogit(healthre ~ 1 | educ + maritals + female + wrkfull, reflevel = 1,
data=chp7)
> summary(mulmodel2c)

Call:
mlogit(formula = healthre ~ 1 | educ + maritals + female + wrkfull,
    data = chp7, reflevel = 1, method = "nr")

Frequencies of alternatives:
        1          2          3          4
0.063001   0.227977   0.487987   0.221036

nr method
6 iterations, 0h:0m:1s
g'(-H)^-1g = 4.62E-05
successive function values within tolerance limits

Coefficients :
                 Estimate   Std. Error   z-value     Pr(>|z|)
2:(intercept)    0.022702     0.459249    0.0494     0.960575
3:(intercept)   -1.199445     0.452359   -2.6515     0.008013 **
4:(intercept)   -3.637155     0.510906   -7.1190    1.087e-12 ***
2:educ           0.084208     0.034834    2.4174     0.015632 *
3:educ           0.194157     0.034029    5.7057    1.159e-08 ***
4:educ           0.315088     0.037290    8.4496    < 2.2e-16 ***
2:maritals       0.386250     0.229941    1.6798     0.093001 .
3:maritals       0.688490     0.219957    3.1301     0.001747 **
4:maritals       0.646024     0.234013    2.7606     0.005769 **
2:female        -0.021563     0.212826   -0.1013     0.919298
3:female         0.169787     0.204131    0.8318     0.405546
4:female         0.156590     0.219206    0.7144     0.475009
2:wrkfull        0.327136     0.238023    1.3744     0.169322
3:wrkfull        0.957817     0.226461    4.2295    2.342e-05 ***
4:wrkfull        0.822496     0.240184    3.4244     0.000616 ***

---
Signif. codes:  0 '***' 0.001 '**' 0.01 '*' 0.05 '.' 0.1 ' ' 1

Log-Likelihood: -2137.6
McFadden R^2:  0.044945
Likelihood ratio test : chisq = 201.19 (p.value = < 2.22e-16)
```

The R output includes the call of the model command, the frequencies of alternatives, the estimation method, the coefficients section, the log likelihood value, the McFadden R^2, and the likelihood ratio test statistic. The call shows the R command for the model. The second section shows the Newton-Raphson method for maximum likelihood estimation. The third section shows the coefficients table including the parameter estimates for the intercept and the four predictor variables. Finally, the log likelihood value, the McFadden R^2, and the likelihood ratio test statistic are displayed at the end.

The coefficients section (labeled `Coefficients:`) displays the parameter estimates for the intercepts and the predictor variables, their standard errors, the Wald z statistics, and the associated p values. It displays the parameter estimates for the three binary logistic models comparing each category with the base category 1. For example, the three coefficients for `educ` are displayed as `2:educ`, `3:educ`, and `4:educ`, respectively.

We use the `coef(mulmodel2c)` command to extract the logit coefficients and use the `confint(mulmodel2c)` command to obtain the confidence intervals. The output is omitted here.

The odds ratios can be obtained with `exp(coef(mulmodel2c))`. We also use the `exp(confint(mulmodel2c))` command to obtain the corresponding confidence intervals. Both results are combined with the `cbind()` function.

```
> exp(coef(mulmodel2c))
2:(intercept)   3:(intercept)   4:(intercept)       2:educ       3:educ
   1.02296127      0.30136144      0.02632714   1.08785483   1.21428738
       4:educ      2:maritals      3:maritals   4:maritals     2:female
   1.37038032      1.47145243      1.99070700   1.90793964   0.97866759
     3:female        4:female       2:wrkfull    3:wrkfull    4:wrkfull
   1.18505292      1.16951655      1.38698996   2.60600205   2.27617406

> exp(confint(mulmodel2c))
                     2.5 %      97.5 %
2:(intercept)  0.415858299  2.51636139
3:(intercept)  0.124176395  0.73136860
4:(intercept)  0.009672079  0.07166175
2:educ         1.016061603  1.16472085
3:educ         1.135942082  1.29803610
4:educ         1.273795029  1.47428918
2:maritals     0.937605127  2.30925811
3:maritals     1.293540013  3.06361947
4:maritals     1.206069519  3.01826190
2:female       0.644877409  1.48522840
3:female       0.794293723  1.76804925
4:female       0.761058277  1.79719347
2:wrkfull      0.869896566  2.21145965
3:wrkfull      1.671902920  4.06198627
4:wrkfull      1.421545886  3.64460156
```

```
> cbind(exp(coef(mulmodel2c)), exp(confint(mulmodel2c)))
                                    2.5 %      97.5 %
2:(intercept)     1.02296127   0.415858299   2.51636139
3:(intercept)     0.30136144   0.124176395   0.73136860
4:(intercept)     0.02632714   0.009672079   0.07166175
2:educ            1.08785483   1.016061603   1.16472085
3:educ            1.21428738   1.135942082   1.29803610
4:educ            1.37038032   1.273795029   1.47428918
2:maritals        1.47145243   0.937605127   2.30925811
3:maritals        1.99070700   1.293540013   3.06361947
4:maritals        1.90793964   1.206069519   3.01826190
2:female          0.97866759   0.644877409   1.48522840
3:female          1.18505292   0.794293723   1.76804925
4:female          1.16951655   0.761058277   1.79719347
2:wrkfull         1.38698996   0.869896566   2.21145965
3:wrkfull         2.60600205   1.671902920   4.06198627
4:wrkfull         2.27617406   1.421545886   3.64460156
```

The `ggpredict()` function in the `ggeffects` package also works with the `mlogit()` function. The output is omitted here.

7.7 MAKING PUBLICATION-QUALITY TABLES

7.7.1 Presenting the Results of the `vglm` Models Using the `texreg` Package

The `stargazer()` function currently cannot directly produce the results table from the `vglm` models, so we use the `screenreg()` and `htmlreg()` functions from the `texreg` package (Leifeld, 2013). Since the package has been installed in preceding chapters, we only need to load the package by typing `library(texreg)`.

After we use the `vglm()` function to fit the single-predictor model `mulmodel1` and the multiple-predictor model `mulmodel2`, we create a table containing the results of the both model with the following command: `screenreg(list(mulmodel1, mulmodel2))`. In the `screenreg()` function, we specify the two model objects to be presented with the `list()` function. The output is a plain text table.

```
> # Presenting the results of the multinomial logistic Models using the texreg
package
> library(texreg)
Version:    1.37.5
Date:       2020-06-17
Author:     Philip Leifeld (University of Essex)

> screenreg(list(mulmodel1, mulmodel2))
```

```
=========================================================
                   Model 1             Model 2
---------------------------------------------------------
(Intercept):1        0.18                0.02
                    (0.43)              (0.46)
(Intercept):2       -0.78               -1.20 **
                    (0.42)              (0.45)
(Intercept):3       -3.28 ***           -3.64 ***
                    (0.48)              (0.51)
educ:1               0.09 **             0.08 *
                    (0.03)              (0.03)
educ:2               0.22 ***            0.19 ***
                    (0.03)              (0.03)
educ:3               0.34 ***            0.32 ***
                    (0.04)              (0.04)
maritals:1                               0.39
                                        (0.23)
maritals:2                               0.69 **
                                        (0.22)
maritals:3                               0.65 **
                                        (0.23)
female:1                                -0.02
                                        (0.21)
female:2                                 0.17
                                        (0.20)
female:3                                 0.16
                                        (0.22)
wrkfull:1                                0.33
                                        (0.24)
wrkfull:2                                0.96 ***
                                        (0.23)
wrkfull:3                                0.82 ***
                                        (0.24)
---------------------------------------------------------
Log Likelihood   -2166.65            -2137.62
DF                5613                5604
Num. obs.         5619                5619
=========================================================
*** p < 0.001; ** p < 0.01; * p < 0.05

> htmlreg(list(mulmodel1, mulmodel2), file="chap7mul.doc", doctype=TRUE,
html.tag=TRUE, head.tag=TRUE)
The table was written to the file 'chap7mul.doc'.
```

We can also use the `htmlreg()` function to create a regression table for the estimated results and save it to a Microsoft Word file named `chap7mul.doc` with the following command: `htmlreg(list(mulmodel1, mulmodel2), file = "chap7mul.doc", doctype = TRUE, html.tag = TRUE, head.tag = TRUE)`. It automatically produces Table 7.3, as shown here in its original format, presenting the results of both the single-predictor and the multiple-predictor multinomial logistic regression models.

TABLE 7.3 ● Results of the Multinomial Logistic Regression Models: Single-Predictor Model and Multiple-Predictor Models (Shown in Original Format Generated by R)

	Model 1	Model 2
(Intercept):1	0.18	0.02
	(0.43)	(0.46)
(Intercept):2	−0.78	-1.20^{**}
	(0.42)	(0.45)
(Intercept):3	-3.28^{***}	-3.64^{***}
	(0.48)	(0.51)
educ:1	0.09^{**}	0.08^{*}
	(0.03)	(0.03)
educ:2	0.22^{***}	0.19^{***}
educ:3	(0.03)	(0.03)
	0.34^{***}	0.32^{***}
maritals:1	(0.04)	(0.04)
		0.39
maritals:2		(0.23)
		0.69^{**}
maritals:3		(0.22)
		0.65^{**}
female:1		(0.23)
		−0.02
female:2		(0.21)
		0.17
female:3		(0.20)
		0.16
wrkfull:1		(0.22)
		0.33

Table 7.3 *(Continued)*

	Model 1	Model 2
wrkfull:2		(0.24)
		0.96***
wrkfull:3		(0.23)
		0.82***
Log Likelihood		(0.24)
	−2166.65	−2137.62
DF	5613	5604
Num. obs.	5619	5619

***$p < 0.001$
**$p < 0.01$
*$p < 0.05$

7.8 REPORTING THE RESULTS

Reporting the results for multinomial logistic regression is similar to that used for binary logistic regression. The following are the generic guidelines for reporting the results. You may need to adjust your writing since your discipline or journals may have different requirements.

First, describe the multinomial logistic regression model, the nominal response variable and the independent variables, and your research hypothesis or the purpose of your study. Include a couple of sentences explaining why this model is appropriate for the analysis.

Second, if available, report the likelihood ratio test statistic for the model and the associated p value, followed by the interpretation on whether the fitted model is better than the null model. If more than one model is developed, then compare models using likelihood ratio test statistics and/or the AIC and BIC statistics.

Third, report the parameter estimates for the predictor variables, their standard errors, the associated p values in a table. Since a multinomial logistic regression model includes $J - 1$ binary comparisons, label them in the table. If more than one model is fitted, then the results of all the competing models need to be presented in a table. In addition, report the odds ratios or relative risk ratios for each predictor in the table or text and interpret the results. The following is an example of summarizing the results for the multinomial logistic regression model illustrated previously.

The multinomial logistic regression analysis was conducted to predict the ordinal outcome variable, health status, from a set of predictor variables, such as marital status, years of education, gender, and working status. Although the multinomial logistic regression model is normally used to estimate the nominal response variables, it is an alternative to estimate ordinal response variables when the proportional odds assumption is violated.

The likelihood ratio test for the fitted model $\chi^2_{(12)} = 151.28$, $p < .001$, which indicated that the four-predictor model provided a better fit than the null model with no independent variables in predicting the logit of being in any other category of health status compared with being in the base category (i.e., poor health).

Table 7.2 displays the parameter estimates for the three binary logistic models comparing each category with the base category since the multinomial logistic regression model is treated as a series of binary logistic regression models. These three equations compare categories 2 with 1, 3 with 1, and 4 with 1, respectively.

For `educ`, the odds ratios of being in category 2 versus category 1, category 3 versus category 1, and category 4 versus category 1 are 1.088, 1.214, and 1.370, respectively. The results indicate that the odds of being in category 2 versus the base category, category 3 versus the base category, and category 4 versus the base category increase by a factor of 1.088, 1.214, and 1.370, respectively, for a one-unit increase in the `educ` predictor when holding all other predictors constant.

For `maritals`, the odds ratios for the three binary comparisons (i.e., categories 2 vs. 1, 3 vs. 1, and 4 vs. 1) are 1.471, 1.991, and 1.908, respectively. OR(2,1) = 1.471, $p > .05$, which indicates that there is no relationship between `maritals` and the odds of being in category 2 versus the base category 1. The odds ratios for the other two comparisons are significant, OR(3,1) = 1.991, OR(4,1) = 1.908, which indicates that the odds of being in category 3 versus the base category and category 4 versus the base category for the married are 1.991 and 1.908 times as large as the odds for the unmarried, respectively, when holding other predictors constant.

The odds ratios for `wrkfull` can be interpreted in the similar way. The odds ratios for three binary comparisons (i.e., categories 2 vs. 1, 3 vs. 1, and 4 vs. 1) are 1.387, 2.606, and 2.276, respectively. OR(2,1) = 1.387, $p > .05$, which indicates that there is no relationship between `wrkfull` and the odds of being in category 2 versus the base category 1. The odds ratios for the other two comparisons are significant. The odds of being in category 3 versus the base category and category 4 versus the base category for those working full time are 2.606 and 2.267 times as large as the odds for those not working full time, respectively, when holding the other predictors constant.

With regard to `female`, none of the odds ratios for the binary comparisons are significant. It indicates that being female does not impact the odds of being in any particular category versus the base category when holding other predictors constant.

7.9 SUMMARY OF R COMMANDS IN THIS CHAPTER

```
# Chap 7 R Script
# Remove all objects
rm(list = ls(all = TRUE))

# The following user-written packages need to be installed first by using install.packages(" ") and then
by loading it with library()
# library(VGAM)          # It is already installed for Chapter 4
# library(rcompanion)    # It is already installed for Chapter 3
# library(ggeffects)     # It is already installed for Chapter 2
# library(texreg)        # It is already installed for Chapter 4
# library(nnet)
# library(mlogit)

# Import the GSS 2016 data
library(foreign)
chp7.mul <- read.dta("C:/CDA/gss2016.dta")
chp7.mul$healthre <- factor(chp7.mul$healthre)
chp7.mul$educ <- as.numeric(chp7.mul$educ)
chp7.mul$wrkfull <- as.numeric(chp7.mul$wrkfull)
chp7.mul$maritals <- as.numeric(chp7.mul$maritals)
attach(chp7.mul)

# One-predictor multinomial logistic regression model with vglm() in VGAM
library(VGAM)
mulmodel1 <- vglm(healthre ~ educ, multinomial(refLevel = 1), data=chp7.mul)
summary(mulmodel1)
coef(mulmodel1, matrix = TRUE)
confint(mulmodel1, matrix = TRUE)
exp(coef(mulmodel1, matrix = TRUE))
exp(confint(mulmodel1, matrix = TRUE))
cbind(exp(coef(mulmodel1)), exp(confint(mulmodel1)))
```

```
# Testing the overall model using the likelihood ratio test
mulmodel0 <- vglm(healthre ~ 1, multinomial(refLevel = 1), data=chp7.mul)
summary(mulmodel0)
lrtest(mulmodel0, mulmodel1)

# Pseudo R2 with nagelkerke()
library(rcompanion)
nagelkerke(mulmodel1)

# Pseudo R2 with equations
LLM1 <- logLik(mulmodel1)
LL0 <- logLik(mulmodel0)
McFadden1 <- 1-(LLM1/LL0)
McFadden1
CS1 <- 1-exp(2*(LL0-LLM1)/1873)
CS1
NG1 <- CS1/(1-exp(2*LL0/1873))
NG1

# AIC and BIC Statistics
AIC(mulmodel1)
BIC(mulmodel1)

# Multiple-predictor multinomial logistic regression model with vglm() in VGAM
mulmodel2 <- vglm(healthre ~ educ + maritals + female + wrkfull, multinomial(refLevel = 1),
data=chp7.mul)
summary(mulmodel2)
coef(mulmodel2, matrix = TRUE)
confint(mulmodel2, matrix = TRUE)
exp(coef(mulmodel2, matrix = TRUE))
exp(confint(mulmodel2, matrix = TRUE))
cbind(exp(coef(mulmodel2)), exp(confint(mulmodel2)))

# Testing the overall model using the likelihood ratio test
lrtest(mulmodel0, mulmodel2)

# Pseudo R2 with nagelkerke()
nagelkerke(mulmodel2)

# Pseudo R2 with equations
LLM2 <- logLik(mulmodel2)
McFadden2 <- 1-(LLM2/LL0)
McFadden2
CS2 <- 1-exp(2*(LL0-LLM2)/1873)
CS2
NG2 <- CS2/(1-exp(2*LL0/1873))
NG2

# AIC and BIC Statistics
AIC(mulmodel2)
BIC(mulmodel2)
```

```
# Model comparison with the likelihood ratio test
lrtest(mulmodel1, mulmodel2)


# Predicted probabilities with ggpredict() in ggeffects
library(ggeffects)
pr.mul2m <- ggpredict(mulmodel2, terms="educ[12, 14, 16]", ci=NA)
pr.mul2m
plot(pr.mul2m)


pr.mul2 <- ggpredict(mulmodel2, terms=c("educ[12, 14, 16]", "maritals"), ci=NA)
pr.mul2
plot(pr.mul2)


# Presenting the results of the multinomial logistic Models using the texreg package
library(texreg)
screenreg(list(mulmodel1, mulmodel2))
htmlreg(list(mulmodel1, mulmodel2), file="chap7mul.doc", doctype=TRUE, html.tag=TRUE, head.tag=TRUE)


# Using multinom() in nnet
library(nnet)
mulmodel2b <- multinom(healthre ~ educ + maritals + female + wrkfull, data=chp7.mul)
summary(mulmodel2b)
coef(mulmodel2b)
z <- summary(mulmodel2b)$coefficients/summary(mulmodel2b)$standard.errors
z
p <- (1-pnorm(abs(z),0,1))*2
p
exp(coef(mulmodel2b))
exp(confint(mulmodel2b))


# Using mlogit() in mlogit
library(mlogit)
chp7 <- mlogit.data(chp7.mul, choice="healthre", shape="wide")
mulmodel2c <- mlogit(healthre ~ 1 | educ + maritals + female + wrkfull, reflevel = 1, data=chp7)
summary(mulmodel2c)
coef(mulmodel2c)
confint(mulmodel2c)
exp(coef(mulmodel2c))
exp(confint(mulmodel2c))
cbind(exp(coef(mulmodel2c)), exp(confint(mulmodel2c)))


detach(chp7.mul)
```

Glossary

The multinomial distribution is an extension of the binomial distribution when the discrete random variable is a nominal variable with more than two categories.

The multinomial logistic regression model is used to estimate nominal response variables with multiple unordered categories. This model is a generalization of binary logistic regression when there are more than two categories in a response variable.

The odds in the multinomial logistic model can be defined as the ratio of the probability of being in a particular category to the probability of being in the base category.

Exercises

Use the GSS 2016 data available at **https://edge.sagepub.com/liu1e** for the following problems.

1. Conduct an analysis for a multinomial logistic regression model and estimate the ordinal response variable `happy` from the four predictor variables, `sex`, `educ`, and `satfin`. Choose category 3 (i.e., not too happy) as the referent category.
2. Interpret the likelihood ratio test for the overall model.
3. In the regression table, identify the logit coefficients for the predictor variable `satfin` across two binary comparisons. Are they both statistically significant? What categories are they comparing?
4. Compute the odds ratios for the predictor variable `satfin` across two binary comparisons.
5. Interpret the relative risk ratios/odd ratios of `satfin`.
6. Based on the parameter estimates in the output, write the two equations for the model.
7. Make a publication-quality table containing the estimated logit.
8. Write a report to summarize the results from the output.

8

POISSON REGRESSION MODELS

OBJECTIVES OF THIS CHAPTER

This chapter introduces Poisson regression models. It first starts with an introduction to the Poisson regression model followed by a discussion of the incidence rates and incidence rate ratios in the model, goodness-of-fit statistics, and how to interpret parameter estimates. After a description of the research example, the data, and the sample, a one-predictor Poisson regression model and a multiple-predictor Poisson regression model are illustrated with the `glm()` function in R. The `vglm()` function in the `VGAM` package is also used to fit the multiple-predictor model. R commands and output are explained in detail. This chapter focuses on fitting the Poisson regression models with R, as well as on interpreting and presenting the results. After reading this chapter, you should be able to:

- Identify when Poisson regression models are used.
- Fit a Poisson regression model using R.
- Interpret the output.
- Interpret the incidence rate ratios and marginal effects.
- Compute, plot, and interpret the predicted counts.
- Compare models using the likelihood ratio test.
- Present results in publication-quality tables using R.
- Write the results for publication.

8.1 POISSON REGRESSION MODELS: AN INTRODUCTION

The Poisson regression model is used to estimate a count response variable. The count response variable can be a count of events, or the number of events during a time period or in a location. For example, researchers are interested in the number of AP courses which students have taken in high school, the number of absences in a college class, the number of publications by new faculty in a year, the number of visits to a doctor in a year, and the number of hospitalized patients. These count response variables are nonnegative integers and follow a Poisson distribution.

The Poisson regression model (Cameron & Trivedi, 2013; Hilbe, 2014; Long & Freese, 2014) can be expressed as follows:

$$\ln(\mu) = \alpha + \beta_1 X_1 + \beta_2 X_2 + \ldots + \beta_p X_p \tag{8.1}$$

where μ is the mean or the expected number of events; α is the intercept; and $\beta_1, \beta_2, \ldots, \beta_p$ are the Poisson coefficients for the predictors. The left side of the equation, $\ln(\mu)$, is the log link function. The right side of the equation is the linear predictor. An important assumption of the Poisson regression model is that the mean of the count response variable is equal to the variance of the variable. $E(Y) = \text{Variance }(Y) = \mu$. By exponentiating both sides of the equation, we get the predicted mean of the count response variable:

$$\mu = \exp\left(\alpha + \beta_1 X_1 + \beta_2 X_2 + \ldots + \beta_p X_p\right) \tag{8.2}$$

When a count response variable is the number of events during a time period or in a location, a count of events can also referred to as an incidence rate. If we define the incidence rate as the expected number of events per unit time or location, μ/t, the Poisson model can also be expressed as follows:

$$\ln(\mu/t) = \alpha + \beta_1 X_1 + \beta_2 X_2 + \ldots + \beta_p X_p \tag{8.3}$$

where μ is the mean count; t is a period of time; μ/t is the incidence rate; and $\ln(\mu/t)$ is the log of the incidence rate or log incidence rate. Since $\ln(\mu/t) = \ln(\mu) - \ln(t)$, the equation can be rewritten as:

$$\ln(\mu) = \ln(t) + \alpha + \beta_1 X_1 + \beta_2 X_2 + \ldots + \beta_p X_p \tag{8.4}$$

where $\ln(t)$ is the offset in the model equation. When $t = 1$, the offset $\ln(t) = 0$, which can be omitted from the equation. For example, if the count response variable (Y) is the number of visits to a zoo in a year in the model, then the mean count (μ) is the average number of visits to a zoo, and the time period (t) is 1. Therefore, the expected number of visits to a zoo is the same as the incidence rate.

8.1.1 The Poisson Distribution

The Poisson probability distribution for a count response variable is expressed as:

$$P(y) = \frac{e^{-u}u^y}{y!} \tag{8.5}$$

where y is a count response variable, μ is the expected or the average number of events, and $y!$ is the y factorial or the factorial of y. $y! = y*(y-1) \ldots 2*1$. In literature, μ is often symbolized as λ. In the Poisson distribution, the mean of a count variable is equal to the variance of the variable.

$$\mu = \text{E}(y) = \text{Variance}(y).$$

The log likelihood function for the Poisson distribution or the Poisson log likelihood function is expressed as:

$$l(u;\, y) = \sum_{i=1}^{n} \{y_i \ln u_i - u_i - \ln(y_i!)\} \tag{8.6}$$

where $l(u; y)$ is the log likelihood function of μ given the values of the count variable y.

8.1.2 Incidence Rate Ratios in Poisson Regression Models

The Poisson regression model estimates the log expected counts of an event or the log incidence rate of the response variable. The incidence rate is defined as the expected number of events during a period of time or in a location. In a simple Poisson regression model with one predictor, $\ln(\mu) = \alpha + \beta X$, where μ is the expected counts of an event or the incidence rate. The estimated coefficient is the Poisson coefficient, which is the coefficient on the scale of the natural logarithm. It can be also referred to as the log coefficient. We estimate the relationship between the predictor variable and the log function of the expected counts of an event or the log incidence rate.

By exponentiating both sides of the equation, we get the expected counts of an event or the incidence rate:

$$\mu = \exp(\alpha + \beta X) \tag{8.7}$$

If the independent variable X is a categorical variable with the values of 0 and 1, the incidence rates of the response variable can be computed as follows.

When $X = 0$, the incidence rate $= \exp(\alpha)$, which is the exponentiated intercept.

When $X = 1$, the incidence rate $= \exp(\alpha + \beta)$, which is the exponentiated sum of the intercept and the Poisson coefficient.

The incidence rate ratio of the group 1 ($X = 1$) to the group 2 ($X = 0$):

$$\text{IRR} = \frac{\exp(\alpha + \beta)}{\exp(\alpha)} = \frac{\exp(\alpha) \times \exp(\beta)}{\exp(\alpha)} = \exp(\beta) \tag{8.8}$$

For a one-unit increase in an independent variable (e.g., from 0 to 1 in the previous example) the change in the incidence rate is the incidence rate ratio, which is the exponentiated Poisson coefficient. When the independent variable is continuous, for a one-unit increase from any value of x to the value of $(x + 1)$, the change in the incidence rate is still the exponentiated Poisson coefficient.

8.1.3 Model Fit Statistics

Same as those discussed in the previous chapters, model fit statistics, such as the log likelihood statistic, the residual deviance, the model chi-square statistic, the AIC and BIC statistics, and the pseudo R^2 statistics, can be computed for the Poisson regression model. The likelihood ratio test and the AIC and BIC statistics can also be used for model comparisons.

8.1.4 Interpretation of Model Parameter Estimates

When the Poisson coefficient is positive, it indicates the relationship between the predictor variable and the log expected counts of an event or the log incidence rate ratio is positive. By exponentiating the log coefficient, we get the incidence rate ratio, which is larger than 1. This means that the expected number of events or the incidence rate of a response variable increases for a one-unit increase in the predictor variable.

When the Poisson coefficient is negative, it indicates that the relationship between the predictor variable and the log expected counts of an event or the log incidence rate is negative. The exponentiated coefficient, the incidence rate ratio, is less than 1. It means that the expected number of events or the incidence rate of a response variable decreases for a one-unit increase in the predictor variable.

When the Poisson coefficient equals 0, the incidence rate ratio equals 1. It indicates that a one-unit increase in the predictor variable does not impact the expected number of events or the incidence rate of a response variable.

When there are multiple predictors in the model, the incidence rate ratio for a predictor can be interpreted as the change in the expected number of events or the incidence rate of a response variable for a one-unit change in a predictor variable when holding other predictor variables constant.

8.1.5 Interpreting an Incidence Rate Ratio as a Percentage Change in an Incidence Rate

Another way of interpreting an incidence rate ratio is the percentage change in an incidence rate. It can be calculated by using (Incidence rate − 1) × 100%. A positive percentage change in an incidence rate indicates there is an increase in the incidence rate, whereas a negative percentage change corresponds to a decrease in the rate. A zero percentage change indicates no change in the rate at all. In other words, the predictor variable does not influence the incidence rate of a response variable.

For example, if an incidence rate ratio for a predictor variable equals 1.7, then the percentage change in the incidence rate can be computed as follows: $(1.7 - 1) \times 100\% = 70\%$. This indicates that each one-unit increase in the predictor variable corresponds to an increase of 70% in the incidence rate of a response variable.

In another example, if an incidence rate ratio = .80, then the percentage change in the incidence rate is $(.80 - 1) \times 100\% = -20\%$. Since the percentage change is negative, it indicates that for each one-unit increase in the predictor variable, there is a decrease of 20% in the incidence rate.

8.1.6 Interpreting Marginal Effects as Changes in Predicted Counts

In Poisson regression, a marginal effect is a change in the expected counts of a response variable related to the change in an independent variable. Mathematically, it is the product of a Poisson coefficient for a particular predictor variable (β_p) and the predicted mean or the expected number of events (μ). Recall that in Equation 8.2 the predicted mean is the exponential of the linear predictor, $\exp(\alpha + \beta_1 X_1 + \beta_2 X_2 + \ldots + \beta_p X_p)$. So a marginal effect $= \beta_p \times \mu = \beta_p \times \exp(\alpha + \beta_1 X_1 + \beta_2 X_2 + \ldots + \beta_p X_p)$. It varies across the values of a predictor variable. In other words, for each predictor variable, we can calculate the marginal effect at each value of that variable. An average marginal effect (AME) is the mean value of the marginal effects at these values.

8.2 RESEARCH EXAMPLE AND DESCRIPTION OF THE DATA AND SAMPLE

We will investigate the relationships between the count response variable, the number of zoo visits in a year, and four predictor variables. Unlike other chapters, however, here the research interest focuses on using Poisson regression to predict the count response variable. The GSS 2016 data are used for the following analyses. The following are the variables used for data analysis in this chapter:

- `vistzoo`: the recoded variable of the number of zoo visits in a year
- `maritals`: the recoded variable of marital (marital status) with 1 = currently married and 0 = not currently married
- `educ`: the highest education completed
- `female`: recoded variable of sex with 1 = female and 0 = male
- `wrkfull`: working full time or not

8.3 FITTING A ONE-PREDICTOR POISSON REGRESSION MODEL WITH R

8.3.1 Packages and Functions for Poisson Regression Models in R

Several packages in R can be used for fitting Poisson regression models. This chapter focuses on the `glm()` function in R and the `vglm()` function in the `VGAM` package (Yee, 2010). The `glm()` function can be used directly since it is included in the `stats` package with the installation of R. However, you need to load the `VGAM` package with `library(VGAM)` if it is installed. The `glm()` function is introduced first.

8.3.2 The `glm()` Function

As introduced in the earlier chapters, the `glm()` function is normally used to fit generalized linear models. The model formula in `glm()` specifies the dependent variable and the predictor variable(s), which are separated by the tilde (~). The plus (+) symbol is used to connect multiple predictor variables. We also need to specify the probability distribution of the outcome variable with the `family` argument. We specify `family = poisson` for a count outcome variable in Poisson regression. For more details on how to use this function, type `help(glm)` in the command prompt.

8.3.3 The Poisson Regression Model: One-Predictor Model

In the following example, the command `PR.1 <- glm(vistzoo ~ educ, family = poisson, data = count)` tells R to predict the count outcome variable `vistzoo` from the independent variable `educ` with Poisson regression by specifying the Poisson family (`family = poisson`). The fitted model is named `PR.1`. The `summary(PR.1)` command displays the output of the fitted model.

```
> # One-predictor Poisson regression model with glm()
> PR.1 <- glm(vistzoo ~ educ, family = poisson, data = count)
> summary(PR.1)

Call:
glm(formula = vistzoo ~ educ, family = poisson, data = count)

Deviance Residuals:
    Min        1Q     Median        3Q       Max
-1.5545   -1.2289   -1.0389    0.3291    4.1534

Coefficients:
             Estimate    Std. Error   z value    Pr(>|z|)
(Intercept)  -1.15403     0.18915     -6.101    1.05e-09 ***
educ          0.06716     0.01295      5.187    2.13e-07 ***
---
Signif. codes: 0 '***' 0.001 '**' 0.01 '*' 0.05 '.' 0.1 ' ' 1

(Dispersion parameter for poisson family taken to be 1)

    Null deviance: 1364.6  on 901  degrees of freedom
Residual deviance: 1337.6  on 900  degrees of freedom
AIC: 2342.6

Number of Fisher Scoring iterations: 5
```

8.3.4 Interpreting the Output

In the R output for the one-predictor Poisson regression model, the first part is the call, which shows the R command for the model. The second part shows the minimum, first quarter, median, third quarter, and maximum values of the deviance residuals. The third part shows the coefficients table including the parameter estimates for the predictor variable and the intercept, their standard errors, the Wald z statistics, and the associated p values. The null hypothesis for the Wald test is that the coefficient of the predictor variable is zero, and the alternative hypothesis is that the coefficient of the predictor variable is significantly different from zero.

The Wald z statistic equals the parameter estimate divided by its standard error.

$$\text{Wald } z = \frac{\hat{\beta}}{\text{SE}(\hat{\beta})} \tag{8.9}$$

where $\hat{\beta}$ is the estimated log coefficient and $\text{SE}(\hat{\beta})$ is the standard error.

$$\text{Confidence intervals for the parameter } \beta = \hat{\beta} \pm z{*}\text{SE}(\hat{\beta})$$

where $\hat{\beta}$ is the estimated log coefficient for a predictor, z is the z-score from a normal distribution for the chosen confidence interval, and $\text{SE}(\hat{\beta})$ is the standard error. For the 95% confidence intervals, $z = 1.96$. So 95% confidence intervals for the parameter $\beta = \hat{\beta} \pm 1.96{*}\text{SE}(\hat{\beta})$.

For the predictor variable `educ`, Wald $z = 5.187$. The associated p value, `Pr(>|z|)` < `.001`, so we reject the null hypothesis. The rejection of the null hypothesis indicates that the predictor variable `educ` is a significant predictor of the count response variable `vistzoo`. For a one-unit increase in education the log expected number of visits to a zoo increases by a factor of .067.

Finally, the fourth part of the output shows the fit statistics including the null deviance, the residual deviance, and the AIC. The null deviance is the deviance for the null model with the intercept only. The residual deviance is the deviance for the fitted model, which is defined as −2(log likelihood of the current model − log likelihood of the saturated model). The difference between the residual deviance and the null deviance can be used to evaluate the significance of the fitted model.

We can extract the coefficients with `coef(PR.1)` and obtain the profiled confidence intervals with the `confint(PR.1)` command. The Wald confidence intervals are extracted with `confint.default(PR.1)`. The profiled confidence intervals are based on the profile likelihood while the Wald confidence intervals are based on the Wald test. Although the results of both confidence intervals are similar in this example, the former is preferred when the sample size is small.

```
> coef(PR.1)
(Intercept)          educ
-1.15403149    0.06715748
> confint(PR.1)
Waiting for profiling to be done...
                   2.5 %          97.5 %
(Intercept)    -1.52744224    -0.78592700
educ            0.04178749     0.09254033
> confint.default(PR.1)
                   2.5 %          97.5 %
(Intercept)    -1.52475464    -0.78330834
educ            0.04178338     0.09253158
```

8.3.5 Interpreting the Incidence Rate Ratios in the One-Predictor Poisson Regression Model

The Poisson regression model estimates the log expected counts of an event. Recall that the exponentiated (β_j) is the incidence rate ratio (IRR) for a one-unit change in a predictor variable. In this model, the IRR for `educ` is 1.069, which indicates that for a one-unit increase in education the incidence rate or the expected number of visits to a zoo increases by a factor of 1.069. In other words, for a one-unit increase in education the expected number of visits to a zoo increases by 6.9%.

The above results can be obtained using the `exp(coef(PR.1))` command. We also use the `exp(confint(PR.1))` command to obtain the corresponding confidence intervals. Both results are combined with the `cbind(exp(PR.1)), exp(confint(PR.1)))` command.

```
> exp(coef(PR.1))
(Intercept)         educ
  0.3153628    1.0694639
> exp(confint(PR.1))
Waiting for profiling to be done...
                 2.5 %       97.5 %
(Intercept)    0.2170902    0.4556971
educ           1.0426729    1.0969574
> cbind(exp(coef(PR.1)), exp(confint(PR.1)))
Waiting for profiling to be done...
                               2.5 %        97.5 %
(Intercept)    0.3153628    0.2170902    0.4556971
educ           1.0694639    1.0426729    1.0969574
```

The standard errors of IRRs can be obtained with `exp(coef(PR.1))*sqrt(diag(vcov(PR.1)))`.

```
> exp(coef(PR.1))*sqrt(diag(vcov(PR.1)))
(Intercept)          educ
 0.05965023    0.01384550
```

8.3.6 Model Fit Statistics

Testing the Overall Model Using the Likelihood Ratio Test

To test if the overall model is significant, we fit a null model with the intercept only and compare the single-predictor model with the null model using the `anova()` function. The null model is fitted using the `glm()` function with 1 as the intercept term in the model formula. The command and the output are displayed below.

```
> # Testing the overall model using the likelihood ratio test
> PR.0 <- glm(vistzoo ~ 1, family = poisson, data = count)
> summary(PR.0)

Call:
glm(formula = vistzoo ~ 1, family = poisson, data = count)

Deviance Residuals:
    Min        1Q    Median       3Q      Max
-1.2731   -1.2731   -1.2731   0.2031   3.6938

Coefficients:
              Estimate   Std. Error   z value   Pr(>|z|)
(Intercept)   -0.21020      0.03699    -5.683   1.32e-08 ***
---
Signif. codes: 0 '***' 0.001 '**' 0.01 '*' 0.05 '.' 0.1 ' ' 1

(Dispersion parameter for poisson family taken to be 1)

    Null deviance: 1364.6  on 901  degrees of freedom
Residual deviance: 1364.6  on 901  degrees of freedom
AIC: 2367.6

Number of Fisher Scoring iterations: 6
```

The `anova(PR.0, PR.1, test = "Chisq"))` command compares the deviance statistics of the fitted model `PR.1` and the null model `PR.0` using the likelihood ratio test.

```
> anova(PR.0, PR.1, test = "Chisq")
Analysis of Deviance Table

Model 1: vistzoo ~ 1
Model 2: vistzoo ~ educ
  Resid. Df  Resid. Dev  Df  Deviance    Pr(>Chi)
1       901      1364.5
2       900      1337.6   1    26.935  2.104e-07 ***
---
Signif. codes: 0 '***' 0.001 '**' 0.01 '*' 0.05 '.' 0.1 ' ' 1
```

Here we first fit the null model and then compare the single-predictor model with the null model. This two-step process can also be simplified to the one-line command with the `update()` function within the `anova()` function. In the `anova(PR.1, update(PR.1, ~1), test = "Chisq")` command, we use `update(PR.1, ~ 1)` to fit the null model.

```
> anova(PR.1, update(PR.1, ~1), test = "Chisq")
Analysis of Deviance Table

Model 1: vistzoo ~ educ
Model 2: vistzoo ~ 1
  Resid. Df  Resid. Dev  Df  Deviance    Pr(>Chi)
1       900      1337.6
2       901      1364.5  -1   -26.935  2.104e-07 ***
---
Signif. codes: 0 '***' 0.001 '**' 0.01 '*' 0.05 '.' 0.1 ' ' 1
```

The null hypothesis of the test for the overall model is that the predictor variable does not contribute to the model, and the alternative hypothesis is that the one-predictor model is better than the null model with no independent variables. The likelihood ratio test statistic $LR\ \chi^2_{(1)} = 26.935$, $p < .001$, which indicates that the overall model with one predictor `educ` is significantly different from zero. Therefore, the one-predictor model provides a better fit than the null model in predicting the log number of visits to a zoo in a year. The output displays a negative value for the likelihood ratio test statistic since the residual deviance for the one-predictor model is smaller than that for the null model.

Pseudo R^2

We use the `nagelkerke()` function in the `rcompanion` package (Mangiafico, 2021) to compute the pseudo R^2 statistics for the single-predictor model. We load the package first with `library(rcompanion)` and then execute the `nagelkerke(PR.1)` command.

```
> # Pseudo R2 with nagelkerke()
> library(rcompanion)
> nagelkerke(PR.1)
$`Models`

Model: "glm, vistzoo ~ educ, poisson, count"
Null:  "glm, vistzoo ~ 1, poisson, count"

$Pseudo.R.squared.for.model.vs.null
                             Pseudo.R.squared
McFadden                            0.0113862
Cox and Snell (ML)                  0.0294198
Nagelkerke (Cragg and Uhler)        0.0317234

$Likelihood.ratio.test
Df.diff   LogLik.diff   Chisq     p.value
     -1       -13.467   26.935   2.1043e-07

$Number.of.observations

Model: 902
Null:  902

$Messages
[1] "Note: For models fit with REML, these statistics are based on refitting with ML"

$Warnings
[1] "None"
```

The McFadden R^2 is .011, the Cox and Snell R^2 is .029, and the Nagelkerke R^2 is .032. The same results can be computed using the equations for these three pseudo R^2 statistics. In the R command below, `LLM1` is the log-likelihood value for the single-predictor model and `LL0` is the log-likelihood value for the null model. The number of the observations is 902. In addition, `McFadden1` is the object name for the McFadden R^2, `CS1` for the Cox and Snell R^2, and `NG1` for the Nagelkerke R^2.

```
> # Pseudo R2 with equations
> LLM1 <- logLik(PR.1)
> LL0 <- logLik(PR.0)
> McFadden1 <- 1-(LLM1/LL0)
> McFadden1
'log Lik.' 0.01138621 (df=2)
> CS1 <- 1-exp(2*(LL0-LLM1)/902)
> CS1
'log Lik.' 0.02941984 (df=1)
> NG1 <- CS1/(1-exp(2*LL0/902))
> NG1
'log Lik.' 0.03172344 (df=1)
```

AIC and BIC Statistics

The AIC and BIC statistics can also be computed from the `AIC()` and `BIC()` functions. The output is shown as follows.

```
> # AIC and BIC Statistics
> AIC(PR.1)
[1] 2342.636
> BIC(PR.1)
[1] 2352.245
```

8.4 FITTING A MULTIPLE-PREDICTOR POISSON REGRESSION MODEL WITH R

8.4.1 The Poisson Regression Model: Multiple-Predictor Model

We still use the `glm()` function for multiple Poisson regression. The `glm(vistzoo ~ educ + maritals + female + wrkfull, family = poisson, data = count)` command tells R to predict the count response variable `vistzoo` from the four independent variables. In the `glm()` function, the model equation is specified as `vistzoo ~ educ + maritals + female + wrkfull`. The `family = poisson` argument specifies that the Poisson family is used to fit the model. The data argument specifies `data = count`. The fitted model is named `PR.2`. The output is shown by the `summary(PR.2)` function.

```
> # Multiple-predictor Poisson regression model with glm()
> PR.2<-glm(vistzoo ~ educ + maritals + female + wrkfull, family = poisson,
data = count)
> summary(PR.2)

Call:
glm(formula = vistzoo ~ educ + maritals + female + wrkfull, family = poisson,
    data = count)

Deviance Residuals:
    Min        1Q      Median      3Q       Max
-1.7880   -1.1659   -0.9280   0.4835   4.4634

Coefficients:
             Estimate  Std. Error  z value  Pr(>|z|)
(Intercept)  -1.34993     0.20084   -6.721  1.80e-11 ***
educ          0.05086     0.01343    3.787  0.000152 ***
maritals      0.21382     0.07494    2.853  0.004328 **
female        0.04837     0.07535    0.642  0.520896
wrkfull       0.53962     0.07792    6.925  4.35e-12 ***
---
Signif. codes: 0 '***' 0.001 '**' 0.01 '*' 0.05 '.' 0.1 ' ' 1

(Dispersion parameter for poisson family taken to be 1)

    Null deviance: 1364.6  on 901  degrees of freedom
Residual deviance: 1275.8  on 897  degrees of freedom
AIC: 2286.8

Number of Fisher Scoring iterations: 6
```

8.4.2 Interpreting R Output

In the R output for the multiple Poisson regression model, the first part is the call, which shows the R command for the model. The second part shows the minimum, first quarter, median, third quarter, and maximum values of the deviance residuals. The third part shows the coefficients table including the parameter estimates for the four predictor variables and the intercept, their standard errors, the Wald z statistics, and the associated p values.

For the predictor variable `educ`, Wald $z = 3.787$. The associated p value, `Pr(>|z|)` `< .001`, so we reject the null hypothesis. The rejection of the null hypothesis indicates that the predictor variable `educ` is a significant predictor of the count response variable `vistzoo`.

For the predictor variable `maritals`, Wald $z = 2.853$. The associated p value, `Pr(>|z|) < .001`, so we reject the null hypothesis. For the predictor variable `wrkfull`, Wald $z = 6.925$. The associated p value, `Pr(>|z|) < .001`, so we also reject the null hypothesis. Therefore, `maritals` and `wrkfull` are significant predictors of the count response variable.

For the predictor variable `female`, the Wald $z = .642$. The associated p value `Pr(>|z|) = .521`, so we fail to reject the null hypothesis and conclude that there is no significant effect of female on the outcome variable. In other words, whether a person is a female or male does not significantly predict the log expected number of visits to a zoo.

Finally, the fourth part of the output shows the fit statistics including the null deviance, the residual deviance, and the AIC.

We use the `coef(PR.2)` command to extract the coefficients. Then we use the `confint(PR.2)` command to compute the corresponding confidence intervals.

```
> coef(PR.2)
(Intercept)          educ      maritals        female       wrkfull
-1.34992531    0.05085899    0.21382421    0.04837019    0.53962440
> confint(PR.2)
Waiting for profiling to be done...
                    2.5 %        97.5 %
(Intercept)   -1.74637649   -0.95905948
educ           0.02452348    0.07716665
maritals       0.06694611    0.36083727
female        -0.09900149    0.19647922
wrkfull        0.38773418    0.69331480
```

We request the IRRs with `exp(coef(PR.2))` and `exp(confint(PR.2))`, respectively. The results are combined with the `cbind()` function.

```
> exp(coef(PR.2))
(Intercept)        educ     maritals      female     wrkfull
  0.2592596   1.0521745   1.2384049   1.0495591   1.7153624
> exp(confint(PR.2))
Waiting for profiling to be done...
                2.5 %     97.5 %
(Intercept) 0.1744048  0.3832532
educ        1.0248267  1.0802221
maritals    1.0692379  1.4345300
female      0.9057414  1.2171100
wrkfull     1.4736380  2.0003353
> cbind(exp(coef(PR.2)), exp(confint(PR.2)) )
Waiting for profiling to be done...
                          2.5 %     97.5 %
(Intercept) 0.2592596 0.1744048  0.3832532
educ        1.0521745 1.0248267  1.0802221
maritals    1.2384049 1.0692379  1.4345300
female      1.0495591 0.9057414  1.2171100
wrkfull     1.7153624 1.4736380  2.0003353
```

To compute the standard errors of the IRRs, we multiply the IRRs by the standard errors of the Poisson coefficients with the `exp(coef(PR.2))*sqrt(diag(vcov(PR.2)))` command. In the syntax, `exp(coef(PR.2))` exponentiates the coefficients and `sqrt(diag(vcov(PR.2)))` computes the square root of the variances of the coefficients which are the diagonal elements in the variance-covariance matrix to obtain the standard errors of the coefficients.

```
> exp(coef(PR.2))*sqrt(diag(vcov(PR.2)))
(Intercept)        educ     maritals      female      wrkfull
 0.05207020  0.01412980   0.09280886  0.07908132   0.13366331
```

8.4.3 Interpreting the Incidence Rate Ratios (IRRs) in the Multiple-Predictor Poisson Model

The incidence rate ratio is the exponentiated coefficients in a Poisson regression model. It is the ratio of two incidence rates. A positive Poisson regression coefficient corresponds to an incidence rate ratio greater than 1, whereas a negative coefficient is associated with an incidence rate ratio less than 1.

The incidence rate ratio for a predictor can be interpreted as the change in the incidence rate or the number of events for a one-unit increase in the predictor variable when holding other predictors constant.

For educ, the incidence rate ratio is 1.052. The result indicates that the incidence rate increases by a factor of 1.052 for a one-unit increase in education when holding all other predictors constant. It can also be interpreted as the change in the number of events as follows. The expected number of visits to a zoo increases by 1.052 for a one-unit increase in education when holding all other predictors constant. In other words, for a one-unit increase in education the expected number of visits to a zoo increases by 5.2%.

For maritals, the incidence rate ratio is 1.238, which indicates that expected number of visits to a zoo for the married is 1.238 times as large as the expected number of visits for the unmarried when holding the other predictors constant. In other words, the expected number of visits to a zoo for the married is 23.8% higher than that for the unmarried.

The incidence rate ratio for wrkfull can be interpreted in the similar way. The incidence rate ratio is 1.715, which indicates that expected number of visits to a zoo for those working full time is 1.715 times as large as that for those not working full time when holding the other predictors constant. In other words, the expected number of visits to a zoo for those working full time is 71.5% higher than that for those not working full time.

With regard to female, the incidence rate ratio is 1.050, which is not significant (see the associated p value in the coefficients table). It indicates that being female does not impact the expected number of visits to a zoo.

8.4.4 Model Fit Statistics

Testing the Overall Model Using the Likelihood Ratio Test

To test if the overall model is significant, we compare the multiple-predictor model with the null model using the anova() function. The anova(PR.0, PR.2, test = "Chisq")) command compares the log-likelihood statistics of the fitted model PR.2 and the null model PR.0 using the likelihood ratio test. The resulting output is displayed below.

```
> anova(PR.2, update(PR.2, ~1), test = "Chisq")
Analysis of Deviance Table

Model 1: vistzoo ~ educ + maritals + female + wrkfull
Model 2: vistzoo ~ 1
     Resid. Df    Resid. Dev   Df    Deviance     Pr(>Chi)
1          897        1275.8
2          901        1364.5   -4     -88.772     < 2.2e-16 ***
---
Signif. codes: 0 '***' 0.001 '**' 0.01 '*' 0.05 '.' 0.1 ' ' 1
```

The same results can be obtained with the anova(PR.2, update(PR.2, ~ 1), test = "Chisq")) command. In the command, we use update(PR.2, ~ 1) to fit the null model.

```
> anova(PR.0, PR.2, test = "Chisq")
Analysis of Deviance Table

Model 1: vistzoo ~ 1
Model 2: vistzoo ~ educ + maritals + female + wrkfull
     Resid. Df    Resid. Dev    Df    Deviance     Pr(>Chi)
1          901        1364.5
2          897        1275.8     4      88.772    < 2.2e-16 ***
---
Signif. codes: 0 '***' 0.001 '**' 0.01 '*' 0.05 '.' 0.1 ' ' 1
```

The likelihood ratio chi-square test $\chi^2_{(4)} = 88.772$, $p < .001$ indicates that the full model with the four predictors provides a better fit than the null model with no independent variables in predicting the count response variable.

Pseudo R^2

The `nagelkerke(PR.2)` command computes the three types of pseudo R^2 statistics and the likelihood ratio test statistic for the overall multiple-predictor model.

```
> # Pseudo R2 with nagelkerke()
> nagelkerke(PR.2)
$`Models`

Model: "glm, vistzoo ~ educ + maritals + female + wrkfull, poisson, count"
Null:  "glm, vistzoo ~ 1, poisson, count"

$Pseudo.R.squared.for.model.vs.null
                              Pseudo.R.squared
McFadden                             0.0375265
Cox and Snell (ML)                   0.0937285
Nagelkerke (Cragg and Uhler)         0.1010680

$Likelihood.ratio.test
Df.diff    LogLik.diff     Chisq       p.value
     -4        -44.386    88.772    2.4012e-18

$Number.of.observations

Model: 902
Null:  902

$Messages
[1] "Note: For models fit with REML, these statistics are based on refitting with ML"

$Warnings
[1] "None"
```

We can also compute the three types of pseudo R^2 statistics with their equations for the multiple-predictor model as follows.

```
> # Pseudo R2 with equations
> LLM2 <- logLik(PR.2)
> McFadden2 <- 1-(LLM2/LL0)
> McFadden2
'log Lik.' 0.03752648 (df=5)
> CS2 <- 1-exp(2*(LL0-LLM2)/902)
> CS2
'log Lik.' 0.0937285 (df=1)
> NG2 <- CS2/(1-exp(2*LL0/902))
> NG2
'log Lik.' 0.1010675 (df=1)
```

In the output, `LLM2` is the log-likelihood value for the multiple-predictor model and `LL0` is the log-likelihood value for the null model. In addition, the number of the observations is 902. The three types of pseudo R^2 statistics are as follows. The McFadden R^2 is .038, the Cox and Snell R^2 is .094, and the Nagelkerke R^2 is .101.

AIC and BIC Statistics

The `AIC(PR.2)` and `BIC(PR.2)` commands produce the AIC and BIC statistics. We also use the `AIC(PR.1, PR.2)` and `BIC(PR.1, PR.2)` commands to compare the AIC and BIC statistics between the two models, respectively.

```
> # AIC and BIC Statistics
> AIC(PR.2)
[1] 2286.799
> BIC(PR.2)
[1] 2310.822
> AIC(PR.1, PR.2)
        df      AIC
PR.1     2 2342.636
PR.2     5 2286.799
> BIC(PR.1, PR.2)
        df      BIC
PR.1     2 2352.245
PR.2     5 2310.822
```

The AIC and BIC statistics for the multiple-predictor model are 2,286.799 and 2,310.822, respectively. Compared with the single-variable model, both AIC and BIC indicate that the multiple-predictor model fits the data better.

8.4.5 Interpreting the Marginal Effects in the Poisson Regression Model

We load the `margins` package (Leeper, 2021) with the `library(margins)` command, compute the average marginal effects with the `margins(PR.2)` command, and name the results `marg.pr`. The summary results are obtained with `summary(marg.pr)` as follows.

```
> # marginal effects
> library(margins)
> marg.pr <- margins(PR.2)
> summary(marg.pr)
   factor     AME      SE       z       p    lower   upper
     educ  0.0412  0.0110  3.7506  0.0002   0.0197  0.0628
   female  0.0392  0.0611  0.6418  0.5210  -0.0805  0.1589
 maritals  0.1733  0.0611  2.8374  0.0045   0.0536  0.2930
  wrkfull  0.4373  0.0652  6.7087  0.0000   0.3096  0.5651
```

The average marginal effect for `educ` is .041. The result indicates that there are on average .041 more visits to a zoo for a one-unit increase in education when holding all other predictors constant.

The average marginal effect for `maritals` is .173. This indicates that the married have on average .173 more visits to the zoo than the unmarried when holding the other predictors constant. The marginal effects of the other two predictor variable can be interpreted in a similar way.

8.4.6 Interpreting the Predicted Counts With the `ggpredict()` Function in the `ggeffects` Package

By using the `ggpredict()` function in the `ggeffects` package (Lüdecke, 2018b), we can compute the predicted number of events of the count response variable at specified values of the predictor variables. We first load the package with `library(ggeffects)` since it has been installed in the previous chapters. The command `pr.ed <- ggpredict(PR.2, terms = "educ[12, 14, 16]")` tells R to compute the predicted counts of the response variable using the `ggpredict()` function. The argument inside the function includes the estimated model, `PR.2`, the `terms = "educ[12, 14, 16]"` option, which specifies the predictor variable `educ` at the values of 12, 14, and 16 when holding other predictor variables at their means. The `terms` option can specify up to four variables, including the second to fourth grouping variables. The output is assigned to the object named `pr.ed`. The `as.data.frame()` function or the `sqrt(diag(vcov()))` function can be used to request the standard errors. The output is omitted here.

```
> library(ggeffects)
> pr.ed <- ggpredict(PR.2, terms = "educ[12, 14, 16]")
> pr.ed
# Predicted counts of vistzoo


educ    |   Predicted    |          95% CI
-----------------------------------------
  12    |        0.70    |    [0.63, 0.77]
  14    |        0.77    |    [0.71, 0.83]
  16    |        0.85    |    [0.78, 0.93]

Adjusted for:
* maritals = 0.44
*   female = 0.56
*  wrkfull = 0.47
> plot(pr.ed)
```

The results table displays the values of `educ`, the predicted counts, and the confidence intervals. When `educ` = 12, and the other predictor variables are held at their means (`maritals` = .44, `female` = .56, and `wrkfull` = .47), the predicted number of visits to a zoo is .70.

When `educ` = 14, and the other three predictor variables are held at their means, the predicted number of visits to a zoo is .77.

When `educ` = 16, and the other predictor variables are held at their means, the predicted number of visits to a zoo is .85.

The predicted counts are plotted using `plot(pr.ed)`. Figure 8.1 shows the predicted number of visits to a zoo when `educ` is at the values of 12, 14, and 16.

The graph shows that with the increase of the years of education, the predicted number of visits to a zoo increases. In other words, people with higher levels of education are associated with having more visits to a zoo.

In the next example, we compute the predicted counts for a continuous variable at given values by different groups. In the following example, we compute the predicted number of visits to a zoo for `educ` at the values of 12, 14, and 16 by the two groups in `wrkfull` when holding other variables at their means. The command is as follows: `pr.ew <- ggpredict(PR.2, terms = c("educ[12, 14, 16]", "wrkfull"))`. In the `ggpredict()` function, the `terms = c("educ[12, 14, 16]", "wrkfull")` option specifies both `educ` and `wrkfull`, with the latter as the grouping variable. The output is assigned to an object named `pr.ew` and is plotted with the `plot(pr.ew)` function.

FIGURE 8.1 ● Predicted Counts for educ at 12, 14, and 16

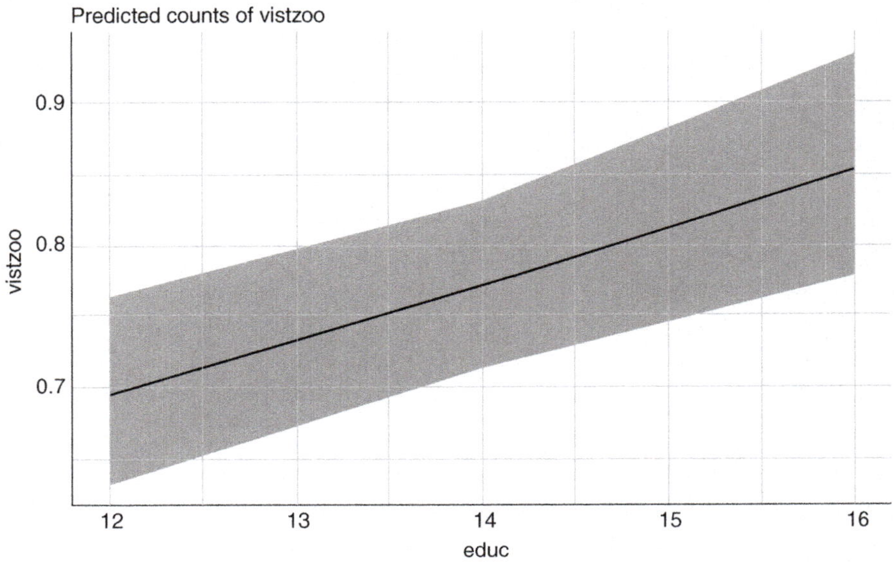


```
> pr.ew <- ggpredict(PR.2, terms = c("educ[12, 14, 16]", "wrkfull"))
> pr.ew
# Predicted counts of vistzoo

# wrkfull = 0

educ    |   Predicted    |         95% CI
-----------------------------------------
  12    |        0.54    |   [0.47, 0.61]
  14    |        0.60    |   [0.53, 0.67]
  16    |        0.66    |   [0.58, 0.75]

# wrkfull = 1

educ    |   Predicted    |         95% CI
-----------------------------------------
  12    |        0.92    |   [0.82, 1.04]
  14    |        1.02    |   [0.93, 1.12]
  16    |        1.13    |   [1.02, 1.25]

Adjusted for:
* maritals = 0.44
*   female = 0.56
> plot(pr.ew)
```

FIGURE 8.2 ● Predicted Counts for educ at 12, 14, and 16 by maritals

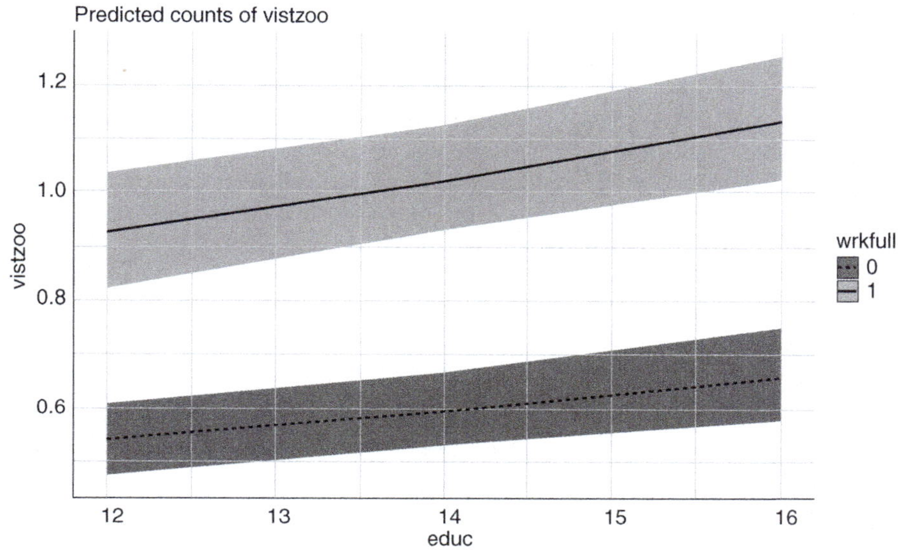

Figure 8.2 shows the predicted counts for `educ` at 12, 14, and 16 by the grouping variable `wrkfull`. As shown in the graph, the predicted number of visits to a zoo increases with the increase of years of education and the predicted counts for the people who work full-time are higher than the predicted counts for those not working full-time.

8.4.7 Model Comparisons Using the Likelihood Ratio Test

The likelihood ratio test, or the deviance difference test, is used to compare the multiple-predictor model and the one-predictor model. In the `anova(PR.1, PR.2, test = "Chisq")` command, `PR.1` and `PR.2` are the two models being compared. The following output is displayed.

```
> # Model comparison with the likelihood ratio test
> anova(PR.1, PR.2, test = "Chisq")
Analysis of Deviance Table

Model 1: vistzoo ~ educ
Model 2: vistzoo ~ educ + maritals + female + wrkfull
    Resid. Df   Resid. Dev   Df   Deviance    Pr(>Chi)
1         900       1337.6
2         897       1275.8    3     61.837   2.381e-13 ***
---
Signif. codes: 0 '***' 0.001 '**' 0.01 '*' 0.05 '.' 0.1 ' ' 1
```

The likelihood ratio test, $\chi^2_{(3)} = 61.837, p < .001$, which indicates that the full model with the four predictor variables fits the data better than the single-predictor model.

8.5 POISSON REGRESSION WITH THE vglm() FUNCTION IN THE VGAM PACKAGE

We can also use the `vglm()` function in the `VGAM` package to fit Poisson regression models. Since `VGAM` has been installed for the previous chapters, you just need to load the package by typing `library(VGAM)`. The syntax for Poisson regression models is similar to that for other models using the `vglm()` function. The `poissonff` family needs to be specified for the `family` argument. For example, the command `vglm(y ~ x, family = poissonff, data = data1)` tells R to fit a Poisson regression model predicting the dependent variable *y* with an independent variable *x*. The argument `family = poissonff` tells R that it is the Poisson family. For more details on how to use this function, type `help(poissonff)` in the command prompt after loading the `VGAM` package.

In the following example, the `pr.v <- vglm(vistzoo ~ educ + maritals + female + wrkfull, family = poissonff, data = count)` command tells R to predict the count response variable `vistzoo` from the four independent variables. In the model formula for the `vglm()` function, the dependent variable `vistzoo` and the four predictor variables are separated by the tilde (~). The four predictor variables include, `educ`, `maritals`, `female`, and `wrkfull` which are connected by plus (+) symbols. We also specify the data arguments `data = count`. The fitted model is named `pr.v`. The following output is shown by the `summary(pr.v)` command.

```
> # Multiple-predictor Poisson regression model with vglm() in VGAM
> library(VGAM)
Loading required package: stats4
Loading required package: splines
> pr.v <- vglm(vistzoo ~ educ + maritals + female + wrkfull, family = poissonff,
data=count)
> summary(pr.v)

Call:
vglm(formula = vistzoo ~ educ + maritals + female + wrkfull,
     family = poissonff, data = count)

Pearson residuals:
                      Min        1Q     Median       3Q     Max
loglink(lambda)    -1.264   -0.8244    -0.6562   0.5319   8.247

Coefficients:
              Estimate   Std. Error    z value    Pr(>|z|)
(Intercept)   -1.34993      0.20084     -6.721    1.80e-11 ***
educ           0.05086      0.01343      3.787    0.000152 ***
maritals       0.21382      0.07494      2.853    0.004328 **
female         0.04837      0.07535      0.642    0.520896
wrkfull        0.53962      0.07792      6.925    4.35e-12 ***
---
Signif. codes:  0 '***' 0.001 '**' 0.01 '*' 0.05 '.' 0.1 ' ' 1

Name of linear predictor: loglink(lambda)
```

```
Residual deviance: 1275.779 on 897 degrees of freedom

Log-likelihood: -1138.399 on 897 degrees of freedom

Number of Fisher scoring iterations: 6

No Hauck-Donner effect found in any of the estimates
```

The R output produced by the `vglm()` function for the Poisson regression model is similar to that by the `glm()` function. It includes the call of the model command, the Pearson residuals, the coefficients, the name of linear predictor, the residual deviance, the log-likelihood, and the number of Fisher scoring iterations.

The first section shows the call, which is the R command for the model. The second section shows the minimum, first quarter, median, third quarter, and maximum values of the Pearson residuals. The third section shows the coefficients table including the parameter estimates for the predictor variable and the intercept, their standard errors, the Wald z statistics, and the associated p values. The fourth section shows the name of linear predictor or the link function, which is the log link for the expected count. The fifth section shows the residual deviance and the degrees of freedom. The sixth section provides the log-likelihood value and the degrees of freedom. Finally, the number of Fisher scoring iterations is displayed at the end.

The Poisson coefficients in the coefficients section (labeled `Coefficients:`) are the same as those produced from the `glm()` function. See the preceding section on the interpretation of the coefficients.

The Poisson coefficients of the predictor variables can be extracted by using `coef(pr.v, matrix = TRUE)`. The confidence intervals are obtained with the `confint(pr.v, matrix = TRUE)` command.

```
> coef(pr.v, matrix = TRUE)
                loglink(lambda)
(Intercept)         -1.34992531
educ                 0.05085899
maritals             0.21382421
female               0.04837019
wrkfull              0.53962440
> confint(pr.v, matrix = TRUE)
                      2.5 %          97.5 %
(Intercept)     -1.74356822     -0.95628240
educ             0.02453836      0.07717963
maritals         0.06694009      0.36070834
female          -0.09930758      0.19604796
wrkfull          0.38690141      0.69234738
```

We use the `exp(coef(pr.v, matrix = TRUE))` and `exp(confint(pr.v, matrix = TRUE))` commands to compute the IRRs and the corresponding confidence intervals, respectively. Both results are combined with the `cbind()` function.

```
> exp(coef(pr.v, matrix = TRUE))
              loglink(lambda)
(Intercept)         0.2592596
educ                1.0521745
maritals            1.2384049
female              1.0495591
wrkfull             1.7153624
> exp(confint(pr.v, matrix = TRUE))
                 2.5 %     97.5 %
(Intercept)  0.1748952   0.384319
educ         1.0248419   1.080236
maritals     1.0692314   1.434345
female       0.9054642   1.216585
wrkfull      1.4724113   1.998401
> cbind(exp(coef(pr.v, matrix = TRUE)), exp(confint(pr.v, matrix = TRUE)))
              loglink(lambda)      2.5 %     97.5 %
(Intercept)         0.2592596  0.1748952   0.384319
educ                1.0521745  1.0248419   1.080236
maritals            1.2384049  1.0692314   1.434345
female              1.0495591  0.9054642   1.216585
wrkfull             1.7153624  1.4724113   1.998401
```

The standard errors of the IRRs are computed with the `exp(coef(pr.v))*sqrt(diag(vcov(pr.v)))` command.

```
> exp(coef(pr.v))*sqrt(diag(vcov(pr.v)))
(Intercept)          educ      maritals        female       wrkfull
 0.05207020    0.01412980    0.09280886    0.07908133    0.13366331
```

We use the `nagelkerke(pr.v)` command to compute the three types of pseudo R^2 statistics and the likelihood ratio test statistic for the multiple-predictor model.

```
> # Pseudo R2 with nagelkerke()
> library(rcompanion)
> nagelkerke(pr.v)
$`Models`

Model: "vglm, vistzoo ~ educ + maritals + female + wrkfull, poissonff, count"
Null:  "vglm, vistzoo ~ 1, poissonff, count"

$Pseudo.R.squared.for.model.vs.null
                             Pseudo.R.squared
McFadden                            0.0375265
Cox and Snell (ML)                  0.0937285
Nagelkerke (Cragg and Uhler)        0.1010680

$Likelihood.ratio.test
Df.diff    LogLik.diff    Chisq      p.value
      4        -44.386   88.772   2.4012e-18
```

```
$Number.of.observations

Model: 902
Null:  902

$Messages
[1] "Note: For models fit with REML, these statistics are based on refitting with ML"

$Warnings
[1] "None"
```

8.6 MAKING PUBLICATION-QUALITY TABLES

8.6.1 Presenting the Results Using the stargazer Package

We can use the `stargazer` package (Hlavac, 2018) to make a table containing the results of the fitted models with the `glm()` function. Since the package has been installed in the preceding chapters, we only need to load the package by typing `library(stargazer)`. After fitting the single-predictor model `PR.1` and the multiple-predictor model `PR.2`, we use the command as follows: `stargazer(PR.1, PR.2, type = "text", align = TRUE, out = "pr2mod.txt")`. In the `stargazer()` function, we first specify the two model objects to be presented and then the type of the table. The option `type = "text"` specifies the table type and the `align = TRUE` option aligns the results of the two models. The `out = "pr2mod.txt"` argument saves the output named `pr2mod.txt`.

```
> library(stargazer)
> stargazer(PR.1, PR.2, type = "text", align = TRUE, out = "pr2mod.txt")

==========================================
                    Dependent variable:
              ----------------------------
                        vistzoo
                   (1)             (2)
------------------------------------------
educ             0.067***        0.051***
                 (0.013)         (0.013)

maritals                         0.214***
                                 (0.075)

female                            0.048
                                 (0.075)

wrkfull                          0.540***
                                 (0.078)

Constant        -1.154***       -1.350***
                 (0.189)         (0.201)

------------------------------------------
Observations       902             902
Log Likelihood  -1,169.318      -1,138.400
Akaike Inf. Crit. 2,342.636     2,286.799
==========================================
Note:          *p<0.1;  **p<0.05;  ***p<0.01
```

TABLE 8.1 ● Results of the Poisson Regression Models: Single-Predictor and Multiple-Predictor Models (Shown in Original Format Generated by R)

	Dependent variable:	
	vistzoo	
	(1)	(2)
educ	0.067***	0.051***
	(0.013)	(0.013)
maritals		0.214***
		(0.075)
female		0.048
		(0.075)
wrkfull		0.540***
		(0.078)
Constant	−1.154***	−1.350***
	(0.189)	(0.201)
Observations	902	902
Log Likelihood	−1,169.318	−1,138.400
Akaike Inf. Crit.	2,342.636	2,286.799

*$p < .1$
**$p < .05$
***$p < .01$

We can also create the table in the HTML format and copy it into Microsoft Word. The command is as follows: `stargazer(PR.1, PR.2, type = "html", align = TRUE, out = "pr2mod.htm")`. It produces Table 8.1, as shown here in its original format, presenting the results of both the single-predictor and the multiple-predictor Poisson regression models.

Presenting the Results of the `vglm` Models Using the `texreg` Package

We can also use the `screenreg()` and `htmlreg()` functions from the `texreg` package (Leifeld, 2013) to present the results. If you have not done so, you need to install the package first by typing `install.packages("texreg")` and then load the package by typing `library(texreg)`. Since the package has been installed in previous chapters, we only need to load the package by typing `library(texreg)`. We use the: `screenreg(list(PR.1, PR.2))` command to display the results. In the `screenreg()` function, we specify the two model objects to be presented with the `list()` function. The output for the resulting plain text table is omitted here.

8.7 REPORTING THE RESULTS

Reporting the results for Poisson regression is similar to that used for binary logistic regression. The following are the generic guidelines for reporting the results. You may need to adjust your writing since your discipline or journals may have different requirements.

First, describe the Poisson regression model, the count response variable and the independent variables, and your research hypothesis or the purpose of your study. Include a couple of sentences justifying your use of this model for the analysis.

Second, report the likelihood ratio test statistic for the model and the associated p value, followed by the interpretation on whether the fitted model is better than the null model. If more than one model is developed, then compare models using likelihood ratio test statistics and/or the AIC and BIC statistics.

Third, report the parameter estimates for the predictor variables, their standard errors, and the associated p values in a table. In addition, report the incidence rate ratio for each predictor in the table or text and interpret the results. The following is an example of summarizing the results for the Poisson regression model illustrated previously.

The Poisson regression analysis was conducted to predict the count outcome variable, the number of zoo visits in a year, from a set of predictor variables, such as marital status, years of education, gender, and working status. The Poisson regression model was fitted since the response variable was a count of the number of visits to a zoo in a year.

The likelihood ratio test for the fitted model $\chi^2_{(4)} = 88.772$, $p < .001$, indicates that the full model with the four predictors provides a better fit than the null model with no independent variables in predicting the count response variable.

Table 8.1 displays the parameter estimates for the multiple-predictor Poisson regression model. The results can be interpreted in terms of the incidence rate ratios which are the exponentiated coefficients.

For `educ`, the incidence rate ratio is 1.052. The result indicates that the incidence rate increases by a factor of 1.052 for a one-unit increase in education when holding all other predictors constant.

For `maritals`, the incidence rate ratio is 1.238, which indicates that expected number of visits to a zoo for the married is 1.238 times as large as the expected number of visits for the unmarried when holding other predictors constant. In other words, the expected number of visits to a zoo for the married is 23.8% higher than that for the unmarried.

The incidence rate ratio for `wrkfull` can be interpreted in the similar way. The incidence rate ratio is 1.715, which indicates that the expected number of visits to a zoo for those working full time is 71.5% higher than that for those not working full time.

With regard to `female`, the incidence rate ratio is 1.050, $p = .521$, which is not significant. It indicates that being female does not impact the expected number of visits to a zoo.

8.8 SUMMARY OF R COMMANDS IN THIS CHAPTER

```
# Chap 8 R Script

# Remove all objects
rm(list = ls(all = TRUE))

# The following user-written packages need to be installed first by using
install.packages(" ") and then by loading it with library()
# library(VGAM)            # It is already installed for Chapter 4
# library(rcompanion)      # It is already installed for Chapter 3
# library(margins)         # It is already installed for Chapter 3
# library(ggeffects)       # It is already installed for Chapter 2
# library(stargazer)       # It is already installed for Chapter 2

# Import the count dataset
library(foreign)
count <- read.dta("C:/CDA/count.dta")

# Convert variables from integer to numeric so they will work well with ggpredict()
count$educ <- as.numeric(count$educ)
count$wrkfull <- as.numeric(count$wrkfull)
count$maritals <- as.numeric(count$maritals)

attach(count)

# One-predictor Poisson regression model with glm()
PR.1 <- glm(vistzoo ~ educ, family = poisson, data = count)
summary(PR.1)
coef(PR.1)
confint(PR.1)
confint.default(PR.1)
exp(coef(PR.1))
```

```
exp(coef(PR.1))*sqrt(diag(vcov(PR.1)))
exp(confint(PR.1))
cbind(exp(coef(PR.1)), exp(confint(PR.1)))

# Testing the overall model using the likelihood ratio test
PR.0 <- glm(vistzoo ~ 1, family = poisson, data = count)
summary(PR.0)
anova(PR.0, PR.1, test = "Chisq")
anova(PR.1, update(PR.1, ~1), test = "Chisq")

# Pseudo R2 with nagelkerke()
library(rcompanion)
nagelkerke(PR.1)

# Pseudo R2 with equations
LLM1 <- logLik(PR.1)
LL0 <- logLik(PR.0)
McFadden1 <- 1-(LLM1/LL0)
McFadden1
CS1 <- 1-exp(2*(LL0-LLM1)/902)
CS1
NG1 <- CS1/(1-exp(2*LL0/902))
NG1

# AIC and BIC Statistics
AIC(PR.1)
BIC(PR.1)

# Multiple-predictor Poisson regression model with glm()
PR.2 <- glm(vistzoo ~ educ + maritals + female + wrkfull, family = poisson, data = count)
summary(PR.2)
coef(PR.2)
confint(PR.2)
exp(coef(PR.2))
exp(confint(PR.2))
cbind(exp(coef(PR.2)), exp(confint(PR.2)) )
exp(coef(PR.2))*sqrt(diag(vcov(PR.2)))

# Testing the overall model using the likelihood ratio test
anova(PR.2, update(PR.2, ~1), test = "Chisq")
anova(PR.0, PR.2, test = "Chisq")

# Pseudo R2 with nagelkerke()
nagelkerke(PR.2)

# Pseudo R2 with equations
LLM2 <- logLik(PR.2)
McFadden2 <- 1-(LLM2/LL0)
McFadden2
CS2 <- 1-exp(2*(LL0-LLM2)/902)
CS2
NG2 <- CS2/(1-exp(2*LL0/902))
NG2
```

```
# AIC and BIC Statistics
AIC(PR.2)
BIC(PR.2)
AIC(PR.1, PR.2)
BIC(PR.1, PR.2)

# Model comparison with the likelihood ratio test
anova(PR.1, PR.2, test = "Chisq")

library(stargazer)
stargazer(PR.1, PR.2, type = "text", align = TRUE, out = "pr2mod.txt")
stargazer(PR.1, PR.2, type = "html", align = TRUE, out = "pr2mod.htm")

# Marginal effects: AME
library(margins)
marg.pr <- margins(PR.2)
summary(marg.pr)

# Predicted counts with ggpredict() in ggeffects
library(ggeffects)
pr.ed <- ggpredict(PR.2, terms = "educ[12, 14, 16]")
pr.ed
plot(pr.ed)
pr.ew <- ggpredict(PR.2, terms = c("educ[12, 14, 16]", "wrkfull"))
pr.ew
plot(pr.ew)

# Multiple-predictor Poisson regression model with vglm() in VGAM
library(VGAM)
pr.v <- vglm(vistzoo ~ educ + maritals + female + wrkfull, family = poissonff,
data=count)
summary(pr.v)
coef(pr.v, matrix = TRUE)
confint(pr.v, matrix = TRUE)
exp(coef(pr.v, matrix = TRUE))
exp(confint(pr.v, matrix = TRUE))
cbind(exp(coef(pr.v, matrix = TRUE)), exp(confint(pr.v, matrix = TRUE)))
exp(coef(pr.v))*sqrt(diag(vcov(pr.v)))

# Pseudo R2 with nagelkerke()
library(rcompanion)
nagelkerke(pr.v)

detach(count)
```

Glossary

A marginal effect in Poisson regression is a change in the expected counts of a response variable related to the change in an independent variable.

Count response variables are nonnegative integers and follow a Poisson distribution.

In the **Poisson distribution**, the mean of a count variable is equal to the variance of the variable.

The incidence rate is defined as the expected number of events during a period of time or in a location.

The Poisson regression model is used to estimate a count response variable. It estimates the log expected counts of an event or the log incidence rate of the response variable.

Exercises

Use the `rwm1984` data (Hilbe, 2014) available at **https://edge.sagepub.com/liu1e** for the following problems.

1. Conduct an analysis for a Poisson regression model and estimate the count response variable `docvis` (the number of visits to a doctor in a year) from the two predictor variables, `outwork` (1 = not working and 0 = working) and `female` (1 = female and 0 = male).
2. Interpret the likelihood ratio test for the overall model.
3. List three measures of pseudo R^2 and the AIC and BIC statistics.
4. In the regression table, identify the coefficients for the predictor variable `outwork` and `female`. Are they both statistically significant?
5. Compute the IRRs and interpret the IRR for the predictor variable `outwork`.
6. Make a publication-quality table containing the estimated coefficients.
7. Write a report to summarize the results from the output.

9

NEGATIVE BINOMIAL REGRESSION MODELS AND ZERO-INFLATED MODELS

OBJECTIVES OF THIS CHAPTER

This chapter introduces negative binomial regression models and zero-inflated models. It first starts with an introduction to the negative binomial regression model followed by a discussion of how to interpret parameter estimates. Next, after a description of the research example, the data, and the sample, a multiple-predictor negative binomial regression model are illustrated with the `glm.nb()` function in the `MASS` package. Then, the `vglm()` function in the `VGAM` package is also used to fit the same multiple-predictor model. Finally, the `zeroinfl()` function in the `pscl` package is used to fit both the zero-inflated Poisson regression model and the zero-inflated negative binomial model. R commands and output are explained in detail. This chapter focuses on fitting the negative binomial regression models and zero-inflated models with R, as well as on interpreting and presenting the results. After reading this chapter, you should be able to:

- Identify when binomial regression models and zero-inflated models are used.
- Fit a negative binomial regression model and a zero-inflated model using R.
- Interpret the output.
- Interpret the incidence rate ratios and marginal effects.
- Compute, plot, and interpret the predicted counts.
- Compare nested models using the likelihood ratio test and compare non-nested models using the Vuong test.
- Present results in publication-quality tables using R.
- Write the results for publication.

9.1 NEGATIVE BINOMIAL REGRESSION MODELS: AN INTRODUCTION

The Poisson regression model which we discussed in the preceding chapter assumes that the mean of the count response variable is equal to the variance of the variable. However, this assumption is often violated. In real data analysis, it is common that the variance could be either greater or less than the mean of the count response variable. Overdispersion occurs when the variance of the count response variable is greater than the mean, whereas underdispersion is present when the variance is less than the mean. Both overdispersion and underdispersion impact the estimation of the standard errors of parameter estimates which result in biased results. This chapter focuses on the issue of overdispersion only since it is more common than underdispersion. To deal with overdispersion in Poisson regression models for count data, we can use several methods to estimate the standard errors. These methods include the use of quasi-Poisson models, robust standard errors, and bootstrapped standard errors. A better option is to use the negative binomial regression model which relaxes the equality of the mean and the variance assumption and allows overdispersion. In Poisson regression, the response variable follows a Poisson distribution with the mean equal to the variance, μ = Variance (Y). In negative binomial regression, the response variable follows a negative binomial distribution which accommodates overdispersion. To relax the equality assumption, the variance of the response variable, Variance (Y), is a function of the mean μ and a dispersion parameter α. Variance (Y) = $\mu(1 + \alpha\mu)$. When the dispersion parameter $\alpha = 0$, the variance is equal to the means. Thus, the negative binomial model is the same as the Poisson model under this circumstance. With both the mean μ and a dispersion parameter α, the negative binomial distribution can be also referred to as the Poisson-Gamma mixture distribution.

As with the Poisson regression model, the negative binomial regression model (Cameron & Trivedi, 2013; Hardin & Hilbe, 2018; Hilbe, 2011; Long & Freese, 2014) can be expressed as follows:

$$\ln(\mu) = \beta_0 + \beta_1 X_1 + \beta_2 X_2 + \cdots + \beta_p X_p + \varepsilon \tag{9.1}$$

where μ is the mean or the expected number of events; β_0 is the intercept; β_1, β_2, ..., β_p are the coefficients for the predictors; and ε is the error term. The left side of the equation, $\ln(\mu)$, is the log link function. The right side of the equation is the linear predictor. The error term ε on the right side of the equation reflects overdispersion. In the negative binomial regression model, the variance of the response variable is a function of the mean μ and a dispersion parameter α. It can be expressed as either a linear form, Variance (Y) = $\mu(1 + \mu)$, or a quadratic form, Variance (Y) = $\mu(1 + \alpha\mu) = \mu + \alpha\mu^2$. The model with a linear form for the variance is called the NB1 negative binomial model, while the model with a quadratic form is referred to as the NB2 negative binomial model in the literature (Hilbe, 2014). The NB2 model is more commonly used and is the focus in this chapter.

We obtain the predicted mean of the count response variable by exponentiating both sides of Equation 9.1. Since we assume that exp(ε) is 1 here, as with the Poisson regression model, the predicted mean of the count response variable in the negative binomial regression model is also expressed as:

$$\mu = \exp\left(\beta_0 + \beta_1 X_1 + \beta_2 X_2 + \cdots + \beta_p X_p\right) \tag{9.2}$$

When a count response variable is the number of events during a time period or in a location, a count of events can also be referred to as an incidence rate. If we define the incidence rate as the expected number of events per unit time or location, μ/t, the negative binomial model can also be expressed as follows:

$$\ln(\mu/t) = \beta_0 + \beta_1 X_1 + \beta_2 X_2 + \cdots + \beta_p X_p + \varepsilon \tag{9.3}$$

where μ is the mean count; t is a period of time; μ/t is the incidence rate; and $\ln(\mu/t)$ is the log of the incidence rate. Since $\ln(\mu/t) = \ln(\mu) - \ln(t)$, the equation can be rewritten as:

$$\ln(\mu) = \ln(t) + \beta_0 + \beta_1 X_1 + \beta_2 X_2 + \cdots + \beta_p X_p + \varepsilon \tag{9.4}$$

where $\ln(t)$ is the offset in the model equation.

9.1.1 The Negative Binomial Distribution

When there is overdispersion, we prefer the negative binomial probability distribution to the Poisson probability distribution. In the negative binomial probability distribution, we count the number of independent Bernoulli trials before a certain number of successes are achievement, so the number of the total trials equals the summation of the two numbers. There are different ways to express the negative binomial probability distribution. We express the distribution in terms of the probability of the number of trials before successes are achieved. With y trials before k successes in a total of $(y + k)$ trials and the success probability p for each individual trial, we express the negative binomial probability distribution as:

$$P(Y = y) = \binom{y+k-1}{y} p^k (1-p)^y \tag{9.5}$$

where $\binom{y+k-1}{y}$ is the negative binomial coefficient, y is the number of trials before a success is achieved, k is the number of successes in $y + k$ trials, and p is the probability of success for an individual trial.

9.1.2 Incidence Rate Ratios in Negative Binomial Regression Models

Like the Poisson regression model, the negative binomial regression model also estimates the log expected counts of an event or the log incidence rate of the response variable. The incidence rate is defined as the expected number of events during a period of time or in a location. In a simple negative binomial regression model with one predictor, ln $(\mu) = \alpha + \beta X + \varepsilon$, where μ is the expected counts of an event or the incidence rate. The estimated coefficient is the negative binomial regression coefficient, which is the coefficient on the scale of the natural logarithm. It can be also referred to as the log coefficient. We estimate the relationship between the predictor variable and the log function of the expected counts of an event or the log incidence rate.

By exponentiating both sides of the equation, we get the expected counts of an event or the incidence rate:

$$\mu = \exp(\alpha + \beta X) \tag{9.6}$$

If the independent variable X is a categorical variable with the values of 0 and 1, what are the incidence rates of the response variable?

When $X = 0$, the incidence rate $= \exp(\alpha)$, which is the exponentiated intercept.

When $X = 1$, the incidence rate $= \exp(\alpha + \beta)$, which is the exponentiated sum of the intercept and the coefficient.

The incidence rate ratio of group 1 ($X = 1$) to group 2 ($X = 0$):

$$\text{IRR} = \frac{\exp(\alpha + \beta)}{\exp(\alpha)} = \frac{\exp(\alpha) \times \exp(\beta)}{\exp(\alpha)} = \exp(\beta) \tag{9.7}$$

For a one-unit increase in an independent variable (e.g., from 0 to 1 in the previous example) the change in the incidence rate is the incidence rate ratio, which is the exponentiated coefficient. When the independent variable is continuous, for a one-unit increase from any value of x to the value of $(x + 1)$, the change in the incidence rate is still the exponentiated coefficient.

9.1.3 Model Fit Statistics

As with those discussed in the previous chapters, model fit statistics, such as the log likelihood statistic, the residual deviance, the model chi-square statistic, the AIC and BIC statistics, and the pseudo R^2 statistics, can be computed for the negative binomial regression model. The likelihood ratio test, the AIC and BIC statistics, and the Vuong test can also be used for model comparisons.

9.1.4 Interpretation of Model Parameter Estimates

Interpreting the coefficients in a negative binomial regression model is the same as that in a Poisson regression model. When there are multiple predictors in the model, the incidence rate ratio for a predictor can be interpreted as the change in the expected number of events or the incidence rate of a response variable for a one-unit change in a predictor variable when holding other predictor variables constant.

We can also interpret an incidence rate ratio as a percentage change in an incidence rate. It can be calculated by using (incidence rate − 1) × 100%. A positive percentage change in an incidence rate indicates there is an increase in the incidence rate, whereas a negative percentage change corresponds to a decrease in the rate. A zero percentage change indicates no change in the rate at all.

9.2 RESEARCH EXAMPLE AND DESCRIPTION OF THE DATA AND SAMPLE

This chapter focuses on the same research problem as that in Chapter 8. We will still investigate the relationships between the count response variable, the number of zoo visits in a year, and four predictor variables. The GSS 2016 data are used for the following analyses. The following are the variables used for data analysis in this chapter:

- `vistzoo`: the recoded variable of the number of zoo visits in a year
- `maritals`: the recoded variable of marital (marital status) with 1 = currently married and 0 = not currently married
- `educ`: the highest education completed
- `female`: recoded variable of sex with 1 = female and 0 = male
- `wrkfull`: working full time or not.

9.3 FITTING A MULTIPLE-PREDICTOR NEGATIVE BINOMIAL REGRESSION MODEL WITH R

9.3.1 Packages and Functions for Negative Binomial Regression Models in R

The `glm.nb()` function in the `MASS` package (Venables & Ripley, 2002) and the `vglm()` function in the `VGAM` package (Yee, 2010) are used for fitting the negative binomial regression model. Since `MASS` is a part of R base distribution, you just need to load the package by typing `library(MASS)`. You also need to load the `VGAM` package with the `library(VGAM)` function if it is installed. The `glm.nb()` function is introduced first.

9.3.2 The `glm.nb()` Function

The `glm.nb()` function in the `MASS` package is used to fit the negative binomial regression model. As with the `glm()` function, the model formula in `glm.nb()` specifies the dependent variable and the predictor variable(s), which are separated by the tilde (~). The plus (+) symbol is used to connect multiple predictor variables. The data argument specifies the data frame. Please note that no `family` argument needs to be specified and the default link function, the log function, can be omitted. For more details on how to use this function, after loading the `MASS` package with the `library(VGAM)` function, type `help(glm.nb)` in the command prompt.

9.3.3 The Negative Binomial Regression Model: Multiple-Predictor Model

In the following example, the `glm.nb(vistzoo ~ educ + maritals + female + wrkfull, data = count)` command tells R to predict the count response variable `vistzoo` from the four independent variables. In the `glm.nb()` function, the model equation is specified as `vistzoo ~ educ + maritals + female + wrkfull`. The data argument specifies `data = count`. The `family` argument is not needed. The fitted model is named `nbr`. The `summary(nbr)` function displays the output of the fitted model.

```
> # Negative binomial regression model with glm.nb() in MASS
> library(MASS)
> nbr <- glm.nb(vistzoo ~ educ + maritals + female + wrkfull, data=count)
> summary(nbr)

Call:
glm.nb(formula = vistzoo ~ educ + maritals + female + wrkfull,
    data = count, init.theta = 1.58331383, link = log)

Deviance Residuals:
    Min        1Q    Median      3Q      Max
-1.4702   -1.0592   -0.8788   0.3811   3.3187

Coefficients:
             Estimate   Std. Error   z value   Pr(>|z|)
(Intercept)  -1.30862     0.24565    -5.327   9.98e-08 ***
educ          0.04982     0.01665     2.992    0.00277 **
maritals      0.19126     0.09337     2.048    0.04052 *
female        0.02412     0.09386     0.257    0.79719
wrkfull       0.53499     0.09525     5.617   1.94e-08 ***
---
Signif. codes: 0 '***' 0.001 '**' 0.01 '*' 0.05 '.' 0.1 ' ' 1
```

```
(Dispersion parameter for Negative Binomial(1.5833) family taken to be 1)

    Null deviance: 931.01  on 901  degrees of freedom
Residual deviance: 873.58  on 897  degrees of freedom
AIC: 2200

Number of Fisher Scoring iterations: 1

              Theta:  1.583
          Std. Err.:  0.251

2 x log-likelihood:  -2188.013
```

9.3.4 Interpreting the Output

The output of the `glm.nb()` function is similar to that of the `glm()` function. The only difference is that the output of the `glm.nb()` function provides the additional estimate of the dispersion parameter. In the output, the first part is the call, which shows the R command for the negative binomial regression model. The second part shows the minimum, first quarter, median, third quarter, and maximum values of the deviance residuals. The third part shows the coefficients table including the parameter estimates for the four predictor variables and the intercept, their standard errors, the Wald z statistics, and the associated p values. The fourth part of the output shows the fit statistics including the null deviance, the residual deviance, and the AIC. Finally, the fifth part provides the dispersion parameter estimate labeled `Theta`, its standard error, and the 2 log likelihood value.

We start from the coefficients section. The coefficients in the negative binomial model can be interpreted in a similar way to those in the Poisson regression model. For the predictor variable `educ`, Wald $z = 2.992$. The associated p value, `Pr(>|z|) < .01`, so we rejected the null hypothesis. The rejection of the null hypothesis indicates that the predictor variable `educ` is a significant predictor of the count response variable `vistzoo`. For a one-unit increase in education the log expected number of visits to a zoo increases by a factor of .050 after controlling for other predictors in the model.

For the predictor variable `maritals`, Wald $z = 2.048$. The associated p value, `Pr(>|z|) < .05`, so we reject the null hypothesis. For the predictor variable `wrkfull`, Wald $z = 5.617$. The associated p value, `Pr(>|z|) < .001`, so we also reject the null hypothesis. Therefore, `maritals` and `wrkfull` are significant predictors of the response variable.

For the predictor variable `female`, the Wald $z = .257$. The associated p value `Pr(>|z|) = .797`, so we fail to reject the null hypothesis and conclude that there is no significant effect of female on the outcome variable. In other words, whether a person is a female or male does not significantly predict the log expected number of visits to a zoo.

Next, the fit statistics section displays the null deviance for the intercept-only model, the residual deviance for the model, and the AIC.

Finally, the dispersion parameter labeled `Theta`, its standard error, and the 2 log likelihood value are shown at the bottom. Recall that the variance of the response variable in the negative binomial regression model is expressed as a function of the mean μ and a dispersion parameter α. In the NB2 model, Variance $(Y) = \mu(1 + \alpha\mu) = \mu + \alpha\mu^2$. The `glm.nb()` function uses a slightly different form, Variance $(Y) = \mu + \mu^2/\theta$. As we see, $\theta = 1/\alpha$. `Theta` = 1.583 and its standard error is .251. We can easily compute $\alpha = 1/1.583 = .632$.

```
> alpha <- 1/1.583
> alpha
[1] 0.6317119
```

2 log likelihood = −2,188.01. We will use it to test the overall model fit later in the chapter.

We extract the coefficients with the `coef(nbr)` command and obtain the profiled confidence intervals with `confint(nbr)` as follows.

```
> coef(nbr)
(Intercept)        educ    maritals      female     wrkfull
-1.30862163  0.04982352  0.19125927  0.02411989  0.53499022
> confint(nbr)
Waiting for profiling to be done...
                  2.5 %      97.5 %
(Intercept) -1.79149450 -0.83255567
educ         0.01694296  0.08288263
maritals     0.00766270  0.37477527
female      -0.15909658  0.20790336
wrkfull      0.34869644  0.72241114
```

We request the incidence rate ratios with the `exp(coef(nbr))` and `exp(confint(nbr))` commands, respectively. Both results are combined with `cbind(exp(nbr)), exp(confint(nbr)))`.

```
> exp(coef(nbr))
(Intercept)       educ   maritals     female    wrkfull
  0.2701922  1.0510856  1.2107733  1.0244131  1.7074315
> exp(confint(nbr))
Waiting for profiling to be done...
```

```
                  2.5 %       97.5 %
(Intercept)   0.1667108    0.4349363
educ          1.0170873    1.0864143
maritals      1.0076921    1.4546645
female        0.8529140    1.2310942
wrkfull       1.4172189    2.0593927

> cbind(exp(coef(nbr)), exp(confint(nbr)))
Waiting for profiling to be done...
                               2.5 %       97.5 %
(Intercept)   0.2701922    0.1667108    0.4349363
educ          1.0510856    1.0170873    1.0864143
maritals      1.2107733    1.0076921    1.4546645
female        1.0244131    0.8529140    1.2310942
wrkfull       1.7074315    1.4172189    2.0593927
```

The standard errors of the incidence rate ratios can be obtained with `exp(coef(nbr))*sqrt(diag(vcov(nbr)))`.

```
> exp(coef(nbr))*sqrt(diag(vcov(nbr)))
(Intercept)           educ      maritals        female       wrkfull
0.06637319      0.01750264    0.11304683    0.09614906    0.16262904
```

9.3.5 Interpreting the Incidence Rate Ratios in the Negative Binomial Regression Model

As with the Poisson regression model, the negative binomial regression model also estimates the log expected counts of an event. Recall that the exponentiated (β_j) is the incidence rate ratio (IRR) for a one-unit change in a predictor variable when holding other predictors constant. A positive Poisson regression coefficient corresponds to an incidence rate ratio greater than 1, whereas a negative coefficient is associated with an incidence rate ratio less than 1.

In this model, for `educ`, the incidence rate ratio is 1.051. The result indicates that the incidence rate increases by a factor of 1.051 for a one-unit increase in education when holding all other predictors constant. It can also be interpreted as the change in the number of events as follows. The expected number of visits to a zoo increases by 1.051 for a one-unit increase in education when holding all other predictors constant. In other words, for a one-unit increase in education the expected number of visits to a zoo increases by 5.1%.

For `maritals`, the incidence rate ratio is 1.211, which indicates that the expected number of visits to a zoo for the married is 1.211 times as high as that for the unmarried when holding the other predictors constant. In other words, the expected number of visits to a zoo for the married is 21.1% higher than that for the unmarried.

The incidence rate ratio for `wrkfull` can be interpreted in a similar way. The incidence rate ratio is 1.707, which indicates that the expected number of visits to a zoo for those working full time is 1.707 times as high as that for those not working full time when holding the other predictors constant. In other words, the expected number of visits to a zoo for those working full time is 70.7% higher than that for those not working full time.

With regard to `female`, the incidence rate ratio is 1.024, which is not significant (see the associated $p = .797$ in the coefficients table). This indicates that being female does not impact the expected number of visits to a zoo.

9.3.6 Interpreting the Marginal Effects in the Negative Binomial Regression Model

We load the `margins` package (Leeper, 2021) with the `library(margins)` command, compute the marginal effects with the `margins(nbr)` command, and obtain the summary results with `summary(marg.nbr)` as follows.

```
> # marginal effects
> library(margins)
> marg.nbr <- margins(nbr)
> summary(marg.nbr)
   factor     AME     SE      z      p    lower   upper
     educ  0.0403  0.0137  2.9452  0.0032   0.0135  0.0672
   female  0.0195  0.0760  0.2570  0.7972  -0.1294  0.1684
 maritals  0.1548  0.0762  2.0330  0.0421   0.0056  0.3041
  wrkfull  0.4330  0.0813  5.3269  0.0000   0.2737  0.5924
```

The average marginal effect for `educ` is .040. The result indicates that there are on average .040 more visits to a zoo for a one-unit increase in education when holding all other predictors constant.

The average marginal effect for `maritals` is .155. This indicates that the married have on average .155 more visits to the zoo than the unmarried when holding the other predictors constant. The average marginal effects for the other two predictor variables can be interpreted in a similar way.

9.3.7 Model Fit Statistics

Testing the Overall Model Using the Likelihood Ratio Test

To test if the overall model is significant, we fit a null model with the intercept only and compare the single-predictor model with the null model using the `anova()` function. The null model is fitted using the `glm.nb()` function with 1 as the

```
> # Testing the overall model using the likelihood ratio test
> nbr.0 <- glm.nb(vistzoo ~ 1, data = count)
> summary(nbr.0)

Call:
glm.nb(formula = vistzoo ~ 1, data = count, init.theta = 1.256798254,
    link = log)

Deviance Residuals:
    Min        1Q    Median        3Q       Max
-1.1184   -1.1184   -1.1184    0.1561    2.4080

Coefficients:
              Estimate   Std. Error   z value    Pr(>|z|)
(Intercept)   -0.21020     0.04744    -4.431    9.37e-06 ***

---
Signif. codes: 0 '***' 0.001 '**' 0.01 '*' 0.05 '.' 0.1 ' ' 1

(Dispersion parameter for Negative Binomial(1.2568) family taken to be 1)

    Null deviance: 870.56  on 901  degrees of freedom
Residual deviance: 870.56  on 901  degrees of freedom
AIC: 2247

Number of Fisher Scoring iterations: 1

            Theta:  1.257
        Std. Err.:  0.177

2 x log-likelihood: -2243.012
```

intercept term in the model formula. The command and the output are displayed below.

The `anova(nbr.0, nbr, test = "Chisq")` command compares the log-likelihood statistics of the fitted model `nbr` and the null model `nbr.0` using the likelihood ratio test.

```
> anova(nbr.0, nbr, test = "Chisq")
Likelihood ratio tests of Negative Binomial Models

Response: vistzoo
                                 Model      theta   Resid. df    2 x log-lik.
1                                    1   1.256798         901       -2243.012
2    educ + maritals + female + wrkfull   1.583314         897       -2188.013

      Test     df     LR stat.        Pr(Chi)
1
2   1 vs 2      4     54.99921   3.250222e-11
```

Here we first fit the null model and then compare the single-predictor model with the null model. This two-step process can also be simplified to the one-line command with the `update()` function within the `anova()` function. In the `anova(nbr, update(nbr, ~ 1), test = "Chisq")` command, we use `update(nbr, ~ 1)` to fit the null model with the intercept only.

```
> anova(nbr, update(nbr, ~1), test = "Chisq")
Likelihood ratio tests of Negative Binomial Models

Response: vistzoo
                                   Model      theta   Resid. df    2 x log-lik.
1                                      1   1.256798         901       -2243.012
2    educ + maritals + female + wrkfull   1.583314         897       -2188.013

      Test    df    LR stat.         Pr(Chi)
1
2    1 vs 2    4    54.99921    3.250222e-11
```

The null hypothesis of the test for the overall model is that the four predictor variables do not contribute to the model, and the alternative hypothesis is that the multiple-predictor model is better than the null model with no independent variables. The likelihood ratio test statistic $LR\ \chi^2_{(4)} = 54.999$, $p < .001$, which indicates that the overall model with the four predictors is significantly different from zero. Therefore, the multiple-predictor model provides a better fit than the null model with no independent variables in predicting the log number of visits to a zoo in a year.

Pseudo R^2

We use the `nagelkerke()` function in the `rcompanion` package (Mangiafico, 2021) to compute the pseudo R^2 statistics for the single-predictor model. We load the package first with `library(rcompanion)` and then use the `nagelkerke(nbr)` command.

```
> # Pseudo R2 with nagelkerke()
> library(rcompanion)
> nagelkerke(nbr)
$`Models`

Model: "glm.nb, vistzoo ~ educ + maritals + female + wrkfull, count, 1.58331383, log"
Null:  "glm.nb, vistzoo ~ 1, count, 1.256798254, log"

$Pseudo.R.squared.for.model.vs.null
                              Pseudo.R.squared
McFadden                             0.0245202
Cox and Snell (ML)                   0.0591530
Nagelkerke (Cragg and Uhler)         0.0645200
```

```
$Likelihood.ratio.test
Df.diff    LogLik.diff    Chisq       p.value
     -4          -27.5   54.999    3.2502e-11

$Number.of.observations
Model: 902
Null:  902

$Messages
[1] "Note: For models fit with REML, these statistics are based on refitting with ML"

$Warnings
[1] "None"
```

McFadden's R^2 is .025, Cox and Snell's R^2 is .059, and Nagelkerke's R^2 is .065. The same results can be computed using the equations for these three pseudo R^2 statistics. In the R command below, `LLM1` is the log-likelihood value for the multiple-predictor model and `LL0` is the log-likelihood value for the null model. The number of observations is 902. In addition, `McFadden1` is the object name for McFadden's R^2, `CS1` for Cox and Snell's R^2, and `NG1` for Nagelkerke's R^2.

```
> # Pseudo R2 with equations
> LLM1 <- logLik(nbr)
> LL0 <- logLik(nbr.0)
> McFadden1 <- 1-(LLM1/LL0)
> McFadden1
'log Lik.' 0.02452024 (df=6)
> CS1 <- 1-exp(2*(LL0-LLM1)/902)
> CS1
'log Lik.' 0.05915298 (df=2)
> NG1 <- CS1/(1-exp(2*LL0/902))
> NG1
'log Lik.' 0.06451996 (df=2)
```

AIC and BIC Statistics

The AIC and BIC statistics can also be computed by the `AIC(nbr)` and `BIC(nbr)` commands. The `AIC(PR.2, nbr)` and `BIC(PR.2, nbr)` commands compare the AIC and BIC statistics between the Poisson regression model and the negative binomial regression model, respectively. The output is as follows.

```
> # AIC and BIC Statistics
> AIC(nbr)
[1] 2200.013
> BIC(nbr)
[1] 2228.841
> PR.2 <- glm(vistzoo ~ educ + maritals + female + wrkfull, family = poisson, data
= count)
```

```
> AIC(PR.2, nbr)
        df       AIC
PR.2     5  2286.799
nbr      6  2200.013

> BIC(PR.2, nbr)
        df       BIC
PR.2     5  2310.822
nbr      6  2228.841
```

The AIC and BIC statistics for the multiple-predictor negative binomial regression model are 2,200.013 and 2,228.841, respectively. Compared with those in the Poisson regression model, both AIC and BIC indicate that the negative binomial regression model fits the data better.

9.3.8 Interpreting the Predicted Counts With the `ggpredict()` Function in the `ggeffects` Package

By using the `ggpredict()` function in the `ggeffects` package (Lüdecke, 2018b), we can compute the predicted number of events of the count response variable at specified values of the predictor variables. We first load the package with `library(ggeffects)` since it has been installed in previous chapters. The command `nbr.ed <- ggpredict(nbr, terms = "educ[12, 14, 16]")` tells R to compute the predicted counts of the response variable using the `ggpredict()` function. The argument inside the function includes the estimated model name, `nbr`, and the `terms = "educ[12, 14, 16]"` argument, which specifies the predictor variable `educ` at the values of 12, 14, and 16 when holding other predictor variables at their means. The `terms` option can specify up to four variables, including the second to fourth grouping variables. The output is assigned to the object named `nbr.ed`. The `as.data.-frame()` function or the `sqrt(diag(vcov()))` function can be used to request the standard errors. The output is omitted here.

```
> # Predicted counts with ggpredict() in ggeffects
> library(ggeffects)
> nbr.ed <- ggpredict(nbr,terms = "educ[12, 14, 16]")
> nbr.ed
# Predicted counts of vistzoo

educ | Predicted |         95% CI
---------------------------------
  12 |      0.70 | [0.62, 0.78]
  14 |      0.77 | [0.70, 0.85]
  16 |      0.85 | [0.76, 0.95]

Adjusted for:
* maritals = 0.44
*   female = 0.56
*  wrkfull = 0.47
> plot(nbr.ed)
```

The results table displays the values of `educ`, the predicted counts, their standard errors, and the confidence intervals. The predicted counts in the negative binomial regression model are similar to those in the Poisson regression model which was introduced in the previous chapter.

> When `educ` = 12, and the other predictor variables are held at their means (`maritals` = .44, `female` = .56, and `wrkfull` = .47), the predicted number of visits to a zoo is .70.
>
> When `educ` = 14, and the other three predictor variables are held at their means, the predicted number of visits to a zoo is .77.
>
> When `educ` = 16, and the other predictor variables are held at their means, the predicted number of visits to a zoo is .85.

The predicted counts are plotted using the `plot(nbr.ed)` function. Figure 9.1 shows the predicted number of visits to a zoo when `educ` is at the values of 12, 14, and 16.

The graph shows that with the increase of the years of education, the predicted number of visits to a zoo increases. In other words, people with higher levels of education are associated with having more visits to a zoo.

We can also compute the predicted counts for a continuous variable at given values by different groups. In the following example, we compute the predicted number of visits to a zoo for `educ` at the values of 12, 14, and 16 by the two groups in `wrkfull`

FIGURE 9.1 ● Predicted Counts for educ at 12, 14, and 16

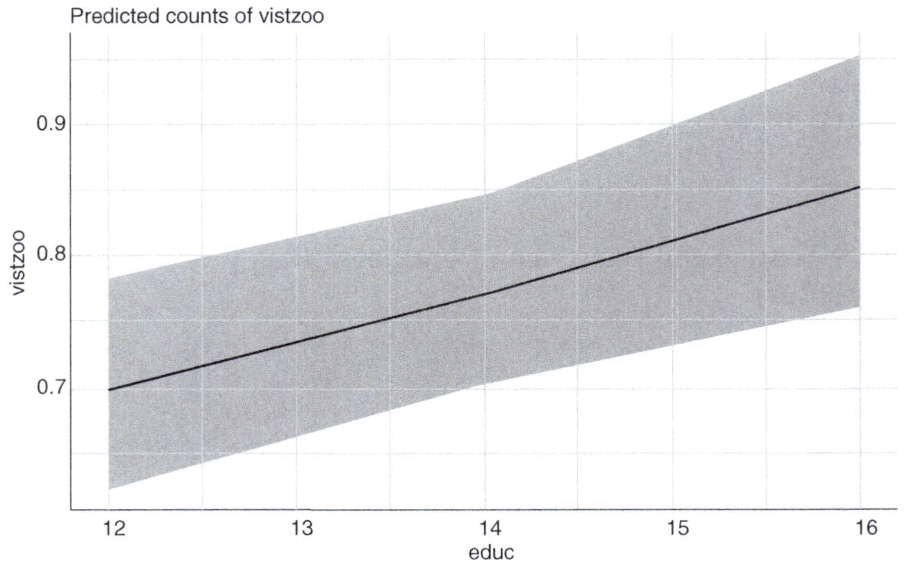

when holding other variables at their means. The command is as follows: nbr.ew <- ggpredict(nbr, terms = c("educ[12, 14, 16]", "wrkfull")). In the ggpredict() function, the terms = c("educ[12, 14, 16]", "wrkfull") option specifies both educ and wrkfull, with the latter as the grouping variable. The output is assigned to an object named nbr.ew and is plotted with plot(nbr.ew).

```
> nbr.ew <- ggpredict(nbr, terms = c("educ[12, 14, 16]", "wrkfull"))
> nbr.ew
# Predicted counts of vistzoo

# wrkfull = 0

educ | Predicted |       95% CI
-------------------------------
  12 |      0.54 | [0.47, 0.63]
  14 |      0.60 | [0.52, 0.69]
  16 |      0.66 | [0.56, 0.77]

# wrkfull = 1

educ | Predicted |       95% CI
-------------------------------
  12 |      0.92 | [0.80, 1.07]
  14 |      1.02 | [0.90, 1.15]
  16 |      1.13 | [0.99, 1.29]

Adjusted for:
* maritals = 0.44
*   female = 0.56
> plot(nbr.ew)
```

Figure 9.2 shows the predicted counts for educ at 12, 14, and 16 by the grouping variable wrkfull. As shown in the graph, the predicted number of visits to a zoo increases with an increase in years of education and the predicted counts for the people who work full-time are higher than the predicted counts for those not working full-time.

9.3.9 Testing the Dispersion Parameter Using the Likelihood Ratio Test

The likelihood ratio test is used to test whether the dispersion parameter is significantly different from zero by comparing the Poisson regression model and the negative binomial regression model. If the dispersion parameter α is significant, then we favor the negative binomial regression model. If it is not significant, then we conclude that there is no overdispersion, so we prefer the Poisson regression model.

We use the lrtest() function in the lmtest package (Zeileis & Hothorn, 2002) to compare these two models since the anova() function is inappropriate here. If you

FIGURE 9.2 ● Predicted Counts for `educ` at 12, 14, and 16 by `wrkfull`

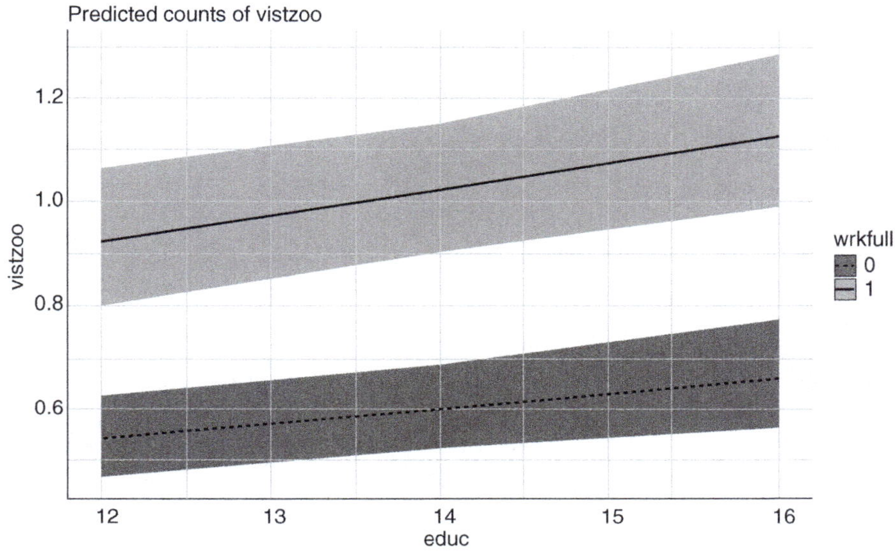

use the `anova()` function, you will get the same degrees of freedom of the deviances for these two models, which will make the model comparison impossible. To use the `lrtest()` function, you need to install the `lmtest` package first by typing `install.packages("lmtest")` and then load it by typing `library (lmtest)`. In the `lrtest(PR.2, nbr)` command, `PR.2` and `nbr` are the Poisson regression model and the negative binomial regression model, respectively. The following output is displayed.

```
> # Model comparison with the likelihood ratio test
> library(lmtest)
> lrtest(PR.2, nbr)
Likelihood ratio test

Model 1: vistzoo ~ educ + maritals + female + wrkfull
Model 2: vistzoo ~ educ + maritals + female + wrkfull
     #Df     LogLik    Df      Chisq     Pr(>Chisq)
1      5    -1138.4
2      6    -1094.0     1     88.786      < 2.2e-16 ***

---
Signif. codes:  0 '***' 0.001 '**' 0.01 '*' 0.05 '.' 0.1 ' ' 1
```

The likelihood ratio test, $\chi^2_{(1)} = 88.79$, $p < .001$, which indicates that the dispersion parameter α is significantly different from zero, so we favor the negative binomial regression model.

9.4 NEGATIVE BINOMIAL REGRESSION WITH THE vglm() FUNCTION IN THE VGAM PACKAGE

We can also use the `vglm()` function in the `VGAM` package to fit negative binomial regression models. Since `VGAM` has been installed for the previous chapters, you just need to load the package by typing `library(VGAM)`. The syntax for negative binomial regression models is similar to that for other models using the `vglm()` function. The `negbinomial` family needs to be specified for the `family` argument. For example, the command `vglm(y ~ x, family = negbinomial, data = data1)` tells R to fit a negative binomial regression model predicting the dependent variable *y* with an independent variable *x*. The `family = negbinomial` argument tells R that it is the negative binomial family. When there are more than one predictor variable in the formula, they are connected by plus (+) symbols. For more details on how to use this function, type `help(negbinomial)` in the command prompt after loading the `VGAM` package.

In the following example, the `nb.v <- vglm(vistzoo ~ educ + maritals + female + wrkfull, family = negbinomial, data = count)` command tells R to predict the count response variable `vistzoo` from the four independent variables. In the model formula for the `vglm()` function, the dependent variable `vistzoo` and the four predictor variables are separated by the tilde (~). The four predictor variables include `educ`, `maritals`, `female`, and `wrkfull`, which are connected by plus (+) symbols. We also specify the data arguments `data = count`. The fitted model is named `nb.v`. The following output is shown by the `summary(nb.v)` command.

```
> # Negative binomial regression model with vglm() in VGAM
> library(VGAM)
> nb.v <- vglm(vistzoo ~ educ + maritals + female + wrkfull, family = negbinomial,
data = count)
> summary(nb.v)

Call:
vglm(formula = vistzoo ~ educ + maritals + female + wrkfull,
    family = negbinomial, data = count)

Pearson residuals:
                    Min       1Q    Median      3Q     Max
loglink(mu)      -0.885  -0.6873   -0.5854  0.4253   7.239
loglink(size)   -13.773  -0.5511   -0.3851  0.9561   1.454
```

```
Coefficients:
                Estimate   Std. Error   z value    Pr(>|z|)
(Intercept):1   -1.30861     0.24566    -5.327    9.98e-08 ***
(Intercept):2    0.45953     0.16463     2.791     0.00525 **
educ             0.04982     0.01665     2.992     0.00277 **
maritals         0.19126     0.09337     2.048     0.04052 *
female           0.02412     0.09386     0.257     0.79722
wrkfull          0.53499     0.09525     5.617    1.95e-08 ***

---
Signif. codes: 0 '***' 0.001 '**' 0.01 '*' 0.05 '.' 0.1 ' ' 1

Names of linear predictors: loglink(mu), loglink(size)

Log-likelihood: -1094.007 on 1798 degrees of freedom

Number of Fisher scoring iterations: 5

No Hauck-Donner effect found in any of the estimates
```

The R output produced by the `vglm()` function for the negative binomial regression model is similar to that for the Poisson regression model. It includes the call of the model command, the Pearson residuals, the coefficients, the names of linear predictors, the log-likelihood, and the number of Fisher scoring iterations.

The first section shows the call, which is the R command for the model. The second section shows the minimum, first quarter, median, third quarter, and maximum values of the Pearson residuals. The third section shows the coefficients table including the parameter estimates for the predictor variable, the intercept, and the dispersion parameter, their standard errors, the Wald z statistics, and the associated p values. The fourth section shows the two names of linear predictors, `loglink(mu)` and `loglink(size)`. The former is the log link for the expected count (μ) and the latter is the log link for the index parameter (k) or the dispersion parameter ($1/k$). The fifth section shows the log-likelihood value and the degrees of freedom. Finally, the number of Fisher scoring iterations is displayed at the end.

The negative binomial coefficients in the coefficients section (labeled `Coefficients:`) are the same as those produced from the `glm.nb()` function. See the preceding section on the interpretation of the coefficients.

Also in the coefficients section, the log dispersion parameter is labeled `(Intercept):2`. To obtain the dispersion parameter, we need to exponentiate the estimate of the `(Intercept):2`. The dispersion parameter $\theta = \exp(.460) = 1.583$. As with the `glm.nb()` function, the `vglm()` function also uses a slightly different form for the NB2 model, Variance $(Y) = \mu + \mu^2/\theta$ rather than Variance $(Y) = \mu + \alpha\mu^2$. Since $\theta = 1/\alpha$, we can compute $\alpha = 1/1.583 = .632$.

```
> exp(.45953)
[1] 1.58333
> alpha.v <- 1/1.583
> alpha.v
[1] 0.6317119
```

The coefficients of the predictor variables can be extracted by using the `coef(nb.v, matrix = TRUE)` command. The confidence intervals are obtained with `confint (nb.v, matrix = TRUE)`.

```
> coef(nb.v, matrix = TRUE)
              loglink(mu)   loglink(size)
(Intercept)   -1.30861425       0.4595301
educ           0.04982324       0.0000000
maritals       0.19125640       0.0000000
female         0.02411682       0.0000000
wrkfull        0.53498927       0.0000000

> confint(nb.v, matrix = TRUE)
                        2.5 %         97.5 %
(Intercept):1     -1.790091654   -0.82713686
(Intercept):2      0.136868259    0.78219189
educ               0.017185431    0.08246105
maritals           0.008256362    0.37425644
female            -0.159844305    0.20807794
wrkfull            0.348303906    0.72167464
```

We use the `exp(coef(nb.v, matrix = TRUE))` and `exp(confint(nb.v, matrix = TRUE))` commands to compute the IRRs and the corresponding confidence intervals, respectively.

```
> exp(coef(nb.v, matrix = TRUE))
              loglink(mu)   loglink(size)
(Intercept)     0.2701942         1.58333
educ            1.0510853         1.00000
maritals        1.2107699         1.00000
female          1.0244100         1.00000
wrkfull         1.7074299         1.00000

> exp(confint(nb.v, matrix = TRUE))
                    2.5 %       97.5 %
(Intercept):1   0.1669449    0.4372995
(Intercept):2   1.1466771    2.1862591
educ            1.0173339    1.0859564
maritals        1.0082905    1.4539099
female          0.8522765    1.2313091
wrkfull         1.4166627    2.0578765
```

The standard errors of the IRRs are computed with the `exp(coef(nb.v))*sqrt(diag(vcov(nb.v)))` command.

```
> exp(coef(nb.v))*sqrt(diag(vcov(nb.v)))
(Intercept):1   (Intercept):2        educ    maritals      female
   0.06637490      0.26065788  0.01750293  0.11304847  0.09615055
      wrkfull
   0.16263165
```

We can use the `nagelkerke(nb.v)` command to compute the three types of pseudo R^2 statistics and the likelihood ratio test statistic for the multiple-predictor model. The output is omitted here.

9.5 ZERO-INFLATED POISSON REGRESSION WITH THE `zeroinfl()` FUNCTION IN THE `pscl` PACKAGE

When we fit Poisson regression and negative binomial models, sometimes we notice there are many zeros in the count response variable. We may wonder whether these zeros are so excessive that they impact both Poisson and negative binomial distributions of the response variable. Recall that in Poisson regression models, we assume the mean is equal to the variance, which is the equidispersion assumption. So having excessive zero counts does impact the Poisson distribution of the response variable. This is also true for the negative binomial regression model although it has a different over-dispersion assumption. If we have an excess of zero counts in the response variable, what are better alternatives to model this type of count data? In this section, we discuss the zero-inflated models which address the issue of excessive zero counts (Friendly & Meyer, 2016; Hilbe, 2014; Lambert, 1992; Long & Freese, 2014). The zero-inflated model can be applied to both Poisson regression and negative binomial regression. The former is referred to as the zero-inflated Poisson regression model and the latter is called the zero-inflated negative binomial regression model.

Having zero counts in the response variable does not mean the variable is zero-inflated. The zero-inflated model can determine whether these zero counts occur by chance or they are associated with other factors. In other words, can a variable or a set of variables predict these zero counts? This question can be addressed by using a logistic regression model. Therefore, the zero-inflated model includes two components. One is the binary logistic regression model component which predicts if the response variable is a zero count or not. The other is the count model component which predicts the count response variable. The count model component can be either a Poisson regression model or a negative binomial regression model. In the binary logistic regression model component, the outcome is the dichotomized count response variable with the zero counts coded as 1 and the positive counts coded as 0, so we predict the probability of having zero counts. If a predictor or a set of predictors can significantly predict the probability of having zero counts, then there are excessive zero counts associated with that predictor or that set of predictors.

The zero counts are included in both the binary logistic regression model component and the count model component, so there is an overlap of zero counts in both components. Therefore, the zero-inflated model is also called the zero-inflated mixture model.

We use the `zeroinfl()` function in the `pscl` package (Jackman, 2020; Zeileis et al., 2008) to fit both the zero-inflated Poisson regression model and the zero-inflated negative binomial model. We introduce the former model first in this section. Since `pscl` is a user-written package, you need to install it first by typing `install.packages("pscl")` and then load the package by typing `library(pscl)`. The syntax for the zero-inflated models with the `zeroinfl()` function is similar to that for other models using the `glm()` function except that we need to specify the additional zero-inflation part in the model formula. The model formula in `zeroinfl()` includes both the count part and the zero-inflation part. It first specifies the dependent variable and the predictor variable(s) in the count part, which are separated by the tilde (~). When there are multiple predictor variables, they are connected by plus (+) symbols. It then specifies the zero-inflation part which is separated from the count part by the vertical line (|). For example, the model formula, `y ~ x1 + x2 | x1 + x2`, includes the two predictor variables, *x*1 and *x*2 in both the count and the zero-inflation parts. When the same predictors are included in both parts, the model formula can be simplified as `y ~ x1 + x2`. You can also specify different predictors in both parts. The `dist = "poisson"` family is used for the Poisson `family` argument. You can omit this family argument since it is the default for the zero-inflated Poisson regression model. For example, the `zeroinfl(y ~ x1 + x2 | x1 + x2, dist = "poisson", data = data1)` command tells R to fit a zero-inflated Poisson regression model predicting the dependent variable *y* with the two independent variables *x*1 and *x*2. The `dist = "poisson"` argument tells R that it is the Poisson family. For more details on how to use this function, type `help(zeroinfl)` in the command prompt after loading the `pscl` package.

In the following example, the `zip <- zeroinfl(vistzoo ~ educ + maritals + female + wrkfull | educ + maritals + female + wrkfull, data = count)` command tells R to predict the count response variable `vistzoo` from the four independent variables. In the model formula for the `zeroinfl()` function, the dependent variable `vistzoo` and the four predictor variables are separated by the tilde (~). The four predictor variables include `educ`, `maritals`, `female`, and `wrkfull`, which are connected by plus (+) symbols in the count part. The same four predictors are specified in the zero-inflation part. We also specify the data arguments `data = count`. The fitted model is named `zip`. The following output is shown by the `summary(zip)` command.

```
> # ZIP model with zeroinfl() in pscl
> library(pscl)
> zip <- zeroinfl(vistzoo ~ educ + maritals + female + wrkfull | educ + maritals +
female + wrkfull, data = count)
> summary(zip)

Call:
zeroinfl(formula = vistzoo ~ educ + maritals + female + wrkfull |
   educ + maritals + female + wrkfull, data = count)

Pearson residuals:
    Min        1Q    Median       3Q      Max
-1.1051   -0.6683   -0.5906   0.4267   7.1068

Count model coefficients (poisson with log link):
              Estimate   Std. Error   z value    Pr(>|z|)
(Intercept)   -1.09095      0.30492    -3.578    0.000346 ***
educ           0.07632      0.02012     3.794    0.000148 ***
maritals       0.16696      0.10394     1.606    0.108187
female         0.04846      0.09851     0.492    0.622779
wrkfull        0.07383      0.11289     0.654    0.513107

Zero-inflation model coefficients (binomial with logit link):
              Estimate   Std. Error   z value    Pr(>|z|)
(Intercept)   -1.43959      0.91564    -1.572       0.116
educ           0.09034      0.06079     1.486       0.137
maritals      -0.13685      0.32803    -0.417       0.677
female         0.03580      0.31216     0.115       0.909
wrkfull       -1.75639      0.41260    -4.257    2.07e-05 ***

---
Signif. codes: 0 '***' 0.001 '**' 0.01 '*' 0.05 '.' 0.1 ' ' 1

Number of iterations in BFGS optimization: 17
Log-likelihood: -1108 on 10 Df
```

The R output produced by the `zeroinfl()` function for the zero-inflated Poisson regression model is similar to that for the Poisson regression model by the `glm()` function. The major difference is that the former function displays the coefficients section for the additional zero-inflation component. It includes the call of the model command, the Pearson residuals, the count model coefficients, the zero-inflation model coefficients, the number of iterations, and the log-likelihood and corresponding degrees of freedom.

Specifically, the first section shows the call, which is the R command for the model. The second section shows the minimum, first quarter, median, third quarter, and maximum values of the Pearson residuals. The third section shows the coefficients table for the Poisson regression model including the parameter estimates for the predictor variables, the intercept, their standard errors, the Wald z statistics, and the associated

p values. The fourth section shows the coefficients table for the zero-inflation model estimating the probability when the response variable equals 0, $P(Y = 0)$. The fourth section shows the number of iterations and the log-likelihood and corresponding degrees of freedom.

In the third section, the coefficients table for the Poisson regression model component is labeled with the title `Count model coefficients (poisson with log link)`. The interpretation of the Poisson coefficients is the same as those in the Poisson regression model and the negative binomial regression model. Each coefficient can be interpreted as the change in the log expected of the incidence rate or the log expected number of events for a one-unit increase in the predictor variable when holding other predictors constant.

For the predictor variable `educ`, Wald $z = 3.794$. The associated p value, `Pr(>|z|)` `< .001`, so we reject the null hypothesis. The rejection of the null hypothesis indicates that `educ` is a significant predictor of the count response variable `vistzoo`.

For the predictor variable `maritals`, Wald $z = 1.606$. The associated p value, `Pr(>|z|)` = `.108`, so we fail to reject the null hypothesis. For the predictor variable `female`, the Wald $z = .492$. The associated p value `Pr(>|z|)` = `.663`, so we also fail to reject the null hypothesis. For the predictor variable `wrkfull`, Wald $z = .654$. The associated p value, `Pr(>|z|)` = `.513`, so we fail to reject the null hypothesis, too. Therefore, `maritals`, `female`, and `wrkfull` are not significant predictors of the count response variable.

In the fourth section, the coefficients table for the zero-inflation model component is labeled with the title `Zero-inflation model coefficients (binomial with logit link)`. Recall that the zero-inflation model component is the logistic regression model for the binary outcome variable comparing zero and non-zero counts. The interpretation of the logit coefficients here is the same as those in the logistic regression model which were introduced in earlier chapters. Each coefficient can be interpreted as the change in the logit of having zero counts for a one-unit increase in the predictor variable when holding other predictors constant.

Among the four predictors in the zero-inflation model component, only `wrkfull` is significant. Wald $z = -4.257$. The associated p value, `Pr(>|z|)` `< .001`, so we reject the null hypothesis and conclude that `wrkfull` is significant in predicting excess zeros.

The coefficients of the predictor variables for both count and zero-inflation models can be extracted by using the `coef(zip)` command. The confidence intervals are obtained with `confint(zip)`. The output is omitted here.

We use the `exp(coef(zip))` and `exp(confint(zip))` commands to compute the IRRs and the corresponding confidence intervals, respectively. The results are combined with the `cbind()` function.

```
> exp(coef(zip))
count_(Intercept)         count_educ   count_maritals      count_female
        0.3358971          1.0793123        1.1817114         1.0496532
    count_wrkfull  zero_(Intercept)        zero_educ     zero_maritals
        1.0766275          0.2370247        1.0945462         0.8721047
      zero_female       zero_wrkfull
        1.0364488          0.1726664

> exp(confint(zip))
                        2.5 %       97.5 %
count_(Intercept)  0.18478087    0.6105982
count_educ         1.03758476    1.1227180
count_maritals     0.96391844    1.4487137
count_female       0.86535012    1.2732094
count_wrkfull      0.86291889    1.3432626
zero_(Intercept)   0.03939143    1.4262163
zero_educ          0.97159626    1.2330548
zero_maritals      0.45851228    1.6587704
zero_female        0.56213136    1.9109879
zero_wrkfull       0.07691409    0.3876233

> cbind(exp(coef(zip)), exp(confint(zip)))
                                   2.5 %       97.5 %
count_(Intercept)  0.3358971  0.18478087    0.6105982
count_educ         1.0793123  1.03758476    1.1227180
count_maritals     1.1817114  0.96391844    1.4487137
count_female       1.0496532  0.86535012    1.2732094
count_wrkfull      1.0766275  0.86291889    1.3432626
zero_(Intercept)   0.2370247  0.03939143    1.4262163
zero_educ          1.0945462  0.97159626    1.2330548
zero_maritals      0.8721047  0.45851228    1.6587704
zero_female        1.0364488  0.56213136    1.9109879
zero_wrkfull       0.1726664  0.07691409    0.3876233
```

The standard errors of the IRRs are computed with the `exp(coef(zip))*sqrt(diag(vcov(zip)))` command.

```
> exp(coef(zip))*sqrt(diag(vcov(zip)))
count_(Intercept)         count_educ   count_maritals      count_female
       0.10242210         0.02171242       0.12282326        0.10340395
    count_wrkfull  zero_(Intercept)        zero_educ     zero_maritals
       0.12154469         0.21702862       0.06654227        0.28607446
      zero_female       zero_wrkfull
       0.32353663         0.07124141
```

We can use the `nagelkerke(zip)` command to compute the three types of pseudo R^2 statistics and the likelihood ratio test statistic for the model. The output is omitted here.

We use the Vuong test to compare the zero-inflated Poisson regression model and the Poisson regression model. This test can be used to compare two non-nested models. A significant z statistic suggests that one model is better than the other. The `vuong()`

function in the `pscl` package function is used for model comparison. In the `vuong(zip,PR.2)` command, `zip` and `PR.2` are the zero-inflated Poisson regression model and the Poisson regression model, respectively. The following output is displayed.

```
> # Vuong test
> vuong(zip, PR.2)
Vuong Non-Nested Hypothesis Test-Statistic:
(test-statistic is asymptotically distributed N(0,1) under the
null that the models are indistinguishible)
-------------------------------------------------------------
              Vuong z-statistic             H_A    p-value
Raw                    3.068161 model1 > model2  0.0010769
AIC-corrected          2.571295 model1 > model2  0.0050659
BIC-corrected          1.377672 model1 > model2  0.0841523
```

The Vuong test $z = 3.068$, $p < .01$, which indicates that the Vuong z statistic is significantly different from zero, so we prefer the zero-inflated Poisson regression model to the Poisson regression model.

Since among the four predictors in the zero-inflation model component only `wrkfull` is significant in predicting zeros, we can remove the other three predictors in the binary zero-inflation model. The simplified model can be fitted using the `zip2 <- zeroinfl(vistzoo ~ educ + maritals + female + wrkfull | wrkfull, data = count)` command. The results of the model are displayed by the `summary(zip2)` command.

```
> # ZIP reduced model
> zip2 <- zeroinfl(vistzoo ~ educ + maritals + female + wrkfull | wrkfull, data =
count)
> summary(zip2)

Call:
zeroinfl(formula = vistzoo ~ educ + maritals + female + wrkfull |
   wrkfull, data = count)

Pearson residuals:
    Min       1Q   Median      3Q     Max
-1.1557  -0.6592  -0.5575  0.4306  7.0883

Count model coefficients (poisson with log link):
            Estimate Std. Error z value Pr(>|z|)
(Intercept) -0.80700    0.23268  -3.468 0.000524 ***
educ         0.05660    0.01524   3.715 0.000203 ***
maritals     0.20464    0.08287   2.469 0.013536 *
female       0.04562    0.08251   0.553 0.580321
wrkfull      0.05033    0.11187   0.450 0.652782

Zero-inflation model coefficients (binomial with logit link):
            Estimate Std. Error z value Pr(>|z|)
(Intercept)  -0.1693     0.1716  -0.987    0.324
wrkfull      -1.8017     0.4457  -4.043 5.29e-05 ***
```

```
---
Signif. codes: 0 '***' 0.001 '**' 0.01 '*' 0.05 '.' 0.1 ' ' 1

Number of iterations in BFGS optimization: 15
Log-likelihood: -1109 on 7 Df
```

We use the likelihood ratio test to compare the zero-inflated Poisson regression model and the simplified model. In the `lrtest(zip2, zip)` command, `zip2` and `zip` are the two models being compared. The following output is displayed.

```
> lrtest(zip2, zip)
Likelihood ratio test

Model 1: vistzoo ~ educ + maritals + female + wrkfull | wrkfull
Model 2: vistzoo ~ educ + maritals + female + wrkfull | educ + maritals +
    female + wrkfull
   #Df   LogLik  Df   Chisq   Pr(>Chisq)
1    7  -1108.8
2   10  -1107.5   3  2.4508       0.4842
```

The likelihood ratio test, $\chi^2_{(3)} = 2.451$, $p = .484$, which is not significant. This indicates that there is no significant difference between these two models. Since the simplified model is more parsimonious than the original model, we prefer the former model with fewer parameters.

9.6 ZERO-INFLATED NEGATIVE BINOMIAL REGRESSION WITH THE `zeroinfl()` FUNCTION IN THE `pscl` PACKAGE

We again use the `zeroinfl()` function in the `pscl` package to fit and the zero-inflated negative binomial model. Since `pscl` has been loaded in the last section with `library(pscl)`, we do not need to load it again. To fit the zero-inflated negative binomial regression model, with all else the same as those in the zero-inflated Poisson regression model, we only need to specify a different family argument, `dist = "negbin"`. The model formula in `zeroinfl()` still includes the negative binomial model and the zero-inflation model components. The `zinb <- zeroinfl(vistzoo ~ educ + maritals + female + wrkfull | educ + maritals + female + wrkfull, dist = "negbin", data = count)` command tells R to predict the count response variable `vistzoo` from the four independent variables with the zero-inflated negative binomial regression model. The fitted model is named `zinb`. The following output is shown by the `summary(zinb)` command.

```
> # Zero-inflated NB model with zeroinfl() in pscl
> zinb <- zeroinfl(vistzoo ~ educ + maritals + female + wrkfull | educ + maritals +
female + wrkfull, dist="negbin", data = count)
> summary(zinb)

Call:
zeroinfl(formula = vistzoo ~ educ + maritals + female + wrkfull |
   educ + maritals + female + wrkfull, data = count, dist = "negbin")

Pearson residuals:
    Min        1Q    Median       3Q      Max
-0.9752   -0.6224   -0.5706   0.4036   6.7580

Count model coefficients (negbin with log link):
             Estimate  Std. Error   z value   Pr(>|z|)
(Intercept)  -1.11891     0.29408    -3.805   0.000142 ***
educ          0.05972     0.01930     3.094   0.001978 **
maritals      0.19616     0.10398     1.887   0.059223 .
female        0.04780     0.10249     0.466   0.640922
wrkfull       0.18744     0.13569     1.381   0.167149
Log(theta)    0.80108     0.21345     3.753   0.000175 ***

Zero-inflation model coefficients (binomial with logit link):
             Estimate  Std. Error   z value   Pr(>|z|)
(Intercept)  -2.02306     1.34445    -1.505      0.132
educ          0.07225     0.08484     0.852      0.394
maritals     -0.01096     0.50484    -0.022      0.983
female        0.19995     0.50826     0.393      0.694
wrkfull      -8.28490    17.01979    -0.487      0.626

---
Signif. codes: 0 '***' 0.001 '**' 0.01 '*' 0.05 '.' 0.1 ' ' 1

Theta = 2.2279
Number of iterations in BFGS optimization: 25
Log-likelihood: -1088 on 11 Df
```

The R output created by the `zeroinfl()` function for the zero-inflated negative binomial regression model is similar to that for the zero-inflated Poisson regression model. The only difference is that the output for the zero-inflated negative binomial regression model displays the additional dispersion parameter `Log(theta)`. It includes the call of the model command, the Pearson residuals, the count model coefficients, the zero-inflation model coefficients, the dispersion parameter theta, the number of iterations, and the log-likelihood and corresponding degrees of freedom.

The fourth section shows the coefficients table for the zero-inflation model estimating the probability when the response variable equals 0, $P(Y = 0)$. The fourth section shows the number of iterations and the log-likelihood and the corresponding degrees of freedom.

The coefficients table for the negative binomial regression model component is labeled with the title `Count model coefficients (negbin with log link).` It shows the estimates for the predictor variables, the intercept, and the dispersion parameter, their standard errors, the Wald z statistics, and the associated p values. The interpretation of the coefficients here is the same as those in the negative binomial regression model. Each coefficient can be interpreted as the change in the log expected of the incidence rate or the log expected number of events for a one-unit increase in the predictor variable when holding other predictors constant.

For the predictor variable `educ`, Wald $z = 3.094$. The associated p value, `Pr(>|z|) < .01`, so we reject the null hypothesis. The rejection of the null hypothesis indicates that `educ` is a significant predictor of the count response variable `vistzoo`.

For the predictor variable `maritals`, Wald $z = 1.887$. The associated p value, `Pr(>|z|) = .059`, so we fail to reject the null hypothesis. For the predictor variable `female`, the Wald $z = .466$. The associated p value `Pr(>|z|) = .640`, so we also fail to reject the null hypothesis. For the predictor variable `wrkfull`, Wald $z = 1.381$. The associated p value, `Pr(>|z|) = .167`, so we also fail to reject the null hypothesis.

Also, in the coefficients section, the log dispersion parameter is labeled `Log(theta)`. Wald $z = 3.753$. The associated p value, `Pr(>|z|) < .01`, which is significant. To obtain the dispersion parameter, we need to exponentiate the estimate of the `Log(theta)`. The dispersion parameter $\theta = \exp(.801) = 2.228$, which is also displayed at the bottom of the output. As with the `glm.nb()` function, the `zeroinfl()` function also uses a slightly different form for the NB2 model, Variance $(Y) = \mu + \mu^2/\theta$ rather than Variance $(Y) = \mu + \alpha\mu^2$. Since $\theta = 1/\alpha$, we can compute $\alpha = 1/2.228 = .449$. The R output for computing θ and α is as follows.

```
> exp(0.80108)
[1] 2.227946
> alpha.zinb <- 1/2.2279
> alpha.zinb
[1] 0.4488532
```

The coefficients table for the zero-inflation model component is labeled with the title `Zero-inflation model coefficients (binomial with logit link).` The interpretation of the logit coefficients here is the same as those in the logistic regression model. Each coefficient can be interpreted as the change in the logit of having zero counts for a one-unit increase in the predictor variable when holding other predictors constant.

Among the four predictors in the zero-inflation model component, none of them are significant in predicting zeros. This suggests that the more parsimonious negative binomial model may be a better choice. We use the Vuong test to compare these two models later in the chapter.

The coefficients of the predictor variables can be extracted with `coef(zinb)`. The confidence intervals are obtained with the `confint(zinb)` command. The output is omitted here.

We use the `exp(coef(zinb))` and `exp(confint(zinb))` commands to compute the IRRs and the corresponding confidence intervals, respectively. The results are combined with the `cbind()` function.

```
> exp(coef(zinb))
count_(Intercept)         count_educ     count_maritals      count_female
     0.3266351837       1.0615383658       1.2167216992      1.0489635783
    count_wrkfull  zero_(Intercept)          zero_educ     zero_maritals
     1.2061628254       0.1322496364       1.0749230663      0.9891001495
      zero_female       zero_wrkfull
     1.2213384461       0.0002522989

> exp(confint(zinb))
                          2.5 %       97.5 %
count_(Intercept)  1.835458e-01 5.812747e-01
count_educ         1.022124e+00 1.102472e+00
count_maritals     9.923939e-01 1.491758e+00
count_female       8.580655e-01 1.282332e+00
count_wrkfull      9.245002e-01 1.573638e+00
zero_(Intercept)   9.484141e-03 1.844128e+00
zero_educ          9.102592e-01 1.269374e+00
zero_maritals      3.677225e-01 2.660482e+00
zero_female        4.510252e-01 3.307282e+00
zero_wrkfull       8.215654e-19 7.747980e+10

> cbind(exp(coef(zinb)), exp(confint(zinb)))
                                        2.5 %       97.5 %
count_(Intercept) 0.3266351837 1.835458e-01 5.812747e-01
count_educ        1.0615383658 1.022124e+00 1.102472e+00
count_maritals    1.2167216992 9.923939e-01 1.491758e+00
count_female      1.0489635783 8.580655e-01 1.282332e+00
count_wrkfull     1.2061628254 9.245002e-01 1.573638e+00
zero_(Intercept)  0.1322496364 9.484141e-03 1.844128e+00
zero_educ         1.0749230663 9.102592e-01 1.269374e+00
zero_maritals     0.9891001495 3.677225e-01 2.660482e+00
zero_female       1.2213384461 4.510252e-01 3.307282e+00
zero_wrkfull      0.0002522989 8.215654e-19 7.747980e+10
```

The standard errors of the IRRs are computed with the `exp(coef(zinb))*sqrt(diag(vcov(zinb)))` command.

```
> exp(coef(zinb))*sqrt(diag(vcov(zinb)))
count_(Intercept)        count_educ     count_maritals      count_female
      0.096055754       0.020492340        0.126513598       0.107508696
    count_wrkfull  zero_(Intercept)          zero_educ     zero_maritals
      0.163663375       0.177802801        0.091191896       0.499336677
      zero_female      zero_wrkfull
      0.620762554       0.004294075
```

We can use the nagelkerke(zinb) command to compute the three types of pseudo R^2 statistics and the likelihood ratio test statistic for the multiple-predictor model. The output is omitted here.

We use the Vuong test to compare the zero-inflated negative binomial regression model and the negative binomial regression model. In the vuong(zinb,nbr) command, zinb and nbr are the zero-inflated negative binomial regression model and the negative binomial regression model, respectively. The following output is displayed.

```
> vuong(zinb, nbr)
Vuong Non-Nested Hypothesis Test-Statistic:
(test-statistic is asymptotically distributed N(0,1) under the
null that the models are indistinguishible)
-------------------------------------------------------------
              Vuong z-statistic             H_A    p-value
Raw                   1.6470053 model1 > model2 0.04977848
AIC-corrected         0.2044316 model1 > model2 0.41900814
BIC-corrected        -3.2610737 model2 > model1 0.00055496
```

The Vuong test $z = 1.647$, $p = .05$; the AIC-corrected test $z = .204$, $p = .419$. The results of both tests are not significantly different from zero, so we conclude that there is no significant difference between the zero-inflated negative binomial regression model and the negative binomial regression model. We prefer the negative binomial regression model since it is more parsimonious than the zero-inflated model. In addition, the BIC-corrected test $z = -3.261$, $p < .001$, which suggests that the negative binomial regression model be preferred.

We also use the Vuong test to compare the negative binomial regression model and the simplified zero-inflated Poisson regression model. In the vuong(nbr,zip2) command, nbr and zip2 are the negative binomial regression model and the simplified zero-inflated Poisson regression model, respectively. The following output is displayed.

```
> vuong(nbr, zip2)
Vuong Non-Nested Hypothesis Test-Statistic:
(test-statistic is asymptotically distributed N(0,1) under the
null that the models are indistinguishable)
-------------------------------------------------------------
              Vuong z-statistic             H_A   p-value
Raw                    1.996957 model1 > model2 0.0229149
AIC-corrected          2.267855 model1 > model2 0.0116690
BIC-corrected          2.918638 model1 > model2 0.0017578
```

The Vuong test $z = 1.997$, $p < .05$, which indicates that the Vuong z statistic is significantly different from zero, so we prefer the negative binomial regression model to the simplified zero-inflated Poisson regression model.

9.7 MAKING PUBLICATION-QUALITY TABLES

9.7.1 Presenting the Results Using the stargazer Package

We can use the `stargazer` package (Hlavac, 2018) to make a table containing the results of the fitted negative binomial model with the `glm.nb()` function. Since the package has been installed in earlier chapters, we only need to load the package by typing `library(stargazer)`. After fitting the negative binomial model `nbr`, we use the command as follows: `stargazer(nbr, type = "text", align = TRUE, out = "nbrmod.txt")`. In the `stargazer()` function, we first specify the model object to be presented and then the type of table. The option `type = "text"` specifies the table type and the `align = TRUE` option aligns the results of the model. The `out = "nbrmod.txt"` argument saves the output named `nbrmod.txt`.

```
> # Presenting the results with stargazer()
> library(stargazer)
> stargazer(nbr, type = "text", align = TRUE, out = "nbrmod.txt")

===================================
                Dependent variable:
              ---------------------
                     vistzoo
-----------------------------------
educ                 0.050***
                     (0.017)

maritals             0.191**
                     (0.093)

female               0.024
                     (0.094)

wrkfull              0.535***
                     (0.095)

Constant            -1.309***
                     (0.246)

-----------------------------------
Observations           902
Log Likelihood     -1,095.007
theta            1.583*** (0.251)
Akaike Inf. Crit.   2,200.013
===================================
Note:       *p<0.1; **p<0.05; ***p<0.01
```

We can also create the table in HTML format and copy it into Microsoft Word. The command is as follows: `stargazer(nbr, type = "html", align = TRUE, out = "nbrmod.htm")`. It produces Table 9.1, as shown here in its original format, presenting the results of the multiple-predictor negative binomial regression model.

TABLE 9.1 ● Results of the Negative Binomial Regression Model (Shown in Original Format Generated by R)

	Dependent variable:
	vistzoo
educ	0.050***
	(0.017)
maritals	0.191**
	(0.093)
female	0.024
	(0.094)
wrkfull	0.535***
	(0.095)
Constant	−1.309***
	(0.246)
Observations	902
Log Likelihood	−1,095.007
Theta	1.583*** (0.251)
Akaike Inf. Crit.	2,200.013

*$p < 0.1$
**$p < 0.05$
***$p < 0.01$.

9.7.2 Presenting the Results Using the `texreg` Package

The results of the reduced zero-inflated Poisson model, the full zero-inflated Poisson model, and the zero-inflated negative binomial regression model can also be displayed in a table by using the `screenreg()` and `htmlreg()` functions from the `texreg` package (Leifeld, 2013). Since the package has been installed in earlier chapters, we only need to load the package by typing `library(texreg)`. We use the following command to display the results: `screenreg(list(zip2, zip, zinb))`. In the `screenreg()` function, we specify the three model objects to be presented with the `list()` function.

```
> screenreg(list(zip2, zip, zinb))

=========================================================
                           Model 1      Model 2     Model 3
---------------------------------------------------------
Count model: (Intercept)   -0.81 ***    -1.09 ***   -1.12 ***
                           (0.23)       (0.30)      (0.29)
Count model: educ           0.06 ***     0.08 ***    0.06 **
                           (0.02)       (0.02)      (0.02)
Count model: maritals       0.20 *       0.17        0.20
                           (0.08)       (0.10)      (0.10)
Count model: female         0.05         0.05        0.05
                           (0.08)       (0.10)      (0.10)
Count model: wrkfull        0.05         0.07        0.19
                           (0.11)       (0.11)      (0.14)
Zero model: (Intercept)    -0.17        -1.44       -2.02
                           (0.17)       (0.92)      (1.34)
Zero model: wrkfull        -1.80 ***    -1.76 ***   -8.28
                           (0.45)       (0.41)     (17.02)
Zero model: educ            0.09         0.07
                                        (0.06)      (0.08)
Zero model: maritals       -0.14        -0.01
                                        (0.33)      (0.50)
Zero model: female          0.04         0.20
                                        (0.31)      (0.51)
Count model: Log(theta)                              0.80 ***
                                                    (0.21)
---------------------------------------------------------
AIC                      2231.50      2235.05     2198.60
Log Likelihood          -1108.75     -1107.52    -1088.30
Num. obs.                  902          902         902
=========================================================
*** p < 0.001; ** p < 0.01; * p < 0.05
```

To create the table in the HTML format and copy it into Microsoft Word, we use the command as follows: `htmlreg(list(zip2, zip, zinb), file = "zipzinb.doc", doctype = TRUE, html.tag = TRUE, head.tag = TRUE)`. The table is omitted here.

9.8 REPORTING THE RESULTS

Reporting the results for the negative binomial regression is similar to that used for Poisson regression. The only difference is that you need to report the additional dispersion parameter. The following are the generic guidelines for reporting the results. You may need to adjust your writing since your discipline or journals may have different requirements.

First, describe the negative binomial regression model, the count response variable and independent variables, and your research hypothesis or the purpose of your study. Include a couple of sentences justifying your use of this model for the analysis.

Second, report the likelihood ratio test statistic for the model and the associated *p* value, followed by the interpretation on whether the fitted model is better than the null model. If more than one model is developed, then compare models using likelihood ratio test statistics and/or the AIC and BIC statistics.

Third, report the parameter estimates for the predictor variables, their standard errors, the associated *p* values, and the dispersion parameter in a table. In addition, report the incidence rate ratio for each predictor in the table or text and interpret the results. The following is an example of summarizing the results for the negative binomial regression model illustrated previously.

The negative binomial regression analysis was conducted to predict the count outcome variable, the number of zoo visits in a year, from a set of predictor variables, such as marital status, years of education, gender, and working status. The negative binomial regression model was fitted since the response variable was a count of the number of visits to a zoo in a year and there was overdispersion in the Poisson regression model.

The likelihood ratio test for the fitted model $\chi^2_{(4)} = 54.999$, $p < .001$, indicates that the full model with the four predictors provides a better fit than the null model with no independent variables in predicting the count response variable.

The likelihood ratio test is used to test whether the dispersion parameter is significantly different from zero by comparing the Poisson regression model and the negative binomial regression model. The likelihood ratio test, $\chi^2_{(1)} = 88.79$, $p < .001$, which indicates that the dispersion parameter α is significantly different from zero, so the use of the negative binomial regression model is justified.

Table 9.1 displays the parameter estimates for the multiple-predictor negative binomial regression model. The results can be interpreted in terms of the incidence rate ratios which are the exponentiated coefficients.

In this model, for `educ`, the incidence rate ratio is 1.051. The result indicates that for a one-unit increase in education the expected number of visits to a zoo increases by 5.1%.

For `maritals`, the incidence rate ratio is 1.211, which indicates that expected number of visits to a zoo for the married is 1.211 times as high as that for the unmarried when holding other predictors constant.

The incidence rate ratio for `wrkfull` can be interpreted in the similar way. The incidence rate ratio is 1.707, which indicates that the expected number of visits to a zoo for those working full time is 1.707 times as high as that for those not working full time when holding the other predictors constant.

With regard to `female`, the incidence rate ratio is 1.024, which is not significant (see the associated $p = .797$ in the coefficients table). This indicates that being female does not impact the expected number of visits to a zoo.

9.9 SUMMARY OF R COMMANDS IN THIS CHAPTER

```
# Chap 9 R Script
# Remove all objects
rm(list = ls(all = TRUE))

# The following user-written packages need to be installed first by using
install.packages(" ") and then by loading it with library()
# library(MASS)
# library(VGAM)              # It is already installed for Chapter 4
# library(rcompanion)        # It is already installed for Chapter 3
# library(margins)           # It is already installed for Chapter 3
# library(ggeffects)         # It is already installed for Chapter 2
# library(texreg)            # It is already installed for Chapter 4
# library(pscl)

# Import the count dataset
library(foreign)
count <- read.dta("C:/CDA/count.dta")

# Convert variables from integer to numeric so they will work well with ggpredict()
count$educ <- as.numeric(count$educ)
count$wrkfull <- as.numeric(count$wrkfull)
count$maritals <- as.numeric(count$maritals)


attach(count)
str(count)

# Negative binomial regression model with glm.nb() in MASS
library(MASS)
nbr <- glm.nb(vistzoo ~ educ + maritals + female + wrkfull, data = count)
summary(nbr)
alpha<-1/1.583
alpha
```

```
coef(nbr)
confint(nbr)
exp(coef(nbr))
exp(confint(nbr))
cbind(exp(coef(nbr)), exp(confint(nbr)))
exp(coef(nbr))*sqrt(diag(vcov(nbr)))

# marginal effects
library(margins)
marg.nbr <- margins(nbr)
summary(marg.nbr)

# Testing the overall model using the likelihood ratio test
nbr.0 <- glm.nb(vistzoo ~ 1, data = count)
summary(nbr.0)
anova(nbr.0, nbr, test = "Chisq")
anova(nbr, update(nbr, ~1), test = "Chisq")

# Pseudo R2 with nagelkerke()
library(rcompanion)
nagelkerke(nbr)

# Pseudo R2 with equations
LLM1 <- logLik(nbr)
LL0 <- logLik(nbr.0)
McFadden1 <- 1-(LLM1/LL0)
McFadden1
CS1 <- 1-exp(2*(LL0-LLM1)/902)
CS1
NG1 <- CS1/(1-exp(2*LL0/902))
NG1

# AIC and BIC Statistics
AIC(nbr)
BIC(nbr)

PR.2 <- glm(vistzoo ~ educ + maritals + female + wrkfull, family = poisson, data =
count)
AIC(PR.2, nbr)
BIC(PR.2, nbr)

# Presenting the results with stargazer()
library(stargazer)
stargazer(nbr, type="text", align=TRUE, out="nbrmod.txt")
stargazer(nbr, type="html", align=TRUE, out="nbrmod.htm")

# Predicted counts with ggpredict() in ggeffects
library(ggeffects)
nbr.ed <- ggpredict(nbr, terms = "educ[12, 14, 16]")
nbr.ed
plot(nbr.ed)

nbr.ew <- ggpredict(nbr, terms = c("educ[12, 14, 16]", "wrkfull"))
nbr.ew
plot(nbr.ew)
```

```
# Negative binomial regression model with vglm() in VGAM
library(VGAM)
nb.v <- vglm(vistzoo ~ educ + maritals + female + wrkfull, family = negbinomial,
data=count)
summary(nb.v)
coef(nb.v, matrix = TRUE)
confint(nb.v, matrix = TRUE)
exp(coef(nb.v, matrix = TRUE))
exp(confint(nb.v, matrix = TRUE))
exp(coef(nb.v))*sqrt(diag(vcov(nb.v)))
exp(.45953)
alpha.v <- 1/1.583
alpha.v
nagelkerke(nb.v)
AIC(nb.v)
BIC(nb.v)

# Model comparison with the likelihood ratio test
library(lmtest)
lrtest(PR.2, nbr)

# ZIP model with zeroinfl() in pscl
# install.packages("pscl")
library(pscl)
zip <- zeroinfl(vistzoo ~ educ + maritals + female + wrkfull | educ+ maritals +
female + wrkfull, data=count)
summary(zip)
coef(zip)
confint(zip)
exp(coef(zip))
exp(confint(zip))
cbind(exp(coef(zip)), exp(confint(zip)))
exp(coef(zip))*sqrt(diag(vcov(zip)))

# Pseudo R2 with nagelkerke()
library(rcompanion)
nagelkerke(zip)

# Vuong test
vuong(zip, PR.2)

# ZIP reduced model
zip2 <- zeroinfl(vistzoo ~ educ + maritals + female + wrkfull | wrkfull,
data=count)
summary(zip2)
lrtest(zip2, zip)
# Zero-inflated NB model with zeroinfl() in pscl
Zinb <- zeroinfl(vistzoo ~ educ + maritals + female + wrkfull | educ + maritals +
female + wrkfull, dist="negbin", data=count)
summary(zinb)
coef(zinb)
confint(zinb)
exp(coef(zinb))
exp(confint(zinb))
cbind(exp(coef(zinb)), exp(confint(zinb)))
exp(coef(zinb))*sqrt(diag(vcov(zinb)))
nagelkerke(zinb)
```

```
vuong(zinb, nbr)
vuong(nbr, zip2)

# Presenting the results with textreg
library(texreg)
screenreg(list(zip2, zip, zinb))
htmlreg(list(zip2, zip, zinb), file="zipzinb.doc", doctype=TRUE, html.tag=TRUE,
head.tag=TRUE)

detach(count)
```

Glossary

Overdispersion occurs when the variance of the count response variable is greater than the mean.

The NB2 negative binomial model is defined when the variance has a quadratic form, $\mu + \alpha\mu^2$.

The negative binomial regression model relaxes the equality of the mean and the variance assumption and allows overdispersion.

The variance of the response variable in the negative binomial regression model is a function of the mean μ and a dispersion parameter α.

The zero-inflated models address the issue of excessive zero counts. The zero-inflated model can be applied to both Poisson regression and negative binomial regression. The former is referred to as the **zero-inflated Poisson regression model** and the latter is called the **zero-inflated negative binomial regression model**.

Exercises

Use the `rwm1984` data (Hilbe, 2014) available at **https://edge.sagepub.com/liu1e** for the following problems.

1. Conduct an analysis for a negative binomial regression model and estimate the count response variable `docvis` (the number of visits to a doctor in a year) from the two predictor variables, `outwork` (1 = not working and 0 = working) and `female` (1 = female and 0 = male).
2. Interpret the likelihood ratio test for the overall model.
3. List three measures of pseudo R^2 and the AIC and BIC statistics.
4. In the regression table, identify the coefficients for the predictor variables `outwork` and `female`. Are they both statistically significant?
5. Identify the dispersion parameter labeled `Theta` and compute the dispersion parameter α.
6. Compute the IRRs and interpret the IRR for the predictor variable `female`.
7. Make a publication-quality table containing the estimated coefficients.
8. Write a report to summarize the results from the output.

10

MULTILEVEL MODELING FOR CONTINUOUS RESPONSE VARIABLES

OBJECTIVES OF THIS CHAPTER

This chapter introduces multilevel modeling for continuous response variables. It starts with an introduction to multilevel modeling, model-building strategies, model fit statistics, centering, and data structure followed by a description of the research questions and data. Then several models, from the unconditional (null) model to the random-intercept model and random-coefficient model to the contextual models, are illustrated using R with step-by-step instructions. R commands are explained, and the output is interpreted for each model in detail. The chapter also illustrates how the results are displayed in publication-quality tables using the R command and reported in text. It focuses on model fitting with R, as well as on interpreting and presenting the results. After reading this chapter, you should be able to:

- Determine when multilevel modeling for continuous variables is used.
- Formulate multilevel models.
- Conduct multilevel modeling analysis for continuous response variables using R.
- Interpret the output.
- Compute and interpret the intraclass correlation coefficient (ICC).
- Be familiar with model fitting strategies.
- Compare models using the likelihood ratio test.
- Present results in publication-quality tables using R.
- Write the results for publication.

10.1 MULTILEVEL MODELING: AN INTRODUCTION

In previous chapters, we have focused on single-level analytic techniques for categorical response variables. Multilevel modeling has been widely used in education, social, and behavioral sciences in recent years, and researchers are increasingly interested in applying this technique to analyze multilevel data in their research. This chapter presents multilevel modeling for continuous outcome variables when the data structure has more than one level.

10.1.1 Multilevel Data Structure

Multilevel data, nested data, or hierarchical structured data have a data format in which observations at lower levels are nested within a higher level. For example, in businesses, employees are nested within companies; in educational research, students are nested in schools; in medical science, patients are nested within hospitals; in political sciences, voters are nested within districts; and in sociology, families are nested within communities. Observations in the same group could be more homogeneous than those across different groups, and thus, the assumption of independence is violated. Another type of multilevel data structure occurs in longitudinal studies in which there are repeated measures for each subject. In this case, measures for multiple time points are nested within a subject. This type of analysis is known as the multilevel analysis for change (Singer & Willett, 2003). The focus of this text is the cross-sectional data structure.

What can multilevel modeling do? There are several advantages to using multilevel modeling. First, in multilevel modeling, variables at higher levels can be included in the model to estimate their relationships with the outcome variable. Second, we can examine whether an effect or slope of a variable at a lower level is allowed to vary among higher level variables. Third, we can also examine whether higher level variables moderate the relationships between lower level variables and the outcome variable.

10.1.2 Intraclass Correlation

With a multilevel data structure, the observations within a group or cluster may violate the assumption of independency. In other words, the observations within the same group or cluster may be more homogeneous than those in other groups or clusters. To justify why multilevel modeling is warranted, we also need to examine how much variance of the outcome variable is accounted for by groups or clusters. The intraclass correlation coefficient (ICC) is used as an index to measure the proportion of variance in the outcome variable explained by groups or clusters (Hox, 2010; Raudenbush & Bryk, 2002; Snijders & Bosker, 2012). It is the ratio of the between-group variance to the total variance. Its range is from 0 to 1. When it is close to 0, it means that using multilevel modeling might not be a good strategy for data analysis. A larger ICC provides strong evidence that this technique is needed.

10.1.3 Overview of a Basic Two-Level Model

Let us look at a basic two-level model with one predictor in each level. In the following example, researchers are interested in estimating the math achievement scores from a student-level variable, math self-efficacy, and a school-level variable (whether a school is public). The level 1 predictor variable is math self-efficacy (`gceffic`). Both the intercept and the slope of `gceffic` are allowed to vary randomly across schools. The level 2 predictor is whether a school is public or private (`public`). Following the convention of model specification by Raudenbush and Bryk (2002), a two-level model can be expressed as:

$$\begin{aligned} \text{Level 1}: Y_{ij} &= \beta_{0j} + \beta_{1j}\texttt{gceffic}_{ij} + r_{ij} \\ \text{Level 2}: \beta_{0j} &= \gamma_{00} + \gamma_{01}\texttt{public}_j + u_{0j} \\ \beta_{1j} &= \gamma_{10} + \gamma_{11}\texttt{public}_j + u_{1j} \end{aligned}$$

where Y_{ij} represents the math achievement score for the *i*th student in the *j*th school, β_{0j} is the level 1 intercept, the average math achievement score in the *j*th school, β_{1j} is the level 1 slope for `gceffic` in the *j*th school, and $\texttt{gceffic}_{ij}$ represents the value of math self-efficacy of the *i*th student in the *j*th school. r_{ij} is the random error, which is the deviation of the individual's math score from the average math score in the school.

The γ_{00} is the overall intercept of the outcome variable across schools. It is the predicted mean math achievement score controlling for the effect of the level 2 predictor (i.e., when the level 2 predictor variables are held constant at 0). γ_{01} represents the effect of the level 2 variable `public` on the intercept. γ_{10} represents the mean of the level 1 slope when the level 2 predictors are held constant at 0, and γ_{11} represents the effect of the level 2 predictor `public`. γ_{11} is the cross-level interaction between `gceffic` and `public`, which moderates the effect of math self-efficacy on math achievement scores. u_{0j} and u_{1j} are the random effects associated with the level 1 intercept and the slope of `gceffic` across schools, respectively. In other words, the level 1 intercept and the slope of `gceffic` are allowed to vary randomly across schools so their respective variances (i.e., between-group variance and slope variance) need to be estimated.

Fix Effects Versus Random Effects

In multilevel modeling, fixed effects are the regression coefficients that estimate the relationships between the predictor variables and the outcome variable from the entire population (West et al., 2014), whereas random effects are the randomly varying parameters across higher level units. For example, the random intercept (u_{0j}) in the previous example is a random deviation from the overall intercept, and the random coefficient (u_{1j}) is a random deviation from the overall fixed effect, the slope of `gceffic`. Variance and covariance components are estimated for random effects. For example, in the two-level model above, the variance and covariance components include the between-group variance for the intercept, the variance for the random coefficient or slope, and the covariance between the intercept and the coefficient. Sometimes, random effects are estimated in terms of standard deviations since they are just the square root of variances.

What if an ordinary least-squares (OLS) regression instead of multilevel modeling is used? In other words, what will happen if random effects are not estimated? When a single-level regression analysis is conducted to analyze multilevel data, the precision of parameter estimates is compromised (Heck & Thomas, 2015). Heck and Thomas (2015) pointed out that multilevel modeling has four advantages over the OLS regression method. These advantages include incorporating regression equations at different levels into a single statistical model, more accurate estimates of standard errors, flexibility in specifying various models, and the capability of estimating different types of response variables.

By using multilevel modeling, we can estimate the influence of both the student-level and school-level predictors on the outcome variable. We can also investigate whether there are cross-level interactions between variables at different levels. In addition, we can estimate random effects by allowing the intercept and slopes of lower level predictors to vary randomly at higher levels. The variance and covariance components of the random effects can also be determined. For example, the estimated error variance for r_{ij} is the within-group variance, the estimated variance for u_{0j} is the intercept variance, which is the between-group variance, and the estimated variance for u_{1j} is the slope variance.

10.1.4 Model-Building Strategies

Although researchers may have their own strategies to build multilevel models, a common practice illustrated by Raudenbush and Bryk (2002), Snijders and Bosker (2012), and other publications (Garson, 2013, 2020; Heck et al., 2010; Heck & Thomas, 2015; Kreft & de Leeuw, 1998; Luke, 2004; West et al., 2014) is to start from a basic model and work up to more complex models. Specifically, this strategy starts with the unconditional means model with no level 1 or level 2 predictors (null model). This model is equivalent to the one-way random-effects analysis of variance (ANOVA) model. This model serves as the baseline model for future model comparisons. The unconditional means model estimates the overall average of the outcome variable across all subjects and the between-group and within-group variances. The variance between groups or clusters estimated from this model can be used to calculate the ICC so that we can decide whether multilevel modeling is needed. The between-group and within-group variances can also be used to compute the proportion of variance explained after the level 1 and level 2 predictors are added to the model. Next, we can add level 1 predictors and build a random-intercept model and a random-coefficient model. In the random-intercept model, only intercepts are allowed to vary freely in higher level clusters and the level 1 slopes are fixed. In the random-coefficient model, both intercepts and coefficients of the level 1 predictors are allowed to vary across higher level clusters. Finally, we add level 2 predictors to the level 2 model so the random-coefficient model includes both level 1 and level 2 predictors. This model is referred to as the contextual model. If the model has more than two levels, then higher level predictors can be added.

Although the earlier simple-to-complex model building strategy is commonly followed by researchers, you can decide whether all the steps need to be followed for your own research.

10.1.5 Model Fit Statistics

As with logistic regression models, several measures of goodness-of-fit statistics, such as the likelihood ratio test, the Akaike information criterion (AIC), and Bayesian information criterion (BIC) statistics, can be applied to multilevel modeling for continuous outcome variables. The following discussion is a brief review of these tests (see Chapter 3 for a more detailed description).

Likelihood Ratio Test

The likelihood ratio test can be used to compare nested models. Models are nested when one model, the reduced model, is a special case of the other one, the full model. For example, more constraints can be put on parameters in one model than the other. A simple case is that one model (model 1) contains predictor variables X_1 and X_2, and the second model (model 2) contains an extra variable X_3. We conclude that model 1 is nested within model 2 since predictors in the former are the subset of the latter. In multilevel modeling, an unconditional model is nested within a random-intercept model, which is then nested within a random-coefficient model and finally a contextual model with both level 1 and level 2 predictor variables.

The likelihood ratio test statistic is expressed as the difference in -2LL between nested models, where LL stands for the log likelihood value for the fitted model with either the full maximum likelihood (ML) estimation or the restricted maximum likelihood (REML) estimation. Since deviance equals -2LL, the likelihood ratio test is also referred to as the difference in deviance, which follows a chi-square distribution, with the degrees of freedom of the distribution equaling the difference in the number of parameters between two nested models. The difference in deviance is often expressed as a generic form: $G =$ Deviance for the reduced model $-$ Deviance for the full model or $D_{\text{Reduced}} - D_{\text{Full}}$, where the reduced model has fewer variables and is nested within the full model. As with logistic regression models in previous chapters, we use the likelihood ratio test to compare nested models from a simple model with one predictor to more complex models with multiple predictors. In multilevel modeling, we can also use the same test to compare a series of nested models from the unconditional means model to the random-intercept model to more complex models, such as the contextual models with level 1 and level 2 predictor variables. A significant likelihood ratio chi-square test statistic indicates that a more complex model fits the data better than a simpler, nested model.

Information Criteria Indices: AIC and BIC

The AIC and the BIC statistics can be used to compare non-nested models. Both AIC and BIC statistics can be applied to multilevel modeling. The AIC statistic adjusts the deviance by the number of parameters. It is expressed as $-2\text{LL} + 2k$ or deviance $+ 2k$, where k is the number of parameters. The BIC statistic is defined as $\text{BIC} = -2\text{LL} + \ln(n) \times k = D_m + \ln(n) \times k$, where n is the sample size and k is the number of parameters. Smaller AIC and BIC statistics indicate a better fit of the model.

10.1.6 Centering

The purpose of centering is to make the results more interpretative. It is often used when a predictor variable does not have a meaningful value of zero. By subtracting the mean of a predictor variable from each value, we obtain a meaningful zero for the predictor variable. Predictors at both levels of the model can be centered. Two types of centering are often used in multilevel modeling. One is grand-mean centering, and the other is group-mean centering. For the grand-mean centering, we subtract the grand mean of the predictor variable from each value of the sample. For example, when we use grand-mean centering of the math efficacy (`efficacy`), we compute the overall mean of this variable and then subtract it from each score of efficacy. For the group-mean centering, we subtract the group mean, which is the mean of each group or cluster where individuals are nested from each score. For example, to group-mean center the predictor variable `efficacy`, we first compute the group mean for each school where a student belongs and then subtract the mean for each school (i.e., group mean) from each score of efficacy.

The choice of grand-mean centering and group-mean centering is complicated, and this topic has been widely discussed in the literature (Enders & Tofighi, 2007; Garson, 2020; Hofmann & Gavin, 1988; Hox, 2010; Kreft et al., 1995; Luke, 2004; Ma et al., 2008; McCoach, 2010; Paccagnella, 2006). The advantage of grand-mean centering is that the subsequent multilevel models with this centering are mathematically equivalent to the models using raw scores without centering. It also makes the computation faster and reduces convergence problems (Hox, 2010). On the other hand, group-mean centering produces a model that is mathematically different from the raw score model. Hox (2010) suggested using group-mean centering with caution for novice users. Enders and Tofighi (2007) suggested that researchers use group-mean centering when level 1 variables and the interactions among them are the research interests, whereas grand-mean centering is a good choice if level 2 variables are the focus after controlling for level 1 variables. Therefore, the decision of using centering methods should be based on research questions or theories.

10.1.7 Sample Size

In multilevel modeling, the sample size needs to be considered at different levels. Theoretically, we would like to have a large sample size for all levels to obtain unbiased estimates for fixed and random effects. Factors such as the complexity of the model, the intraclass correlation, cross-level interactions, and power considerations impact sample size determination. Although there is no definite number to define a sufficient sample size in the literature, researchers have suggested several rules-of-thumb for a two-level model. Kreft and de Leeuw (1998) recommended a sample size of more than 20 groups for cross-level interactions. Maas and Hox (2005) conducted simulations and the results suggested that a sample size of 50 or more groups is needed to obtain unbiased estimates of the standard errors at level 2. They also found that the standard errors were underestimated at level 2 with a sample size of 30 groups. Hox (2010) suggested a 50/20 rule for cross-level interactions, that is, 50 level 2 groups with 20 level 1 subjects. In addition, a 100/10 rule (i.e., 100 level 2 groups with 10 level 1 subjects) was suggested when the focus was on random effects. For a more detailed review of other simulation studies, refer to Garson (2020). Sometimes, when the group number at level 2 is small, multilevel modeling can still be a useful tool for analyzing nested data, but the results should be interpreted with caution.

10.1.8 Data Structure for Model Fitting

In R, the data structure for multilevel modeling is a single dataset containing variables at different levels. For example, a two-level model needs a single dataset with both student-level and school-level variables and the former variables are nested with the latter. If the original student-level and school-level variables are saved in two separate datasets, then they need to be merged into one dataset in a format where students are nested within schools. This stacked data format requires that each school have multiple records, one for each student. For example, when 50 students are selected from a school, in the dataset, 50 students with different IDs (with each one having a row) are nested within the same school ID. Such a dataset needs to be created before model fitting.

10.2 MULTILEVEL MODELING FOR CONTINUOUS OUTCOME VARIABLES

10.2.1 Research Example and Research Questions

In the following example, researchers are interested in examining the relationships between high-school students' mathematics achievement and mathematics self-efficacy, school type, and school climate using the Educational Longitudinal Study of 2002 (ELS: 2002) data. The student-level predictor variable is students' mathematics self-efficacy, and the two school-level predictor variables are school type and school climate. The following research questions will be addressed:

1. Can high-school students' mathematics scores be predicted by students' mathematics self-efficacy?
2. Do school characteristics, such as school type and school climate, impact math achievement?
3. Do mathematics scores vary across schools?
4. Does the relationship between mathematics self-efficacy and mathematics achievement vary across schools?
5. Are there any interaction effects between the two school-level variables (i.e., school type and school climate) and math self-efficacy? In other words, does school type or school climate moderate or influence the relationship between mathematics self-efficacy and mathematics achievement? Put it another way: Does the effect of mathematics self-efficacy on mathematics achievement vary across school type and school climate?

10.2.2 Description of the Data and Sample

The ELS:2002 base-year data are used for the following analyses. The variables are listed as follows:

- `mathach`: mathematics item response theory (IRT) estimated scores of high-school students
- `gceffic`: math self-efficacy (grand-mean centered)
- `public`: school type (1= public, 0 = private and others)
- `csclimat`: school climate (grand-mean centered).

10.3 MULTILEVEL MODELING FOR CONTINUOUS RESPONSE VARIABLES WITH R

10.3.1 The `lme()` Function in the `nlme` Package

The `lme()` function in the `nlme` package (Pinheiro et al., 2021) is used for multilevel models with continuous response variables. Since `nlme` is a user-written package, you need to install it first by typing `install.packages("nlme")` and then load the package by typing `library(nlme)`.

The basic syntax for the model formula of `lme()` includes two components. The first component or argument specifies the dependent variable and the predictor variable(s) for the fixed effects component, which are separated by the tilde (~). When there are multiple predictor variables in the formula, they are connected by plus (+) symbols. This is the fixed effects part of the model, which looks the same as the model formula for any linear regression in `lm()`. The predictor variables from different levels are specified here, but the command itself does not tell the specific levels within which the variables belong to. Next, the `random` argument specifies the random effects of the model and the grouping variable which are separated by a vertical line (|). A predictor variable or a list of predictor variables that have random coefficients is specified first, followed by a vertical line (|), and then, an identifier variable at a higher level as the grouping variable. Sometimes the grouping variable may be omitted. In addition to the model formula, several optional arguments, such as the `data` argument for the data frame, `method = "ML"` for the maximum likelihood estimation, and `na.action = "na.omit"` for removing missing data, can be specified. For example, the command `lme(y ~ x, random = ~ 1| schid, method = "ML")` tells R to fit a multilevel model to estimate a continuous outcome variable `y` with a predictor variable `x`, and `random = ~ 1| schid` specifies the random intercepts varying across schools with `1` as the intercept and `schid` as the identifier variable. The `method = "ML"` argument requests the full maximum likelihood estimation rather than the default restricted maximum likelihood (method = "REML"). For more details on how to use this function, type help(lme) in the command prompt after loading the nlme package.

10.3.2 Unconditional Means Model (Model 1: Null Model)

The unconditional means model or the null model is known as the one-way random-effects ANOVA. Neither level 1 nor level 2 predictor variables are included in the model. This model can be expressed as follows:

$$\begin{aligned}\text{Level 1}: Y_{ij} &= \beta_{0j} + r_{ij}\\ \text{Level 2}: \beta_{0j} &= \gamma_{00} + u_{0j}\end{aligned}$$

The m1 <- lme(mathach ~ 1, random = ~1|SCH_ID, na.action = "na.omit", method = "ML", data = chp10) command is used to fit the unconditional model (model 1). In the lme() function, the fixed effects part of the model is specified first. Because this is the unconditional model without any predictor variables, the continuous outcome variable mathach and the intercept 1 are specified as mathach ~1. The random effects part of the model is then specified after the fixed part as random = ~1 which is separated from the grouping variable SCH_ID by a vertical line (|). SCH_ID is the grouping variable or identifier variable. Since no random coefficients for any predictor variables are specified in this model, only 1 is specified as the intercept. The method = "ML" argument requests the maximum likelihood estimation method. The na.action = "na.omit" argument removes the missing data if there are any. The fitted model is named m1 and the output is displayed by the summary(m1) command.

```
> library(foreign)
> chp10<-read.dta("C:/CDA/els2002.dta")
> chp10  <-  chp10[!is.na(chp10$mathach)&!is.na(chp10$efficacy)&!is.na(chp10$
public)&!is.na(chp10$sclimate), ]
> chp10$csclimat <- chp10$sclimate-mean(chp10$sclimate, na.rm=TRUE)
> chp10$gceffic <- chp10$efficacy-mean(chp10$efficacy, na.rm=TRUE)
> attach(chp10)
> # install.packages("nlme")
> library(nlme)
> # Null model with lme()
> m1 <- lme(mathach ~ 1, random = ~1|SCH_ID, na.action="na.omit", method="ML",
data=chp10)
> summary(m1)
Linear mixed-effects model fit by maximum likelihood
Data: chp10
      AIC        BIC       logLik
75354.68    75376.27    -37674.34


Random effects:
 Formula: ~1 | SCH_ID
             (Intercept)    Residual
StdDev:           5.6338    10.45764
```

```
Fixed effects: mathach ~ 1
                  Value   Std.Error   DF    t-value   p-value
(Intercept)    39.12354   0.2537879  9249  154.1584         0

Standardized Within-Group Residuals:
        Min           Q1          Med           Q3          Max
-3.08523311  -0.71736082   0.02771707   0.74094481   2.99690652

Number of Observations: 9866
Number of Groups: 617
```

Interpreting the Output

The R output for the `lme()` function includes the model fit statistics such as the AIC, the BIC, and the log-likelihood value, the random effects, the fixed effects, the standardized within-group residuals, and the number of observations and groups.

First, the statement "`Linear mixed-effects model fit by maximum likelihood`" tells us the linear mixed model is fitted with the ML estimation rather than the REML estimation method. The AIC and BIC statistics are 75,354.68 and 75,376.27, respectively. The log likelihood value is −37,674.34. These fit statistics will be used for model comparisons in the following sections.

Then, the random effects section contains the model formula of the random effects and the standard deviations of the intercept and residual. The column (labeled `Intercept`) reports the standard deviation at level 2 (i.e., schools) and the column (labeled `Residual`) reports the within-school standard deviation. The between-school standard deviation is 5.634 and the within-school standard deviation is 10.458. To request the variance components of the model, we need to either square the standard deviations or use the `VarCorr(m1)` command. The between-school variance (τ_{00}) is 31.740, and the within-school variance (σ^2) is 109.362.

```
> VarCorr(m1)
SCH_ID = pdLogChol(1)
             Variance    StdDev
(Intercept)   31.7397   5.63380
Residual     109.3623  10.45764
```

Next, the fixed effects section contains the model formula of the fixed effects and the estimate for the intercept, its standard error, the degrees of freedom, the t statistic, and the associated p value. Since no predictors are included in the model, this section only displays the estimate for the intercept. The intercept γ_{00} (labeled `Intercept`) is 39.124, which is significant (`p = .000`). This means that the average math achievement score across all schools is 39.124.

Since currently the `confint()` function does not work with `lme()` function, we use the `intervals(m1)` command to extract the 95% confidence intervals of the estimates. The results are displayed as follows.

```
> confint(m1)
Error in confint.lme(m1) :
  not (yet) implemented. Contributions are welcome; use intervals() instead (for
now)
> intervals(m1)
Approximate 95% confidence intervals
 Fixed effects:
                 lower       est.     upper
(Intercept)   38.62609   39.12354   39.621

attr(,"label")
[1] "Fixed effects:"
 Random Effects:
  Level: SCH_ID
                     lower      est.      upper
sd((Intercept))   5.246673   5.6338   6.049491

 Within-group standard error:
   lower      est.      upper
10.30785   10.45764   10.60962
```

The ICC is defined as the proportion of total variance in the outcome variable ($\sigma^2 + \tau_{00}$) explained by the between-group variance (τ_{00}). It is expressed as ICC $= \tau_{00}/(\sigma^2 + \tau_{00})$.

From the earlier output, we compute

$$\text{ICC} = 31.740/(31.740 + 109.362) = .225,$$

which indicates that 22.5% of the total variance is accounted for by schools in level 2.

Finally, the output shows the minimum, first quartile, median, third quartile, and maximum values of the standardized within-group residuals. These residuals are normally distributed. In addition, the number of observations and the number of groups are provided. A total of 9,866 observations in level 1 is nested in 617 groups (i.e., schools) in level 2.

10.3.3 Random-Intercept Model (Model 2)

Next, we include the predictor variable `gceffic` (math self-efficacy) to the level 1 equation, with all other parts of the level 1 equation the same. The model is referred to as the random-intercept model since the intercept is allowed to vary across schools. This model can be expressed as follows:

$$\begin{aligned}\text{Level 1}: Y_{ij} &= \beta_{0j} + \beta_{1j}\texttt{gceffic}_{ij} + r_{ij}\\ \text{Level 2}: \beta_{0j} &= \gamma_{00} + u_{0j}\\ \beta_{1j} &= \gamma_{10}\end{aligned}$$

In the level 1 equation, the predictor variable is math self-efficacy (`gceffic`) and the outcome variable is math achievement (`mathach`). The predictor variable is grand-mean centered. The level 2 equations express the random intercepts (β_{0j}) and the fixed slopes (β_{1j}).

The command `m2 <- lme(mathach ~ gceffic, random = ~1|SCH_ID, na.action = "na.omit", method = "ML", data = chp10)` is used to fit the random-intercept model (model 2) after a predictor variable `gceffic` is added to the fixed part of the model. The fitted model is named `m2`. Just as in the unconditional model, no random coefficients are specified in the random part of the model. The following output is displayed by the `summary(m2)` command.

```
> # Random-intercept model
> m2 <- lme(mathach ~ gceffic, random = ~1|SCH_ID, na.action="na.omit", method="ML",
data=chp10)
> summary(m2)
Linear mixed-effects model fit by maximum likelihood
 Data: chp10
      AIC        BIC       logLik
73960.62    73989.41    -36976.31

Random effects:
 Formula: ~1 | SCH_ID
            (Intercept)     Residual
StdDev:        5.243221     9.743885

Fixed effects: mathach ~ gceffic
                Value     Std.Error     DF     t-value    p-value
(Intercept)  39.12132     0.2362576   9248    165.5875          0
gceffic       4.67671     0.1207781   9248     38.7215          0
Correlation:
           (Intr)
gceffic        0

Standardized Within-Group Residuals:
         Min           Q1          Med           Q3          Max
-3.96543934  -0.66900059   0.05549942   0.71325893   3.30894729

Number of Observations: 9866
Number of Groups: 617
```

Interpreting the Output

In the fixed-effects section, the estimated intercept is 39.121, and the coefficient for `gceffic` is 4.677. Both estimates are significant ($p < .001$). The intercept can be interpreted as follows: The average math achievement score is 39.121 for students with a value of math self-efficacy at 0. The coefficient for `gceffic` is 4.677, $t = 38.722$,

$p < .001$, which indicates that for a one-unit increase in math self-efficacy, there is an increase of 4.677 points in math achievement scores.

We also use the `intervals(m2)` command to extract the 95% confidence intervals of the parameter estimates. The results are displayed as follows.

```
> intervals(m2)
Approximate 95% confidence intervals

 Fixed effects:
                 lower       est.      upper
(Intercept)   38.65825  39.121317  39.584387
gceffic        4.43998   4.676707   4.913435
attr(,"label")
[1] "Fixed effects:"

 Random Effects:
  Level: SCH_ID
                   lower       est.      upper
sd((Intercept))  4.882778  5.243221  5.630272

 Within-group standard error:
   lower      est.     upper
9.604314  9.743885  9.885484
```

To request the variance components of the model, we need to either square the standard deviations or use the `VarCorr(m2)` command. The between-school variance (τ_{00}) is 27.491, and the within-school variance (σ^2) is 94.943.

```
> VarCorr(m2)
SCH_ID = pdLogChol(1)
             Variance    StdDev
(Intercept)  27.49137  5.243221
Residual     94.94330  9.743885
```

After the level 1 predictor is entered in the model, the variance for the random intercept has decreased to 27.491, compared with the original 31.740 in the unconditional model.

Likelihood Ratio Test Comparing the Unconditional Model and the Random Intercept Model

```
> # Model comparison
> anova(m1, m2)
   Model df      AIC      BIC    logLik   Test  L.Ratio p-value
m1     1  3 75354.68 75376.27 -37674.34
m2     2  4 73960.62 73989.41 -36976.31 1 vs 2 1396.056  <.0001
```

The likelihood ratio test is used to compare the unconditional model (model 1) and the random-intercept model (model 2). The `anova(m1, m2)` command compares the log-likelihood statistics of these two models using the likelihood ratio test. The likelihood ratio test $\chi^2_{(1)} = 1{,}396.056$, $p < .001$, which indicates that the random-intercept model fits the data better than the unconditional model.

10.3.4 Random-Coefficient Model: Random-Intercept and Slope Model With Level 1 Variable (Model 3)

In addition to the random intercept, level 1 slopes (i.e., the coefficients of the level 1 predictors) can also be specified to be random. In other words, a predictor may have a random slope across clusters. For example, we may allow the effect of math self-efficacy on math achievement to vary across different schools. This model can be expressed as follows:

$$\begin{aligned} \text{Level 1}: \; Y_{ij} &= \beta_{0j} + \beta_{1j}\texttt{gceffic}_{ij} + r_{ij} \\ \text{Level 2}: \; \beta_{0j} &= \gamma_{00} + u_{0j} \\ \beta_{1j} &= \gamma_{10} + u_{1j} \end{aligned}$$

The level 1 equation for the random-coefficient model is the same as that for the random-intercept model. The math self-efficacy (`gceffic`) is still the only predictor variable, and math achievement is the outcome variable. Unlike the random-intercept model, in the random-coefficient model, the level 2 equations specify that both the intercept and the coefficient at level 1 are random across schools.

The command `m3 <- lme(mathach ~ gceffic, random = ~gceffic| SCH_ID, na.action = "na.omit", method = "ML", data = chp10)` is used to fit the random-coefficient model (model 3) after a predictor variable `gceffic` is added to the random part of the model. The fitted model is named `m3`. The `summary(m3)` command displays the output of the fitted model.

```
> # Random-coefficient model
> m3 <- lme(mathach ~ gceffic, random = ~gceffic|SCH_ID, na.action="na.omit",
method="ML", data=chp10)
> summary(m3)
Linear mixed-effects model fit by maximum likelihood
 Data: chp10
      AIC          BIC        logLik
73953.39    73996.57    -36970.7

Random effects:
 Formula: ~gceffic | SCH_ID
 Structure: General positive-definite, Log-Cholesky parametrization
                    StdDev       Corr
(Intercept)       5.237083      (Intr)
gceffic           1.003340       0.389
Residual          9.708897
```

```
Fixed effects: mathach ~ gceffic
                Value      Std.Error   DF    t-value     p-value
(Intercept)   39.09802   0.2360843   9248   165.61046          0
gceffic        4.64856   0.1281118   9248    36.28516          0
 Correlation:
          (Intr)
gceffic    0.115

Standardized Within-Group Residuals:
        Min            Q1           Med            Q3           Max
-3.88739725   -0.66933019   0.05610879   0.71053443   3.31730425

Number of Observations: 9866
Number of Groups: 617
```

The fixed effects in the output look similar to those in the random-intercept model (model 2). We use the `intervals(m3, which = "fixed")` command to extract the 95% confidence intervals of the parameter estimates of the fixed effects. Without the `which = "fixed"` argument, an error message will occur since the confidence intervals cannot be computed for the variance and covariance components. The results are displayed as follows.

```
> intervals(m3)
Error in intervals.lme(m3) :
  cannot get confidence intervals on var-cov components: Non-positive definite
approximate variance-covariance
 Consider 'which = "fixed"'
> intervals(m3, which="fixed")
Approximate 95% confidence intervals

 Fixed effects:
                 lower        est.       upper
(Intercept)  38.635295   39.098025   39.560755
gceffic       4.397457    4.648559    4.899661
attr(,"label")
[1] "Fixed effects:"
```

To request the variance and covariance components of the model, we use the `VarCorr(m3)` command. The results are displayed as follows.

```
> VarCorr(m3)
SCH_ID = pdLogChol(gceffic)
               Variance      StdDev      Corr
(Intercept)   27.427040    5.237083    (Intr)
gceffic        1.006691    1.003340     0.389
Residual      94.262689    9.708897

> .389*1.003*5.237
[1] 2.043305
```

Interpreting the Output

The `VarCorr(m3)` command reports the unstructured variance–covariance components. The between-school variance (τ_{00}) is 27.427 and the within-school variance (σ^2) is 94.263.

The variance for the random coefficient of `gceffic` or the slope variance (labeled `var(gceffic)`) is 1.007. The output reports the correlation between `gceffic` and the intercept rather than the covariance between them. The correlation coefficient for `gceffic` and the intercept is .389. We can compute the covariance between `gceffic` and the intercept with the correlation between them and their standard deviations. The covariance equals the product of the correlation and the two standard deviations, so we compute .389*1.003*5.237 = 2.043.

The within-school variance or the level 1 residual variance (labeled `var(Residual)`) is 94.263, which is similar to that of the random-intercept model.

Model Comparisons Using the Likelihood Ratio Test

```
> anova(m2, m3)
     Model  df       AIC       BIC     logLik    Test    L.Ratio  p-value
m2       1   4  73960.62  73989.41  -36976.31
m3       2   6  73953.39  73996.57  -36970.70  1 vs 2  11.23146   0.0036
```

The likelihood ratio test is used to compare the random-intercept model (model 2) and the random-coefficient model with a level 1 predictor (model 3). The `anova(m2, m3)` command compares the log likelihood statistics of these two models. The likelihood ratio test $\chi^2_{(2)} = 11.231$, $p < .01$, which indicates that the random-coefficient model has a significantly better fit than the random-intercept model. Therefore, allowing a random coefficient in the model is justified.

10.3.5 Contextual Model With Level 1 and Level 2 Variables (Model 4)

The contextual model is a special case of the random-coefficient model when both level 1 and level 2 predictor variables are included. The two level 2 equations specify the intercept and slope from the level 1 equation to be random. In this example, two school-level variables, `public` and `csclimat`, are added to the level 2 equation. This model can be expressed as follows:

$$\begin{aligned}
\text{Level 1}: \; Y_{ij} &= \beta_{0j} + \beta_{1j}\texttt{gceffic}_{ij} + r_{ij} \\
\text{Level 2}: \; \beta_{0j} &= \gamma_{00} + \gamma_{01}\texttt{public}_j + \gamma_{02}\texttt{csclimat}_j + u_{0j} \\
\beta_{1j} &= \gamma_{10} + u_{1j}
\end{aligned}$$

With the level 1 equation the same as that for the previous random-coefficient model (model 3), the two school-level predictor variables are added to the equation for the random intercept at level 2. On the other hand, the equation for the random slope β_{1j} contains no predictor variables.

The command `m4 <- lme(mathach ~ gceffic + public + csclimat, random = ~gceffic|SCH_ID, na.action = "na.omit", method = "ML", data = chp10)` is used to fit the contextual model (model 4) after the two level 2 predictor variables `public` and `csclimat` are added to the model.

```
> # Contextual model with predictor variables in both levels
> m4 <- lme(mathach ~ gceffic + public + csclimat, random = ~gceffic|SCH_ID,
na.action = "na.omit", method = "ML", data = chp10)
> summary(m4)
Linear mixed-effects model fit by maximum likelihood
 Data: chp10
      AIC        BIC       logLik
73806.56   73864.14   -36895.28

Random effects:
 Formula: ~gceffic | SCH_ID
 Structure: General positive-definite, Log-Cholesky parametrization
              StdDev     Corr
(Intercept)  4.447067   (Intr)
gceffic      1.037690    0.565
Residual     9.713777

Fixed effects: mathach ~ gceffic + public + csclimat
                 Value   Std.Error    DF    t-value   p-value
(Intercept)   41.39652   0.4556953  9248   90.84255         0
gceffic        4.62698   0.1282449  9248   36.07922         0
publicpublic  -2.89695   0.5248300   614   -5.51979         0
csclimat       3.02747   0.3249720   614    9.31610         0
 Correlation:
              (Intr)   gceffc   pblcpb
gceffic        0.065
publicpublic  -0.890    0.014
csclimat      -0.304   -0.008    0.350

Standardized Within-Group Residuals:
        Min          Q1         Med          Q3         Max
-3.88077064  -0.67411813  0.05771525  0.70333885  3.41379144

Number of Observations: 9866
Number of Groups: 617
```

Interpreting the Output

The fixed-effects table displays the intercept and coefficients for both the level 1 and level 2 predictor variables. The coefficient for `gceffic` is 4.627, $t = 36.079$, $p < .001$, which indicates that students with higher mathematics self-efficacy tend to have higher mathematics achievement when holding other predictors constant. The effects of two school-level predictor variables are significant. The coefficient for `public` is −2.897, $t = -5.520$, $p < .001$, which indicates that students' mathematics scores in public schools tend to be lower than those in private schools. The coefficient for `csclimat` is 3.027, $t = 9.316$, $p < .001$, which indicates that schools with a better social climate tend to have higher mathematics scores.

We use the `intervals(m4, which = "fixed")` command to extract the 95% confidence intervals of the parameter estimates of the fixed effects. The results are displayed as follows.

```
> intervals(m4, which = "fixed")
Approximate 95% confidence intervals

 Fixed effects:
                    lower        est.       upper
(Intercept)     40.503438   41.396520   42.289602
gceffic          4.375639    4.626976    4.878313
publicpublic    -3.927424   -2.896953   -1.866483
csclimat         2.389408    3.027470    3.665532

attr(,"label")
[1] "Fixed effects:"
```

To request the variance and covariance components of the model, we use the `VarCorr(m4)` command. The results are displayed as follows.

```
> VarCorr(m4)
SCH_ID = pdLogChol(gceffic)
               Variance     StdDev     Corr
(Intercept)   19.776405   4.447067   (Intr)
gceffic        1.076801   1.037690    0.565
Residual      94.357456   9.713777

> .565*1.038*4.447
[1] 2.608032
```

The variance for the random coefficient of `gceffic` or the slope variance (labeled `gceffic)`) is 1.077, the between-group variance (labeled `Intercept)` is 19.776. Since the correlation coefficient for `gceffic` and the intercept is .565 and the standard deviations of `gceffic` and the intercept are 1.038 and 4.447, respectively, we can compute the covariance between gceffic and the intercept as follows: .565*1.038*4.447 = 2.608. Both the slope variance and the covariance in model 4 look similar to those in the random-coefficient model (model 3). After two school-level predictors are included in the model, the between-group variance decreases from 27.427 to 19.776. The percentage decrease is computed as (27.427−19.776)/27.427 = 27.9%. In other words, the school-level variables explain 27.9% of the between-group variance.

```
> (27.427-19.776)/27.427
[1] 0.2789587
```

The within-school variance or the level 1 residual variance (labeled Residual) is 94.357, which is similar to that of the random-coefficient model.

Model Comparisons Using the Likelihood Ratio Test

```
> anova(m3, m4)
   Model df      AIC      BIC    logLik   Test  L.Ratio p-value
m3     1  6 73953.39 73996.57 -36970.70
m4     2  8 73806.56 73864.14 -36895.28 1 vs 2 150.8293  <.0001
```

The anova(m3, m4) command compares the log-likelihood statistics of the contextual model with both level 1 and level 2 predictor variables (model 4: m4) and the random-intercept and slope model with a level 1 predictor (model 3: m3). A comparison between these two models yields the value of the likelihood ratio test $\chi^2_{(2)} = 150.829$, $p < .001$, which indicates that the contextual model fits the data better.

10.3.6 Contextual Model With Cross-Level Interactions (Model 5)

We can also add cross-level interactions to the model by including level 2 predictor variables in the equation for the random slope (β_{1j}). This model can be expressed as follows:

$$
\begin{aligned}
\text{Level 1}: \; Y_{ij} &= \beta_{0j} + \beta_{1j}\texttt{gceffic}_{ij} + r_{ij} \\
\text{Level 2}: \; \beta_{0j} &= \gamma_{00} + \gamma_{01}\texttt{public}_j + \gamma_{02}\texttt{csclimat}_j + u_{0j} \\
\beta_{1j} &= \gamma_{10} + \gamma_{11}\texttt{public}_j + \gamma_{12}\texttt{csclimat}_j + u_{1j}
\end{aligned}
$$

In R, we need to add the interaction term between public and gceffic and the interaction term between csclimat and gceffic in the model formula. The command m5 <- lme(mathach ~ gceffic + public + csclimat + public*gceffic + csclimat*gceffic, random = ~gceffic|SCH_ID, method = "ML", data = chp10) is used to fit the contextual model with cross-level interactions (model 5) after the cross-level interactions, `public*gceffic` and `csclimat*gceffic`, are added to the model. The fitted model is named `m5`. The `summary(m5)` command displays the resulting output.

```
> # Contextual model with cross-level interactions
> m5 <- lme(mathach ~ gceffic + public + csclimat + public*gceffic + csclimat*gceffic,
random = ~gceffic|SCH_ID, method="ML", data=chp10)
> summary(m5)
Linear mixed-effects model fit by maximum likelihood
 Data: chp10
       AIC        BIC      logLik
  73806.63   73878.6   -36893.31

Random effects:
 Formula: ~gceffic | SCH_ID
 Structure: General positive-definite, Log-Cholesky parametrization
             StdDev     Corr
(Intercept)  4.453850   (Intr)
gceffic      1.001821   0.583
Residual     9.713346

Fixed effects: mathach ~ gceffic + public + csclimat + public * gceffic + csclimat *
gceffic
                        Value  Std.Error    DF    t-value  p-value
(Intercept)          41.31079  0.4598111  9246   89.84297   0.0000
gceffic               4.22730  0.2792129  9246   15.14006   0.0000
publicpublic         -2.77430  0.5312182   614   -5.22252   0.0000
csclimat              3.00094  0.3297024   614    9.10198   0.0000
gceffic:publicpublic  0.52294  0.3237919  9246    1.61505   0.1063
gceffic:csclimat     -0.10002  0.2066567  9246   -0.48400   0.6284
 Correlation:
                      (Intr)  gceffc  pblcpb  csclmt  gcffc:p
gceffic                0.140
publicpublic          -0.892  -0.125
csclimat              -0.307  -0.058   0.351
gceffic:publicpublic  -0.125  -0.889   0.148   0.060
gceffic:csclimat      -0.056  -0.333   0.058   0.162   0.368

Standardized Within-Group Residuals:
        Min          Q1          Med          Q3          Max
-3.90127367 -0.67502516  0.05711133  0.70307008  3.42698496

Number of Observations: 9866
Number of Groups: 617
```

Interpreting the Output

The fixed-effects section displays the intercept and the coefficients for the level 1 and level 2 predictor variables and two interaction terms. Let us take a look at the coefficients for the two interaction terms first since we are interested in the cross-level interactions. The coefficient for `gceffic:public` (γ_{11}) = .523, t = 1.615, $p > .05$, and the coefficient for gceffic:csclimat (γ_{12}) = −.100, t = −.484, p > .05, which means both interactions are not significant. What if there is a cross-level effect? How can it be interpreted? For example, if the interaction between public and gceffic is significant, then this effect can be interpreted as follows: The relationship between mathematics self-efficacy and mathematics achievement varies across school type.

We use the intervals(m5, which = "fixed") command to extract the 95% confidence intervals of the parameter estimates of the fixed effects. The results are displayed as follows.

```
> intervals(m5, which="fixed")
Approximate 95% confidence intervals

 Fixed effects:
                            lower        est.       upper
(Intercept)            40.4097378  41.3107948  42.2118519
gceffic                 3.6801490   4.2273014   4.7744538
publicpublic           -3.8172044  -2.7742967  -1.7313890
csclimat                2.3536604   3.0009446   3.6482288
gceffic:publicpublic   -0.1115691   0.5229414   1.1574518
gceffic:csclimat       -0.5049920  -0.1000226   0.3049469

attr(,"label")
[1] "Fixed effects:"
```

To request the variance and covariance components of the model, we use the VarCorr(m5) command. The results are displayed as follows.

```
> VarCorr(m5)
SCH_ID = pdLogChol(gceffic)
             Variance    StdDev    Corr
(Intercept)  19.836777  4.453850  (Intr)
gceffic       1.003644  1.001821  0.583
Residual     94.349097  9.713346

> .583*1.002*4.454
[1] 2.601875
```

The variance and covariance components for the random effects for model 5 look similar to those for model 4, so they will not be interpreted here.

Model Comparisons Using the Likelihood Ratio Test

```
> anova(m4, m5)
   Model df      AIC      BIC    logLik   Test  L.Ratio p-value
m4     1  8 73806.56 73864.14 -36895.28
m5     2 10 73806.63 73878.60 -36893.31 1 vs 2 3.933261  0.1399
```

After two interaction terms are added to the model, we again use the `anova(m4, m5)` command to compare the log-likelihood statistics of the contextual models with or without interactions (model 5 vs. model 4). The likelihood ratio test $\chi^2_{(2)} = 3.93$, $p = .140$, which indicates that the contextual model with cross-level interactions does not fit the data better than the contextual model without interactions. Therefore, we should remove the interaction terms from the model.

10.4 MULTILEVEL MODELING FOR CONTINUOUS RESPONSE VARIABLES WITH THE `lmer()` FUNCTION IN THE `lme4` PACKAGE

10.4.1 The `lmer()` Function in the `lme4` Package

The `lmer()` function in the `lme4` package (Bates et al., 2015) can also be used for multilevel models for continuous response variables. Since `lme4` is a user-written package, you need to install it first by typing `install.packages("lme4")` and then load the package by typing `library(lme4)`. The model formula syntax of `lmer()` is similar to that of `lme()` introduced in the last section, but they are slightly different in specifying the random effects. The model formula in `lmer()` also includes two components connected by a plus (+) sign. The first component specifies the dependent variable and the predictor variable(s), which are separated by the tilde (~). When there are more than one predictor variable in the formula, they are connected by plus (+) symbols. This is the fixed effects component of the model. Next, the random effects component of the model is specified within parentheses, which is different from the `random` argument specified in `lme()` for linear mixed effects models. Within parentheses, a predictor variable or a list of predictor variables that have random coefficients is specified first, followed by a vertical line (|), and then, an identifier variable as the grouping variable at a higher level. The `REML = FALSE` argument requests the full maximum likelihood estimation rather than the default restricted likelihood (`REML = TRUE`). For example, the command `lmer(y ~ x + (1|schid), REML = FALSE, data = data1)` tells R to fit a multilevel model to estimate a continuous outcome variable `y` on a predictor variable `x` with random intercepts varying across schools by specifying `(1|schid)`, where `1` indicates the random intercept on the left of the vertical line (|) and `schid` is the identifier variable on the right. The `REML = FALSE` argument requests the maximum likelihood estimation. For more details on how to use this function, type `help(lmer)` in the command prompt after loading the `lme4` package.

We use the `lmerTest` package (Kuznetsova et al., 2017) to request the associated *p* values for the *t* tests of the parameter estimates. Since `lmerTest` is a user-written package, you need to install it first by typing `install.packages("lmerTest")` and then load the package by typing `library(lmerTest)`.

In the following example, the `m4.b <- lmer(mathach ~ gceffic + public + csclimat + (gceffic|SCH_ID), data = chp10, REML = FALSE)` command tells R to fit the same contextual model (model 4) in the previous section after the two level 2 predictor variables `public` and `csclimat` are added to the model. The fitted model is named `m4.b`. The following output is displayed by the `summary(m4.b)` command.

```
> # Contextual model with lmer()
> library(lmerTest)
> library(lme4)
> m4.b <- lmer(mathach ~ gceffic + public + csclimat + (gceffic|SCH_ID),
data=chp10, REML = FALSE)
> summary(m4.b)
Linear mixed model fit by maximum likelihood . t-tests use Satterthwaite's
  method [lmerModLmerTest]
Formula: mathach ~ gceffic + public + csclimat + (gceffic | SCH_ID)
  Data: chp10

     AIC       BIC      logLik    deviance   df.resid
 73806.6   73864.1   -36895.3    73790.6       9858

Scaled residuals:
     Min        1Q    Median       3Q       Max
-3.8808   -0.6741    0.0577   0.7033   3.4138

Random effects:
  Groups          Name     Variance    Std.Dev.   Corr
  SCH_ID    (Intercept)      19.777       4.447
                gceffic       1.077       1.038   0.56
Residual                     94.357       9.714
Number of obs: 9866, groups: SCH_ID, 617
Fixed effects:
               Estimate   Std. Error         df   t value    Pr(>|t|)
(Intercept)     41.3965       0.4556   581.5800    90.860   < 2e-16 ***
gceffic          4.6270       0.1282   536.0261    36.086   < 2e-16 ***
publicpublic    -2.8969       0.5247   573.6036    -5.521  5.12e-08 ***
csclimat         3.0275       0.3249   570.2441     9.318   < 2e-16 ***
---
Signif. codes: 0 '***' 0.001 '**' 0.01 '*' 0.05 '.' 0.1 ' ' 1

Correlation of Fixed Effects:
              (Intr)   gceffc   pblcpb
gceffic        0.065
publicpublc   -0.890    0.014
csclimat      -0.304   -0.008    0.350
```

The R output produced by the `lmer()` function is similar to that by the `lme()` function introduced in the previous section. It includes the model fit statistics, the scaled residuals, the random effects, the number of observations and the number of groups, the fixed effects, and the correlations of the fixed effects.

The random effects section provides the variances and standard deviations for within-group, between-group, and the random coefficient of gceffic. It also provides the correlation coefficient for gceffic and the intercept. The results are the same as those provided by the VarCorr(m4) command in the previous section.

The fixed effects section looks the same as that produced by the lme() function. We use the confint(m4.b) command rather than the intervals() function to extract the 95% confidence intervals of the parameter estimates. The results are displayed as follows.

```
> confint(m4.b)
Computing profile confidence intervals ...
                    2.5 %      97.5 %
.sig01          4.1147172    4.805720
.sig02          0.2734630    1.000000
.sig03          0.4181463    1.493107
.sigma          9.5715169    9.859538
(Intercept)    40.4992243   42.291286
gceffic         4.3740861    4.880153
publicpublic   -3.9287910   -1.863101
csclimat        2.3894054    3.665517
```

The 95% confidence intervals for both the random effects and the fixed effects are provided. Please note that the confidence intervals for the random effects component are based on the standard deviations rather than the variances. The 95% confidence intervals for the fixed effects component are the same as those provided by the intervals(m4, which = "fixed") command introduced in the previous section.

10.4.2 Interpreting the Predicted Values With the ggpredict() Function in the ggeffects Package

By using the ggpredict() function in the ggeffects package (Lüdecke, 2018b), we can compute the predicted values of the continuous response variable at specified values of the predictor variables. We first load the package with library(ggeffects) since it has been installed in previous chapters. The command m4.b.pub <- ggpredict(m4.b, terms = "public") tells R to compute the predicted values of the response variable using the ggpredict() function. The argument inside the function includes the estimated model, m4.b, and the terms = "public" option which specifies the predictor variable public. The other predictor variables are held either at their means or at the reference level. The output is assigned to an object named m4.b.pub. The as.data.frame() function is used to request the standard errors.

```
> library(ggeffects)
> m4.b.pub <- ggpredict(m4.b, terms = "public")
> m4.b.pub
```

```
# Predicted values of mathach
public                    |  Predicted   |          95% CI
-------------------------------------------------------
catholic or other private |    41.40     |  [40.50, 42.29]
public                    |    38.50     |  [38.03, 38.97]

Adjusted for:
* gceffic = 0.00
* csclimat = -0.00
* SCH_ID = 0 (population-level)
> as.data.frame(m4.b.pub)
              x  predicted  std.error  conf.low  conf.high  group
1   catholic or   41.39651  0.4556121  40.50352   42.28949      1
  other private
2        public   38.49957  0.2395262  38.03010   38.96903      1
> plot(m4.b.pub)
Loading required namespace: ggplot2
```

The output displays the predicted math scores for both Catholic or other private schools and public schools and other predictor variables are fixed at their means. The last section under the title "`Adjusted for`" lists the means of the other three variables. For Catholic or other private schools, the predicted math scores = 41.397 and for public schools, the predicted math scores = 38.500.

The predicted math scores are plotted with `plot(m4.b.pub)`. Figure 10.1 shows the predicted math scores for both Catholic or other private schools and public schools.

FIGURE 10.1 ● Predicted Math Scores for `public` at 0 and 1 With Others Fixed at Their Means

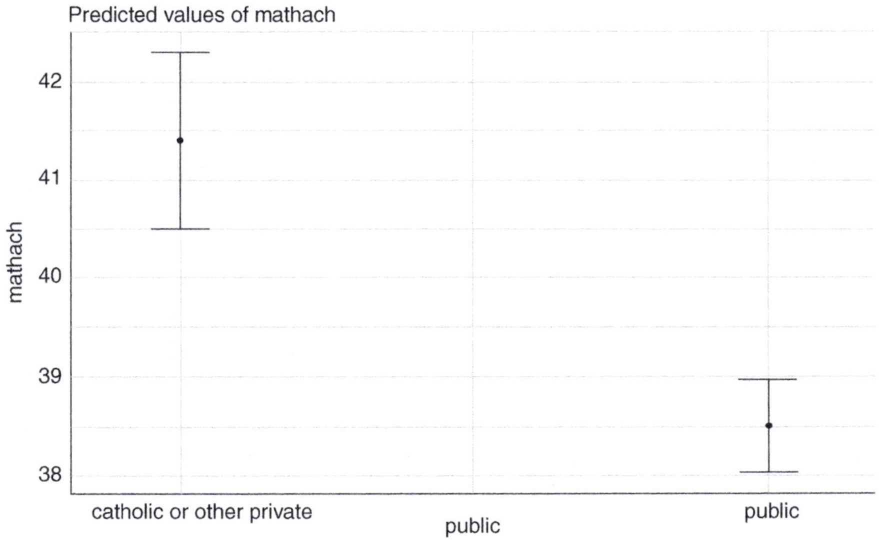

As shown in the graph, the predicted math scores for students in Catholic or other private schools are higher than those in public schools.

We can compute the predicted scores for a continuous variable at given values by different groups. In the following example, we compute the predicted math scores for `gceffic` at different values by the two groups in `public` when holding other variables at their means. The command is as follows: `m4.efpub <- ggpredict(m4.b, terms = c("gceffic", "public"))`. In the `ggpredict()` function, the `terms = c("gceffic", "public")` option specifies both `gceffic` and `public`, with the latter as the grouping variable. The output is assigned to an object named `m4.efpub` and is plotted with the `plot(m4.efpub)` command.

```
> m4.efpub <- ggpredict(m4.b, terms = c("gceffic", "public"))
> m4.efpub
# Predicted values of mathach
# public = catholic or other private

gceffic   |   Predicted   |           95% CI
--------------------------------------------
-1.60     |       33.99   |   [33.04, 34.95]
-1.00     |       36.77   |   [35.86, 37.68]
-0.60     |       38.62   |   [37.72, 39.52]
 0.00     |       41.40   |   [40.50, 42.29]
 0.60     |       44.17   |   [43.26, 45.09]
 1.60     |       48.80   |   [47.80, 49.80]

# public = public

gceffic   |   Predicted   |           95% CI
--------------------------------------------
-1.60     |       31.10   |   [30.53, 31.67]
-1.00     |       33.87   |   [33.38, 34.37]
-0.60     |       35.72   |   [35.25, 36.19]
 0.00     |       38.50   |   [38.03, 38.97]
 0.60     |       41.28   |   [40.76, 41.79]
 1.60     |       45.90   |   [45.24, 46.57]

Adjusted for:
* csclimat = -0.00
*   SCH_ID = 0 (population-level)
> plot(m4.efpub)
```

Figure 10.2 shows the predicted math scores for `gceffic` at different values by the grouping variable `public`. As shown in the graph, the predicted math scores increase with the increase of students' self-efficacy and the predicted math scores for students in Catholic or other private schools are higher than those in public schools.

In the example above, we disregard the random effects component and the predicted math scores are on the population level. To account for the variability of the random effects, we need to specify the `type = "re"` option. With this option, the predicted math scores are still on the population level and stay the same, but the confidence intervals change after accounting for the variances of the random effects. The command

FIGURE 10.2 ● Predicted Math Scores for gceffic by public

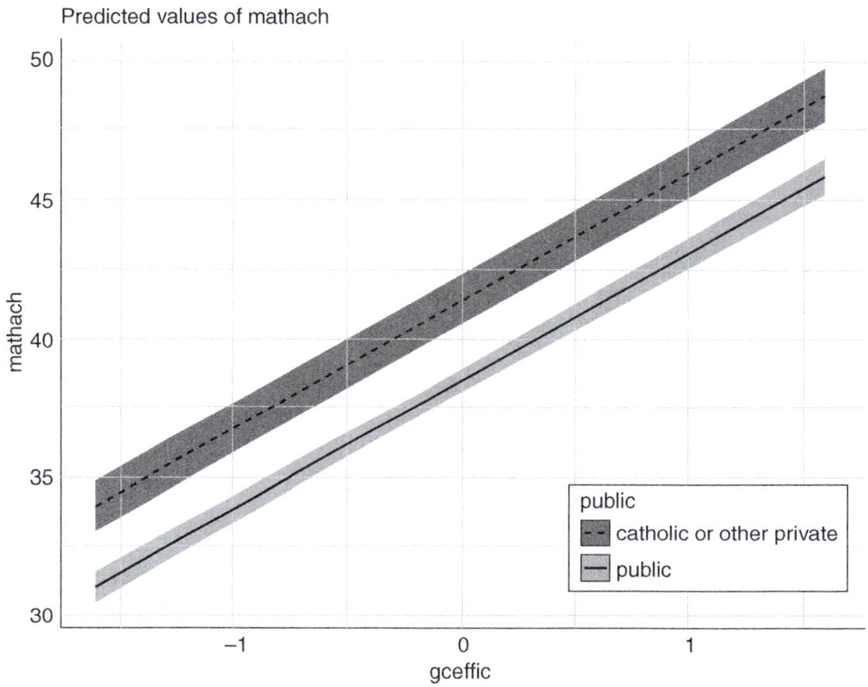

is as follows: `m4.efpub.r <- ggpredict(m4.b, terms = c("gceffic", "public"), type = "re")`. In the `ggpredict()` function, everything stays the same exception the added `type = "re"` option. The output is assigned to an object named `m4.efpub.r` and is plotted with the `plot(m4.efpub.r)` command.

```
> m4.efpub.r <- ggpredict(m4.b, terms = c("gceffic", "public"), type = "re")
> m4.efpub.r

# Predicted values of mathach

# public = catholic or other private
gceffic   |   Predicted   |          95% CI
--------------------------------------------
 -1.60    |       33.99   |   [14.93, 53.06]
 -1.00    |       36.77   |   [17.71, 55.83]
 -0.60    |       38.62   |   [19.56, 57.68]
  0.00    |       41.40   |   [22.34, 60.46]
  0.60    |       44.17   |   [25.11, 63.23]
  1.60    |       48.80   |   [29.73, 67.86]
```

```
# public = public
gceffic    |   Predicted    |           95% CI
----------------------------------------------
-1.60      |       31.10    |   [12.05, 50.14]
-1.00      |       33.87    |   [14.83, 52.92]
-0.60      |       35.72    |   [16.68, 54.77]
 0.00      |       38.50    |   [19.46, 57.54]
 0.60      |       41.28    |   [22.23, 60.32]
 1.60      |       45.90    |   [26.85, 64.95]

Adjusted for:
* csclimat = -0.00
*   SCH_ID = 0 (population-level)
Intervals are prediction intervals.
> plot(m4.efpub.r)
```

Figure 10.3 shows the predicted math scores for `gceffic` by the grouping variable `public` with random effects. As shown in the graph, the predicted math scores are the same as those in Figure 10.2, but the confidence intervals are wider than those in Figure 10.2.

FIGURE 10.3 ● Predicted Math Scores for **`gceffic`** by **`public`** With Random Effects

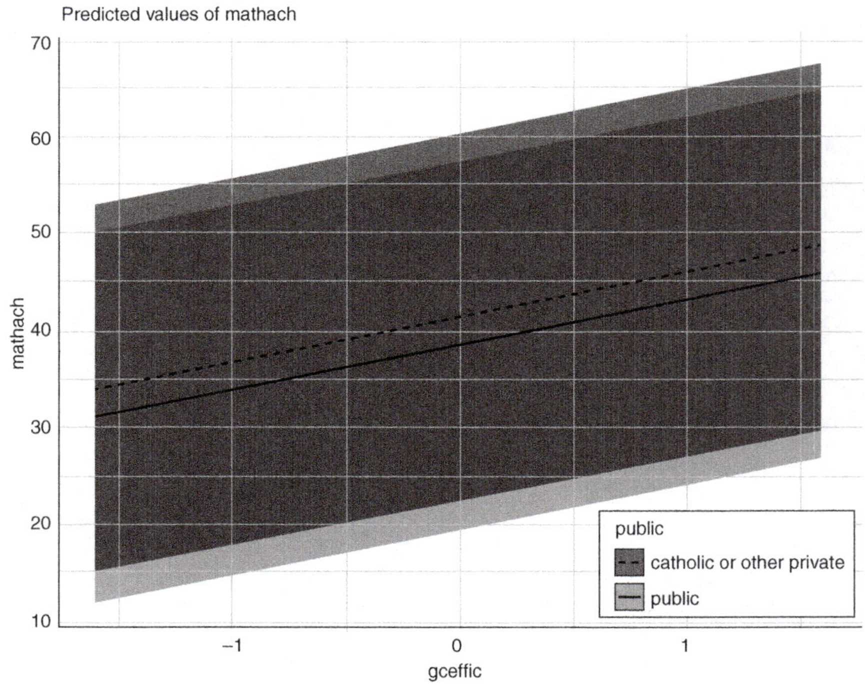

10.5 MAKING PUBLICATION-QUALITY TABLES

10.5.1 Presenting the Results Using the stargazer Package

For illustration purposes, we will make a table containing the results for only two models, the unconditional model and the contextual model with cross-level interactions. The commands for creating this table are explained as follows. By following this example, readers should easily be able to create a table containing the results for all the fitted models.

We can use the `stargazer` package (Hlavac, 2018) to make a table containing the results of the two fitted models. After fitting the `m1` and `m4` models with the `lme()` function, we use the command as follows: `stargazer(m1, m4, type = "text", align = TRUE, out = "mul.txt")`. In the `stargazer()` function, we first specify the two model objects to be presented and then the type of table. The option `type = "text"` specifies the table type and the `align = TRUE` option aligns the results of the model. The `out = "mul.txt"` argument saves the output named `mul.txt`. Please note that only the fixed effects are presented in the table.

```
> library(stargazer)
Please cite as:
Hlavac, Marek (2018). stargazer: Well-Formatted Regression and Summary Statistics
Tables.
R package version 5.2.2. https://CRAN.R-project.org/package=stargazer

> stargazer(m1, m4, type="text", align=TRUE, out="mul.txt")
=====================================================
                          Dependent variable:
                       ------------------------------
                                mathach
                           (1)              (2)
-----------------------------------------------------
gceffic                                  4.627***
                                         (0.128)

publicpublic                            -2.897***
                                         (0.525)

csclimat                                 3.027***
                                         (0.325)

Constant                39.124***       41.397***
                         (0.254)         (0.456)

-----------------------------------------------------
Observations              9,866           9,866
Log Likelihood         -37,674.340     -36,895.280
Akaike Inf. Crit.       75,354.680      73,806.560
Bayesian Inf. Crit.     75,376.270      73,864.140
=====================================================
Note:                *p<0.1; **p<0.05;   ***p<0.01
```

We can also create the table in the HTML format and copy it into Microsoft Word. The command is as follows: `stargazer(m1, m4, type = "html", align = TRUE, out = "mul.htm")`. It produces Table 10.1, as shown here in its original format, presenting the results for the unconditional model and the contextual model without cross-level interactions.

We can add the variances and covariance directly from the output of the unconditional model (model 1) and the contextual model (model 4). The edited table is displayed as Table 10.2.

10.5.2 Presenting the Results Using the `texreg` Package

The results can also be displayed in a table using the `screenreg()` and `htmlreg()` functions from the `texreg` package (Leifeld, 2013). Since `texreg`

TABLE 10.1 ● Results of the Two Multilevel Models: The Unconditional Model and the Contextual Model (Shown in Original Format Generated by R)

	Dependent variable:	
	mathach	
	(1)	(2)
gceffic		4.627***
		(0.128)
publicpublic		−2.897***
		(0.525)
csclimat		3.027***
		(0.325)
Constant	39.124***	41.397***
	(0.254)	(0.456)
Observations	9,866	9,866
Log Likelihood	−37,674.340	−36,895.280
Akaike Inf. Crit.	75,354.680	73,806.560
Bayesian Inf. Crit.	75,376.270	73,864.140

*$p < .1$
**$p < .05$
***$p < .01$.

TABLE 10.2 ● Results of the Two Multilevel Models: The Unconditional Model and the Contextual Model (Edited)

	Dependent variable:	
	mathach	
	Unconditional Model (Model 1)	Contextual Model (Model 4)
	Coefficient (SE)	Coefficient (SE)
Fixed Effects		
gceffic		4.627***
		(0.128)
publicpublic		−2.897***
		(0.525)
csclimat		3.027***
		(0.325)
Constant	39.124***	41.397***
	(0.254)	(0.456)
Random Effects		
Slope variance (`gceffic`)	–	1.077
Within-school variance (σ^2)	109.362	94.357
Between-group variance (τ_{00})	31.740	19.776
Covariance	–	2.608
Observations	9,866	9,866
Log Likelihood	−37,674.340	−36,895.280
Akaike Inf. Crit.	75,354.680	73,806.560
Bayesian Inf. Crit.	75,376.270	73,864.140

*$p < .1$
**$p < .05$
***$p < .01$

has been installed for previous chapters, we load the package by typing `library (texreg)`. We use the following command: `screenreg(list(m1, m4))`.

```
> library(texreg)
Version:    1.36.23
Date:       2017-03-03
Author:     Philip Leifeld (University of Glasgow)

Please cite the JSS article in your publications - see citation("texreg").
> screenreg(list(m1, m4))
==========================================
                    Model 1       Model 2
------------------------------------------
(Intercept)         39.12 ***     41.40 ***
                    (0.25)        (0.46)
gceffic                            4.63 ***
                                  (0.13)
publicpublic                      -2.90 ***
                                  (0.52)
csclimat                           3.03 ***
                                  (0.32)
------------------------------------------
AIC                 75354.68      73806.56
BIC                 75376.27      73864.14
Log Likelihood     -37674.34     -36895.28
Num. obs.            9866          9866
Num. groups           617           617
==========================================
*** p < 0.001,      ** p < 0.01,      * p < 0.05
```

To create the table in the HTML format and copy it into Microsoft Word, we use the command as follows: `htmlreg(list(m1, m4), file = "mul.doc", doctype = TRUE, html.tag = TRUE, head.tag = TRUE)`. The table is omitted here.

10.6 REPORTING THE RESULTS

Since multilevel models estimate the fixed effects and random effects, the results of both need to be reported. The following are the basic guidelines for reporting. Several common reporting guidelines provided in previous chapters can also be applied to the reporting for the multilevel modeling. Please note that what needs to be reported in the Results section in a research article and the formats for displaying results vary across disciplines and journals.

First, as with other research examples, describe the purpose of your study and explain why the multilevel modeling is needed for the analysis.

Second, if a series of nested models is fitted, then report model-building steps and briefly describe each model. Report and interpret the intraclass correlation coefficient.

Third, if necessary, report the results of the fitted models in a table including both the parameter estimates for the fixed effects and the variances and covariances for the random effects. If available, include deviance statistics (i.e., −2LL) and AIC and BIC statistics for these models in the table.

Fourth, report and interpret the fixed effects of the predictor variables and variances and covariances in the final model. The following is an example of summarized results for the unconditional model and the contextual model.

Multilevel modeling was used to examine the relationships between high-school students' mathematics achievement and mathematics self-efficacy, school type, and school climate. Five models, from the unconditional (null) model to the contextual model with cross-level interactions, were fitted. Table 10.2 presents the parameter estimates for the fixed effects and random effects for the fitted models. For illustration purposes, the following interpretations only focused on the results of the unconditional model (model 1) and the contextual model without cross-level interactions (model 4).

Results for the Unconditional Model

The between-school variance (τ_{00}) was 31.740, and the within-school variance (σ^2) was 109.362. The ICC = 31.740/(31.740 + 109.362) = .225, which indicated that 22.5% of the total variance was explained by schools in level 2. This empirical evidence showed that it was appropriate to use multilevel modeling for data analysis.

Results for the Contextual Model Without Cross-Level Interactions

The coefficient for `gceffic` was 4.627, $t = 36.079$, $p < .001$, which indicated that students with higher mathematics self-efficacy tended to have higher mathematics achievement when holding other predictors constant. The effects of two school-level predictor variables were also significant. The coefficient for `public` was −2.897, $t = -5.520$, $p < .01$, which indicated that students' mathematics scores in public schools tended to be lower than those in private schools. The coefficient for `csclimat` was 3.027, $t = 9.316$, $p < .001$, which indicated that schools with better social climate tended to have higher mathematics scores.

Regarding the random effects, the variance and covariance components are displayed in Table 10.2. After two school-level variables were included in the contextual model (model 4), the between-school variance (τ_{00}) decreased from 27.427 to 19.776 when compared with that for the random coefficient model (model 3): (27.427 − 19.776)/27.427 = 27.9%, which indicated that there was a decrease of 27.9% in the between-school variance from the random-coefficient model (model 3) to the contextual model without the cross-level interactions (model 4) after the two school-level variables were included.

10.7 SUMMARY OF R COMMANDS IN THIS CHAPTER

```
# Chap 10 R Script

# Remove all objects
rm(list = ls(all = TRUE))

# The following user-written packages need to be installed first by using
install.packages(" ") and then by loading it with library()
# library(nlme)           # It is part of R base distribution
# library(lmerTest)
# library(lme4)
# library(ggeffects)      # It is already installed for Chapter 2
# library(stargazer)      # It is already installed for Chapter 2
# library(texreg)         # It is already installed for Chapter 4


# Import the dataset
library(foreign)
chp10 <- read.dta("C:/CDA/els2002.dta")
chp10 <- chp10[!is.na(chp10$mathach)&!is.na(chp10$efficacy)&!is.na(chp10$
public)&!is.na(chp10$sclimate), ]
chp10$csclimat <- chp10$sclimate-mean(chp10$sclimate, na.rm=TRUE)
chp10$gceffic <- chp10$efficacy-mean(chp10$efficacy, na.rm=TRUE)
attach(chp10)

library(nlme)

# Null model with lme()
m1 <- lme(mathach ~ 1, random = ~1|SCH_ID, na.action="na.omit", method="ML",
data=chp10)
summary(m1)
intervals(m1)
VarCorr(m1)

# Random-intercept model
m2 <- lme(mathach ~ gceffic, random = ~1|SCH_ID, na.action="na.omit",
method="ML", data=chp10)
summary(m2)
intervals(m2)
VarCorr(m2)
```

```
# Model comparison
anova(m1, m2)

# Random-coefficient model
m3 <- lme(mathach ~ gceffic, random = ~gceffic|SCH_ID, na.action="na.omit",
method="ML", data=chp10)
summary(m3)
intervals(m3, which="fixed")
VarCorr(m3)
anova(m2, m3)

# Contextual model with predictor variables in both levels
m4 <- lme(mathach ~ gceffic + public + csclimat, random = ~gceffic|SCH_ID,
na.action="na.omit", method="ML", data=chp10)
summary(m4)
intervals(m4, which="fixed")
VarCorr(m4)
anova(m3, m4)

# Contextual model with cross-level interactions
m5 <- lme(mathach ~ gceffic + public + csclimat + public*gceffic + csclimat*gceffic,
random = ~gceffic|SCH_ID, method="ML", data=chp10)
summary(m5)
intervals(m5, which="fixed")
VarCorr(m5)

anova(m4, m5)

# Presenting the results with stargazer()
library(stargazer)
stargazer(m1, m4, type="text", align=TRUE, out="mul.txt")
stargazer(m1, m4, type="html", align=TRUE, out="mul.htm")

# Presenting the results with texreg()
library(texreg)
screenreg(list(m1, m4))
htmlreg(list(m1, m4), file="mul.doc", doctype=TRUE, html.tag=TRUE, head.tag=TRUE)

# Contextual model with lmer()
library(lmerTest)
library(lme4)
m4.b <- lmer(mathach ~ gceffic + public + csclimat + (gceffic|SCH_ID), data=chp10,
REML = FALSE)
summary(m4.b)
confint(m4.b)
confint(m4.b, method="Wald")

# Marginal effects/Predicted values with ggpredict() in ggeffects
library(ggeffects)
m4.b.pub <- ggpredict(m4.b, terms = "public")
m4.b.pub
as.data.frame(m4.b.pub)
plot(m4.b.pub)

m4.efpub <- ggpredict(m4.b, terms = c("gceffic", "public"))
m4.efpub
plot(m4.efpub)

m4.efpub.r <- ggpredict(m4.b, terms = c("gceffic", "public"), type = "re")
m4.efpub.r
plot(m4.efpub.r)

detach(chp10)
```

Glossary

Cross-level interactions in a two-level model involve including level 2 predictor variables in the equation for the random slope. In other words, there are interaction terms between level 1 and level 2 variables.

Fixed effects are the regression coefficients that estimate the relationships between the predictor variables and the outcome variable from the entire population.

Grand-mean centering involves subtracting the grand mean of the predictor variable from each value of the sample.

In group-mean centering we subtract the group mean from each value, which is the mean of each group or cluster where individuals are nested from each score.

In the random-intercept model the intercept is allowed to vary across groups or clusters.

Multilevel data, nested data, or hierarchical structured data have a data format in which observations at lower levels are nested within a higher level.

Random-coefficient models include both random intercept and random slopes. In addition to the random intercept, level 1 slopes (i.e., the coefficients of the level 1 predictors) can also be specified to be random.

Random effects are the randomly varying parameters across higher level units.

The contextual model is a special case of the random-intercept model or the random-coefficient model when both level 1 and level 2 predictor variables are included.

The intraclass correlation coefficient (ICC) is used to measure the proportion of variance in the outcome variable explained by groups or clusters. It is the ratio of the between-group variance to the total variance.

The purpose of centering is to make the results more interpretative. By subtracting the mean of a predictor variable from each value, we obtain a meaningful zero for the predictor variable.

The unconditional means model or the null model is known as the one-way random-effects ANOVA. Neither level 1 nor level 2 predictor variables are included in the model.

Exercises

Use the ELS:2002 data available at **https://edge.sagepub.com/liu1e** for the following problems. The following variables are used for the multilevel modeling.

`mathach`: mathematics IRT scores of high-school students

`gender`: gender (1 = female; 0 = male)

`byses`: socioeconomic status composite for base-year data.

We will conduct a study investigating the relationships between students' math achievement and the two student-level predictors. The outcome variable is `mathach`, and the predictor variables are `gender` and `byses`. The multilevel modeling for the continuous response variable will be used for data analysis. Answer the following questions or perform the following analyses:

1. Fit an unconditional model and obtain the between-group variance. Compute the ICC. What does it tell us?
2. Fit a two-level, random-intercept model with a random intercept and two student-level predictor variables `gender` and `byses`. Use the grand-mean centering for `byses` before fitting the model.
3. Interpret the fixed effects of the two predictor variables.
4. Conduct a likelihood ratio test comparing the random-intercept model and the unconditional model. Interpret the results.

11

MULTILEVEL MODELING FOR BINARY RESPONSE VARIABLES

OBJECTIVES OF THIS CHAPTER

This chapter introduces multilevel modeling for binary response variables. It starts with an introduction to the model specification, the odds and odds ratios in the model, and a description of the research questions and data. Then several models, from the unconditional (null) model to the random-intercept model and random-coefficient model to the contextual models, are illustrated using R with step-by-step instructions. The two packages, `lme4` and `ordinal`, are used for fitting multilevel logistic regression models. R commands are explained, and the output is interpreted for each model in detail. This chapter also illustrates how the results are displayed in publication-quality tables using R and reported in text. It focuses on model fitting with R, as well as on interpreting and presenting the results. After reading this chapter, you should be able to:

- Determine when multilevel modeling for binary response variables is used.
- Formulate multilevel models.
- Conduct the multilevel modeling analysis for binary response variables using R.
- Interpret the output.
- Compute and plot the predicted probabilities using the `ggeffects()` function.
- Be familiar with model fitting strategies.
- Compare models using the likelihood ratio test.
- Present results in publication-quality tables using R.
- Write the results for publication.

11.1 MULTILEVEL MODELING FOR BINARY OUTCOME VARIABLES

11.1.1 Model Specification

In the previous chapter, we introduced multilevel models for continuous outcome variables. Multilevel modeling can be extended for categorical and count response variables that follow a distribution from the exponential family when data are nested or hierarchical. These models are referred to as multilevel generalized linear models or generalized linear mixed models. In this section, multilevel models for binary response variables (Goldstein, 2003; Hedeker & Gibbon, 1994; Hox, 2010; O'Connell et al., 2008; Raudenbush & Bryk, 2002) are introduced.

Recall that in the single-level binary logistic regression model introduced in Chapter 3, the response variable only has two values (i.e., 1 and 0). Rather than estimating the binary outcome variable directly, we estimate the probability of success or of having an event via a logistic transformation, logit(π) or ln(odds). The logit transformation is defined as follows:

$$\eta = \text{logit}(\pi) = \ln\frac{\pi}{1-\pi} \tag{11.1}$$

The generic form for the multiple logistic regression model is then expressed as:

$$\eta = \ln\left(\frac{\pi(x)}{1-\pi(x)}\right) = \beta_0 + \beta_1 X_1 + \beta_2 X_2 + \ldots + \beta_p X_p \tag{11.2}$$

where β_0 is the intercept; $X_1, X_2, \ldots, X_p$ are the predictor variables, and $\beta_1, \beta_2, \ldots, \beta_p$ are the logit coefficients of these predictors. This model estimates the relationship between a set of independent variables and the binary outcome variable on a scale of the logit or log odds. The link function transforms the original outcome variable Y and connects it with the linear predictor η.

In multilevel modeling for binary response variables, we have a nested data structure that allows us to estimate the relationships between predictor variables from different levels and the binary response variable.

Let us look at a basic two-level logistic regression model with only one predictor variable at each level of the equations. For example, researchers are interested in estimating whether students are proficient in mathematics from a student-level variable, student's major language, and a school-level variable, public or private. The outcome variable is whether students are capable of doing simple arithmetical operations on whole numbers. The level 1 predictor variable is whether a student's native language is English (`stlang`). The level 2 predictor is whether the school is public or private (`public`). Both the intercept and the slope of the level 1 predictor variable `stlang` are allowed to vary randomly across schools.

The level 1 equation is similar to that for the single-level logistic regression model, whereas in multilevel modeling, students are nested with schools so we have a subscript i to represent students and a subscript j to represent schools. As with a single-level logistic regression model, no error term is specified in the level 1 equation:

$$\begin{aligned}
\text{Level 1: } \eta_{ij} &= \text{logit}\left[\pi\left(x_{ij}\right)\right] = \ln\left(\frac{\pi\left(x_{ij}\right)}{1-\pi\left(x_{ij}\right)}\right) = \beta_{0j} + \beta_{1j}\texttt{stlang}_{ij} \\
\text{Level 2: } \beta_{0j} &= \gamma_{00} + \gamma_{01}\texttt{public}_j + u_{0j} \\
\beta_{1j} &= \gamma_{10} + \gamma_{11}\texttt{public}_j + u_{1j}
\end{aligned} \quad (11.3)$$

The equations at level 2 look the same as those for the multilevel models for continuous outcome variables. Both the intercept and the coefficient (i.e., the slope of `stlang`) are allowed to vary randomly across schools at level 2 equations, where they are specified as the outcome variables.

In the level 1 equation, $\text{logit}[\pi(x_{ij})]$ is the logit link for the probability of success for the ith student in the jth school, where $\pi(x_{ij})$ can also be expressed as π_{ij}. In level 2 equations, γ_{00}, γ_{01}, γ_{10}, and γ_{11} are level 2 coefficients, which are the fixed effects. γ_{00} is the overall average logit or log odds of reaching math proficiency across schools. γ_{10} is the overall average slope for `stlang` when the school-level predictor `public` is zero (`public` $= 0$). u_{0j} is an error term associated with the random intercept, the random variation for the intercept, and u_{1j} is the error term associated with the random slope, the random variation for the level 1 slope across schools. γ_{01} is the fixed effect of the level 2 predictor `public` on the log odds of reaching math proficiency; and γ_{11} is the cross-level interaction between `stlang` and `public`.

Estimation of the level 2 random effects results in the variance and covariance matrix, which includes the intercept variance, the slope variance, and the covariance between the intercept and the slope.

11.1.2 Odds and Odds Ratios in Multilevel Logistic Regression Models

As with the single-level conventional binary logistic regression model, the odds and odds ratios in multilevel models for binary response variables can be interpreted in a similar way. The odds of success or of having an event are still the ratio of the probability of success or of having an event (p) to the probability of failure or of not having that event ($1 - p$).

The odds ratios are the exponentiated logit coefficients. They are interpreted as the change in odds of success or of having an event for a one-unit increase in an independent variable. When an odds ratio (OR) is larger than 1, the odds of success or of having an event increases for a one-unit increase in the predictor variable; when an OR is less than 1, the odds of success or of having an event decreases for a one-unit increase in the predictor variable. In addition, an OR of 1 indicates that there is no relationship between the predictor and the odds.

11.2 RESEARCH EXAMPLE AND DESCRIPTION OF THE DATA AND SAMPLE

11.2.1 Research Example and Research Questions

In this research example, using the ELS:2002, researchers are interested in predicting whether high-school students can do simple arithmetical operations on whole numbers from three predictor variables: whether English is students' native language, school type, and school climate. The first predictor is a student-level variable, and the other two are school-level variables. The following research questions will be addressed:

1. Can the probability of being in mathematics proficient level 1 (i.e., doing simple arithmetical operations on whole numbers) be predicted by whether English is the students' native language?
2. Do school characteristics, such as school type and school climate, impact mathematics proficiency?
3. Does students' mathematics proficiency level 1 vary across schools?
4. Does the relationship between students' native language and the probability of being in mathematics proficient level 1 vary across schools?
5. Are there any interaction effects between the two school-level variables (i.e., school type and school climate) and student language? In other words, does the relationship between students' native language and the probability of being in mathematics proficiency level 1 vary across school type and school climate?

11.2.2 Description of the Data and Sample

The ELS:2002 base-year data are used for the following analyses. The variables are listed as follows:

- `Profmath1`: mathematics proficiency level 1
- `stlang`: student language is English (1 = yes, 0 = no)
- `public`: school type (1 = public, 0 = private and others)
- `sclimat`: school climate
- `csclimat`: school climate (grand-mean centered)

For demonstration purposes, the missing values of the variables are deleted. The descriptive statistics are summarized and the school-level variable `sclimat` is grand-mean centered.

11.3 MULTILEVEL MODELING FOR BINARY RESPONSE VARIABLES WITH R

11.3.1 Packages and Functions for Multilevel Modeling for Binary Response Variables in R

The `glmer()` function in the `lme4` package (Bates et al., 2015) and the `clmm()` function in the `ordinal` package (Christensen, 2019) are used for fitting multilevel models with binary outcome variables. The `glmer()` function is the major focus, so it is introduced first.

11.3.2 The `glmer()` Function in the `lme4` Package

The `glmer()` function in the `lme4` package is used for multilevel models with binary outcome variables or mixed-effects logistic regression models. Since `lme4` is a user-written package, you need to install it first by typing `install.packages("lme4")` and then load the package by typing `library(lme4)`. If you have installed the package for the previous chapter, you just need to load it. The model formula syntax of `glmer()` is similar to that of `lme()` introduced in the last chapter, but they are slightly different in specifying the random effects. The model formula in `glmer()` also includes two components connected by a plus (+) sign. The first component specifies the dependent variable and the predictor variable(s), which are separated by the tilde (~).When there are multiple predictor variables, they are connected by plus (+) symbols. This is the fixed effects part of the model. Next, the random effects part of the model is specified within parentheses, which is different from the `random=` argument specified in `lme()` for linear mixed effects models. Within parentheses, a predictor variable or a list of predictor variables that have random coefficients is specified first, followed by a vertical line (|), and then, an identifier variable at a higher level as the grouping variable. In addition to the model formula, we also need to specify the `family = binomial` argument since the outcome variable is dichotomous. For example, the command `glmer(y ~ x + (1|schid), family = binomial, nAGQ = 7)` tells R to fit a multilevel model to estimate a binary outcome variable `y` on a predictor variable `x` with random intercepts varying across schools by specifying `(1|schid)`, where `1` indicates the random intercept on the left of the vertical line (|) and `schid` is the identifier variable on the right. The `nAGQ = 7` option specifies the number of quadrature points for adaptive Gauss-Hermite quadrature. The default number is 1, which is used for the Laplace approximation. The maximum number of quadrature points is 25 in the current version of the `lme4` package (Version 1.1-27.1). It takes longer to fit the model with the increase of the integration points, so reducing the number of `nAGQ` can speed up the computation process. As of the writing, the adaptive Gauss-Hermite quadrature method is currently only for models with a single random-effects term, such as the null model or the random-intercept model. When there is more than a single random effects term in the model, only the Laplace approximation with `nAGQ = 1` is available. For more details on how to use this function, type `help(glmer)` in the command prompt after loading the `lme4` package.

11.3.3 Unconditional Model or Null Model (Model 1)

The unconditional model, or null model, for binary response variables includes no predictor variable at any level. This model estimates the overall probability of success or of having an event and the variability in the probability of success between groups (e.g., schools). It can be expressed as follows:

$$\begin{aligned} \text{Level 1: } \operatorname{logit}\left[\pi\left(x_{ij}\right)\right] &= \beta_{0j} \\ \text{Level 2: } \beta_{0j} &= \gamma_{00} + u_{0j} \end{aligned} \quad (11.4)$$

Recall that in binary logistic regression logit(π) is the logistic transformation of the probability of success of an event occurrence. In multilevel logistic regression models, logit(π_{ij}) is the logit link for the probability of success for the *i*th student in the *j*th school. It can be expressed as $\ln\frac{\pi_{ij}}{1-\pi_{ij}}$. In the level 2 equation, γ_{00} is the overall average logit or log odds of reaching math proficiency across schools and u_{0j} is an error term at the school level.

The `glmer()` Command and Output

The command `melr.1 <- glmer(Profmath1 ~ + (1|SCH_ID), data = chp11, family = binomial, nAGQ = 7)` is used to fit the unconditional model (model 1). This command tells R to fit a mixed logistic regression model for the binary response variable `Profmath1`. Since this is the unconditional model, no predictors are specified on the right-hand side of the tilde (~). We can specify either no predictors or `1` as the intercept. The random effects part of the model specifies the random effects at the school level within parentheses, `(1|SCH_ID)`, where `1` indicates the random intercept and `SCH_ID` is the identifier variable. No random coefficients for any predictor variables are specified in this model. The model is assigned to an object named `melr.1`. Next, the `summary()` function is used to display the output of the model.

```
> library(lme4)
> # Unconditional model/Null model with glmer()
> melr.1 <- glmer(Profmath1 ~ + (1|SCH_ID), data = chp11, family = binomial, nAGQ = 7)
> summary(melr.1)
Generalized linear mixed model fit by maximum likelihood (Adaptive
  Gauss-Hermite Quadrature, nAGQ = 7) [glmerMod]
 Family: binomial  ( logit )
Formula: Profmath1 ~ + (1 | SCH_ID)
   Data: chp11

   AIC      BIC     logLik   deviance   df.resid
4862.1   4877.0   -2429.0     4858.1      12791

Scaled residuals:
    Min       1Q    Median      3Q      Max
-5.8103   0.1333   0.1439   0.2349   0.6861
```

```
Random effects:
 Groups Name          Variance  Std.Dev.
 SCH_ID (Intercept)   1.334     1.155
Number of obs: 12793, groups: SCH_ID, 619

Fixed effects:
            Estimate  Std. Error  z value  Pr(>|z|)
(Intercept)  3.47168     0.08592    40.41   <2e-16 ***
---
Signif. codes: 0 '***' 0.001 '**' 0.01 '*' 0.05 '.' 0.1 ' ' 1
```

Interpreting the Output

The R output produced by the `glmer()` function is similar to that by the `lmer()` function introduced in the previous section. It includes the note for the model estimation method, the distribution family, the model formula, and the data, the model fit statistics such as the AIC, the BIC, the log-likelihood value, and the deviance with its degrees of freedom, the scaled residuals, the random effects, the number of observations and groups, and the fixed effects.

The maximum likelihood estimation with numerical integration is used to fit the multilevel logistic regression model. The numerical integration method in the model is the adaptive Gauss–Hermite quadrature method with 7 integration points. When `nAGQ = 1`, the method is the Laplace approximation. With the increase of the integration points, the estimation gets more accurate, but it is time consuming to compute and may have converging issues.

The random effects section provides the variance and the standard deviation for the intercepts across all schools. This is the between-school variance (τ_{00}), which is 1.334. The ICC in multilevel logistic regression is also defined as the proportion of total variance in the outcome variable explained by the between-group variance. Since the variance of the logistic distribution is $\pi^2/3 = 3.29$, we compute ICC $= \tau_{00}/(\tau_{00} + 3.29) = 1.334/(1.334 + 3.29) = .288$, which indicates that 28.8% of the total variance is accounted for by schools in level 2. This is a large ICC, so we can justify that using multilevel logistic regression is appropriate for the analysis.

The fixed effects section displays the estimate for the intercept, its standard error, the Wald z statistic, and the associated p value. The intercept γ_{00} (labeled `Intercept`) is 3.472, which is significant (`z = 40.41, P(>|z|) < 2e-16`). This means that the overall average logit or log odds of being proficient in mathematics across all schools is 3.472. The estimated logit can be transformed to the odds ratio, which will be explained next.

The 95% confidence interval of the estimate is computed as follows.

```
> coef <- fixef(melr.1)
> se <- sqrt(diag(vcov(melr.1)))
> LL <- fixef(melr.1)-1.96*se
> UL <- fixef(melr.1)+1.96*se
> coeftab <- cbind(coef, LL, UL)
> coeftab
                coef        LL        UL
(Intercept)  3.471677  3.303271  3.640084
```

The same results can be obtained with the `confint(melr.1, method = "Wald")` command.

```
> confint(melr.1, method = "Wald")
                  2.5 %      97.5 %
.sig01              NA          NA
(Intercept)   3.303274    3.640081
```

With `exp(fixef(melr.1))`, we can get the odds ratios for the predictor variables. Since this is the unconditional model, only the odds ratio for the intercept is reported. OR = 32.191, which is the overall average odds of being proficient in math. The confidence interval for the odds ratio can be obtained with the `exp(coeftab)` command.

```
> exp(fixef(melr.1))
(Intercept)
   32.19069
> exp(coeftab)
                  coef          LL          UL
(Intercept)   32.19069    27.20146    38.09504
```

11.3.4 Random-Intercept Model (Model 2)

Next, we include the predictor variable, whether student's native language is English (`stlang`), to the level 1 equation. The model is referred to as the random-intercept model since the intercept is allowed to vary freely across schools. This model can be expressed as follows:

$$\begin{aligned} \text{Level 1: } \operatorname{logit}\left[\pi(x_{ij})\right] &= \beta_{0j} + \beta_{1j}\texttt{stlang}_{ij} \\ \text{Level 2: } \beta_{0j} &= \gamma_{00} + u_{0j} \\ \beta_{1j} &= \gamma_{10} \end{aligned} \tag{11.5}$$

The level 1 equation includes the predictor variable, whether student's native language is English (`stlang`), and the binary response variable, whether students are capable of doing simple arithmetical operations on whole numbers (`Profmath1`). The level 2 equation for the intercept (β_{0j}) is specified to be random, whereas the slope for `stlang` (β_{1j}) is constrained to be fixed.

The `glmer()` Command and Output

The command `melr.2 <- glmer(Profmath1 ~ stlang + (1|SCH_ID), data = chp11, family = binomial, nAGQ = 7)` is used to fit the random-intercept model (model 2) after a predictor variable `stlang` is added to the fixed part of the model immediately after the tilde (~). Just like the unconditional model, no random coefficients are specified in the random part of the model. Again, the `summary()` function is used to display the output of the fitted random-intercept model named `melr.2`. The following output is displayed.

```
> # Random-intercept model
> melr.2 <- glmer(Profmathl ~ stlang + (1|SCH_ID), data = chp11, family = binomial, nAGQ = 7)
> summary(melr.2)
Generalized linear mixed model fit by maximum likelihood (Adaptive
  Gauss-Hermite Quadrature, nAGQ = 7) [glmerMod]
 Family: binomial  ( logit )
Formula: Profmathl ~ stlang + (1 | SCH_ID)
   Data: chp11


   AIC       BIC     logLik    deviance    df.resid
4814.4    4836.8    -2404.2      4808.4       12790


Scaled residuals:
    Min         1Q     Median         3Q        Max
-6.5650    0.1298    0.1523     0.2298     0.9754


Random effects:
  Groups Name           Variance    Std.Dev.
  SCH_ID (Intercept)       1.191       1.091
Number of obs: 12793, groups:  SCH_ID, 619


Fixed effects:
              Estimate    Std. Error     z value     Pr(>|z|)
(Intercept)     2.8094        0.1180      23.805      < 2e-16 ***
stlangYes       0.7808        0.1070       7.294     3.02e-13 ***
---
Signif. codes:  0 '***' 0.001 '**' 0.01 '*' 0.05 '.' 0.1 ' ' 1
Correlation of Fixed Effects:
             (Intr)
stlangYes    -0.709
```

Interpreting the Output

In the random effects section, the row (labeled `Intercept`) reports that the intercept variance (τ_{00}) is 1.191. In the fixed effects section, the estimation table reports the coefficients for `stlang` and the intercept. The intercept γ_{00} (labeled `Intercept`) = 2.809. It is significantly different from 0 (`z = 23.805`, `P(>|z|) = 2e-16`), which means that the estimated logit or log odds of being proficient in doing simple arithmetical operations on whole numbers for students whose native language is not English (`stlang` = 0) in a typical school is 2.809. The logit coefficient for `stlang` (γ_{10}) = .781, $z = 7.294$, $p < .001$, which indicates that the logit or log odds of being proficient in math for students with English as their native language is .781 times as great as that for those students whose native language is not English.

The 95% confidence intervals of the estimates are computed as follows.

```
> coef <- fixef(melr.2)
> se <- sqrt(diag(vcov(melr.2)))
> LL <- fixef(melr.2)-1.96*se
> UL <- fixef(melr.2)+1.96*se
> coeftab <- cbind(coef, LL, UL)
> coeftab
                   coef          LL          UL
(Intercept)   2.8093522   2.5780426   3.0406618
stlangYes     0.7807545   0.5709418   0.9905672
```

The same results can be obtained with `confint(melr.2, method = "Wald")`.

```
> confint(melr.2, method = "Wald")
                  2.5 %      97.5 %
.sig01               NA          NA
(Intercept)   2.5780469   3.0406576
stlangYes     0.5709457   0.9905633
```

With the `exp(fixef(melr.2))` command, we can get the odds ratios for the predictor variables. Interpretation of odds ratios in multilevel logistic regression is the same as that for binary logistic regression. OR = 2.183. It can be interpreted as meaning that the odds of being proficient in math for students whose native language is English are 2.183 times as large as the odds for those whose native language is not English.

The odds ratio for the intercept is 16.599, which is the overall average odds of being proficient in math. The confidence interval for the odds ratio can be obtained with `exp(coeftab)`.

```
> exp(fixef(melr.2))
(Intercept)   stlangYes
  16.599162    2.183119
> exp(coeftab)
                   coef          LL          UL
(Intercept)   16.599162   13.171332   20.919083
stlangYes      2.183119    1.769933    2.692761
```

Likelihood Ratio Test Comparing the Unconditional Model and the Random-Intercept Model

```
> anova(melr.1, melr.2)
Data: chp11
Models:
melr.1: Profmath1 ~ +(1 | SCH_ID)
melr.2: Profmath1 ~ stlang + (1 | SCH_ID)
         npar     AIC      BIC    logLik   deviance    Chisq   Df   Pr(>Chisq)
melr.1      2  4862.1   4877.0   -2429.0     4858.1
melr.2      3  4814.4   4836.8   -2404.2     4808.4   49.611    1   1.875e-12 ***
---
Signif. codes: 0 '***' 0.001 '**' 0.01 '*' 0.05 '.' 0.1 ' ' 1
```

The likelihood ratio test is used to compare the unconditional model and the random-intercept model. The `anova(melr.1, melr.2)` command compares the log-likelihood statistics of these two models: $\chi^2_{(1)} = 49.611, p < .001$, which indicates that the random-intercept model fits the data better than the unconditional model, so we are in favor of the former model with a varying intercept.

11.3.5 Random-Coefficient Model With a Level-1 Variable (Model 3)

Just like the random-coefficient model for linear mixed modeling, we can also fit a random-coefficient logistic regression model. Although the expressions for the level 1 equation are different between the two models, both level 2 equations for the random intercept and random coefficient are the same. This model is referred to as the random-coefficient model since both the intercept and coefficient in the level 1 equation are allowed to vary across schools at level 2. This model also estimates variance–covariance components, including variances in intercepts and coefficients/slopes and the covariance between them. The random coefficient logistic regression model can be expressed as follows:

$$\begin{aligned} \text{Level 1: } \operatorname{logit}\left[\pi\left(x_{ij}\right)\right] &= \beta_{0j} + \beta_{1j}\texttt{stlang}_{ij} \\ \text{Level 2: } \beta_{0j} &= \gamma_{00} + u_{0j} \\ \beta_{1j} &= \gamma_{10} + u_{1j} \end{aligned} \tag{11.6}$$

The level 1 equation includes the predictor variable, students' native language (`stlang`), and the binary response variable (`Profmath1`). The level 2 equations for the intercept and the slope for `stlang` are both specified to be random with level 2 residuals u_{0j} and u_{1j}, respectively.

The `glmer()` Command and Output

The command `melr.3 <- glmer(Profmath1 ~ stlang + (stlang| SCH_ID), data = chp11, family = binomial, nAGQ = 1)` is used to fit the random-coefficient model (model 3). The fixed part of the model is the same as that of the random-intercept model. Different from the random-intercept model, the predictor variable `stlang` is specified in the random part of the model, `(stlang| SCH_ID)`, which indicates that the slope of the level 1 predictor `stlang` is allowed to vary randomly across schools. The `nAGQ = 1` argument specifies that the number of quadrature points is 1 since only the Laplace approximation is available in the random-coefficient model with more than one random-effects term. An error message will be displayed in the output if `nAGQ` is set more than 1. The next command `summary(melr.3)` displays the detailed output of the fitted model.

```
> # Random-coefficient model
> melr.3 <- glmer(Profmath1 ~ stlang + (stlang|SCH_ID), data = chp11, family = binomial,
nAGQ = 1)
> summary(melr.3)
Generalized linear mixed model fit by maximum likelihood (Laplace
  Approximation) [glmerMod]
 Family: binomial  ( logit )
Formula: Profmath1 ~ stlang + (stlang | SCH_ID)
   Data: chp11


   AIC        BIC      logLik     deviance     df.resid
4804.6    4841.9     -2397.3       4794.6        12788

Scaled residuals:
    Min         1Q     Median          3Q        Max
-6.1628     0.1237     0.1479     0.2217     0.8101

Random effects:
 Groups     Name            Variance     Std.Dev.     Corr
 SCH_ID     (Intercept)     1.2936       1.1374
            stlangYes       0.7614       0.8726       -0.34
Number of obs: 12793, groups:  SCH_ID, 619

Fixed effects:
                 Estimate    Std. Error     z value     Pr(>|z|)
(Intercept)        2.7939        0.1586      17.614      < 2e-16 ***
stlangYes          0.8753        0.1783       4.909     9.16e-07 ***
---
Signif. codes:  0 '***' 0.001 '**' 0.01 '*' 0.05 '.' 0.1 ' ' 1

Correlation of Fixed Effects:
             (Intr)
stlangYes    -0.833
```

Interpreting the Output

Regarding the random effects in the model, the intercept variance (τ_{00}) is 1.294 and the slope variance for `stlang` is .761. The correlation between the intercept and the `stlang` slope is −.34. The covariance between two variables equals the product of the correlation coefficient and the two standard deviations. To compute the covariance between the intercept and the `stlang` slope, we use the following equation: $1.137*.873*(-.34) = -.337$ since it equals the product of the correlation and the two standard deviations.

The estimated logit coefficients for the intercept and `stlang` in the output are similar to those for the previous random-intercept model. They can also be interpreted in a similar way.

The 95% confidence intervals of the estimates are computed as follows.

```
> coef <- fixef(melr.3)
> se <- sqrt(diag(vcov(melr.3)))
> LL <- fixef(melr.3)-1.96*se
> UL <- fixef(melr.3)+1.96*se
> coeftab <- cbind(coef, LL, UL)
> coeftab
                 coef          LL          UL
(Intercept)  2.7939497   2.4830599   3.104839
stlangYes    0.8752682   0.5257977   1.224739
```

The same results can be obtained with `confint(melr.3, method = "Wald")`.

```
> confint(melr.3, method = "Wald")
                 2.5 %      97.5 %
.sig01              NA          NA
.sig02              NA          NA
.sig03              NA          NA
(Intercept)  2.4830656    3.104834
stlangYes    0.5258041    1.224732
```

With `exp(fixef(melr.3))`, we request the odds ratios for the intercept and predictor variables. The odds ratio for `stlang` is 2.400, which indicates that the odds of being proficient in math for students whose native language is English are 2.400 times as great as the odds for those whose native language is not English. In other words, being students whose native language is English increases the odds of being proficient in math by a factor of 2.400.

The odds ratio for the intercept is 16.345, which is the overall average odds of being proficient in math when the predictor variable is held constant at 0. The confidence interval for the odds ratio can be obtained with the `exp(coeftab)` command.

```
> exp(fixef(melr.3))
(Intercept)    stlangYes
  16.345451     2.399519
> exp(coeftab)
                 coef          LL          UL
(Intercept)  16.345451   11.977859   22.305636
stlangYes     2.399519    1.691808    3.403277
```

Model Comparisons Using the Likelihood Ratio Test

```
> anova(melr.2, melr.3)
Data: chp11
Models:
melr.2: Profmath1 ~ stlang + (1 | SCH_ID)
melr.3: Profmath1 ~ stlang + (stlang | SCH_ID)
        npar    AIC     BIC    logLik  deviance   Chisq  Df  Pr(>Chisq)
melr.2     3  4814.4  4836.8  -2404.2    4808.4
melr.3     5  4804.6  4841.9  -2397.3    4794.6  13.818   2   0.0009988 ***
---
Signif. codes: 0 '***' 0.001 '**' 0.01 '*' 0.05 '.' 0.1 ' ' 1
```

The likelihood ratio test compares the random-intercept model (model 2) and the random-coefficient model (model 3). The `anova(melr.2, melr.3)` command compares the log-likelihood statistics of these two models: $\chi^2_{(2)} = 13.818$, $p < .001$, which indicates that we are in favor of the random-coefficient model, so having a random coefficient in model 3 is justified.

11.3.6 Contextual Model With Level-2 Variables (Model 4)

Just like the contextual model in the linear mixed modeling, the contextual model for the binary response variable also includes both level 1 and level 2 predictor variables. The two level 2 equations specify the random intercept and random slope. In this example, two school-level variables `public` and `csclimat` are added to the level 2 equation for the random intercept. In the level 2 equation for the random slope, no school-level variables are specified. This model is expressed as follows:

$$\begin{aligned} \text{Level 1: } & \text{logit}\left[\pi(x_{ij})\right] = \beta_{0j} + \beta_{1j}\texttt{stlang}_{ij} \\ \text{Level 2: } & \beta_{0j} = \gamma_{00} + \gamma_{01}\texttt{public}_j + \gamma_{02}\texttt{csclimat}_j + u_{0j} \\ & \beta_{1j} = \gamma_{10} + u_{1j} \end{aligned} \tag{11.7}$$

The `glmer()` Command and Output

The command `melr.4 <- glmer(Profmath1 ~ stlang + csclimat + public + (stlang|SCH_ID), data = chp11, family = binomial, nAGQ = 1)` is used to fit the contextual model (model 4). With other parts of the equations the same as those of the random-coefficient model (model 3), the fixed part of the model includes both the level 1 predictor `stlang` and the two school-level predictor variables `public` and `csclimat`. Since only the Laplace approximation is available for models with more than one random effects term, we specify the `nAGQ = 1` option. Again, the results of the fitted model are displayed with the `summary(melr.4)` command.

```
> # Contextual model/Random-coefficient model with predictor variables in both levels
> melr.4 <- glmer(Profmath1 ~ stlang + csclimat + public + (stlang|SCH_ID), data = chp11,
family = binomial, nAGQ = 1)
> summary(melr.4)
Generalized linear mixed model fit by maximum likelihood (Laplace
  Approximation) [glmerMod]
 Family: binomial  ( logit )
Formula: Profmath1 ~ stlang + csclimat + public + (stlang | SCH_ID)
   Data: chp11


   AIC       BIC      logLik    deviance    df.resid
4698.9    4751.1    -2342.5      4684.9       12786

Scaled residuals:
     Min        1Q    Median        3Q       Max
-10.2502    0.1153    0.1655    0.2273    0.8859

Random effects:
 Groups          Name    Variance    Std.Dev.     Corr
 SCH_ID    (Intercept)     1.2409      1.1140
             stlangYes     0.9728      0.9863    -0.59
Number of obs: 12793, groups:  SCH_ID, 619

Fixed effects:
                Estimate   Std. Error    z value    Pr(>|z|)
(Intercept)      3.98201      0.24574     16.204     < 2e-16 ***
stlangYes        0.68224      0.16459      4.145    3.40e-05 ***
csclimat         0.60673      0.09461      6.413    1.43e-10 ***
publicpublic    -1.26803      0.21425     -5.918    3.25e-09 ***
---
Signif. codes:  0 '***' 0.001 '**' 0.01 '*' 0.05 '.' 0.1 ' ' 1

Correlation of Fixed Effects:
               (Intr)    stlngY    csclmt
stlangYes      -0.530
csclimat       -0.024    -0.090
publicpublc    -0.787     0.012     0.213
```

Interpreting the Output

In the random effects section, the residual variance in the intercepts (τ_{00}) is 1.241 and the slope variance for `stlang` is .973. The covariance between the intercept and the `stlang` slope equals the product of the correlation coefficient and the two standard deviations and is computed as follows: $1.114{*}.986{*}(-.59) = -.648$.

The fixed-effects section reports the estimated logit coefficients for both level 1 and level 2 predictor variables and the intercept. The logit coefficients for all predictor variables are significant. The 95% confidence intervals of the estimates are computed as follows.

```
> coef <- fixef(melr.4)
> se <- sqrt(diag(vcov(melr.4)))
> LL <- fixef(melr.4)-1.96*se
> UL <- fixef(melr.4)+1.96*se
> coeftab <- cbind(coef, LL, UL)
> coeftab
                    coef          LL          UL
(Intercept)    3.9820131   3.5003542   4.4636721
stlangYes      0.6822375   0.3596484   1.0048266
csclimat       0.6067300   0.4212852   0.7921748
publicpublic  -1.2680262  -1.6879607  -0.8480917
```

The same results can be obtained with the `confint(melr.4, method = "Wald")` command.

```
> confint(melr.4, method = "Wald")
                   2.5 %      97.5 %
.sig01                NA          NA
.sig02                NA          NA
.sig03                NA          NA
(Intercept)    3.5003630   4.4636633
stlangYes      0.3596544   1.0048207
csclimat       0.4212886   0.7921714
publicpublic  -1.6879530  -0.8480995
```

We use the `exp(fixef(melr.4))` command to request the odds ratios for the predictor variables. The confidence intervals for the odds ratios can be obtained with `exp(coeftab)`.

```
> exp(fixef(melr.4))
 (Intercept)    stlangYes     csclimat  publicpublic
  53.6248803    1.9782993    1.8344230     0.2813865
> exp(coeftab)
                    coef          LL          UL
(Intercept)   53.6248803  33.1271823  86.8056861
stlangYes      1.9782993   1.4328256   2.7314336
csclimat       1.8344230   1.5239189   2.2081935
publicpublic   0.2813865   0.1848962   0.4282313
```

The odds ratio for `stlang` is 1.978, which indicates that the odds of being proficient in math for students whose native language is English are 1.978 times as great as the odds for those whose native language is not English.

The odds ratio for `public` is .281, which indicates that the odds of being proficient in math for students in public schools are .281 times the odds for those in private schools. In other words, being students in public schools decreases the odds of being proficient in math by 71.9%.

The odds ratio for csclimat is 1.834, which indicates that a one-unit increase in school climate corresponds to an increase of 1.834 points in the odds of being proficient in math. In other words, being students in schools with a more positive school climate increases the odds of being proficient in math.

Model Comparisons Using the Likelihood Ratio Test

```
> anova(melr.3, melr.4)
Data: chp11
Models:
melr.3: Profmath1 ~ stlang + (stlang | SCH_ID)
melr.4: Profmath1 ~ stlang + csclimat + public + (stlang | SCH_ID)
        npar    AIC    BIC   logLik  deviance   Chisq  Df  Pr(>Chisq)
melr.3     5 4804.6 4841.9  -2397.3    4794.6
melr.4     7 4698.9 4751.1  -2342.5    4684.9  109.71   2  < 2.2e-16 ***
---
Signif. codes: 0 '***' 0.001 '**' 0.01 '*' 0.05 '.' 0.1 ' ' 1
```

The likelihood ratio test compares the contextual model with both the level 1 and level 2 predictor variables (model 4: ranslop2) with the random-coefficient model with a level 1 predictor only (model 3: ranslop1). With the anova(melr.3, melr.4) command, we compare the log-likelihood statistics of these two models. It yields a value of $\chi^2_{(2)} = 109.71$, $p < .001$, which indicates that the contextual model fits the data better.

11.3.7 Contextual Model With Cross-Level Interactions (Model 5)

The contextual model with cross-level interactions is used to investigate whether there are interaction effects between level 1 and level 2 predictor variables. For example, if we are interested in whether the two school-level variables public and csclimat interact with the level 1 predictor stlang, we can include two interaction terms in the model. The two interaction terms need to be created for the model equation. One interaction public:stlang is created for two categorical variables stlang and public; the other interaction csclimat:stlang is for the categorical variable stlang and the continuous variable csclimat. This model can be expressed as follows:

$$\begin{aligned} \text{Level 1: } \operatorname{logit}\left[\pi(x_{ij})\right] &= \beta_{0j} + \beta_{1j}\texttt{stlang}_{ij} \\ \text{Level 2: } \beta_{0j} &= \gamma_{00} + \gamma_{01}\texttt{public}_j + \gamma_{02}\texttt{csclimat}_j + u_{0j} \\ \beta_{1j} &= \gamma_{10} + \gamma_{11}\texttt{public}_j + \gamma_{12}\texttt{csclimat}_j + u_{1j} \end{aligned} \tag{11.8}$$

The glmer() Command and Output

The melr.5 <- glmer(Profmath1 ~ stlang + csclimat + public + public:stlang + csclimat:stlang + (stlang|SCH_ID), data = chp11, family = binomial, nAGQ = 1) command is used to fit the contextual model with cross-level interactions (model 5). With other parts of the equations the same as those of the contextual model (model 4), the fixed part of the model includes two extra variables `public:stlang` and `csclimat:stlang`, which are the cross-level interactions. The `summary(melr.5)` command displays the output of the fitted model.

```
> # Contextual model with cross-level interactions
> melr.5 <- glmer(Profmath1 ~ stlang + csclimat + public + public:stlang +
csclimat:stlang + (stlang|SCH_ID), data = chp11, family = binomial, nAGQ = 1)
Warning message:
In checkConv(attr(opt, "derivs"), opt$par, ctrl = control$checkConv, :
  Model failed to converge with max|grad| = 0.136969 (tol = 0.002, component 1)
> summary(melr.5)
Generalized linear mixed model fit by maximum likelihood (Laplace
  Approximation) [glmerMod]
 Family: binomial  ( logit )
Formula:
Profmath1 ~ stlang + csclimat + public + public:stlang + csclimat:stlang +
    (stlang | SCH_ID)
  Data: chp11


   AIC       BIC      logLik    deviance    df.resid
4701.0    4768.1    -2341.5     4683.0       12784


Scaled residuals:
     Min         1Q     Median        3Q        Max
-10.4073    0.1117    0.1662    0.2295    0.7972


Random effects:
 Groups            Name      Variance     Std.Dev.     Corr
 SCH_ID      (Intercept)       1.0483       1.0239
               stlangYes       0.7809       0.8837    -0.49
Number of obs: 12793, groups:  SCH_ID, 619


Fixed effects:
                            Estimate    Std. Error     z value    Pr(>|z|)
(Intercept)                   4.0262        0.4575       8.800    < 2e-16 ***
stlangYes                     0.6182        0.4993       1.238     0.21565
csclimat                      0.4202        0.1522       2.760     0.00577 **
publicpublic                 -1.4172        0.4555      -3.111     0.00186 **
stlangYes:publicpublic        0.1996        0.4981       0.401     0.68867
stlangYes:csclimat            0.2550        0.1696       1.504     0.13267
---
Signif. codes:  0 '***' 0.001 '**' 0.01 '*' 0.05 '.' 0.1 ' ' 1


Correlation of Fixed Effects:
               (Intr)     stlngY     csclmt     pblcpb     stlngYs:p
stlangYes      -0.890
csclimat       -0.009     -0.004
publicpublc    -0.945      0.838      0.128
stlngYs:pbl     0.837     -0.940     -0.100     -0.881
stlngYs:csc    -0.001     -0.013     -0.787     -0.101         0.145
convergence code: 0
Model failed to converge with max|grad| = 0.136969 (tol = 0.002, component 1)
```

Interpreting the Output

In the output, let us look at the coefficients for these two interaction terms in the fixed effects section. The coefficient for `public:stlang` (γ_{11}) = .200, z = .401, $p > .05$, and the coefficient for `csclimat:stlang` (γ_{12}) = .255, z = 1.504, $p > .05$. So both interactions are not significant, which indicates that the effect of `stlang` does not differ in public or private schools. In addition, the effect of `stlang` is not influenced by school climate, either. How can a cross-level effect be interpreted if it is identified in the model? For example, if `csclimat:stlang` is significant, then this effect can be interpreted as follows: The relationship between students' native language and the probability of being in mathematics proficiency level 1 varies across school climate.

Since the variance and covariance components for the random effects for model 5 are similar to those for model 4, their interpretation is omitted here.

The 95% confidence intervals of the estimates for the fixed effects are computed as follows.

```
> coef <- fixef(melr.5)
> se <- sqrt(diag(vcov(melr.5)))
> LL <- fixef(melr.5)-1.96*se
> UL <- fixef(melr.5)+1.96*se
> coeftab <- cbind(coef, LL, UL)
> coeftab
                              coef           LL          UL
(Intercept)              4.0262165   3.12943563   4.9229975
stlangYes                0.6182003  -0.36039229   1.5967929
csclimat                 0.4202059   0.12183283   0.7185789
publicpublic            -1.4172172  -2.30996835  -0.5244661
stlangYes:publicpublic   0.1995631  -0.77666953   1.1757957
stlangYes:csclimat       0.2549905  -0.07738362   0.5873646
```

The same results can be obtained with the `confint(melr.5, method = "Wald")` command.

```
> confint(melr.5, method = "Wald")
                               2.5 %       97.5 %
.sig01                            NA           NA
.sig02                            NA           NA
.sig03                            NA           NA
(Intercept)               3.12945211    4.9229810
stlangYes                -0.36037431    1.5967749
csclimat                  0.12183831    0.7185734
publicpublic             -2.30995195   -0.5244825
stlangYes:publicpublic   -0.77665159    1.1757777
stlangYes:csclimat       -0.07737751    0.5873585
```

We use the `exp(fixef(melr.5))` command to request the odds ratios for the predictor variables. The confidence interval for the odds ratio can be obtained with `exp(coeftab)`.

```
> exp(fixef(melr.5))
          (Intercept)                  stlangYes              csclimat
           56.0484530                  1.8555855             1.5222749
         publicpublic     stlangYes:publicpublic    stlangYes:csclimat
            0.2423876                  1.2208692             1.2904493
> exp(coeftab)
                                coef            LL            UL
(Intercept)               56.0484530   22.86107384   137.4138901
stlangYes                  1.8555855    0.69740269     4.9371728
csclimat                   1.5222749    1.12956525     2.0515157
publicpublic               0.2423876    0.09926439     0.5918713
stlangYes:publicpublic     1.2208692    0.45993526     3.2407204
stlangYes:csclimat         1.2904493    0.92553473     1.7992404
```

Model Comparison Using the Likelihood Ratio Test

```
> anova(melr.4, melr.5)
Data: chp11
Models:
melr.4: Profmath1 ~ stlang + csclimat + public + (stlang | SCH_ID)
melr.5: Profmath1 ~ stlang + csclimat + public + public:stlang + csclimat:stlang +
melr.5:     (stlang | SCH_ID)
        npar     AIC     BIC   logLik   deviance    Chisq   Df   Pr(>Chisq)
melr.4     7  4698.9  4751.1  -2342.5     4684.9
melr.5     9  4701.0  4768.1  -2341.5     4683.0   1.8989    2       0.3869
```

The likelihood ratio test is again used to compare the contextual models with or without interactions (model 5 vs. model 4). The `anova(melr.4, melr.5)` command compares the log-likelihood statistics of these two models: $\chi^2_{(2)} = 1.899, p > .05$, which indicates that we are in favor of the null hypothesis. We can conclude that the contextual model with cross-level interactions does not fit the data better the contextual model without interactions. Therefore, the interaction terms can be removed from model 5.

We use the `anova(melr.1, melr.2, melr.3, melr.4, melr.5)` command to compare the log-likelihood statistics of all the five models fitted above. The output is as follows.

```
> # Model comparisons
> anova(melr.1, melr.2, melr.3, melr.4, melr.5)
Data: chp11
Models:
melr.1: Profmath1 ~ +(1 | SCH_ID)
melr.2: Profmath1 ~ stlang + (1 | SCH_ID)
melr.3: Profmath1 ~ stlang + (stlang | SCH_ID)
melr.4: Profmath1 ~ stlang + csclimat + public + (stlang | SCH_ID)
melr.5: Profmath1 ~ stlang + csclimat + public + public:stlang + csclimat:stlang +
melr.5:     (stlang | SCH_ID)
       npar    AIC    BIC  logLik deviance    Chisq Df Pr(>Chisq)
melr.1    2 4862.1 4877.0 -2429.0   4858.1
melr.2    3 4814.4 4836.8 -2404.2   4808.4  49.6109  1  1.875e-12 ***
melr.3    5 4804.6 4841.9 -2397.3   4794.6  13.8179  2  0.0009988 ***
melr.4    7 4698.9 4751.1 -2342.5   4684.9 109.7084  2  < 2.2e-16 ***
melr.5    9 4701.0 4768.1 -2341.5   4683.0   1.8989  2  0.3869458
---
Signif. codes: 0 '***' 0.001 '**' 0.01 '*' 0.05 '.' 0.1 ' ' 1
```

The results of the likelihood ratio test indicate that the contextual model without cross-level interactions (`melr.4`) fits the data best among all the five models. The AIC and BIC statistics of the fourth model are also the smallest, which supports our conclusion.

11.3.8 Interpreting the Predicted Probabilities With the `ggpredict()` Function in the `ggeffects` Package

By using the `ggpredict()` function in the `ggeffects` package (Lüdecke, 2018b), we can compute the predicted probabilities for predictor variables at specified values in the multilevel logistic regression model. Since the interaction terms have been removed from model 5, the `ggpredict()` function is executed for the contextual model without interactions (model 4).

We first load the package with `library(ggeffects)` since it has been installed in previous chapters. The command `melr.4.pub <- ggpredict(melr.4, terms = "public")` tells R to compute the predicted probabilities using the `ggpredict()` function. The argument inside the function includes the estimated model, `melr.4`, and the `terms = "public"` option which specifies the predictor variable `public`. The other predictor variables are held constant. The output is assigned to an object named `melr.4.pub`. The `as.data.frame()` function is used to request the standard errors.

```
> # Marginal effects/Predicted probabilities with ggpredict()
> library(ggeffects)
> melr.4.pub <- ggpredict(melr.4, terms = "public")
> melr.4.pub
# Predicted probabilities of Profmath1

public                     | Predicted |       95% CI
-----------------------------------------------------
catholic or other private  |      0.98 | [0.97, 0.99]
public                     |      0.94 | [0.92, 0.95]
```

```
Adjusted for:
*   stlang =   No
* csclimat = 0.00
*   SCH_ID = 0 (population-level)

> as.data.frame(melr.4.pub)
                           x  predicted  std.error   conf.low  conf.high  group
1   catholic or other private  0.9816933  0.2457444  0.9706981  0.9886111      1
2                      public  0.9378470  0.1529411  0.9179058  0.9531913      1
> plot(melr.4.pub)
Loading required namespace: ggplot2
```

The output displays the predicted probabilities for both Catholic or other private schools and public schools when other predictor variables are fixed at zeros. The last section under the title "`Adjusted for`" lists the names and the values of the other variables. For Catholic or other private schools, the predicted probability = .982 and for public schools, the predicted probability = .938.

To visualize the results, the predicted probabilities are plotted using the `plot(melr.4.pub)` function. Figure 11.1 shows the predicted probabilities of being proficient in doing simple arithmetical operations on whole numbers for `public` at 0 and 1 with others fixed constant at the mean or at the reference category.

FIGURE 11.1 ● Predicted Probabilities for the Binary Predictor Variable `public`

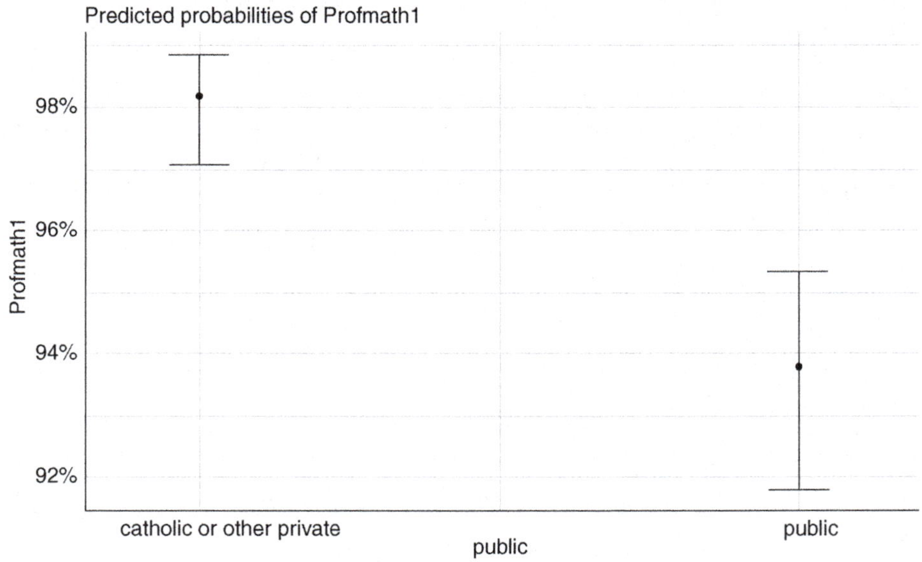

As shown in the graph, the predicted probabilities for students in Catholic or other private schools are higher than those in public schools. In other words, students in Catholic or other private schools are more likely than students in public schools to be proficient in doing simple arithmetical operations on whole numbers.

We can compute the predicted probabilities for a continuous predictor variable at given values by different groups. In the following example, we compute the predicted probabilities for `csclimat` at different values by the two groups in `public` when holding other variables constant. The command is as follows: `me4.efpub <- ggpredict(melr.4, terms = c("csclimat", "public"))`. In the `ggpredict()` function, the `terms = c("csclimat", "public")` option specifies both `csclimat` and `public`, with the latter as the grouping variable. The output is assigned to an object named `me4.efpub` and is plotted with `plot(me4.efpub)`.

```
> me4.efpub <- ggpredict(melr.4, terms = c("csclimat", "public"))
Data were 'prettified'. Consider using `terms="csclimat [all]"` to get smooth plots.
> me4.efpub
# Predicted probabilities of Profmath1

# public = catholic or other private

csclimat    |   Predicted    |         95% CI
---------------------------------------------
   -2.80    |        0.91    |   [0.83, 0.95]
   -2.20    |        0.93    |   [0.88, 0.96]
   -1.60    |        0.95    |   [0.92, 0.97]
   -1.00    |        0.97    |   [0.95, 0.98]
   -0.20    |        0.98    |   [0.97, 0.99]
    1.00    |        0.99    |   [0.98, 0.99]

# public = public

csclimat    |   Predicted    |         95% CI
---------------------------------------------
   -2.80    |        0.73    |   [0.62, 0.82]
   -2.20    |        0.80    |   [0.72, 0.86]
   -1.60    |        0.85    |   [0.80, 0.89]
   -1.00    |        0.89    |   [0.86, 0.92]
   -0.20    |        0.93    |   [0.91, 0.95]
    1.00    |        0.97    |   [0.95, 0.98]

Adjusted for:
* stlang = No
* SCH_ID = 0 (population-level)
> plot(me4.efpub)
```

FIGURE 11.2 ● Predicted Probabilities for `csclimat` by `public`

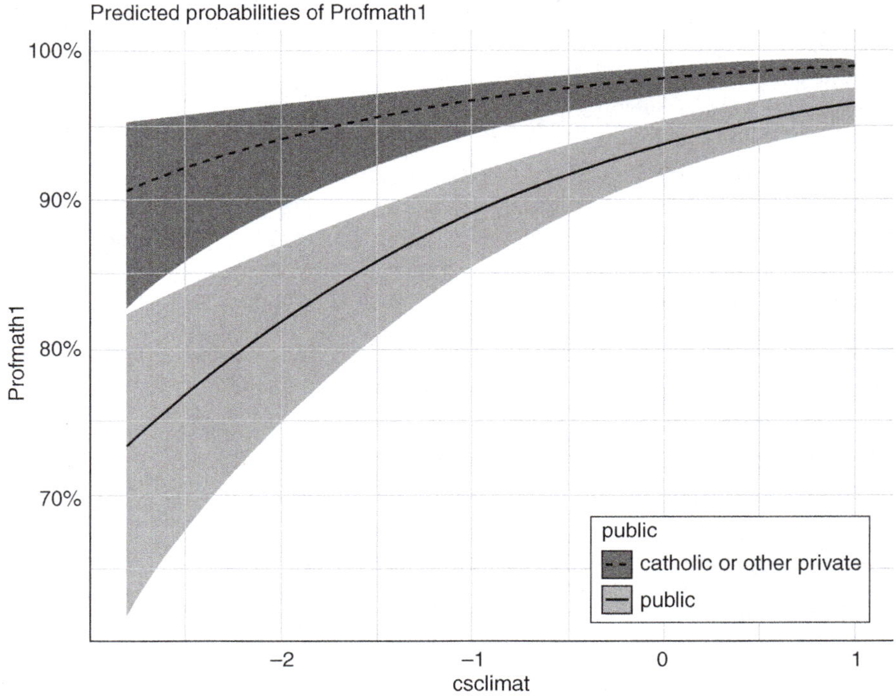

Figure 11.2 shows the predicted probabilities for `csclimat` at different values by the grouping variable `public`. As shown in the graph, the predicted probabilities increase with the increase of school climate and the predicted probabilities for students in Catholic or other private schools are higher than those in public schools.

In the example above, we disregard the random effects component and the predicted probabilities are on the population level. To account for the variability of the random effects, we need to specify the `type = "re"` option. With this option, the predicted probabilities are still on the population level and stay the same, but the confidence intervals change after accounting for the variances of the random effects. The command is as follows: `me4.efpub.r <- ggpredict(me4.b, terms = c("csclimat", "public"), type = "re")`. In the `ggpredict()` function, everything stays the same exception the added `type = "re"` option. The output is assigned to an object named `me4.efpub.r` and is plotted with the `plot(me4.efpub.r)` command.

```
> me4.efpub.r <- ggpredict(melr.4, terms = c("csclimat", "public"), type = "re")
Data were 'prettified'. Consider using `terms="csclimat [all]"` to get smooth plots.
> me4.efpub.r

# Predicted probabilities of Profmath1

csclimat    |    Predicted    |           95% CI
-------------------------------------------------
   -2.80    |         0.91    |     [0.55, 0.99]
   -2.20    |         0.93    |     [0.64, 0.99]
   -1.60    |         0.95    |     [0.73, 0.99]
   -1.00    |         0.97    |     [0.79, 1.00]
   -0.20    |         0.98    |     [0.86, 1.00]
    1.00    |         0.99    |     [0.93, 1.00]

# public = public

csclimat    |    Predicted    |           95% CI
-------------------------------------------------
   -2.80    |         0.73    |     [0.27, 0.95]
   -2.20    |         0.80    |     [0.35, 0.97]
   -1.60    |         0.85    |     [0.44, 0.98]
   -1.00    |         0.89    |     [0.53, 0.98]
   -0.20    |         0.93    |     [0.65, 0.99]
    1.00    |         0.97    |     [0.79, 1.00]

Adjusted for:
* stlang = No
* SCH_ID = 0 (population-level)

Standard errors are on link-scale (untransformed).
Intervals are prediction intervals.
> plot(me4.efpub.r)
```

Figure 11.3 shows the predicted probabilities for `csclimat` by the grouping variable `public` with random effects. As shown in the graph, the predicted probabilities are the same as those in Figure 11.2 but the confidence intervals are wider than those in Figure 11.2.

11.4 MULTILEVEL MODELING FOR BINARY OUTCOME VARIABLES WITH THE `clmm()` FUNCTION IN THE `ordinal` PACKAGE

Although the `clmm()` function in the `ordinal` package is normally used for ordinal response variables, it can also be used for multilevel models for binary response variables. Since the user-written `ordinal` package has been installed for proportional odds models introduced in Chapter 4, you only need to load the package by typing `library(ordinal)`. If it is not already installed, you need to install it first by typing `install.packages("ordinal")` and then load the package.

FIGURE 11.3 ● Predicted Probabilities for `csclimat` by `public` With Random Effects

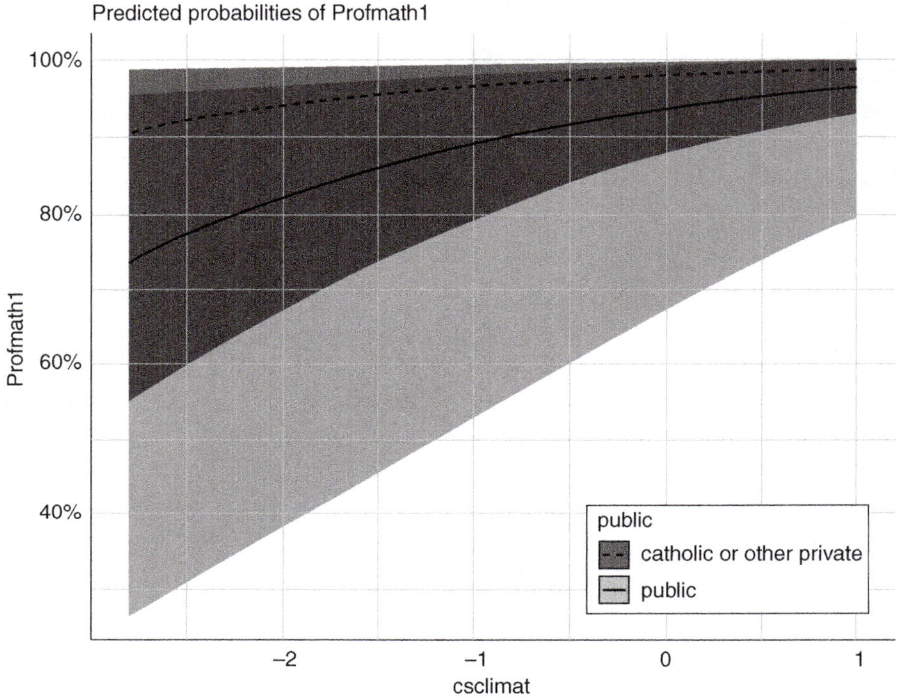

The syntax for `clmm()` is similar to that for `glmer()` for mixed-effects logistic regression models. The model formula in `clmm()` also includes two components including the fixed effects and random effects, which are connected by a plus (+) sign. The fixed effects part of the model specifies the dependent variable and the predictor variable(s) with the dependent variable on the left-hand side of the tilde (~). The predictor variables on the right-hand side are connected by plus (+) symbols. Next, the random effects part of the model is specified within parentheses. Similar to the random effects part of `glmer()`, the random coefficients term is specified first, followed by a vertical line (|), and then, an identifier variable at a higher level as the grouping variable. For example, the command `clmm(y ~ x + (1|schid), Hess = TRUE, nAGQ = 7)` tells R to fit a multilevel model to estimate a response variable `y` on a predictor variable `x` with a randomly varying intercept across schools by specifying `(1| schid)`, where `1` indicates the random intercept on the left of the vertical line (|) and `schid` is the identifier variable on the right. The `Hess = TRUE` argument specifies the Hessian so the output provides variance and covariance estimates of random effects. The `nAGQ = 7` argument specifies the number of quadrature points for the adaptive Gauss-Hermite quadrature method. The default number is 1, which is used for the Laplace approximation. A larger number of quadrature points increases the precision but slows the speed of computation. A negative number can be used for non-adaptive Gauss-Hermite quadrature. When there is more than a single random effects term in the model, such as the random-coefficient model with varying intercepts and coefficients, only the Laplace approximation with `nAGQ = 1` is available. For more details on how to use this function, type `help(clmm)` in the command prompt after loading the `ordinal` package.

In the following example, the command `melr.4b <- clmm(as.factor(Profmath1) ~ stlang + csclimat + public + (stlang|SCH_ID), data = chp11, na.action = "na.omit", Hess = TRUE, nAGQ = 1)` tells R to fit the same contextual model (model 4) in the previous section after the two level 2 predictor variables `public` and `csclimat` are added to the model. The fitted model is named `melr.4b`. The following output is displayed by the `summary(melr.4b)` command.

```
> library(ordinal)
> melr.4b <- clmm(as.factor(Profmath1) ~ stlang + csclimat + public + (stlang|SCH_ID),
data = chp11, na.action = "na.omit", Hess = TRUE, nAGQ = 1)
> summary(melr.4b)
Cumulative Link Mixed Model fitted with the Laplace approximation


formula: as.factor(Profmath1) ~ stlang + csclimat + public + (stlang |
    SCH_ID)
data:    chp11


link     threshold    nobs     logLik       AIC        niter      max.grad      cond.H
logit     flexible    12793    -2342.46    4698.92    441(2681)    3.29e-04    9.3e+01


Random effects:
 Groups   Name          Variance     Std.Dev.    Corr
 SCH_ID   (Intercept)   1.2412       1.1141
          stlangYes     0.9731       0.9864      -0.593
Number of groups:  SCH_ID 619


Coefficients:
                 Estimate    Std. Error    z value    Pr(>|z|)
stlangYes         0.68218       0.16470      4.142    3.44e-05 ***
csclimat          0.60676       0.09463      6.412    1.43e-10 ***
publicpublic     -1.26801       0.21423     -5.919    3.24e-09 ***
---
Signif. codes: 0 '***' 0.001 '**' 0.01 '*' 0.05 '.' 0.1 ' ' 1


Threshold coefficients:
       Estimate     Std. Error     z value
0|1     -3.9821         0.2458       -16.2
```

The R output produced by the `clmm()` function is similar to that by the `glmer()` function introduced in the previous section. It includes the note for the estimation method, the model formula, the data frame, the model fit statistics such as the log-likelihood value and the AIC, the random effects, the number groups, the fixed effects, and the threshold coefficients or the intercepts.

The random effects section (labeled `Random effects:`) provides the variances and the standard deviations for the `Intercept` (between-group) and for the random coefficient of `gceffic`. It also provides the correlation coefficient for `gceffic` and the intercept. The results are the same as those provided by the `glmer()` function in the previous section.

The fixed effects section (labeled `Coefficients:`) looks similar to that produced by the `glmer()` function. The last section (labeled `Threshold coefficients:`) displays the estimate for the intercept. While multiple intercepts or thresholds can be estimated for ordinal response variables which are introduced in the following chapter, only one intercept is estimated for the binary response variable here. Please note that the intercept provided by the `glmer()` function in the previous section is 3.982 whereas it is −3.982 by the `clmm()` function. This difference is due to the fact that the `clmm()` function estimates the log odds of being at or below a category versus being above that category. In this example, the binary model compares the categories 0 and 1 and the intercept is labeled `0|1`. Therefore, we need to reverse the sign before the intercept when we use the `clmm()` function for the multilevel models for binary response variables.

We use the `confint(melr.4b)` command to extract the 95% confidence intervals of the parameter estimates. The results are displayed as follows.

```
> confint(melr.4b)
                    2.5 %       97.5 %
0|1             -4.4638356   -3.5003992
stlangYes        0.3593808    1.0049775
csclimat         0.4212963    0.7922222
publicpublic    -1.6878851   -0.8481322
```

We combine the results of the fixed effects and the confidence intervals with `cbind(coef(melr.4b)), confint(melr.4b))`.

```
> cbind(coef(melr.4b), confint(melr.4b))
                                 2.5 %       97.5 %
0|1             -3.9821174   -4.4638356   -3.5003992
stlangYes        0.6821792    0.3593808    1.0049775
csclimat         0.6067593    0.4212963    0.7922222
publicpublic    -1.2680086   -1.6878851   -0.8481322
```

We use the `exp(coef(melr.4b))` command to obtain the odds ratios and use the `exp(confint(melr.4b))` command to produce the corresponding confidence intervals. The results are combined using `cbind(exp(coef(melr.4b)), exp(confint(melr.4b)))`. The following output is created.

```
> exp(coef(melr.4b))
        0|1      stlangYes      csclimat    publicpublic
0.01864612     1.97818388    1.83447668      0.28139142
> exp(confint(melr.4b))
                    2.5 %        97.5 %
0|1             0.0115181    0.03018533
stlangYes       1.4324422    2.73184593
csclimat        1.5239358    2.20829824
publicpublic    0.1849102    0.42821401
> cbind(exp(coef(melr.4b)),exp(confint(melr.4b)))
                                 2.5 %        97.5 %
0|1             0.01864612   0.0115181    0.03018533
stlangYes       1.97818388   1.4324422    2.73184593
csclimat        1.83447668   1.5239358    2.20829824
publicpublic    0.28139142   0.1849102    0.42821401
```

The odds ratios for the three predictor variables are the same as those from the results of the `glmer()` function in the previous section. Please note that we need to take the multiplicative inverse of the odds ratio for the intercept to get the correct value. The odds ratio for the intercept is 1/.0186 = 53.76. The same value can be obtained if we take the exponential of the intercept 3.982.

The `ggeffects` package also supports the `clmm()` function in the `ordinal` package. You can use the `ggpredict()` function to obtain the predicted probabilities for predictor variables at specified values in the multilevel logistic regression model.

11.5 MAKING PUBLICATION-QUALITY TABLES

11.5.1 Presenting the Results Using the `stargazer` Package

We can use the `stargazer` package (Hlavac, 2018) to make a table containing the results of the fitted unconditional model and the contextual model with cross-level interactions. After fitting these two models, `melr.1` and `melr.5`, with the `glmer()` function, we use the command as follows: `stargazer(melr.1, melr.5, type = "text", align = TRUE, out = "melr.txt")`. In the `stargazer()` function, we first specify the two model objects to be presented and then the type of the table. The `type = "text"` argument specifies the table type and the `align = TRUE` argument aligns the results of the models. The `out = "melr.txt"` argument saves the output object named `melr.txt`.

```
> # Presenting the results with stargazer()
> library(stargazer)

Please cite as:

 Hlavac, Marek (2018). stargazer: Well-Formatted Regression and Summary Statistics
Tables.
 R package version 5.2.2. https://CRAN.R-project.org/package=stargazer

> stargazer(melr.1, melr.5, type = "text", align = TRUE, out = "melr.txt")

================================================
                        Dependent variable:
                     ---------------------------
                             Profmath1
                     (1)                (2)
------------------------------------------------
stlangYes                               0.618
                                       (0.499)

csclimat                                0.420***
                                       (0.152)

publicpublic                           -1.417***
                                       (0.455)

stlangYes:publicpublic                  0.200
                                       (0.498)

stlangYes:csclimat                      0.255
                                       (0.170)

Constant                  3.472***      4.026***
                         (0.086)       (0.458)

------------------------------------------------
Observations              12,793        12,793
Log Likelihood           -2,429.029    -2,341.511
Akaike Inf. Crit.         4,862.058     4,701.021
Bayesian Inf. Crit.       4,876.971     4,768.131
================================================
Note:               *p<0.1;  **p<0.05;  ***p<0.01
> stargazer(melr.1, melr.5, type = "html", align = TRUE, out = "melr.htm")
```

We can also create the table in the HTML format and copy it into Microsoft Word. The command is as follows: `stargazer(melr.1, melr.5, type = "html", align = TRUE, out = "melr.htm")`. The table is omitted here.

11.5.2 Presenting the Results Using the `texreg` Package

We can also use the `screenreg()` and `htmlreg()` functions from the `texreg` package (Leifeld, 2013) to display the results. We load the package first by typing `library(texreg)`.

The same results of the unconditional model and the contextual model with cross-level interactions can also be displayed in a table with the following command: `screenreg(list(melr.1, melr.5))`.

```
> # Presenting the results with texreg()
> library(texreg)
Version:    1.37.5
Date:       2020-06-17
Author:     Philip Leifeld (University of Essex)

Please cite the JSS article in your publications - see citation("texreg").
> screenreg(list(melr.1, melr.5))


===============================================================
                                        Model 1       Model 2
---------------------------------------------------------------
(Intercept)                              3.47 ***      4.03 ***
                                        (0.09)        (0.46)
stlangYes                                              0.62
                                                      (0.50)
csclimat                                               0.42 **
                                                      (0.15)
publicpublic                                          -1.42 **
                                                      (0.46)
stlangYes:publicpublic                                 0.20
                                                      (0.50)
stlangYes:csclimat                                     0.25
                                                      (0.17)
---------------------------------------------------------------
AIC                                     4862.06       4701.02
BIC                                     4876.97       4768.13
Log Likelihood                         -2429.03      -2341.51
Num. obs.                              12793         12793
Num. groups: SCH_ID                      619           619
Var: SCH_ID (Intercept)                    1.33          1.05
Var: SCH_ID stlangYes                                    0.78
Cov: SCH_ID (Intercept) stlangYes                       -0.44
===============================================================
*** p < 0.001, ** p < 0.01, * p < 0.05
> htmlreg(list(melr.1, melr.5), file = "melr.doc", doctype = TRUE, html.tag = TRUE,
head.tag = TRUE)
The table was written to the file 'melr.doc'.
```

To create the table in the HTML format and copy it into Microsoft Word, we use the command as follows: `htmlreg(list(melr.1, melr.5), file = "melr.doc", doctype = TRUE, html.tag = TRUE, head.tag = TRUE)`. Table 11.1, as shown here in its original format, presents the results for the unconditional model and the contextual model with cross-level interactions.

With edits on the model titles, the added odds ratios, and the added labels for the fixed and random effects, the final table is displayed as in Table 11.2.

TABLE 11.1 ● Results of the Two Multilevel Logistic Regression Models: Unconditional Model and Contextual Model (Shown in Original Format Generated by the `texreg` Package)

	Model 1	Model 2
(Intercept)	3.47***	4.03***
	(0.09)	(0.46)
stlangYes		0.62
		(0.50)
csclimat		0.42**
		(0.15)
publicpublic		−1.42**
		(0.46)
stlangYes:publicpublic		0.20
		(0.50)
stlangYes:csclimat		0.25
		(0.17)
AIC	4862.06	4701.02
BIC	4876.97	4768.13
Log Likelihood	−2429.03	−2341.51
Num. obs.	12,793	12,793
Num. groups: SCH_ID	619	619
Var: SCH_ID (Intercept)	1.33	1.05
Var: SCH_ID stlangYes		0.78
Cov: SCH_ID (Intercept) stlangYes		−0.44

***$p < .001$
**$p < .01$
*$p < .05$

TABLE 11.2 ● Results of the Two Multilevel Logistic Regression Models: Unconditional Model and Contextual Model (Edited)

	Unconditional Model (Model 1)		Contextual Model (Model 5)	
	Coefficient (SE)	OR	Coefficient (SE)	OR
Fixed Effects				
(Intercept)	3.47***	32.19	4.03***	56.05
	(0.09)		(0.46)	
Stlang			0.62	1.86
			(0.50)	
Csclimat			0.42**	1.52
			(0.15)	
Public			−1.42**	.24
			(0.46)	
stlang:public			0.20	1.22
			(0.50)	
stlang:csclimat			0.25	1.29
			(0.17)	
AIC	4,862.06		4,701.02	
BIC	4,876.97		4,768.13	
Log Likelihood	−2,429.03		−2,341.51	
Num. obs.	12,793		12,793	
Number of groups: SCH_ID	619		619	
Random Effects				
Between-School Var: SCH_ID (Intercept)	1.33		1.05	
Slope Var: SCH_ID stlang			0.78	
Cov: SCH_ID (Intercept) stlang			−0.44	

Note: Standard errors are shown in parentheses.
***$p < .001$
**$p < .01$
*$p < .05$

11.6 REPORTING THE RESULTS

The same reporting guidelines for multilevel modeling for continuous response variables can be applied to the reporting for the multilevel logistic regression modeling. The only difference is that the odds ratios may need to be reported and interpreted in the latter models. The following is an example of summarizing the results of a contextual multilevel logistic regression model with cross-level interactions.

Multilevel logistic regression models were fitted to predict whether high-school students were capable of doing simple arithmetical operations on whole numbers from three predictor variables: whether English is the students' native language, the school type, and the school climate. The first predictor was a student-level variable, and the other two were school-level variables. Multilevel modeling for binary response variables was used since the outcome variable was dichotomous. Five models, from the unconditional (null) model to the contextual model with cross-level interactions, were fitted. To save space, only the results for the final contextual model with cross-level interactions (model 5) were reported as follows.

Results for the Contextual Model With Cross-Level Interactions (Model 5)

The logit coefficient for `stlang` (γ_{10}) = .618, z = 1.238, p = .216. The odds ratio for `stlang` was 1.856, which indicated that the odds of being proficient in math for students whose native language is English were 1.856 times the odds for those whose native language is not English. However, this effect was not significant.

The logit coefficient for `public` (γ_{01}) = −1.417, $z = -3.111$, $p < .01$. The odds ratio for `public` is .242, which indicated that the odds of being proficient in math for students in public schools were .242 times the odds for those in private schools. In other words, being students in public schools decreased the odds of being proficient in math by 75.5%.

The logit coefficient for `csclimat` (γ_{02}) = .420, z = 2.760, $p < .01$. OR = 1.522, which indicated that for a one-unit increase in school climate, the odds of being above a particular category of math proficiency increased by a factor of 1.522. So the odds of being in higher math proficiency levels were associated with schools with a more positive school climate.

With regard to cross-level interactions, the coefficient for `public:stlang` (γ_{11}) = .200, z = .401, $p > .05$, and the coefficient for `csclimat:stlang` (γ_{12}) = .255, z = 1.504, $p > .05$. So both interactions were not significant, which indicated that the effect of `stlang` does not differ in public or private schools. In addition, the effect of `stlang` was not influenced by school climate either.

Regarding the variance and covariance components for the random effects, the between-group variance, the intercept variance (τ_{00}), was 1.048. The slope variance (τ_{10}) for `stlang` was .781. The covariance between the intercept and the slope was −.440.

After two school-level variables and two cross-level interactions were included in the contextual model (model 5), the between-school variance (τ_{00}) decreased from 1.294 to 1.048 compared with that for the random-coefficient model (model 3): (1.294 − 1.048)/1.294 = 19%. This indicated that there was a decrease of 19% in the between-school variance from the random coefficient model (model 3) to the contextual model (model 5) after the two school-level variables were included.

The likelihood ratio test was used to compare the contextual models with or without interactions (model 5 vs. model 4): $\chi^2_{(2)} = 1.899$, $p > .05$, which indicated that there was no improvement in deviance from the contextual model without cross-level interactions to the contextual model without interactions. Therefore, the interaction terms should be removed from model 5.

11.7 SUMMARY OF R COMMANDS IN THIS CHAPTER

```
# Chap 11 R Script

# Remove all objects
rm(list = ls(all = TRUE))

# The following user-written packages need to be installed first by using
install.packages(" ") and then by loading it with library()
# library(lme4)            # It is already installed for Chapter 10
# library(margins)         # It is already installed for Chapter 3
# library(ggeffects)       # It is already installed for Chapter 2
# library(stargazer)       # It is already installed for Chapter 2
# library(ordinal)         # It is already installed for Chapter 4
```

```
# Import the dataset
library(foreign)
chp11 <- read.dta("C:/CDA/els2002.dta")
chp11 <-
chp11[!is.na(chp11$Profmath1)&!is.na(chp11$stlang)&!is.na(chp11$public)&!is.na(chp11$
sclimate), ]
attach(chp11)
chp11$csclimat <- sclimate-mean(sclimate, na.rm=TRUE)
summary(chp11$csclimat)
table(Profmath1)
table(stlang)
mean(sclimate)
table(public)


library(lme4)
# Unconditional model/Null model with glmer()
melr.1 <- glmer(Profmath1 ~ + (1|SCH_ID), data = chp11, family = binomial, nAGQ = 7)
summary(melr.1)
coef <- fixef(melr.1)
se <- sqrt(diag(vcov(melr.1)))
LL <- fixef(melr.1)-1.96*se
UL <- fixef(melr.1)+1.96*se
coeftab <- cbind(coef, LL, UL)
coeftab
exp(fixef(melr.1))
exp(coeftab)
ortab <- cbind(or=exp(fixef(melr.1)), ORLL=exp(LL), ORUL=exp(UL))
ortab
# confint(melr.1)
confint(melr.1, method = "Wald")


# Random-intercept model
melr.2 <- glmer(Profmath1 ~ stlang + (1|SCH_ID), data = chp11, family = binomial, nAGQ = 7)
summary(melr.2)
coef <- fixef(melr.2)
se <- sqrt(diag(vcov(melr.2)))
LL <- fixef(melr.2)-1.96*se
UL <- fixef(melr.2)+1.96*se
coeftab <- cbind(coef, LL, UL)
coeftab
exp(fixef(melr.2))
exp(coeftab)
ortab <- cbind(or=exp(fixef(melr.2)), ORLL=exp(LL), ORUL=exp(UL))
ortab
# confint(melr.2)
confint(melr.2, method = "Wald")


anova(melr.1, melr.2)


# Random-coefficient model
melr.3 <- glmer(Profmath1~ stlang + (stlang|SCH_ID), data = chp11, family = binomial,
nAGQ = 1)
summary(melr.3)
coef <- fixef(melr.3)
se <- sqrt(diag(vcov(melr.3)))
LL <- fixef(melr.3)-1.96*se
UL <- fixef(melr.3)+1.96*se
coeftab <- cbind(coef, LL, UL)
coeftab
exp(fixef(melr.3))
```

```
exp(coeftab)
ortab <- cbind(or=exp(fixef(melr.3)), ORLL=exp(LL), ORUL=exp(UL))
ortab
confint(melr.3, method = "Wald")
anova(melr.2, melr.3)

# Contextual model/Random-coefficient model with predictor variables in both levels
# It will take several minutes to fit the model
melr.4 <- glmer(Profmath1 ~ stlang + csclimat + public + (stlang|SCH_ID), data = chp11,
family = binomial, nAGQ = 1)
summary(melr.4)
coef <- fixef(melr.4)
se <- sqrt(diag(vcov(melr.4)))
LL <- fixef(melr.4)-1.96*se
UL <- fixef(melr.4)+1.96*se
coeftab <- cbind(coef, LL, UL)
coeftab
confint(melr.4, method = "Wald")
exp(fixef(melr.4))
exp(coeftab)
ortab <- cbind(or=exp(fixef(melr.4)), ORLL=exp(LL), ORUL=exp(UL))
ortab
anova(melr.3, melr.4)

# Contextual model with cross-level interactions
# It will take several minutes to fit the model
melr.5 <- glmer(Profmath1 ~ stlang + csclimat + public + public:stlang + csclimat:stlang
+ (stlang|SCH_ID), data = chp11, family = binomial, nAGQ = 1)
summary(melr.5)
coef <- fixef(melr.5)
se <- sqrt(diag(vcov(melr.5)))
LL <- fixef(melr.5)-1.96*se
UL <- fixef(melr.5)+1.96*se
Coeftab <- cbind(coef, LL, UL)
coeftab
confint(melr.5, method = "Wald")
exp(fixef(melr.5))
exp(coeftab)
ortab <- cbind(or=exp(fixef(melr.5)), ORLL=exp(LL), ORUL=exp(UL))
ortab
anova(melr.4, melr.5)

# Model comparisons
anova(melr.1, melr.2, melr.3, melr.4, melr.5)

# Marginal effects/Predicted probabilities with ggpredict()
library(ggeffects)
melr.4.pub <- ggpredict(melr.4, terms = "public")
melr.4.pub
as.data.frame(melr.4.pub)
plot(melr.4.pub)

me4.efpub <- ggpredict(melr.4, terms = c("csclimat", "public"))
me4.efpub
plot(me4.efpub)
```

```
me4.efpub.r <- ggpredict(melr.4, terms = c("csclimat", "public"), type = "re")
me4.efpub.r
plot(me4.efpub.r)

# Presenting the results with stargazer()
library(stargazer)
stargazer(melr.1, melr.5, type="text", align=TRUE, out="melr.txt")
stargazer(melr.1, melr.5, type="html", align=TRUE, out="melr.htm")

# Presenting the results with texreg()
library(texreg)
screenreg(list(melr.1, melr.5))
htmlreg(list(melr.1, melr.5), file="melr.doc", doctype=TRUE, html.tag=TRUE,
head.tag=TRUE)

# Use the clmm() function in the ordinal package
# Install the ordinal package by using install.packages()
library(ordinal)
melr.4b <- clmm(as.factor(Profmath1) ~ stlang + csclimat + public + (stlang|SCH_ID),
data=chp11, na.action="na.omit", Hess=TRUE, nAGQ=1)
summary(melr.4b)
confint(melr.4b)
cbind(coef(melr.4b), confint(melr.4b))
exp(coef(melr.4b))
exp(confint(melr.4b))
cbind(exp(coef(melr.4b)), exp(confint(melr.4b)))

detach(chp11)
```

Glossary

Cross-level interactions are interaction effects between level 1 and level 2 predictor variables.

In **a random-coefficient logistic regression model**, both the intercept and coefficient in the level 1 equation are allowed to vary across groups at level 2.

In **a random-intercept logistic regression model**, the intercept is allowed to vary freely across groups.

The contextual model for a binary response variable includes both level 1 and level 2 predictor variables.

The unconditional model, or null model, for binary response variables includes no predictor variable at any level. This model estimates the overall probability of success or of having an event and the variability in the probability of success between groups (e.g., schools).

Exercises

Use the ELS:2002 data available at **https://edge.sagepub.com/liu1e** for the following problems. The following variables are used for the multilevel modeling.

`Profmath2`: math proficiency level 2 (1 = capable of doing simple operations with decimals, fractions, powers, and root; 0 = not achieving this level)

`gender`: gender (1 = female; 0 = male)

`byses`: socioeconomic status composite for base-year data

`urban`: urban schools or not (1 = urban; 0 = suburban or rural areas) (school-level variable)

Multilevel logistic regression models will be fitted to estimate whether students are capable of doing simple operations with decimals, fractions, powers, and root. The binary outcome variable is `Profmath2`, and the predictor variables are `gender`, `byses`, and `urban`. Perform the analysis and answer the following questions:

1. Fit an unconditional model and obtain the between-group variance. Compute the ICC and interpret it.
2. Fit a two-level, random-intercept model with a random intercept and the two student-level predictor variables `gender` and `byses`. Use the grand-mean centering for `byses` before fitting the model.
3. Interpret the odds ratios of the two predictor variables.
4. Conduct a likelihood ratio test comparing the random-intercept model and the unconditional model. Interpret the results.
5. Fit a two-level, contextual model with `gender` and `byses` as the student-level predictor variables and `urban` as the school-level predictor variable. Specify a random intercept. No cross-level interactions are included.
6. Interpret the odds ratio of `urban` in the contextual model.
7. Among the three fitted models, which one fits the data best? Why?
8. Make a publication-quality table containing the estimated logit coefficients for the contextual model.
9. Write a report to summarize the results of the contextual model.

12

MULTILEVEL MODELING FOR ORDINAL RESPONSE VARIABLES

OBJECTIVES OF THIS CHAPTER

This chapter introduces multilevel modeling for ordinal response variables. It starts with an introduction to the model followed by a discussion of the model specification, the odds and odds ratios in the model, and the likelihood ratio test, as well as a description of the research questions and data. Then several models are illustrated using R with step-by-step instructions, including the unconditional (null) model, the random-intercept model, the random-coefficient model with a level 1 variable, the contextual model with both level 1 and level 2 variables, and the contextual model with cross-level interactions. The two packages, `ordinal` and `mixor`, are used for fitting multilevel ordinal logistic regression models. R commands are explained, and the output is interpreted for each model in detail. This chapter also illustrates how the results are displayed in publication-quality tables using R and reported in the text. It focuses on model fitting with R, as well as on interpreting and presenting the results. After reading this chapter, you should be able to:

- Determine when multilevel modeling for ordinal response variables is used.
- Formulate multilevel models for ordinal response variables.
- Fit multilevel models for ordinal response variables using the `clmm()` function in the `ordinal` package and the `mixor()` function in the `mixor` package.
- Interpret the output.
- Compute and plot the predicted probabilities using the `ggeffects()` function.
- Be familiar with model fitting strategies.
- Compare models using likelihood ratio tests and AIC and BIC statistics.
- Present results in publication-quality tables using R.
- Write the results for publication.

12.1 MULTILEVEL MODELING FOR ORDINAL RESPONSE VARIABLES: AN INTRODUCTION

Although several types of ordinal logistic regression models have been introduced in the early chapters, this chapter focuses on multilevel proportional odds (PO) models only as multilevel modeling for other ordinal models has not been well developed. Multilevel PO models (Goldstein, 2003; Hedeker, 2007; Hedeker & Gibbon, 1994; Hox, 2010; O'Connell et al., 2008; Rabe-Hesketh & Skrondal, 2012; Raudenbush & Bryk, 2002; Snijders & Bosker, 2012) are an extension of multilevel binary logistic regression models when the response variables are ordinal and are often referred to as multilevel ordinal logistic regression models. They are also a generalization of single-level proportional odds models when the data structure is nested or hierarchical. Just like the proportional odds models, they estimate the cumulative odds of being at or below a particular category of the response variable, or the inversed odds, the odds of being above that particular category. The cumulative odds of being at or below a particular ordinal category are the probability of being at or below a category divided by the probability of being above that category. On the other hand, the cumulative odds of being above a particular category compare the probability of being above a category with the probability of being at or below that particular category. Since below and above a particular category are two opposite directions, these two probabilities are complementary, and thus, these two odds are reciprocal.

With the nested data structure, the multilevel modeling for ordinal response variables allows us to estimate the relationships between predictor variables at different levels and the ordinal response variable. Similar to multilevel logistic regression models, we can specify varying intercepts and slopes in level 2 models, so the multilevel PO models also estimate variance–covariance components for the random effects.

12.1.1 Model Specification

The level 1 equation for the multilevel PO model can be expressed as:

$$\begin{aligned}\text{Level 1: } \text{logit}\left[\pi_{kij}(Y \leq k)\right] &= \ln\left(\frac{\pi\left(Y_{ij} \leq k \middle| x_1, x_2, \ldots, x_p\right)}{\pi\left(Y_{ij} > k \middle| x_1, x_2, \ldots, x_p\right)}\right) \\ &= \alpha_k + \left(-\beta_{1j}X_{1ij} - \beta_{2j}X_{2ij} - \cdots - \beta_{qj}X_{qij}\right)\end{aligned} \tag{12.1}$$

where α_k are the cut points with $k = 1, 2, \ldots, K - 1$; $X_{1ij}, X_{2ij}, \ldots, X_{pij}$ are the predictor variables for the *i*th individual in the *j*th cluster; and $\beta_{1j}, \beta_{2j}, \ldots, \beta_{pj}$ are the logit coefficients of these predictors in the *j*th cluster.

The model specification for multilevel PO models is more complex than those for multilevel modeling for continuous and binary response variables. Detailed explanations are as follows:

First, the subscripts i and j mean the ith individual in the jth cluster. For example, they mean the ith student in the jth school when students are nested within schools. Without the subscripts, this model is just the single-level PO model introduced in Chapter 4, which estimates the relationship between a set of independent variables and the ordinal outcome variable on a scale of the logit or log odds.

Second, let us look at the three parts of the equation with these subscripts. The left side of the equation $\text{logit}[\pi_{kij}(Y \leq k)]$ is the link function. It transforms the original ordinal response variable Y for an ith individual in the jth cluster into the transformed outcome, logit or log odds of the cumulative probability of being at or below a particular category k. The middle part is the log odds, where the cumulative odds of being at or below a particular category equal the ratio of the probability of being at or below a category to the probability of being above that category. The right side of the equation is the linear combination of the predictor variables with each logit coefficient associated with cluster j.

Just like with the single-level PO model, the multilevel PO model also assumes that the logit coefficients for each predictor variable are the same across the categories of the ordinal outcome variable. In other words, the logit coefficients for the underlying binary logistic regression models that dichotomize the ordinal response variable are the same. This is known as the proportional odds or parallel lines assumption.

Unlike multilevel modeling for binary response variables where an intercept is estimated, the multilevel PO models estimate a series of cut points for the underlying binary logistic models for the dichotomized ordinal response variable. When the ordinal response variable has k levels, $K - 1$ cut points will be estimated. As with the randomly varying intercept in the multilevel modeling for binary response variables, the cut points in the multilevel PO models may also be allowed to vary across clusters.

Since the odds of being above a particular category are the multiplicative inverse of the odds of being at or below that particular category of the response variable, the level 1 equation can also be expressed as:

$$\begin{aligned}\text{Level 1}: \text{logit}\left[\pi_{kij}(Y > k)\right] &= \ln\left(\frac{\pi\left(Y_{ij} > k \middle| x_1, x_2, \ldots, x_p\right)}{\pi\left(Y_{ij} \leq k \middle| x_1, x_2, \ldots, x_p\right)}\right) \\ &= -\alpha_k + \left(\beta_{1j}X_{1ij} + \beta_{2j}X_{2ij} + \cdots + \beta_{qj}X_{qij}\right)\end{aligned} \tag{12.2}$$

Please be aware that the signs before the cut points and the coefficient in Equation 12.2 are reversed from those in Equation 12.1.

Model Specification: An Example of a Basic Two-Level PO Model

To understand the specification of multilevel modeling with both level 1 and level 2 equations, let us specify a basic two-level PO model with one predictor variable at each level of the equations. For example, researchers are interested in estimating the relationships between an ordinal response variable, students' mathematics proficiency, and a student-level variable, students' native language, and a school-level variable, `public`.

The level 1 predictor variable is whether students' native language is English (`stlang`), and the level 2 predictor is whether a school is public or private (`public`). Both the cut points and the slope of the level 1 predictor variable `stlang` are allowed to vary randomly across schools.

The level 1 equation is similar to that for the single-level PO model, whereas in multilevel modeling, students are nested with schools so we have a subscript i to represent students and a subscript j to represent schools. As with the single-level PO model, no error term is specified in the level 1 equation:

$$\text{Level 1}: \ \eta_{ij} = \text{logit}\left[\pi_{kij}(Y \leq k)\right] = \ln\left(\frac{\pi\left(Y_{ij} \leq k\right)}{\pi\left(Y_{ij} > k\right)}\right) = \alpha_k - \left(\beta_{0j} + \beta_{1j}\texttt{stlang}_{ij}\right) \tag{12.3}$$

$$\text{Level 2}: \ \beta_{0j} = \gamma_{00} + \gamma_{01}\left(\texttt{public}_j\right) + u_{0j} \tag{12.4}$$

$$\beta_{1j} = \gamma_{10} + \gamma_{11}\left(\texttt{public}_j\right) + u_{1j} \tag{12.5}$$

The equations at level 2 look the same as those for the multilevel models for continuous outcome variables and binary response variables. Both the intercept and the coefficient (i.e., the slope of `stlang`) are allowed to vary randomly across schools at level 2 equations, where they are specified as the outcome variables.

In the level 1 equation, logit(π_{kij}) is the logit link for the cumulative probability of being at or below a particular category k for the ith student in the jth school.

In the level 2 equation, γ_{00}, γ_{01}, γ_{10}, and γ_{11} are level 2 coefficients, which are the fixed effects. γ_{00} is the intercept, the overall logit or log odds of being at or below a particular math proficiency across schools, and u_{0j} is an error term at the school level. The overall intercept is fixed to zero for identification purposes and is removed from the model, so u_{0j} is an error term associated with the cut points, the random variation for the cut points; γ_{10} is the overall average slope for `stlang` when the school-level predictor `public` is zero (`public` = 0), and u_{1j} is the error term associated with the random slope, the random variation for the level 1 slope across schools; γ_{01} is the fixed effect of the level 2 predictor `public` on the log odds of being at or below a particular ordinal category; and γ_{11} is the cross-level interaction between `stlang` and `public`.

The level 1 and level 2 equations can be expressed in a combined form as follows:

$$\begin{aligned}\eta_{ij} &= \text{logit}\left[\pi_{kij}\left(Y_{ij} \leq k\right)\right] = \ln\left(\frac{\pi\left(Y_{ij} \leq k\right)}{\pi\left(Y_{ij} > k\right)}\right) \\ &= \alpha_k - \left(\gamma_{00} + \gamma_{01}\texttt{public}_j + \gamma_{10}\texttt{stlang}_{ij} + \gamma_{11}\texttt{public}_j \times \texttt{stlang}_{ij}\right. \\ &\quad \left. + u_{0j} + u_{1j}\texttt{stlang}_{ij}\right)\end{aligned} \tag{12.6}$$

The intercept γ_{00} is set to be zero and the cut points or thresholds are estimated. After removing the intercept, we can rewrite this composite equation as follows:

$$\eta_{ij} = \text{logit}\left[\pi_{kij}\left(Y_{ij} \leq k\right)\right] = \left(\alpha_k - u_{0j}\right) - \left(\gamma_{01}\texttt{public}_j + \gamma_{10}\texttt{stlang}_{ij} + \gamma_{11}\texttt{public}_j \times \texttt{stlang}_{ij} + u_{1j}\texttt{stlang}_{ij}\right) \quad (12.7)$$

We can see that α_k are the estimated cut points or thresholds and that u_{0j} is the random effect associated with the cut points. When the cut points are allowed to vary randomly across schools, u_{0j} represents the deviation from the cut points across schools.

12.1.2 Odds and Odds Ratios in Multilevel PO Models

As with the single-level conventional PO model, the odds and odds ratios in multilevel PO models can be interpreted in a similar way. The cumulative odds of being at or below a category are the cumulative probability of being at or below a category divided by the probability of being above that category, where the cumulative probability $\pi(Y_{ij} \leq k)$ equals the sum of the probabilities of all categories at or below that category. The corresponding odds ratio of being above a particular category in the PO model is the exponentiated logit coefficient $\exp(\beta)$. This is interpreted as the change in the odds of above a particular category for each one-unit change in a predictor variable.

The odds ratio of being at or below a particular category in the PO model is the exponentiated negative coefficient $\exp(-\beta)$. This is interpreted as the change in the odds of at or below a particular category for each one-unit change in a predictor variable.

12.1.3 Likelihood Ratio Test

As with the multilevel binary logistic regression, the likelihood ratio test can be used to compare nested models since the Laplace approximation or Gauss-Hermite quadrature can be used as numerical integration methods to estimate log likelihood values. The difference in -2 log likelihood (-2LL) or deviance is expressed as $G = D_{\text{Reduced}} - D_{\text{Full}}$. It is the difference between the deviance for the reduced model and that for the full model. When model A is nested within model B, the log likelihood ratio test $G = D_{\text{ModelA}} - D_{\text{ModelB}}$. The likelihood ratio test statistic follows a chi-square distribution with the degrees of freedom of the distribution equal to the difference in the number of parameters between these two models.

In addition to the likelihood ratio test, the Akaike information criterion (AIC) and Bayesian information criterion (BIC) statistics can be used to compare nested models. One advantage of these two statistics is that they can also be used to compare non-nested models.

12.2 RESEARCH EXAMPLE: RESEARCH PROBLEM AND QUESTIONS

Using the Educational Longitudinal Study of 2002 (ELS:2002), researchers are interested in investigating the relationships between an ordinal response variable, students' math proficiency, and a student-level variable, student's native language, and two school-level variables, school type and school climate. The research questions are as follows:

1. Can the cumulative odds of being at or below a particular mathematics proficient level be predicted by whether English is the students' native language?
2. Do school characteristics, such as school type and school climate, impact the odds of being at or below a particular level of mathematics proficiency?
3. Does the relationship between the students' native language and the odds of being at or below a particular level of mathematics proficiency vary across schools?
4. Are there any interaction effects between the two school-level variables (i.e., school type and school climate) and students' native language? In other words, does the relationship between students' native language and the cumulative odds of being at or below a particular mathematics proficiency level vary across school type and school climate?

12.2.1 Description of the Data and Sample

The ELS:2002 base-year data are used for the following analyses. The variables are listed as follows:

- `Profmath`: mathematics proficiency with six levels from 0 to 5
- `stlang`: whether English is the students' native language
- `public`: school type (1 = public, 0 = private and others)
- `sclimat`: school climate
- `csclimat`: school climate (grand-mean centered).

12.3 MULTILEVEL MODELING FOR ORDINAL RESPONSE VARIABLES WITH R

12.3.1 Packages and Functions for Multilevel Modeling for Ordinal Response Variables in R

The `clmm()` function in the `ordinal` package (Christensen, 2019) and the `mixor()` function in the `mixor` package (Archer et al., 2018) are used for fitting multilevel models with ordinal outcome variables. Both packages can fit multilevel PO models with the cumulative odds link which are expressed in Equation 12.1. The `clmm()` function is the major focus, and so it is introduced first. Readers who are interested in multilevel continuation ratio models and multilevel adjacent categories models may refer to the `brms` package (Bürkner, 2017) for the Bayesian approach.

12.3.2 The `clmm()` Function in the `ordinal` Package

The `clmm()` function in the `ordinal` package is used for mixed-effects ordinal logistic regression models or multilevel models for ordinal response variables. The `clmm()` function stands for the function for cumulative link mixed models and is an extension of the `clm()` function for cumulative link models. Since the user-written `ordinal` package has been installed for proportional odds models introduced in Chapter 4, you only need to load the package by typing `library(ordinal)`. If it has not already been installed, you need to install it first by typing `install.packages("ordinal")` and then load the package.

The syntax of `clmm()` for mixed-effects ordinal logistic regression models is similar to that of `glmer()` for mixed-effects logistic regression models. The model formula in `clmm()` also includes two components including the fixed effects and random effects, which are connected by a plus (+) sign. The fixed-effects part of the model specifies the dependent variable and the predictor variable(s) with the dependent variable on the left-hand side of the tilde (~). The predictor variables on the right-hand side are connected by plus (+) symbols. Next, the random-effects part of the model is specified within parentheses. Similar to the random-effects part of `glmer()`, the random coefficients term is specified first, followed by a vertical line (|), and then, an identifier variable at a higher level as the grouping variable. For example, the command `clmm(y ~ x + (1|schid), Hess = TRUE, nAGQ = 7)` tells R to fit a multilevel model to estimate an ordinal response variable `y` on a predictor variable `x` with a randomly varying intercept across schools by specifying `(1|schid)`, where `1` indicates the random intercept on the left of the vertical line (|) and `schid` is the identifier variable on the right. The `Hess = TRUE` argument specifies the Hessian so the output provides variance and covariance estimates of random effects. The `nAGQ = 7` argument specifies the number of quadrature points for the adaptive Gauss-Hermite quadrature method. The default number is 1, which is used for the Laplace approximation. A larger number of quadrature points increases the precision but slows the speed of computation and may have converging issues. A negative number is used for nonadaptive Gauss-Hermite quadrature approximation. When there is more than a single random-effects term in the model, such as the random-coefficient model with varying intercepts and coefficients, only the Laplace approximation with `nAGQ = 1` is currently available. For more details on how to use this function, type `help(clmm)` in the command prompt after loading the `ordinal` package.

The `clmm()` function uses the maximum likelihood estimation with numerical integration for multilevel models for ordinal response variables. This method provides more accurate estimates than the quasi-likelihood methods, such as the marginal quasi-likelihood and the penalized quasi-likelihood methods. However, when using the numerical integration method, you may have converging problems. Sometimes it takes longer to converge or it does not converge at all. The integration methods in the `clmm()` function include both the nonadaptive and adaptive Gauss-Hermite quadrature approximations and the Laplace approximation, so the likelihood ratio test can be used for model comparisons with the `anova()` function. Discussion of the technical details of the estimation methods is beyond the scope of the text.

12.3.3 Unconditional Model or Null Model (Model 1)

The unconditional model or null model for the ordinal response variable includes no predictor variable at any levels. This model estimates the overall cumulative probabilities of being at or below a particular category and the variability in the cumulative probabilities between groups (e.g., schools). It can be expressed as follows:

$$\text{Level 1}: \ \text{logit}\left[\pi_{kij}(Y \leq k)\right] = \ln\left(\frac{\pi(Y_{ij} \leq k)}{\pi(Y_{ij} > k)}\right) = \alpha_k - \beta_{0j}$$

$$\text{Level 2}: \ \beta_{0j} = \gamma_{00} + u_{0j}$$

Since γ_{00} is set to zero, the composite model is $\eta_{ij} = \text{logit}[\pi_{kij}(Y_{ij} \leq k)] = \alpha_k - u_{0j}$.

In multilevel ordinal logistic regression models, logit(π_{kij}) is the logit link for the cumulative probability of being at or below a particular category k for the ith student in the jth school.

In the level 2 equation, γ_{00} is the overall logit or log odds of being at or below a particular math proficiency across schools, and u_{0j} is an error term at the school level. The intercept γ_{00} is set to zero since the cut points or thresholds α_k are estimated.

The `clmm()` Command and Output

The command `mpo.1 <- clmm(Profmath ~ 1 + (1|SCH_ID), data = chp11, na.action = "na.omit", Hess = TRUE, nAGQ = 7)` is used to fit the unconditional model (model 1). The model formula specifies the fixed effects of the model first. The ordinal response variable `Profmath` is on the left-hand side of the tilde (~). It needs to be a factor variable. Otherwise, `clmm()` does not run. You may either set it as a factor variable with the `factor(Profmath, ordered = TRUE)` function before model fitting or use the `as.factor(Profmath)` function in the model formula. We use the first method in this example with the `Profmath <- factor(Profmath, ordered = TRUE)` command. If you use the second method, the `as.factor(Profmath)` function tells R that the outcome variable is a factor or a categorical variable. Since this is the unconditional model, no predictor variables are specified on the right-hand side except the `1` as the intercept. The random effects part of the model specifies the randomly varying intercept across schools `(1|SCH_ID)`, where `1` indicates the random intercept and `SCH_ID` is the identifier variable. No random coefficients for any predictor variables are specified in this model.

The model is assigned to an object named `mpo.1`. To display the output of the model, we use the `summary()` function.

```
> library(ordinal)
> # Unconditional model/Null model with clmm()
> mpo.1 <- clmm(Profmath ~ 1 + (1|SCH_ID), data=chp12, na.action="na.omit",
Hess=TRUE, nAGQ=7)
> summary(mpo.1)
Cumulative Link Mixed Model fitted with the adaptive Gauss-Hermite
quadrature approximation with 7 quadrature points

formula: Profmath ~ 1 + (1 | SCH_ID)
data:    chp12

link   threshold  nobs   logLik     AIC       niter      max.grad  cond.H
logit  flexible   12793  -18809.26  37630.53  486(1948)  3.86e-02  9.3e+01

Random effects:
  Groups   Name          Variance   Std.Dev.
  SCH_ID   (Intercept)   0.8883     0.9425

Number of groups:  SCH_ID 619

No Coefficients

Threshold coefficients:
       Estimate     Std. Error    z value
0|1    -3.2917117   0.0576003    -57.148
1|2    -1.0699575   0.0438138    -24.421
2|3    -0.0002482   0.0427721     -0.006
3|4     1.5163383   0.0450078     33.691
4|5     5.3191981   0.1104598     48.155
```

Interpreting the Output

The output for the `clmm()` command displays the numerical integration method, the model formula, and the estimation information at the top. It then displays the random effects and fixed effects of the mixed-effects ordinal logistic regression model.

The numerical integration method for this model is adaptive Gauss-Hermite quadrature with seven quadrature points. The link function is logit. The thresholds or cut points are flexibly estimated. There are a total of 12,793 observations in level 1, which are nested in 619 schools in level 2. The log likelihood value is −18,809.26 and the AIC is 37,630.53. `niter` reports the number of Newton iterations. `max.grad` and `cond.H` display the maximum absolute gradient of the log likelihood function and the condition number of the Hessian matrix, respectively. These two values look small, so the model estimation shows no issues.

In the random-effects section, the variances and standard deviations of the random variables are reported. Since no random variables are specified in the model, only the between-school variance (τ_{00}) is reported. The column (labeled `Intercept`) reports that the between-school variance (τ_{00}) is .888, which is the intercept variance across all schools.

The intraclass correlation coefficient (ICC) in the multilevel ordinal logistic regression can be computed in the same way as that for the multilevel logistic regression. Since the variance of the logistic distribution is $\pi^2/3 = 3.29$, ICC $= \tau_{00}/(\tau_{00} + 3.29) = .888/(.888 + 3.29) = .213$, which indicates that 21.3% of the total variance is accounted for by schools in level 2. Therefore, using multilevel ordinal logistic regression is appropriate for the analysis.

The fixed-effects section reports the fixed effects of the predictors and the cut points. Since this is the null model with no predictors, only the coefficients of the cut points are reported. The output displays the estimates for the cut points, their standard errors, and the z values.

Five cut points or thresholds (labeled `0|1`, `1|2`, `2|3`, `3|4`, and `4|5`) are reported. As with the single-level proportional odds models, the cut points are used to differentiate the adjacent levels of the latent variable Y^* (i.e., math proficiency). For example, the latent variable falls at or below the first cut point α_1 when the response category is 0; it falls between the first cut point α_1 and the second cut point α_2 when the response category is 1.

With the `exp(coef(mpo.1))` command, we can obtain the odds ratios for the fixed effects. Since this is the unconditional model, only the odds ratios for the intercepts are reported. The results are displayed as follows.

```
> exp(coef(mpo.1))
        0|1          1|2          2|3          3|4            4|5
0.03719013   0.34302308   0.99975187   4.55551373   204.22005833
```

12.3.4 Random-Intercept Model (Model 2)

In the random-intercept model, we include the predictor variable `stlang` (whether a student's native language is English) to the level 1 equation. The model is referred to as the random-intercept model since the intercept is allowed to vary randomly across schools. This model can be expressed as follows:

$$\text{Level 1}: \ \text{logit}\left[\pi_{kij}(Y \leq k)\right] = \ln\left(\frac{\pi(Y_{ij} \leq k)}{\pi(Y_{ij} > k)}\right) = \alpha_k - \left(\beta_{0j} + \beta_{1j}\texttt{stlang}_{ij}\right)$$

$$\text{Level 2}: \ \beta_{0j} = \gamma_{00} + u_{0j}$$

$$\beta_{1j} = \gamma_{10}$$

Substituting the level 2 equations into the level 1 equation, the composite model is expressed as follows:

$$\eta_{ij} = \text{logit}\left[\pi_{kij}\left(Y_{ij} \leq k\right)\right] = \left(\alpha_k - u_{0j}\right) - \left(\gamma_{10}\texttt{stlang}_{ij}\right)$$

The level 1 equation includes the predictor variable, student's native language (`stlang`), and the ordinal response variable (`Profmath`). The level 2 equation for the intercept is specified to be random, and the slope for `stlang` is constrained to be fixed.

The `clmm()` Command and Output

The `mpo.2 <- clmm(Profmath ~ stlang + (1|SCH_ID), data = chp12, Hess = TRUE, nAGQ = 7)` command is used to fit the random-intercept model (model 2). The fixed part of the model includes a predictor variable `stlang` immediately after the ordinal response variable. Just like with the unconditional model, no random coefficients are specified in the random part of the model within the parenthesis. We assign `mpo.2` as the object name of the model so that it can be accessed by the following functions. Then, the `summary()` function is used to display the detailed output of the fitted random-intercept model. The following output is displayed.

```
> # Random-intercept model
> mpo.2 <- clmm(Profmath ~ stlang + (1|SCH_ID), data=chp12, Hess=TRUE, nAGQ=7)
> summary(mpo.2)
Cumulative Link Mixed Model fitted with the adaptive Gauss-Hermite
quadrature approximation with 7 quadrature points

formula: Profmath ~ stlang + (1 | SCH_ID)
data:    chp12

  link   threshold  nobs    logLik     AIC       niter       max.grad  cond.H
  logit  flexible   12793   -18783.19  37580.38  673(2697)   1.32e-02  1.5e+02

Random effects:
  Groups    Name          Variance    Std.Dev.
  SCH_ID    (Intercept)   0.8664      0.9308

Number of groups:  SCH_ID 619

Coefficients:
             Estimate    Std. Error    z value    Pr(>|z|)
stlangYes     0.37875       0.05243      7.225    5.03e-13 ***
---
Signif. codes:  0 '***' 0.001 '**' 0.01 '*' 0.05 '.' 0.1 ' ' 1

Threshold coefficients:
       Estimate    Std. Error    z value
0|1    -2.97675       0.07163    -41.560
1|2    -0.74889       0.06201    -12.077
2|3     0.32361       0.06168      5.246
3|4     1.84192       0.06359     28.967
4|5     5.64342       0.11926     47.319
```

Interpreting the Output

The random-effects section (labeled `Random effects`) reports that the intercept variance (τ_{00}), which is labeled `Intercept`, is .866. The intercept variance here is the variance associated with the thresholds or cut points.

The fixed-effects section (labeled `Coefficients`) reports the coefficient for `stlang` and the threshold coefficients section (labeled `Threshold coefficients`) reports the five cut points or thresholds. The logit coefficient for `stlang` (γ_{10}) = .379, z = 7.225, $p < .001$, which indicates that the cumulative logit or log odds of being above a particular math proficiency level for students with English as their native language is .379 times as high as that for those whose native language is not English. The odds ratio for `stlang` is reported with the `exp(coef(mpo.2))` command, which is interpreted below.

We use the `coef(mpo.2)` command to request the coefficients of the fixed effects and use the `confint(mpo.2)` function to extract the 95% confidence intervals of the parameter estimates. The results are displayed as follows.

```
> coef(mpo.2)
        0|1         1|2         2|3         3|4         4|5   stlangYes
 -2.9767539  -0.7488906   0.3236055   1.8419188   5.6434240   0.3787539
> confint(mpo.2)
                 2.5 %      97.5 %
0|1         -3.1171372  -2.8363706
1|2         -0.8704226  -0.6273586
2|3          0.2027076   0.4445035
3|4          1.7172898   1.9665479
4|5          5.4096707   5.8771772
stlangYes    0.2760009   0.4815069
```

We combine the results of the fixed effects and the confidence intervals with the `cbind(coef(mpo.2)), confint(mpo.2))` command.

```
> cbind(coef(mpo.2), confint(mpo.2))
                                2.5 %      97.5 %
0|1         -2.9767539  -3.1171372  -2.8363706
1|2         -0.7488906  -0.8704226  -0.6273586
2|3          0.3236055   0.2027076   0.4445035
3|4          1.8419188   1.7172898   1.9665479
4|5          5.6434240   5.4096707   5.8771772
stlangYes    0.3787539   0.2760009   0.4815069
```

We use the `exp(coef(mpo.2))` command to request the odds ratios and use the `exp(confint(mpo.2))` command to obtain the corresponding confidence intervals. The results are combined using `cbind(exp(coef(mpo.2)), exp(confint(mpo.2)))`. The following output is created.

```
> cbind(exp(coef(mpo.2)), exp(confint(mpo.2)))
                                   2.5 %         97.5 %
0|1              0.05095798     0.04428376     0.0586381
1|2              0.47289088     0.41877454     0.5340004
2|3              1.38210198     1.22471428     1.5597155
3|4              6.30863182     5.56941369     7.1459650
4|5            282.42808640   223.55796642   356.8006332
stlangYes        1.46046355     1.31784907     1.6185114
```

Interpretation of odds ratios in multilevel ordinal logistic regression is the same as that for single-level ordinal logistic regression. Odds ratio (OR) for `stlang` = 1.460, which means that the odds of being above a particular math proficiency level for students whose native language is English are 1.460 times as high as the odds for those whose native language is not English.

Likelihood Ratio Test Comparing the Unconditional Model and Random-Intercept Model

```
> anova(mpo.1, mpo.2)
Likelihood ratio tests of cumulative link models:
         formula:                              link: threshold:
mpo.1    Profmath ~ 1 + (1 | SCH_ID)           logit flexible
mpo.2    Profmath ~ stlang + (1 | SCH_ID)      logit flexible

         no.par     AIC     logLik     LR.stat    df    Pr(>Chisq)
mpo.1         6   37631     -18809
mpo.2         7   37580     -18783      52.143     1     5.161e-13 ***
---
Signif. codes: 0 '***' 0.001 '**' 0.01 '*' 0.05 '.' 0.1 ' ' 1
```

We can compare the random-intercept model (model 2) and the unconditional model (model 1) using the likelihood ratio test with the `anova(mpo.1, mpo.2)` command: $\chi^2_{(1)} = 52.143, p < .001$, which indicates that the random-intercept model fits the data better than the unconditional model, so we are in favor of the random-intercept model.

12.3.5 Random-Coefficient Model With a Level 1 Variable (Model 3)

Both intercepts and coefficients at level 1 are allowed to vary across schools at level 2 so the model estimates variance–covariance components. This model is referred to as the random-coefficient model since both the level 1 intercept and slope are allowed to vary across schools. When the variance–covariance components are specified as unstructured, the variances in intercepts and coefficients/slopes and the covariance between them will be estimated. This model can be expressed as follows:

$$\text{Level 1}: \ \text{logit}\left[\pi_{kij}(Y \le k)\right] = \ln\left(\frac{\pi(Y_{ij} \le k)}{\pi(Y_{ij} > k)}\right) = \alpha_k - \left(\beta_{0j} + \beta_{1j}\texttt{stlang}_{ij}\right)$$

$$\text{Level 2}: \ \beta_{0j} = \gamma_{00} + u_{0j}$$

$$\beta_{1j} = \gamma_{10} + u_{1j}$$

Since R set γ_{00} to zero, the composite model is expressed as follows:

$$\eta_{ij} = \text{logit}\left[\pi_{kij}(Y_{ij} \le k)\right] = (\alpha_k - u_{0j}) - \left(\gamma_{10}\texttt{stlang}_{ij} + u_{1j}\texttt{stlang}_{ij}\right)$$

The level 1 equation includes the predictor variable, student's native language (`stlang`), and the ordinal response variable (`Profmath`). In the level 2 equations, the level 1 intercept and the slope for `stlang` are both specified to be random with level 2 residuals u_{0j} and u_{1j}, respectively.

The `clmm()` Command and Output

The `mpo.3 <- clmm(Profmath ~ stlang + (stlang|SCH_ID), data = chp12, Hess = TRUE, nAGQ = 1)` command is used to fit the random-coefficient model (model 3). The fixed part of the model is the same as that of the random-intercept model. Unlike the random-intercept model, the predictor variable `stlang` is specified in the random part of the model, which indicates that the slope of the level 1 predictor `stlang` is allowed to vary randomly across schools. The model is assigned to an object named `mpo.3`. Please note that at the time of writing only the Laplace approximation is allowed in the random-coefficient model with more than one random-effects term. You may either set `nAGQ = 1` or leave the option empty since it is the default. Next, the `summary(mpo.3)` command tells R to display the detailed output of the fitted model.

```
> # Random-coefficient model
> # It will take several minutes to fit the model
> mpo.3 <- clmm(Profmath ~ stlang + (stlang|SCH_ID), data=chp12, Hess=TRUE, nAGQ=1)
> summary(mpo.3)
Cumulative Link Mixed Model fitted with the Laplace approximation

formula: Profmath ~ stlang + (stlang | SCH_ID)
data:    chp12

link  threshold nobs  logLik    AIC      niter     max.grad cond.H
logit flexible  12793 -18773.44 37564.88 861(3449) 2.01e-02 2.1e+02
```

```
Random effects:
   Groups    Name          Variance   Std.Dev.   Corr
   SCH_ID    (Intercept)   1.354      1.1637
             stlangYes     0.334      0.5779     -0.650
Number of groups:  SCH_ID 619

Coefficients:
            Estimate   Std. Error   z value   Pr(>|z|)
stlangYes    0.41382     0.06364     6.503     7.9e-11 ***
---
Signif. codes: 0 '***' 0.001 '**' 0.01 '*' 0.05 '.' 0.1 ' ' 1

Threshold coefficients:
      Estimate    Std. Error    z value
0|1   -2.96533      0.08102    -36.600
1|2   -0.71918      0.07341     -9.796
2|3    0.35905      0.07347      4.887
3|4    1.88280      0.07539     24.973
4|5    5.69358      0.12619     45.120
```

Interpreting the Output

We are interested in the variance and covariance components for the random effects since the slope of `stlang` is random across schools. The row (labeled `Intercept`) reports that the intercept variance (τ_{00}) is 1.354. The row (labeled `stlangYes`) reports that the slope variance (τ_{10}) for `stlang` is .334. As demonstrated in the previous chapters, the covariance between two variables equals the product of the correlation coefficient and the two standard deviations. Since the correlation coefficient for the intercept and the slope is $-.650$, we compute the covariance between them as $1.164^* \ .578^* \ (-.650) = -.437$.

In the coefficients and the threshold coefficients sections for the fixed effects, the coefficient for `stlang` and five cut points or thresholds are similar to those for the random-intercept model (model 2). The logit coefficient for `stlang` (γ_{10}) $= .414$, $z = 6.503$, $p < .001$, which indicates that the cumulative logit or log odds of being above a particular category of math proficiency for students with English as their native language is .414 times as great as that for those whose native language is not English.

We can use the `coef(mpo.3)` command to request the coefficients of the fixed effects and use the `confint(mpo.3)` command to extract the 95% confidence intervals of the parameter estimates. The results are omitted here.

We combine the results of the fixed effects and the confidence intervals with the `cbind(coef(mpo.3)), confint(mpo.3))` command.

```
> cbind(coef(mpo.3), confint(mpo.3))
                            2.5 %       97.5 %
0|1        -2.9653269  -3.1241227  -2.8065310
1|2        -0.7191838  -0.8630714  -0.5752963
2|3         0.3590502   0.2150601   0.5030403
3|4         1.8828038   1.7350348   2.0305728
4|5         5.6935792   5.4462539   5.9409045
stlangYes   0.4138193   0.2890873   0.5385513
```

We use the `exp(coef(mpo.3))` command and the `exp(confint(mpo.3))` command to obtain the odds ratios and the corresponding confidence intervals, respectively. The results are combined with `cbind(exp(coef(mpo.3)), exp(confint(mpo.3)))`. The following output is created.

```
> cbind(exp(coef(mpo.3)), exp(confint(mpo.3)))
                                2.5 %        97.5 %
0|1          0.05154362    0.0439755     0.0604142
1|2          0.48714968    0.4218644     0.5625382
2|3          1.43196864    1.2399364     1.6537415
3|4          6.57190517    5.6691249     7.6184488
4|5        296.95457346  231.8878517   380.2787341
stlangYes    1.51258379    1.3352083     1.7135227
```

The odds ratio for `stlang` is 1.513, which indicates that being students whose native language is English increases the odds of being above a particular math proficiency level by a factor of 1.513.

Model Comparison Using the Likelihood Ratio Test

```
> anova(mpo.2, mpo.3)
Likelihood ratio tests of cumulative link models:

        formula:                              link: threshold:
mpo.2   Profmath ~ stlang + (1 | SCH_ID)      logit flexible
mpo.3   Profmath ~ stlang + (stlang | SCH_ID) logit flexible

        no.par    AIC    logLik   LR.stat   df   Pr(>Chisq)
mpo.2        7  37580    -18783
mpo.3        9  37565    -18773    19.501    2    5.826e-05 ***
---
Signif. codes: 0 '***' 0.001 '**' 0.01 '*' 0.05 '.' 0.1 ' ' 1
```

The likelihood ratio test is used to compare the random-intercept model (model 2) and random-coefficient model with a level 1 predictor (model 3). The `anova(mpo.2, mpo.3)` function is used to compare these two models. The likelihood ratio test $\chi^2_{(2)} = 19.501$, $p < .01$, which indicates that the random-coefficient model has a significantly better fit than the random-intercept model.

12.3.6 Contextual Model With Both Level 1 and Level 2 Variables (Model 4)

The contextual model is a random-coefficient model with both level 1 and level 2 predictor variables. The level 1 predictor `stlang` remains unchanged. Two school-level variables `public` and `csclimat` are added to the level 2 equation:

$$\text{Level 1}: \ \text{logit}\left[\pi_{kij}(Y \leq k)\right] = \ln\left(\frac{\pi\left(Y_{ij} \leq k\right)}{\pi\left(Y_{ij} > k\right)}\right) = \alpha_k - \left(\beta_{0j} + \beta_{1j}\texttt{stlang}_{ij}\right)$$

$$\text{Level 2}: \ \beta_{0j} = \gamma_{00} + \gamma_{01}\texttt{public}_j + \gamma_{02}\texttt{csclimat}_j + u_{0j}$$

$$\beta_{1j} = \gamma_{10} + u_{1j}$$

The level 1 equation includes the predictor variable, the student's native language (`stlang`), and the ordinal response variable (`Profmath`). The level 2 equations for the level 1 intercept and the slope for `stlang` are both specified as random with level 2 residuals u_{0j} and u_{1j}, respectively. In level 2 equations, two school-level predictor variables are added to the equation for the level 1 random intercept, whereas the equation for the random slope contains no predictor variables. The composite model is expressed as follows:

$$\eta_{ij} = \text{logit}\left[\pi_{kij}\left(Y_{ij} \leq k\right)\right] = \left(\alpha_k - u_{0j}\right) - \left(\gamma_{01}\texttt{public}_j + \gamma_{02}\texttt{csclimat}_j + \gamma_{10}\texttt{stlang}_{ij} + u_{1j}\texttt{stlang}_{ij}\right)$$

The `clmm()` Command and Output

The `mpo.4 <- clmm(Profmath ~ stlang + public + csclimat + (stlang|SCH_ID), data = chp12, Hess = TRUE, nAGQ = 1)` command is used to fit the contextual model (model 4). The fixed part of the model includes one level 1 predictor, `stlang`, and two level 2 predictor variables, `public` and `csclimat`. The command itself does not distinguish between the level 1 and level 2 predictor variables. The predictor variable `stlang` is specified in the random part of the model `(stlang|SCH_ID)`, which indicates that the slope of the level 1 predictor `stlang` is allowed to vary randomly across schools. Only the Laplace approximation with the default `nAGQ = 1` is available for the model, so this argument is optional and can be omitted in the command. The `summary(mpo.4)` command presents the detailed results of the model.

```
> # Contextual model/Random-coefficient model with predictor variables in both
levels
> # It will take several minutes to fit the model
> mpo.4 <- clmm(Profmath ~ stlang + public + csclimat + (stlang|SCH_ID),
data=chp12, Hess=TRUE, nAGQ=1)
> summary(mpo.4)
Cumulative Link Mixed Model fitted with the Laplace approximation

formula: Profmath ~ stlang + public + csclimat + (stlang | SCH_ID)
data:    chp12

link   threshold  nobs   logLik     AIC       niter        max.grad  cond.H
logit  flexible   12793  -18686.83  37395.65  1500(6003)   6.47e-03  3.7e+02

Random effects:
  Groups   Name         Variance   Std.Dev.   Corr
  SCH_ID   (Intercept)  1.0644     1.0317
           stlangYes    0.3802     0.6166     -0.707
Number of groups:  SCH_ID 619

Coefficients:
                Estimate   Std. Error   z value   Pr(>|z|)
stlangYes        0.40330    0.06403      6.298    3.01e-10 ***
publicpublic    -0.57179    0.08877     -6.441    1.19e-10 ***
csclimat         0.53429    0.05580      9.575     < 2e-16 ***
---
Signif. codes:  0 '***' 0.001 '**' 0.01 '*' 0.05 '.' 0.1 ' ' 1

Threshold coefficients:
      Estimate    Std. Error    z value
0|1   -3.41121     0.10475     -32.566
1|2   -1.16816     0.09883     -11.820
2|3   -0.09007     0.09854      -0.914
3|4    1.43401     0.09940      14.426
4|5    5.23965     0.14133      37.075
```

Interpreting the Output

Regarding the variance and covariance components for the random effects, the row labeled `Intercept` reports that the intercept variance (τ_{00}) is 1.064. The row (labeled `stlangYes`) reports that the slope variance (τ_{10}) for `stlang` is .380. Since the covariance equals the product of the correlation between the two variables and their standard deviations, the covariance between the intercept and the slope can be computed with the equation: $1.032^* .617^* (-.707) = -.450$.

After two school-level variables are included in the contextual model (model 4), the between-school variance (τ_{00}) decreases from 1.354 to 1.064 when compared with that for the random-coefficient model (model 3): (1.354 − 1.064)/1.354 = 21%. This indicates that there is a decrease of 21% in the between-school variance from the random-coefficient model (model 3) to the contextual model (model 4) after the two school-level predictor variables are included.

The coefficients section for the fixed-effects reports the coefficients for `stlang` and the two new school-level predictor variables `public` and `csclimat`. The threshold coefficients section displays the five cut points or thresholds. The logit coefficient for `stlang` (γ_{10}) = .403, z = 6.298, p < .001. The result is similar to that in the random-coefficient table (model 3) so the interpretation for this variable is omitted here. Our interpretation focuses on the other two newly added level 2 predictor variables.

The logit coefficient for `public` (γ_{01}) = −.572, z = −6.441, p < .001, which indicates that being in public schools decreases the logit of being above a particular category of math proficiency by a factor of .572. The logit coefficient for `csclimat` (γ_{02}) = .534, z = 9.575, p < .001. This indicates that the logit of being above a particular category of math proficiency increases by a factor of .534 for a one-unit increase in school climate. The corresponding odds ratio can be obtained by exponentiating the coefficient with `exp(coef(mpo.4))`, which is discussed later in this section.

We use the `coef(mpo.4)` command and the `confint(mpo.4)` command to extract the parameter estimates of the fixed effects and the corresponding 95% confidence intervals, respectively. The results are omitted here.

We combine the results of the fixed effects and the confidence intervals with `cbind(coef(mpo.4)), confint(mpo.4)).`

```
> cbind(coef(mpo.4), confint(mpo.4))
                                    2.5 %      97.5 %
0|1              -3.41120874   -3.6165123   -3.2059052
1|2              -1.16816326   -1.3618704   -0.9744561
2|3              -0.09007254   -0.2832037    0.1030587
3|4               1.43401027    1.2391829    1.6288377
4|5               5.23965417    4.9626582    5.5166501
stlangYes         0.40330254    0.2777971    0.5288080
publicpublic     -0.57178827   -0.7457802   -0.3977963
csclimat          0.53428503    0.4249170    0.6436531
```

By exponentiating the coefficients with `exp(coef(mpo.4))`, we obtain the odds ratios. We also use the `exp(confint(mpo.4))` command to request the corresponding confidence intervals. The results are combined using the `cbind(exp (coef(mpo.4)), exp(confint(mpo.4)))` command. The following output is created.

```
> cbind(exp(coef(mpo.4)), exp(confint(mpo.4)))
                                    2.5 %         97.5 %
0|1                 0.03300129     0.02687625     0.0405222
1|2                 0.31093753     0.25618117     0.3773976
2|3                 0.91386489     0.75336628     1.1085564
3|4                 4.19549054     3.45279087     5.0979458
4|5               188.60486585   142.97335002   248.8001814
stlangYes           1.49675966     1.32021825     1.6969084
publicpublic        0.56451503     0.47436406     0.6717988
csclimat            1.70622791     1.52946347     1.9034215
```

The interpretation of the odds ratios focuses on the two newly added level 2 predictor variables. The odds ratio for `public` = .564, which indicates that a one-unit increase in public is associated with a decrease of .572 points in the odds of being above a particular category of math proficiency. In other words, the odds of being above a particular category for students in a public school are .564 times as high as the odds for students in a private school when holding all the other predictors constant. The odds ratio for `csclimat` = 1.708, which indicates that for a one-unit increase in school climate, the odds of being above a particular category of math proficiency increase by a factor of 1.708. So the odds of being in higher math proficiency levels are associated with schools with a more positive school climate.

Model Comparisons Using the Likelihood Ratio Test

```
> anova(mpo.3, mpo.4)
Likelihood ratio tests of cumulative link models:

        formula:                                                  link:
mpo.3   Profmath ~ stlang + (stlang | SCH_ID)                     logit
mpo.4   Profmath ~ stlang + public + csclimat + (stlang | SCH_ID) logit
        threshold:
mpo.3   flexible
mpo.4   flexible


        no.par    AIC    logLik    LR.stat    df    Pr(>Chisq)
mpo.3        9  37565    -18773
mpo.4       11  37396    -18687     173.23     2    < 2.2e-16 ***
--
Signif. codes: 0 '***' 0.001 '**' 0.01 '*' 0.05 '.' 0.1 ' ' 1
```

The likelihood ratio test is still used to compare the contextual model (model 4) and random-coefficient model (model 3) with the anova(mpo.3, mpo.4) command. The likelihood ratio test $\chi^2_{(2)} = 173.23$, $p < .001$, which indicates that we are in favor of the contextual model (model 4).

12.3.7 Contextual Model With Cross-Level Interactions (Model 5)

The contextual model with cross-level interactions is used to investigate whether there are interaction effects between level 1 and level 2 predictor variables. For example, if we are interested in whether the two school-level variables public and csclimat interact with the level 1 predictor stlang, we can include two interaction terms in the model. The two interaction terms need to be created in the model formula. One interaction public:stlang is created for two categorical variables stlang and public; and the other interaction csclimat:stlang is for the categorical variable stlang and the continuous variable csclimat. This model can be expressed as follows:

$$\text{Level 1}: \ \text{logit}\left[\pi_{kij}(Y \le k)\right] = \ln\left(\frac{\pi\left(Y_{ij} \le k\right)}{\pi\left(Y_{ij} > k\right)}\right) = \alpha_k - \left(\beta_{0j} + \beta_{1j}\mathbf{stlang}_{ij}\right)$$

$$\text{Level 2}: \ \beta_{0j} = \gamma_{00} + \gamma_{01}\mathbf{public}_j + \gamma_{02}\mathbf{csclimat}_j + u_{0j}$$

$$\beta_{1j} = \gamma_{10} + \gamma_{11}\mathbf{public}_j + \gamma_{12}\mathbf{csclimat}_j + u_{1j}$$

The expression of this model (model 5) is similar to that of the previous contextual model (model 4). The only difference is that the level 2 equation for the slope of stlang now includes two predictor variables public and csclimat.

The composite model is expressed as follows:

$$\eta_{ij} = \text{logit}\left[\pi_{kij}\left(Y_{ij} \le k\right)\right] = \left(\alpha_k - u_{0j}\right) - \left(\gamma_{01}\mathbf{public}_j + \gamma_{02}\mathbf{csclimat}_j + \gamma_{10}\mathbf{stlang}_{ij} + \gamma_{11}\mathbf{public}_j \times \mathbf{stlang}_{ij} + \gamma_{12}\mathbf{csclimat}_j \times \mathbf{stlang}_{ij} + u_{1j}\ \mathbf{stlang}_{ij}\right)$$

The clmm() Command and Output

The mpo.5 <- clmm(Profmath ~ stlang + public + csclimat + public:stlang + csclimat:stlang + (stlang|SCH_ID), data = chp12, Hess = TRUE, nAGQ = 1) command is used to fit the contextual model with cross-level interactions (model 5). The fixed part of the model includes one level 1 predictor, stlang, two level 2 predictor variables, public and csclimat, and two interaction terms, public:stlang and csclimat:stlang. The random part of the model and the options are the same as those for model 4. The Laplace approximation again is used in the model. The `summary(mpo.5)` command presents the results of the fitted model.

```
> # Contextual model with cross-level interactions
> mpo.5 <- clmm(Profmath ~ stlang + public + csclimat + public:stlang + csclimat:
stlang + (stlang|SCH_ID), data=chp12, Hess=TRUE, nAGQ=1)
> summary(mpo.5)
Cumulative Link Mixed Model fitted with the Laplace approximation

formula:
Profmath ~ stlang + public + csclimat + public:stlang + csclimat:stlang +
  (stlang | SCH_ID)
data:    chp12

link   threshold  nobs   logLik    AIC      niter      max.grad  cond.H
logit  flexible   12793  -18685.87 37397.74 1695(6769) 7.55e-03  1.5e+03

Random effects:
   Groups  Name         Variance   Std.Dev.   Corr
   SCH_ID  (Intercept)  1.0695     1.0341
           stlangYes    0.3941     0.6278     -0.708
Number of groups:  SCH_ID 619

Coefficients:
                           Estimate    Std. Error    z value    Pr(>|z|)
stlangYes                   0.23732     0.15866       1.496       0.135
publicpublic               -0.75939     0.18970      -4.003    6.25e-05 ***
csclimat                    0.57955     0.10395       5.575    2.47e-08 ***
stlangYes:publicpublic      0.19603     0.17737       1.105       0.269
stlangYes:csclimat         -0.04915     0.09677      -0.508       0.612
---
Signif. codes:  0 '***' 0.001 '**' 0.01 '*' 0.05 '.' 0.1 ' ' 1

Threshold coefficients:
       Estimate   Std. Error   z value
0|1     -3.5727      0.1741    -20.516
1|2     -1.3282      0.1698     -7.824
2|3     -0.2498      0.1694     -1.475
3|4      1.2743      0.1698      7.506
4|5      5.0804      0.1970     25.785
```

Interpreting the Output

In the random effects section, since the variance and covariance components for model 5 look similar to those for model 4, the interpretation is omitted here.

The coefficients section for the fixed effects displays the logit coefficients for the level 1 and level 2 predictor variables and the two interaction terms. The logit coefficient for public:stlang (γ_{11}) = .196, z = 1.105, $p > .05$, and the logit coefficient for csclimat:stlang (γ_{12}) = –.049, z = −.508, $p > .05$. So both interactions are not significant.

We use coef(mpo.5) and confint(mpo.5) to extract the parameter estimates of the fixed effects and the corresponding 95% confidence intervals, respectively. The results are omitted here. We can combine the results of the fixed effects and the confidence intervals with cbind(coef(mpo.5)), confint(mpo.5)).

We use the exp(coef(mpo.5)) command to obtain the odds ratios for the fixed effects by exponentiating the coefficients and use the exp(confint(mpo.5)) command to compute the corresponding confidence intervals. The results are combined with cbind(exp(coef(mpo.5)), exp(confint(mpo.5))). The following output is created.

```
> cbind(exp(coef(mpo.5)), exp(confint(mpo.5)))
                                                   2.5 %          97.5 %
0|1                          0.02808048       0.01996064      0.03950342
1|2                          0.26495609       0.18996458      0.36955166
2|3                          0.77894987       0.55892156      1.08559579
3|4                          3.57614819       2.56395840      4.98792643
4|5                        160.83966485     109.31613039    236.64758070
stlangYes                    1.26784442       0.92899509      1.73028844
publicpublic                 0.46795407       0.32264822      0.67869897
csclimat                     1.78523613       1.45616230      2.18867638
stlangYes:publicpublic       1.21656139       0.85931467      1.72232789
stlangYes:csclimat           0.95203790       0.78756392      1.15086043
```

Model Comparisons Using the Likelihood Ratio Test

```
> anova(mpo.4, mpo.5)
Likelihood ratio tests of cumulative link models:

     formula:
mpo.4 Profmath ~ stlang + public + csclimat + (stlang | SCH_ID)
mpo.5 Profmath ~ stlang + public + csclimat + public:stlang + csclimat:stlang +
(stlang | SCH_ID)
        link:    threshold:
mpo.4   logit    flexible
mpo.5   logit    flexible


        no.par     AIC    logLik    LR.stat    df    Pr(>Chisq)
mpo.4       11   37396    -18687
mpo.5       13   37398    -18686     1.9176     2        0.3833
```

We again use the likelihood ratio test to compare the contextual model with cross-level interactions (model 5) and the contextual model without interactions (model 4). With `anova(mpo.4, mpo.5)`, we compare the log-likelihood statistics of these two models. The likelihood ratio chi-square test $\chi^2_{(2)} = 1.918$, $p = .383$, which indicates that there is no significant improvement from model 4 to model 5. Therefore, the two interaction terms can be removed from the model.

12.3.8 Model Comparisons Using the AIC and the Log-Likelihood Statistics

We use the `anova(mpo.1, mpo.2, mpo.3, mpo.4, mpo.5)` command to compare the AIC and the log-likelihood statistics of all five models fitted above. The output is as follows.

```
> anova(mpo.1, mpo.2, mpo.3, mpo.4, mpo.5)
Likelihood ratio tests of cumulative link models:


      formula:
mpo.1 Profmath ~ 1 + (1  | SCH_ID)
mpo.2 Profmath ~ stlang + (1 | SCH_ID)
mpo.3 Profmath ~ stlang + (stlang | SCH_ID)
mpo.4 Profmath ~ stlang + public + csclimat + (stlang | SCH_ID)
mpo.5 Profmath ~ stlang + public + csclimat + public:stlang + csclimat:stlang +
      (stlang | SCH_ID)
      link:    threshold:
mpo.1 logit    flexible
mpo.2 logit    flexible
mpo.3 logit    flexible
mpo.4 logit    flexible
mpo.5 logit    flexible


      no.par    AIC    logLik    LR.stat   df    Pr(>Chisq)
mpo.1      6  37631    -18809
mpo.2      7  37580    -18783    52.1429    1     5.161e-13 ***
mpo.3      9  37565    -18773    19.5012    2     5.826e-05 ***
mpo.4     11  37396    -18687   173.2284    2     < 2.2e-16 ***
mpo.5     13  37398    -18686     1.9176    2        0.3833
---
Signif. codes: 0 '***' 0.001 '**' 0.01 '*' 0.05 '.' 0.1 ' ' 1
```

In the output, we can see that the AIC and the log-likelihood values decrease dramatically from the null model (model 1: `mpo.1`) to the contextual model without interactions (model 4: `mpo.4`), whereas both statistics increase slightly from model 4 (labeled `mpo.4`) to model 5 (labeled `mpo.5`). For example, the AIC for model 4 is 37,396 and that for model 5 is 37,398. The difference in AIC between these two models is only 2, which is small. Although there is no test to compare the difference in the AIC statistic, this provides additional information to confirm with the likelihood ratio test that there is no significant improvement from model 4 to model 5 ($\chi^2_{(2)}$ = 1.918, p = .383).

12.3.9 Interpreting the Predicted Probabilities With the `ggpredict()` Function in the `ggeffects` Package

By using the `ggpredict()` function in the `ggeffects` package (Lüdecke, 2018b), we can compute the predicted probabilities of being in each category of the ordinal response variable for predictor variables at specified values in the multilevel ordinal logistic regression model. Since the interaction terms have been removed from model 5, we use the `ggpredict()` function for the contextual model without interactions (model 4: `mpo.4`).

We first load the package with `library(ggeffects)` since it has been installed in previous chapters. We can compute the predicted probabilities for a continuous predictor variable at given values by different groups. In the following example, we compute the predicted probabilities of being in each of the six categories of the ordinal response variable for `csclimat` at different values by the two groups in `public` when holding other variables constant. The command is as follows: `mpo4.clpub <- ggpredict(mpo.4, terms = c("csclimat", "public"))`. In the `ggpredict()` function, the `terms = c("csclimat ", "public")` option specifies both `csclimat` and `public`, with the latter as the grouping variable. The output is assigned to an object named `mpo4.clpub` and is plotted with `plot(mpo4.clpub)`. The `as.data.frame()` function can be used to request the standard errors, and the output is omitted here.

```
> mpo4.clpub <- ggpredict(mpo.4, terms = c("csclimat", "public"))
Data were 'prettified'. Consider using `terms="csclimat [all]"` to get smooth
plots.
> mpo4.clpub

# Predicted probabilities of Profmath

# Response Level = 1
# public = catholic or other private

csclimat  | Predicted  |         95% CI
-----------------------------------------
   -2.80  |       0.11 | [0.07, 0.15]
   -1.80  |       0.07 | [0.05, 0.09]
   -1.00  |       0.04 | [0.03, 0.05]
    1.00  |       0.02 | [0.01, 0.02]
```

```
# Response Level = 2
# public = catholic or other private

csclimat  | Predicted |         95% CI
-----------------------------------
   -2.80  |      0.42 | [0.36, 0.48]
   -1.80  |      0.33 | [0.28, 0.38]
   -1.00  |      0.26 | [0.22, 0.30]
    1.00  |      0.12 | [0.10, 0.13]

# Response Level = 3
# public = catholic or other private

csclimat  | Predicted |         95% CI
-----------------------------------
   -2.80  |      0.24 | [0.21, 0.26]
   -1.80  |      0.26 | [0.25, 0.27]
   -1.00  |      0.26 | [0.24, 0.27]
    1.00  |      0.17 | [0.16, 0.19]

# Response Level = 4
# public = catholic or other private

csclimat  | Predicted |         95% CI
-----------------------------------
   -2.80  |      0.17 | [0.12, 0.22]
   -1.80  |      0.24 | [0.20, 0.28]
   -1.00  |      0.29 | [0.27, 0.32]
    1.00  |      0.36 | [0.35, 0.37]

# Response Level = 5
# public = catholic or other private

csclimat  | Predicted |         95% CI
-----------------------------------
   -2.80  |      0.06 | [0.04, 0.08]
   -1.80  |      0.10 | [0.07, 0.12]
   -1.00  |      0.14 | [0.12, 0.17]
    1.00  |      0.32 | [0.29, 0.36]

# Response Level = 6
# public = catholic or other private

csclimat  | Predicted |         95% CI
-----------------------------------
   -2.80  |      0.00 | [0.00, 0.00]
   -1.80  |      0.00 | [0.00, 0.00]
   -1.00  |      0.00 | [0.00, 0.01]
    1.00  |      0.01 | [0.01, 0.01]
```

```
# Response Level = 1
# public = public

csclimat  | Predicted  |        95% CI
-----------------------------------------
   -2.80  |      0.18  | [0.13, 0.22]
   -1.80  |      0.11  | [0.09, 0.13]
   -1.00  |      0.08  | [0.07, 0.09]
    1.00  |      0.03  | [0.02, 0.03]

# Response Level = 2
# public = public

csclimat  | Predicted  |        95% CI
-----------------------------------------
   -2.80  |      0.49  | [0.46, 0.51]
   -1.80  |      0.43  | [0.40, 0.46]
   -1.00  |      0.36  | [0.33, 0.38]
    1.00  |      0.18  | [0.16, 0.21]

# Response Level = 3
# public = public

csclimat  | Predicted  |        95% CI
-----------------------------------------
   -2.80  |      0.19  | [0.16, 0.22]
   -1.80  |      0.23  | [0.22, 0.25]
   -1.00  |      0.26  | [0.25, 0.27]
    1.00  |      0.23  | [0.21, 0.24]

# Response Level = 4
# public = public

csclimat  | Predicted  |        95% CI
-----------------------------------------
   -2.80  |      0.11  | [0.08, 0.14]
   -1.80  |      0.17  | [0.14, 0.19]
   -1.00  |      0.22  | [0.20, 0.24]
    1.00  |      0.34  | [0.33, 0.36]

# Response Level = 5
# public = public

csclimat  | Predicted  |        95% CI
-----------------------------------------
   -2.80  |      0.04  | [0.02, 0.05]
   -1.80  |      0.06  | [0.05, 0.07]
   -1.00  |      0.09  | [0.08, 0.10]
    1.00  |      0.21  | [0.19, 0.24]
```

```
# Response Level = 6
# public = public

csclimat  |  Predicted  |         95% CI
----------------------------------------
   -2.80  |       0.00  |  [0.00, 0.00]
   -1.80  |       0.00  |  [0.00, 0.00]
   -1.00  |       0.00  |  [0.00, 0.00]
    1.00  |       0.01  |  [0.00, 0.01]

Adjusted for:
* stlang =        No
* SCH_ID = 2751.00
> plot(mpo4.clpub)
```

Figure 12.1 shows the predicted probabilities of being in each category of math proficiency for `csclimat` at different values by the grouping variable `public`.

As shown in the graph, the predicted probabilities of being in proficiencies 4 and 5 increase with the increase of school climate and the predicted probabilities for students in Catholic or other private schools are higher than those in public schools.

As of the time of writing, the predicted probabilities for ordinal response variables by the `ggpredict()` function are on the population level since the variability of the random effects is not accounted for. Even if we use the `type = "re"` argument, the results are the same as those with the `type = "fixed"` argument.

12.4 MULTILEVEL MODELING FOR ORDINAL RESPONSE VARIABLES WITH THE `mixor()` FUNCTION IN THE `mixor` PACKAGE

12.4.1 The `mixor()` Function in the `mixor` Package

The `mixor()` function in the `mixor` package can also be used for multilevel models for ordinal response variables. Compared with `clmm()`, `mixor()` is normally faster. Since `mixor` is a user-written package, you need to install it first by typing `install.packages("mixor")` and then load the package by typing `library(mixor)`.

Similar to that in `clmm()`, the model formula in `mixor()` also includes two components connected by a plus (+) sign, but it is different from that of `clmm()` in specifying the random effects. The first component specifies the dependent variable and the predictor variable(s), which are separated by the tilde (~). This is the fixed-effects component of the model. Next, the random-effects component of the model is specified with the `id` argument and the `which.random.slope` argument, which is different from the random effects part of `clmm()` and `glmer()` within parentheses. The `id` argument specifies an identifier variable at a higher level as the grouping or cluster variable. The `which.random.slope` argument specifies a predictor variable

FIGURE 12.1 ● Predicted Probabilities of Being in Each Category of Math Proficiency for `csclimat` by `public`

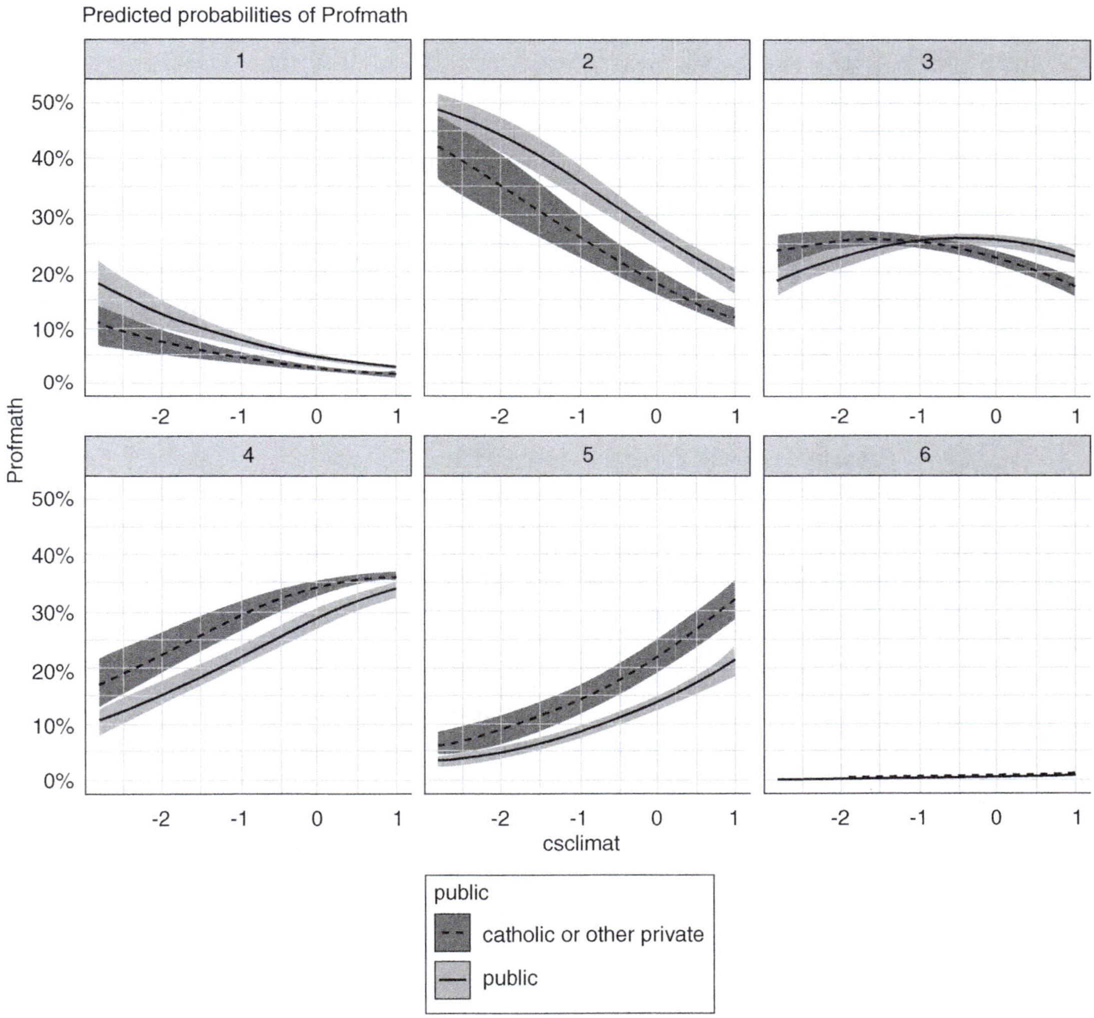

or a list of predictor variables that have random coefficients. For example, `which.random.slope = 1` specifies that the first predictor variable in the model formula has a random slope and `which.random.slope = c(1, 2)` specifies that the first and second predictor variables in the model formula have random slopes. The `nAGQ` argument specifies the number of quadrature points for adaptive Gauss-Hermite quadrature. Larger quadrature points for numerical integration increase the precision but slow the speed of computation. The `adaptive.quadrature = FALSE` specifies nonadaptive Gauss-Hermite quadrature. If not specified, the model uses the default adaptive quadrature method. The `link = "logit"` argument specifies the logit link. If omitted, the default link is `link = "probit"`. For example, the `mixor(y ~ x1 + x2, id = schid, data = data1, which.random.slope = 1, nAGQ = 7, link = "logit")` command tells R to fit a multilevel model to estimate the ordinal response variable `y` on the two predictor variables `x1` and `x2` with random intercepts varying across schools for the first predictor variable `x1` by specifying `which.random.slope = 1`. The `id = schid` specifies the level 2 identifier variable. The `nAGQ = 7` argument requests seven quadrature points although the default number is 11. For more details on how to use this function, type `help(mixor)` in the command prompt after loading the `mixor` package.

12.4.2 Multilevel Model for Ordinal Response Variables With Nonadaptive Gauss-Hermite Quadrature Using the `mixor()` Function

In the following two examples, we first fit one model with nonadaptive Gauss-Hermite quadrature and then fit the other model with the adaptive Gauss-Hermite quadrature. In the first example, the `mpo.4b <- mixor(Profmath ~ stlang + public + csclimat, id = SCH_ID, data = chp12, which.random.slope = 1, adaptive.quadrature = FALSE, link = "logit")` command tells R to fit the same contextual model (model 4) in the previous section after the two level 2 predictor variables `public` and `csclimat` are added to the model. In the function, `id = SCH_ID` specifies the level 2 identifier variable or cluster variable, `which.random.slope = 1` specifies that the first predictor variable `stlang` in the model formula has a random slope, and `adaptive.quadrature = FALSE` specifies the nonadaptive quadrature method for model estimation. The fitted model is named `mpo.4b`. The following output is displayed by the `summary(mpo.4b)` command.

```
> # Contextual model/Random-coefficient model with level 2 variables with mixor()
> # Install the mixor package by using install.packages("mixor")
> library(mixor)
> # Model with non-adaptive Gauss-Hermite quadrature
> mpo.4b <- mixor(Profmath ~ stlang + public + csclimat, id=SCH_ID, data=chp12,
which.random.slope=1, adaptive.quadrature = FALSE, link="logit")
Warning messages:
1: In if (class(Y) == "Surv") { :
   the condition has length > 1 and only the first element will be used
2: In model.response(mf, "numeric") :
   using type = "numeric" with a factor response will be ignored
> summary(mpo.4b)

Call:
mixor(formula = Profmath ~ stlang + public + csclimat, data = chp12,
  id = SCH_ID, which.random.slope = 1, adaptive.quadrature = FALSE,
  link = "logit")


Deviance =          37367.91
Log-likelihood =   -18683.95
RIDGEMAX =          0
AIC =              -18694.95
SBC =              -18719.31
```

```
                           Estimate   Std. Error    z value      P(>|z|)
(Intercept)                3.414720   0.104366      32.7187   < 2.2e-16 ***
stlangYes                  0.400605   0.064280       6.2322   4.599e-10 ***
publicpublic              -0.569141   0.085451      -6.6604   2.730e-11 ***
csclimat                   0.539604   0.056658       9.5238   < 2.2e-16 ***
(Intercept) (Intercept)    1.088290   0.140403       7.7512   9.104e-15 ***
(Intercept) stlangYes     -0.476764   0.122256      -3.8997   9.630e-05 ***
stlangYes stlangYes        0.412918   0.127091       3.2490    0.001158 **
Threshold1                 2.244028   0.038652      58.0577   < 2.2e-16 ***
Threshold2                 3.322575   0.041698      79.6828   < 2.2e-16 ***
Threshold3                 4.847313   0.045142     107.3802   < 2.2e-16 ***
Threshold4                 8.657841   0.109510      79.0596   < 2.2e-16 ***

---
Signif. codes: 0 '***' 0.001 '**' 0.01 '*' 0.05 '.' 0.1 ' ' 1
```

The R output produced by the `mixor()` function includes the call for the model command, the model fit statistics such as the deviance, the log-likelihood value, the AIC, and the SBC (i.e., BIC), and the estimates of the fixed effects, the random effects, and the threshold coefficients.

In the parameter estimates section, the estimates of the fixed effects, the random effects, and the threshold coefficients are ordered in sequence. The coefficients for the fixed effects include the coefficients for the intercept, `stlang`, and the two new school-level predictors, `public` and `csclimat`. The logit coefficient for `stlang` (γ_{10}) = .401, $z = 6.232$, $p < .001$. The logit coefficient for `public` (γ_{01}) = $-.569$, $z = -6.660$, $p < .001$. The logit coefficient for `csclimat` (γ_{02}) = .540, $z = 9.524$, $p < .001$. These results are similar to those estimated by the `clmm()` function in the previous section.

Regarding the variance and covariance components for the random effects, the row (labeled `(Intercept) (Intercept)`) reports that the intercept variance (τ_{00}) is 1.088. The row (labeled `stlangYes stlangYes`) reports that the slope variance (τ_{10}) for `stlang` is .413. The covariance between the intercept and the slope is reported in the row (labeled `(Intercept) stlangYes`), which is $-.477$. These results are also close to those in the random-effects section in model 4 provided by the `clmm()` function in the previous section.

The threshold coefficients are displayed at the end of the estimation table. The intercept and thresholds in the `mixor()` function are parameterized differently from those in the `clmm()` function. In the `clmm()` function, only the thresholds are estimated, whereas in the `mixor()` function, both the intercept and thresholds are estimated. These four thresholds in the output above show the deviations from the intercept. To obtain the same cutpoints or thresholds as those from the `clmm()` function, we need to subtract each threshold from the intercept. For example, the second cutpoint = 3.415 – 2.244 = 1.171. We can use the `Contrasts()` function to compute the cutpoints and this function is introduced later in the section.

12.4.3 Multilevel Model for Ordinal Response Variables With Adaptive Gauss-Hermite Quadrature Using the mixor() Function

In the second example, we fit the same model with the adaptive Gauss-Hermite quadrature method, which provides more accurate results than those from the last model with nonadaptive quadrature. The `mpo.4c <- mixor(Profmath ~ stlang + public + csclimat, id = SCH_ID, data = chp12, which.random.slope = 1, nAGQ = 7, link = "logit")` command tells R to fit the same contextual model (model 4). In the function, with everything else the same, `nAGQ = 7` specifies the adaptive quadrature method with the seven quadrature points for model estimation. The fitted model is named `mpo.4c`. The following output is displayed by the `summary(mpo.4c)` command.

```
> mpo.4c <- mixor(Profmath ~ stlang + public + csclimat, id=SCH_ID, data=chp12,
which.random.slope=1, nAGQ=7, link="logit")
> summary(mpo.4c)

Call:
mixor(formula = Profmath ~ stlang + public + csclimat, data = chp12,
  id = SCH_ID, which.random.slope = 1, nAGQ = 7, link = "logit")

Deviance =              37368.31
Log-likelihood =       -18684.16
RIDGEMAX =              0
AIC =                  -18695.16
SBC =                  -18719.51

                          Estimate   Std. Error   z value     P(>|z|)
(Intercept)               3.409932   0.105612    32.2872   < 2.2e-16 ***
stlangYes                 0.406871   0.065035     6.2562   3.945e-10 ***
publicpublic             -0.571980   0.085948    -6.6549   2.835e-11 ***
csclimat                  0.535154   0.057534     9.3016   < 2.2e-16 ***
(Intercept) (Intercept)   1.098554   0.139982     7.8478   4.219e-15 ***
(Intercept) stlangYes    -0.485904   0.123329    -3.9399   8.152e-05 ***
stlangYes stlangYes       0.422032   0.128816     3.2762    0.001052 **
Threshold1                2.244951   0.038676    58.0452   < 2.2e-16 ***
Threshold2                3.323592   0.041723    79.6594   < 2.2e-16 ***
Threshold3                4.848243   0.045178   107.3132   < 2.2e-16 ***
Threshold4                8.656617   0.109800    78.8401   < 2.2e-16 ***
---
Signif. codes: 0 '***' 0.001 '**' 0.01 '*' 0.05 '.' 0.1 ' ' 1
```

We use the `coef(melr.4c)` command and the `confint(melr.4c)` command to extract the coefficients and the 95% confidence intervals of the parameter estimates, respectively. The results are combined with `cbind(coef(melr.4c)), confint(melr.4c))`. The output is omitted here.

We use the `exp(coef(melr.4c))` command to obtain the odds ratios and use the `exp(confint(melr.4c))` command to produce the corresponding confidence intervals. The results are combined using the `cbind(exp(coef(melr.4c)), exp(confint(melr.4c)))` command. The following output is created.

```
> exp(coef(mpo.4c))
            (Intercept)                stlangYes            publicpublic
             30.2631895                1.5021109               0.5644070
      csclimat (Intercept) (Intercept)   (Intercept) stlangYes
              1.7077105                2.9998256               0.6151406
    stlangYes stlangYes               Threshold1              Threshold2
              1.5250575                9.4399483              27.7598984
             Threshold3               Threshold4
            127.5161953             5748.0567784
> exp(confint(mpo.4c))
                                  2.5 %         97.5 %
(Intercept)                  24.6046534     37.2230659
stlangYes                     1.3223426      1.7063182
publicpublic                  0.4769062      0.6679621
csclimat                      1.5256028      1.9115560
(Intercept) (Intercept)       2.2800445      3.9468325
(Intercept) stlangYes         0.4830549      0.7833436
stlangYes stlangYes           1.1847807      1.9630640
Threshold1                    8.7508181     10.1833478
Threshold2                   25.5801784     30.1253552
Threshold3                  116.7103511    139.3225187
Threshold4                 4635.1025289   7128.2472226
> cbind(exp(coef(mpo.4c)), exp(confint(mpo.4c)))
                                                2.5 %         97.5 %
(Intercept)                  30.2631895    24.6046534     37.2230659
stlangYes                     1.5021109     1.3223426      1.7063182
publicpublic                  0.5644070     0.4769062      0.6679621
csclimat                      1.7077105     1.5256028      1.9115560
(Intercept) (Intercept)       2.9998256     2.2800445      3.9468325
(Intercept) stlangYes         0.6151406     0.4830549      0.7833436
stlangYes stlangYes           1.5250575     1.1847807      1.9630640
Threshold1                    9.4399483     8.7508181     10.1833478
Threshold2                   27.7598984    25.5801784     30.1253552
Threshold3                  127.5161953   116.7103511    139.3225187
Threshold4                 5748.0567784  4635.1025289   7128.2472226
```

Although the output provides the odds ratios for both the fixed effects and the random effects, we are only interested in those for the fixed effects.

We can use the `Contrasts()` function to compute the cutpoints. Since the cutpoints are computed as the differences between the intercept and the thresholds, we need to construct a matrix for the contrasts. In the `Contrasts(mpo.4c, contrast.matrix = cm)` command, `mpo.4c` is the model object and `contrast.matrix = cm` specifies the matrix for the contrast. In the matrix, we code 1 for the four thresholds and –1 for the intercept to create the contrasts.

```
> cm <- matrix(c(-1, 0, 0, 0, 0, 0, 0, 0, 0, 0, 0, -1,0, 0, 0, 0, 0, 0, 1, 0, 0, 0,
+               -1,0, 0, 0, 0, 0, 0, 0, 1, 0, 0,-1,0, 0, 0, 0, 0, 0, 0, 1, 0,
+               -1,0, 0, 0, 0, 0, 0, 0, 0, 0, 1), ncol=5)
> Contrasts(mpo.4c, contrast.matrix = cm)
                            1   2   3   4   5
(Intercept)                -1  -1  -1  -1  -1
stlangYes                   0   0   0   0   0
publicpublic                0   0   0   0   0
csclimat                    0   0   0   0   0
(Intercept) (Intercept)     0   0   0   0   0
(Intercept) stlangYes       0   0   0   0   0
stlangYes stlangYes         0   0   0   0   0
Threshold1                  0   1   0   0   0
Threshold2                  0   0   1   0   0
Threshold3                  0   0   0   1   0
Threshold4                  0   0   0   0   1

     Estimate    Std. Error    z value    P(>|z|)
1   -3.409932    0.105612    -32.2872    <2e-16 ***
2   -1.164982    0.098540    -11.8224    <2e-16 ***
3   -0.086340    0.098684     -0.8749    0.3816
4    1.438311    0.098483     14.6046    <2e-16 ***
5    5.246685    0.137209     38.2387    <2e-16 ***
---
Signif. codes:  0 '***' 0.001 '**' 0.01 '*' 0.05 '.' 0.1 ' ' 1
```

The cutpoints can also be computed manually from the output produced by the `summary(mpo.4b)` command.

Cutpoint 1 = `Intercept` = 3.410.

Cutpoint 2 = `Intercept` − `Threshold1` = 3.410 − 2.245 = 1.165.

Cutpoint 3 = `Intercept` − `Threshold2` = 3.410 − 3.324 = .086.

Cutpoint 4 = `Intercept` − `Threshold3` = 3.410 − 4.848 = −1.438.

Cutpoint 5 = `Intercept` − `Threshold4` = 3.410 − 8.657= −5.247.

Please note that the cutpoints which we compute manually and those from the output with the `Contrasts()` function have opposite signs since the former cutpoints are for the model estimating the cumulative logit odds of being above a particular category, whereas the latter cutpoints are for the model estimating the cumulative logit odds of being at or below that category. If we would like to obtain the cutpoints with the same signs from the output with the `Contrasts()` function, we could code −1 for the four thresholds and 1 for the intercept to create the contrasts in the matrix.

The `ggeffects` package also supports the `mixor` package. You can use the `ggpredict()` function to obtain the predicted probabilities for predictor variables at specified values in the multilevel ordinal logistic regression model.

12.5 MAKING PUBLICATION-QUALITY TABLES USING THE texreg PACKAGE

Currently, the `stargazer` package does not work with `clmm()` in the `ordinal` package, so we use the `screenreg()` and `htmlreg()` functions from the `texreg` package (Leifeld, 2013) to display the results from the unconditional model to the contextual model. We load the package first by typing `library(texreg)`.

The results of the four models, including the unconditional model, the random-intercept model, the random-coefficient model, and the contextual model, can be displayed in a table with the following command: `screenreg(list(mpo.1, mpo.2, mpo.3, mpo.4))`.

```
> # Presenting the results with texreg()
> library(texreg)
Version:     1.37.5
Date:        2020-06-17
Author:      Philip Leifeld (University of Essex)

Please cite the JSS article in your publications -- see citation("texreg").
> screenreg(list(mpo.1, mpo.2, mpo.3, mpo.4))

=====================================================================
                                   Model 1     Model 2     Model 3     Model 4
---------------------------------------------------------------------
0|1                               -3.29 ***   -2.98 ***   -2.97 ***   -3.41 ***
                                  (0.06)      (0.07)      (0.08)      (0.10)
1|2                               -1.07 ***   -0.75 ***   -0.72 ***   -1.17 ***
                                  (0.04)      (0.06)      (0.07)      (0.10)
2|3                               -0.00        0.32 ***    0.36 ***   -0.09
                                  (0.04)      (0.06)      (0.07)      (0.10)
3|4                                1.52 ***    1.84 ***    1.88 ***    1.43 ***
                                  (0.05)      (0.06)      (0.08)      (0.10)
4|5                                5.32 ***    5.64 ***    5.69 ***    5.24 ***
                                  (0.11)      (0.12)      (0.13)      (0.14)
stlangYes                                      0.38 ***    0.41 ***    0.40 ***
                                              (0.05)      (0.06)      (0.06)
publicpublic                                                          -0.57 ***
                                                                      (0.09)
csclimat                                                               0.53 ***
                                                                      (0.06)
---------------------------------------------------------------------
Log Likelihood                  -18809.26   -18783.19   -18773.44   -18686.83
AIC                              37630.53    37580.38    37564.88    37395.65
BIC                              37675.27    37632.58    37631.99    37477.68
Num. obs.                        12793       12793       12793       12793
Groups (SCH_ID)                    619         619         619         619
Variance: SCH_ID: (Intercept)        0.89        0.87        1.35        1.06
Variance: SCH_ID: stlangYes                                0.33        0.38
=====================================================================
*** p < 0.001; ** p < 0.01; * p < 0.05
> htmlreg(list(mpo.1,mpo.2, mpo.3,mpo.4), file="mpo.doc", doctype=TRUE, html.tag=TRUE,
head.tag=TRUE)
The table was written to the file 'mpo.doc'.
```

To create the table in the HTML format and copy it into Microsoft Word, we use the command as follows: `htmlreg(list(mpo.1,mpo.2, mpo.3,mpo.4), file = "mpo.doc", doctype = TRUE, html.tag = TRUE, head.tag = TRUE)`. The original table presenting the results from the unconditional model to the contextual model is omitted here. With edits on the model titles and the added labels for the fixed and random effects, the final table is displayed as in Table 12.1.

TABLE 12.1 ● Results of the Multilevel PO Models: From the Unconditional Model to the Contextual Model (Edited)

	Model 1 Unconditional Model	Model 2 Random-Intercept Model	Model 3 Random-Coefficient Model	Model 4 Contextual Model
	Coefficient (SE)	Coefficient (SE)	Coefficient (SE)	Coefficient (SE)
Fixed Effects				
0|1	−3.29***	−2.98***	−2.97***	−3.41***
	(0.06)	(0.07)	(0.08)	(0.10)
1|2	−1.07***	−0.75***	−0.72***	−1.17***
	(0.04)	(0.06)	(0.07)	(0.10)
2|3	−0.00	0.32***	0.36***	−0.09
	(0.04)	(0.06)	(0.07)	(0.10)
3|4	1.52***	1.84***	1.88***	1.43***
	(0.05)	(0.06)	(0.08)	(0.10)
4|5	5.32***	5.64***	5.69***	5.24***
	(0.11)	(0.12)	(0.13)	(0.14)
stlang		0.38***	0.41***	0.40***
		(0.05)	(0.06)	(0.06)
public				−0.57***
				(0.09)
csclimat				0.53***
				(0.06)

Table 12.1 *(Continued)*

	Model 1 Unconditional Model	Model 2 Random-Intercept Model	Model 3 Random-Coefficient Model	Model 4 Contextual Model
	Coefficient (SE)	Coefficient (SE)	Coefficient (SE)	Coefficient (SE)
Log Likelihood	−18,809.26	−18,783.19	−18,773.44	−18,686.83
AIC	37,630.53	37,580.38	37,564.88	37,395.65
BIC	37,675.27	37,632.58	37,631.99	37,477.68
Num. obs.	12,793	12,793	12,793	12,793
Number of Groups (SCH_ID)	619	619	619	619
Random Effects				
Between-School Variance: SCH_ID: (Intercept)	0.89	0.87	1.35	1.06
Slope Variance: SCH_ID: stlang			0.33	0.38

Note: Standard errors are shown in parentheses.
***$p < .001$
**$p < .01$
*$p < .05$

12.6 REPORTING THE RESULTS

Writing the results of the multilevel PO models is similar to that of the multilevel models for the continuous and binary response variables.

First, describe the purpose of your study and explain why the multilevel PO models are needed for the analysis.

Second, report the model-building steps and briefly describe each model. Report and interpret the intraclass correlation coefficient, which is computed from the between-group variance in the unconditional model.

Third, report the results of a series of fitted models in a table including both the parameter estimates for the fixed effects and the variances and covariances for the random effects. If available, include the log likelihood and the deviance statistics (i.e., −2LL) and the AIC and BIC statistics for these models in the table.

Fourth, as with the conventional PO model, report the parameter estimates for the fixed effects of the predictor variables in the final contextual model and interpret the corresponding odds ratios. In addition, report the variances and covariances for the random effects in text. The following is an example of summarizing the results of the unconditional model and the contextual model fitted earlier.

Multilevel modeling for the ordinal response variable was used to estimate the relationships between an ordinal response variable, students' math proficiency, and a student-level variable, the student's native language, and two school-level variables, the school type and the school climate. Five models, from the unconditional (null) model to the contextual model with cross-level interactions, were fitted. Table 12.1 presents the parameter estimates for the fixed effects and random effects for the fitted models. Since no cross-level interactions were identified in the contextual model, the interaction terms were removed from the model. For illustration purposes, the following presentations only focused on the results of the unconditional model and the contextual model without cross-level interactions.

Results for the Unconditional Model (Model 1)

There was a significant between-group variance: $\tau_{00} = .888$, which is the intercept variance across all schools. The ICC in the unconditional model was .213, which indicated that 21.3% of the total variance was accounted for by schools in level 2. Therefore, using multilevel ordinal logistic regression was justified.

Results for the Contextual Model Without Cross-Level Interactions (Model 4)

The logit coefficient for `stlang` (γ_{10}) = .403, $z = 6.298$, $p < .001$. OR = 1.320, which indicated that the odds of being above a particular mathematics proficiency level for students whose native language was English were 1.320 times as high as the odds for those whose native language was not English.

The logit coefficient for `public` (γ_{01}) = −.571, $z = -6.441$, $p < .001$, which indicated that being in public schools decreased the logit of being above a particular category of mathematics proficiency by a factor of .571. The corresponding odds ratio was .474, which indicated that the odds of being above a particular category for students in a public school were .474 times as high as the odds for students in a private school when holding all the other predictors constant.

The logit coefficient for `csclimat` (γ_{02}) = .534, $z = 9.575$, $p < .001$. OR = 1.529, which indicated that for a one-unit increase in the school climate the odds of being above a particular category of mathematics proficiency increased by a factor of 1.529.

Regarding the variance and covariance components for the random effects, the between-group variance, the intercept variance (τ_{00}), was 1.064. The slope variance (τ_{10}) for `stlang` was .421. The covariance between the intercept and the slope was −.450.

After two school-level variables were included in the contextual model (model 4), the between-school variance (τ_{00}) decreased from 1.354 to 1.064 compared with that for the random-coefficient model (model 3): (1.354 − 1.064)/1.354 = 21%, which indicated that there was a decrease of 21% in the between-school variance from the random-coefficient model (model 3) to the contextual model (model 4) after the two school-level predictor variables were included.

12.7 SUMMARY OF R COMMANDS IN THIS CHAPTER

```
# Chap 12 R Script

# Remove all objects
rm(list = ls(all = TRUE))

# The following user-written packages need to be installed first by using
install.packages(" ") and then by loading it with library()
# library(ordinal)       # It is already installed for Chapter 4
# library(ggeffects)     # It is already installed for Chapter 2
# library(texreg)        # It is already installed for Chapter 4
# library(mixor)
# library(emmeans)

# Import the dataset
library(foreign)
chp12<-read.dta("C:/CDA/els2002.dta")
chp12 <- chp12[!is.na(chp12$Profmath)&!is.na(chp12$stlang)&!is.na(chp12$public)
&!is.na(chp12$sclimate), ]
chp12$Profmath <- factor(chp12$Profmath, ordered=TRUE)
attach(chp12)
table(Profmath)
table(stlang)
mean(sclimate)
table(public)
chp12$csclimat <- sclimate-mean(sclimate, na.rm=TRUE)
summary(csclimat)
```

```
library(ordinal)
# Unconditional model/Null model with clmm()
mpo.1 <- clmm(Profmath ~ 1 + (1|SCH_ID), data=chp12, na.action="na.omit", Hess=
TRUE, nAGQ=7)
summary(mpo.1)
coef(mpo.1)
confint(mpo.1)
exp(coef(mpo.1))
exp(confint(mpo.1))

# Random-intercept model
# It will take several minutes to fit the model
mpo.2 <- clmm(Profmath ~ stlang + (1|SCH_ID), data=chp12, Hess=TRUE, nAGQ=7)
summary(mpo.2)
coef(mpo.2)
confint(mpo.2)
cbind(coef(mpo.2), confint(mpo.2))
exp(coef(mpo.2))
exp(confint(mpo.2))
cbind(exp(coef(mpo.2)), exp(confint(mpo.2)))
anova(mpo.1, mpo.2)

# Random-coefficient model
# It will take several minutes to fit the model
mpo.3 <- clmm(Profmath ~ stlang + (stlang|SCH_ID), data=chp12, Hess=TRUE,
nAGQ=1)
summary(mpo.3)
coef(mpo.3)
confint(mpo.3)
cbind(coef(mpo.3), confint(mpo.3))
exp(coef(mpo.3))
exp(confint(mpo.3))
cbind(exp(coef(mpo.3)), exp(confint(mpo.3)))
anova(mpo.2, mpo.3)

# Contextual model/Random-coefficient model with predictor variables in both levels
# It will take several minutes (normally around 5 minutes) to fit the model
mpo.4 <- clmm(Profmath ~ stlang + public + csclimat + (stlang|SCH_ID), data=
chp12, Hess=TRUE, nAGQ=1)
summary(mpo.4)
coef(mpo.4)
confint(mpo.4)
cbind(coef(mpo.4), confint(mpo.4))
exp(coef(mpo.4))
exp(confint(mpo.4))
cbind(exp(coef(mpo.4)), exp(confint(mpo.4)))
anova(mpo.3, mpo.4)

# Contextual model with cross-level interactions
# It will take several minutes (normally around 5 minutes) to fit the model
mpo.5 <- clmm(Profmath ~ stlang + public + csclimat + public:stlang + csclimat:
stlang + (stlang|SCH_ID), data=chp12, Hess=TRUE, nAGQ=1)
summary(mpo.5)
coef(mpo.5)
confint(mpo.5)
```

```
cbind(coef(mpo.5), confint(mpo.5))
exp(coef(mpo.5))
exp(confint(mpo.5))
cbind(exp(coef(mpo.5)), exp(confint(mpo.5)))
anova(mpo.4, mpo.5)

# Model comparisons
anova(mpo.1, mpo.2, mpo.3, mpo.4, mpo.5)

# Marginal effects/Predicted probabilities with ggpredict()
# install.packages(emmeans)
# install.packages("ggeffects")
library(ggeffects)
mpo4.clpub <- ggpredict(mpo.4, terms = c("csclimat", "public"))
mpo4.clpub
plot(mpo4.clpub)

# Presenting the results with texreg()
library(texreg)
screenreg(list(mpo.1, mpo.2, mpo.3, mpo.4))
htmlreg(list(mpo.1, mpo.2, mpo.3, mpo.4), file="mpo.doc", doctype=TRUE, html.
tag=TRUE, head.tag=TRUE)

# Contextual model/Random-coefficient model with level 2 variables with mixor()
# Install the mixor package by using install.packages("mixor")
library(mixor)
# Model with non-adaptive Gauss-Hermite quadrature
mpo.4b <- mixor(Profmath ~ stlang + public + csclimat, id=SCH_ID, data=chp12,
which.random.slope=1, adaptive.quadrature = FALSE, link="logit")
summary(mpo.4b)

# Model with adaptive Gauss-Hermite quadrature
mpo.4c <- mixor(Profmath ~ stlang + public + csclimat, id=SCH_ID, data=chp12,
which.random.slope=1, nAGQ=7, link="logit")
summary(mpo.4c)
coef(mpo.4c)
confint(mpo.4c)
cbind(coef(mpo.4c), confint(mpo.4c))
exp(coef(mpo.4c))
exp(confint(mpo.4c))
cbind(exp(coef(mpo.4c)), exp(confint(mpo.4c)))
cm <- matrix(c(-1, 0, 0, 0, 0, 0, 0, 0, 0, 0, 0, -1,0, 0, 0, 0, 0, 0, 1, 0, 0, 0,
               -1,0, 0, 0, 0, 0, 0, 0, 1, 0, 0,-1,0, 0, 0, 0, 0, 0, 0, 0, 1, 0,
               -1,0, 0, 0, 0, 0, 0, 0, 0, 0, 1), ncol=5)
Contrasts(mpo.4c, contrast.matrix = cm)

detach(chp12)
```

Glossary

A series of **cut points** for the underlying binary logistic models for the dichotomized ordinal response variable are estimated. When the ordinal response variable has k levels, $K - 1$ cut points will be estimated.

Cross-level interactions are interaction effects between level 1 and level 2 predictor variables.

In **a random-coefficient PO model**, both the intercept and coefficient in the level 1 equation are allowed to vary across groups at level 2.

In **a random-intercept PO model**, the intercept is allowed to vary freely across groups.

Multilevel proportional odds (PO) models are an extension of multilevel binary logistic regression models when the response variables are ordinal.

Multilevel PO models also assume that **the logit coefficients** for each predictor variable are the same across the categories of the ordinal outcome variable. This is known as the proportional odds or parallel lines assumption.

The contextual model for an ordinal response variable includes both level 1 and level 2 predictor variables.

The unconditional model, or null model, for an ordinal response variable includes no predictor variable at any level. This model estimates the overall cumulative probabilities of being at or below a particular category and the variability in the cumulative probabilities between groups (e.g., schools).

Exercises

Use the ELS:2002 data available at **https://edge.sagepub.com/liu1e** with the following variables for the multilevel ordinal logistic regression models.

`Profmath`: mathematics proficiency level with six ordinal categories from 0 to 5

`gender`: gender (1 = female; 0 = male)

`byses`: socioeconomic status composite for base-year data

`urban`: urban schools or not (1 = urban; 0 = suburban or rural areas) (school-level variable).

This study investigates how student-level and school-level predictors influence students' mathematics proficiency. The ordinal outcome variable is `Profmath` with six levels, and the predictor variables are `gender` and `byses`. Conduct the multilevel modeling for the ordinal response variable and then answer the following questions:

1. Use the `clmm()` function to fit an unconditional model and obtain the between-group variance. Compute the ICC and interpret it.

2. Fit a two-level, random-intercept model with a random intercept and the two student-level predictor variables `gender` and `byses`. Use the grand-mean centering for `byses` before fitting the model.
3. Interpret the odds ratios of the two predictor variables.
4. Conduct a likelihood ratio test comparing the random-intercept model and the unconditional model. Interpret the results.
5. Fit a two-level, contextual model with `gender` and `byses` as the student-level predictor variables and `urban` as the school-level predictor variable. Use the grand-mean centering for `byses`. Specify a random intercept. No cross-level interactions are included.
6. Interpret the odds ratio of `urban` in the contextual model.
7. Conduct the likelihood ratio test among the three fitted models. Which one fits the data best? Why?
8. Use the AIC statistic to compare the three fitted models. Interpret the results. Do they support the finding from the likelihood ratio tests?
9. Use the `mixor()` function to fit the same contextual model shown earlier.
10. Present a table containing the estimated logit coefficients for the contextual model.
11. Write a report to summarize the results of the contextual model.

13

MULTILEVEL MODELING FOR COUNT RESPONSE VARIABLES

OBJECTIVES OF THIS CHAPTER

This chapter introduces multilevel modeling for count response variables. It starts with an introduction to the model specification, the incidence rate ratios in the model, and a description of the research questions and data. Then several models, from the unconditional (null) model to the random-intercept model and random-coefficient model to the contextual models, are illustrated using R with step-by-step instructions. The three packages, `lme4`, `glmmTMB`, and `MASS`, are used for fitting multilevel Poisson regression models. In addition, the `glmmTMB` package is used to fit multilevel negative binomial regression models. R commands are explained, and the output is interpreted for each model in detail. This chapter also illustrates how the results are displayed in publication-quality tables using R and reported in text. It focuses on model fitting with R, as well as on interpreting and presenting the results. After reading this chapter, you should be able to:

- Determine when multilevel modeling for count response variables is used.
- Formulate multilevel models for count response variables.
- Conduct the multilevel modeling analysis for count response variables using R.
- Interpret the output.
- Compute and plot the predicted counts using the `ggeffects()` function.
- Be familiar with model fitting strategies and compare models using the likelihood ratio test.
- Present results in publication-quality tables using R.
- Write the results for publication.

13.1 MULTILEVEL MODELING FOR COUNT RESPONSE VARIABLES

13.1.1 Model Specification

In the previous sections, we introduced the two types of generalized linear mixed models, multilevel models for binary and ordinal response variables. Multilevel modeling can be extended for count response variables. In this section, multilevel Poisson models (Goldstein, 2003; Hox, 2010; Raudenbush & Bryk, 2002; Snijders & Bosker, 2012) and multilevel negative binomial models for count response variables are introduced.

Recall that in the single-level Poisson regression model introduced in Chapter 8, the response variable is a count of events, which is a non-negative integer. The Poisson regression model estimates the relationship between a set of independent variables and the log expected counts of an event or the log incidence rate of the response variable. Here the log expected counts of an event is the simplified term for the natural logarithm of expected counts of an event. The model is expressed as follows:

$$\eta = \ln(\mu) = \beta_0 + \beta_1 X_1 + \beta_2 X_2 + \ldots + \beta_p X_p \tag{13.1}$$

where μ is the mean or the expected number of events; β_0 is the intercept; X_1, X_2, ..., X_p are the predictor variables; and β_1, β_2, ..., β_p are the Poisson regression coefficients for the predictor variables on the scale of the natural logarithm, which are also referred to as log coefficients. Poisson regression assumes that a count response variable follows the Poisson distribution and the mean of the count variable is equal to the variance. $E(Y) = \text{variance}\ (Y) = \mu$. The log link function transforms the expected counts of an event or the incidence rate μ and connects it with the linear predictor η.

In multilevel modeling for count response variables, we have a nested data structure that allows us to estimate the relationships between predictor variables from different levels and the count response variable. Let us look at a basic two-level Poisson regression model with only one predictor variable at each level of the equations. For example, researchers are interested in estimating the number of absences of high school students in a semester from a student-level variable, gender, and a school-level variable, a public or a private school. The outcome variable is the number of absences in a semester. The level 1 predictor variable is gender (`bygender`). The level 2 predictor is whether the school is public or private (`public`). Both the intercept and the slope of the level 1 predictor variable `bygender` are allowed to vary randomly across schools.

The level 1 equation is similar to that for the single-level Poisson regression model, whereas in multilevel modeling, students are nested with schools so we have a subscript i to represent students and a subscript j to represent schools. As with a single-level Poisson regression model, no error term is specified in the level 1 equation:

$$\begin{aligned} \text{Level 1}: \ \eta_{ij} &= \ln(\mu_{ij}) = \beta_{0j} + \beta_{1j}\texttt{bygender}_{ij} \\ \text{Level 2}: \ \beta_{0j} &= \gamma_{00} + \gamma_{01}\texttt{public}_j + u_{0j} \\ \beta_{1j} &= \gamma_{10} + \gamma_{11}\texttt{public}_j + u_{1j} \end{aligned} \tag{13.2}$$

The equations at level 2 look the same as those for the multilevel models for continuous outcome variables. Both the intercept and the coefficient (i.e., the slope of `bygender`) are allowed to vary randomly across schools at level 2 equations, where they are specified as the outcome variables.

In the level 1 equation, $\ln(\mu_{ij})$ is the log link for the expected counts of an event or the incidence rate for the *i*th student in the *j*th school. In level 2 equations, γ_{00}, γ_{01}, γ_{10}, and γ_{11} are level 2 coefficients, which are the fixed effects. γ_{00} is the overall average of the log expected counts of absences across schools. γ_{10} is the overall average slope for `bygender` when the school-level predictor `public` is zero (`public` = 0). u_{0j} is an error term associated with the random intercept, the random variation for the intercept, and u_{1j} is the error term associated with the random slope, the random variation for the level 1 slope across schools; γ_{01} is the fixed effect of the level 2 predictor `public` on the log expected counts of absences; and γ_{11} is the cross-level interaction between `bygender` and `public`.

Estimation of the level 2 random effects results in the variance and covariance matrix, which includes the intercept variance, the slope variance, and the covariance between the intercept and the slope.

13.1.2 Incidence Rate Ratios in Multilevel Poisson Regression Models

As with the single-level Poisson regression model, the incidence rate ratios in multilevel models for count response variables can be interpreted in a similar way. As defined in Chapter 8, the incidence rate is the expected number of events during a period of time or in a location. When a specified period of time is 1, the incidence rate is equal to the expected number of events.

The incidence rate ratios are the exponentiated log coefficients. They are interpreted as the change in the incidence rate or the expected number of events for a one-unit increase in an independent variable. When an incidence rate ratio (IRR) is larger than 1, the incidence rate or the expected number of events increases for a one-unit increase in the predictor variable; when an IRR is less than 1, the incidence rate decreases for a one-unit increase in the predictor variable. In addition, an IRR of 1 indicates that there is no relationship between the predictor and the incidence rate.

13.2 RESEARCH EXAMPLE AND DESCRIPTION OF THE DATA AND SAMPLE

13.2.1 Research Example and Research Questions

In this research example, using the ELS:2002, researchers are interested in predicting number of school absences from the four predictor variables: math scores, gender, school type, and school climate. The first two predictor variables are student-level variables, and the other two are school-level variables. The following research questions will be addressed:

1. Can the number of absences be predicted by math scores and gender?
2. Do school characteristics, such as school type and school climate, impact the number of absences?
3. Do students' counts of absences vary across schools?
4. Does the relationship between gender and the number of absences vary across schools?

13.2.2 Description of the Data and Sample

The ELS:2002 base-year data are used for the following analyses. The variables are listed as follows:

- `byabsen`: number of absences from school in the first semester of ninth grade
- `mathach`: mathematics item response theory (IRT) estimated scores of high-school students
- `cmathach`: mathematics item response theory (IRT) estimated scores of high-school students (grand-mean centered)
- `bygender`: gender (1 = male, 0 = female)
- `public`: school type (1 = public, 0 = private and others)
- `sclimat`: school climate
- `csclimat`: school climate (grand-mean centered)

For demonstration purposes, the missing values of the variables are deleted. The student-level variable `mathach` and the school-level variable `sclimat` are grand-mean centered. The `descr()` function from the `sjmisc` package is used to display the descriptive statistics as follows. Here we just need to load the `sjmisc` package with `library(sjmisc)` since we have installed it for the earlier chapters.

```
> library(foreign)
> mecount <- read.dta("C:/CDA/mecount.dta")
> attach(mecount)
> library(sjmisc)
> descr(mecount, byabsen, cmathach, bygender, public, csclimate)
```

```
## Basic descriptive statistics

       var        type      label     n  NA.prc  mean     sd    se    md  trimmed
   byabsen     numeric    byabsen 12183       0  0.64   0.91  0.01  0.00     0.45
  cmathach     numeric   cmathach 12183       0  0.00  11.96  0.11  0.29     0.09
  bygender     numeric   bygender 12183       0  0.50   0.50  0.00  1.00     0.50
    public categorical     public 12183       0  1.78   0.41  0.00  2.00     1.85
 csclimate     numeric  csclimate 12183       0  0.00   0.68  0.01  0.08     0.03
               range   skew
             4 (0-4)   1.78
 57.2 (-25.97-31.23)  -0.05
             1 (0-1)   0.00
             1 (1-2)  -1.35
    3.8 (-2.72-1.08)  -0.53
```

13.3 MULTILEVEL MODELING FOR COUNT RESPONSE VARIABLES WITH R

13.3.1 Packages and Functions for Multilevel Modeling for Count Response Variables in R

The `glmer()` function in the `lme4` package (Bates et al., 2015), the `glmmTMB()` function in the `glmmTMB` package (Brooks et al., 2017), and the `glmmPQL()` function in the `MASS` package (Venables & Ripley, 2002) are used for fitting multilevel models for count outcome variables. The `glmer()` function is introduced first.

13.3.2 The `glmer()` Function in the `lme4` Package

In the previous two chapters, we use the `glmer()` function in the `lme4` package to fit multilevel models for binary and ordinal outcome variables, respectively. In this chapter, the same function is used for fitting multilevel models for count outcome variables. Here you just need to load the package since it has been installed for previous chapters. As introduced in the preceding chapters, the model formula syntax of `glmer()` is similar to that of `lme()`, but they are slightly different in specifying the random effects. The model formula in `glmer()` also includes two components connected by a plus (+) sign. The first component specifies the dependent variable and the fixed effects part of the model, which are separated by the tilde (~). The second component is the random effects part of the model, which is specified within parentheses. Within parentheses, a predictor variable or a list of predictor variables that have random coefficients is specified first, followed by a vertical line (|), and then, an identifier variable at a higher level as the grouping variable. In addition to the model formula, we also need to specify the `family = poisson` argument since the outcome variable is a count variable. For example, the command `glmer(y ~ x + (1|schid), family = poisson, nAGQ = 7)` tells R to fit a multilevel model to estimate a count variable `y` on a predictor variable `x` with random intercepts varying across schools by specifying `(1|schid)`, where `1` indicates the random intercept on the left of the vertical line (|) and `schid` is the identifier variable on the right. The `nAGQ = 7` option specifies the number of quadrature points for adaptive Gauss-Hermite quadrature. The default number is 1, which is used for the Laplace approximation. The maximum number of quadrature points is 25 in the current version of the `lme4` package (Version 1.1-27.1). Although increasing the number of quadrature points can improve precision, but it takes longer to fit the model and may have converging issues. At the time of this writing, the adaptive Gauss-Hermite quadrature method is only for models with a single random effects term, such as the null model or the random-intercept model. When there is more than a single random effects term in the model such as the random-coefficient model, only the Laplace approximation with `nAGQ = 1` is available. In the following four examples, for simplicity and consistency, we use the Laplace approximation with `nAGQ = 1` as the numerical integration method for model estimation. For more details on how to use this function, type `help(glmer)` in the command prompt after loading the `lme4` package.

13.3.3 Unconditional Model or Null Model (Model 1)

The unconditional model, or null model, for count response variables includes no predictor variable at any level. This model estimates the overall average of the log expected counts of events or the log predicted mean count and the variability in the predicted log mean count between groups (e.g., schools). It can be expressed as follows:

$$\begin{aligned} \text{Level 1}: \ln(\mu_{ij}) &= \beta_{0j} \\ \text{Level 2}: \beta_{0j} &= \gamma_{00} + u_{0j} \end{aligned}$$

Recall in Poisson regression that $\ln(\mu)$ is the log transformation of the expected number of events or the predicted mean count. In multilevel Poisson regression models, $\ln(\mu_{ij})$ is the log link for the predicted mean count for the *i*th student in the *j*th school. In the level 2 equation, γ_{00} is the overall average log expected count of absences across schools and u_{0j} is an error term at the school level.

The `glmer()` Command and Output

The command `mepois.1 <- glmer(byabsen ~ 1 + (1|SCH_ID), data = mecount, family = poisson, nAGQ = 1)` is used to fit the unconditional model (model 1). This command tells R to fit a multilevel model for the count response variable `byabsen`. Since this is the unconditional model, no predictors are specified on the right-hand side of the tilde (~). The random effects part of the model specifies the random effects at the school level within parentheses, `(1|SCH_ID)`, where `1` indicates the random intercept and `SCH_ID` is the identifier variable. No random coefficients for any predictor variables are specified in this model. The model is assigned to an object named `mepois.1`. Next, the `summary()` function is used to display the output of the model.

```
> library(lme4)
> # Unconditional Model/Null model with glmer()
>mepois.1<-glmer(byabsen~1+(1|SCH_ID),data=mecount, family=poisson, nAGQ=1)
> summary(mepois.1)
Generalized linear mixed model fit by maximum likelihood (Laplace
  Approximation) [glmerMod]
 Family: poisson  ( log )
Formula: byabsen ~ 1 + (1 | SCH_ID)
   Data: mecount

    AIC      BIC    logLik  deviance  df.resid
26500.1  26514.9  -13248.0   26496.1     12181

Scaled residuals:
    Min       1Q   Median      3Q     Max
-1.0974  -0.7685  -0.6600  0.5254  5.6664

Random effects:
 Groups  Name        Variance  Std.Dev.
 SCH_ID  (Intercept) 0.1089    0.33
Number of obs: 12183, groups: SCH_ID, 617

Fixed effects:
             Estimate  Std. Error  z value  Pr(>|z|)
(Intercept)  -0.50738    0.01829   -27.74    <2e-16 ***
---
Signif. codes: 0 '***' 0.001 '**' 0.01 '*' 0.05 '.' 0.1 ' ' 1
```

Interpreting the Output

The R output produced by the `glmer()` function is similar to that by the `lmer()` function. It includes the note for the model estimation method, the distribution family, the model formula, and the data, the model fit statistics such as the AIC and BIC statistics, the log-likelihood value, and the deviance and its degrees of freedom, the scaled residuals, the random effects, the number of observations and groups, and the fixed effects.

The maximum likelihood estimation with numerical integration is used to fit the multilevel Poisson regression model. The estimation method for the model is the Laplace approximation since `nAGQ = 1`.

The random effects section provides the variance and the standard deviation for the intercept across all schools. This is the between-school variance (τ_{00}), which is 0.109. Since the intraclass correlation is not well developed for multilevel modeling for count response variables, the computation is omitted here.

The fixed effects section displays the estimate for the intercept, its standard error, the Wald z statistic, and the associated p value. The intercept γ_{00} (labeled `Intercept`) is –.507, which is significant (`z = -27.74, P(>|z|) < 2e-16`). This means that the overall average log expected number of absences across all schools is –.507. The estimated log counts can be transformed to the expected number of events or the incidence rate, which will be explained next.

The 95% confidence interval of the estimate is computed as follows.

```
> coef <- fixef(mepois.1)
> se <- sqrt(diag(vcov(mepois.1)))
> LL <- fixef(mepois.1)-1.96*se
> UL <- fixef(mepois.1)+1.96*se
> coeftab <- cbind(coef, LL, UL)
> coeftab
                   coef          LL          UL
(Intercept)  -0.5073773  -0.5432224  -0.4715323
```

The same results can be obtained with the `confint(mepois.1, method = "Wald")` command.

```
> confint(mepois.1, method = "Wald")
                   2.5 %       97.5 %
.sig01                NA           NA
(Intercept)  -0.5432217    -0.471533
```

With the `exp(fixef(mepois.1))` command, we can get the incidence rate ratios (IRR) for the predictor variables. Since this is the unconditional model, only the IRR for the intercept is reported. IRR =.602, which is the overall average incidence rate or the overall expected number of absences. The confidence interval for the IRR can be obtained with the `exp(coeftab)` command.

```
> exp(fixef(mepois.1))
(Intercept)
 0.6020725
> exp(coeftab)
                  coef         LL         UL
(Intercept)  0.6020725  0.5808735  0.6240453
```

The same confidence interval for the IRR can also be obtained with the `irtab <- cbind(ir=exp(fixef(mepois.1)), IRLL=exp(LL), IRUL=exp(UL))` command or the `exp(confint(mepois.1, method = "Wald"))` command.

```
> irtab <- cbind(ir=exp(fixef(mepois.1)), IRLL=exp(LL), IRUL=exp(UL))
> irtab
                     ir       IRLL       IRUL
(Intercept)  0.6020725  0.5808735  0.6240453

> exp(confint(mepois.1, method = "Wald"))
                 2.5 %        97.5 %
.sig01              NA            NA
(Intercept)  0.5808738     0.6240449
```

In the Unconditional Model above, we use the Laplace approximation as the numerical integration method. We can also use the adaptive Gauss-Hermite quadrature with more integration points. In the following example, we specify the `nAGQ = 7` argument in the syntax, so the number of quadrature points for adaptive Gauss–Hermite quadrature is 7. The resulting output is as follows.

```
>mepois.1b<-glmer(byabsen~1+ (1|SCH_ID), data=mecount, family=poisson, nAGQ=7)
> summary(mepois.1b)
Generalized linear mixed model fit by maximum likelihood (Adaptive
 Gauss-Hermite Quadrature, nAGQ = 7) [glmerMod]
 Family: poisson  ( log )
Formula: byabsen ~ 1 + (1 | SCH_ID)
   Data: mecount


    AIC       BIC    logLik  deviance  df.resid
14618.4  14633.3   -7307.2   14614.4     12181


Scaled residuals:
    Min        1Q   Median       3Q      Max
-1.0985  -0.7685  -0.6596   0.5256   5.6706


Random effects:
 Groups  Name        Variance  Std.Dev.
 SCH_ID  (Intercept) 0.1096    0.3311
Number of obs: 12183, groups: SCH_ID, 617


Fixed effects:
              Estimate  Std. Error  z value  Pr(>|z|)
(Intercept)   -0.50763     0.01838   -27.62   <2e-16 ***
---
Signif. codes: 0 '***' 0.001 '**' 0.01 '*' 0.05 '.' 0.1 ' ' 1
```

The results from the unconditional model with the adaptive Gauss-Hermite quadrature method and those from the model with the Laplace approximation look similar although the former estimation method provides more accurate results. Nevertheless, there are differences in the output which are worth noting. The AIC and BIC statistics, the log likelihood values, and the deviances are different between these two models with different estimation methods. Therefore, to compare multilevel Poisson regression models with `glmer()`, it is a good practice to use the same estimation method.

13.3.4 Random-Intercept Model (Model 2)

Next, we include the two predictor variables, math scores (`cmathach`) and gender (`bygender`), to the level 1 equation. The model is referred to as the random-intercept model since the intercept is allowed to vary freely across schools. This model can be expressed as follows:

$$
\begin{aligned}
\text{Level 1}: \ln\left(\mu_{ij}\right) &= \beta_{0j} + \beta_{1j}\texttt{cmathach}_{ij} + \beta_{2j}\texttt{bygender}_{ij} \\
\text{Level 2}: \beta_{0j} &= \gamma_{00} + u_{0j} \\
\beta_{1j} &= \gamma_{10} \\
\beta_{2j} &= \gamma_{20}
\end{aligned}
$$

The level 1 equation includes the predictor variables, math scores (`cmathach`) and gender (`bygender`), and the count response variable, number of absences (`byabsen`). The level 2 equation for the intercept (β_{0j}) is specified to be random, and the slope for `cmathach` (β_{1j}) and `bygender` (β_{2j}) are constrained to be fixed.

The `glmer()` Command and Output

The command `mepois.2 <- glmer(byabsen ~ cmathach + bygender + (1|SCH_ID), data = mecount, family = poisson, nAGQ = 1)` is used to fit the random-intercept model (model 2) after the two predictor variables `cmathach` and `bygender` are added to the fixed part of the model immediately after the tilde (~). Just like the unconditional model, no random coefficients are specified in the random part of the model. Again, the `summary()` function is used to display the output of the fitted random-intercept model named `mepois.2`. The following output is displayed.

```
> # Random-intercept model
> mepois.2 <- glmer(byabsen ~ cmathach + bygender + (1|SCH_ID), data=mecount,
family=poisson, nAGQ=1)
> summary(mepois.2)
Generalized linear mixed model fit by maximum likelihood (Laplace
  Approximation) [glmerMod]
 Family: poisson  ( log )
Formula: byabsen ~ cmathach + bygender + (1 | SCH_ID)
   Data: mecount

     AIC      BIC   logLik deviance df.resid
 25683.6  25713.2 -12837.8  25675.6    12179
```

```
Scaled residuals:
    Min      1Q  Median      3Q     Max
-1.4242 -0.7357 -0.5613  0.5567  7.2371

Random effects:
 Groups  Name        Variance Std.Dev.
 SCH_ID  (Intercept) 0.1107   0.3328
Number of obs: 12183, groups: SCH_ID, 617

Fixed effects:
              Estimate Std. Error z value Pr(>|z|)
(Intercept) -0.854910   0.024088  -35.49   <2e-16 ***
cmathach    -0.016926   0.001038  -16.31   <2e-16 ***
bygender     0.583579   0.024281   24.04   <2e-16 ***
---
Signif. codes: 0 '***' 0.001 '**' 0.01 '*' 0.05 '.' 0.1 ' ' 1

Correlation of Fixed Effects:
         (Intr) cmthch
cmathach  0.126
bygender -0.641 -0.049
convergence code: 0
Model is nearly unidentifiable: very large eigenvalue
 - Rescale variables?
```

Interpreting the Output

In the random effects section, the row (labeled `Intercept`) reports that the intercept variance (τ_{00}) is .111. In the fixed effects section, the estimation table reports the coefficients for `cmathach`, `bygender`, and the intercept. The intercept γ_{00} (labeled `Intercept`) = –.855. It is significantly different from 0 (`z = -35.49`, `P(>|z|) = 2e-16`), which means that the log expected number of absences for female students (`bygender` = 0) whose math scores are zero (`cmathach` = 0) in a typical school is –.855. The log coefficient for `cmathach` (γ_{10}) = –.017, $z = -16.31$, $p < .001$, which indicates that the log expected number of absences decreases by a factor of .017 for a one-unit increase in math scores. The log coefficient for `bygender` (γ_{20}) = .584, $z = 24.04$, $p < .001$, which indicates that the log expected number of absences for male students is .584 times as great as that for female students.

The 95% confidence intervals of the estimates are computed as follows.

```
> coef <- fixef(mepois.2)
> se <- sqrt(diag(vcov(mepois.2)))
> LL <- fixef(mepois.2)-1.96*se
> UL <- fixef(mepois.2)+1.96*se
> coeftab <- cbind(coef, LL, UL)
> coeftab
```

```
                    coef          LL            UL
(Intercept)  -0.85491024  -0.90212344  -0.80769704
cmathach     -0.01692627  -0.01896027  -0.01489227
bygender      0.58357886   0.53598866   0.63116907
```

The same results can be obtained with the `confint(mepois.2, method = "Wald")` command.

```
> confint(mepois.2, method = "Wald")
                  2.5 %       97.5 %
.sig01               NA           NA
(Intercept)  -0.90212257  -0.80769791
cmathach     -0.01896024  -0.01489231
bygender      0.53598953   0.63116819
```

With the `exp(fixef(mepois.2))` command, we can get the IRRs for the predictor variables. The confidence intervals for the IRRs can be obtained with the `exp(coeftab)` command.

```
> exp(fixef(mepois.2))
(Intercept)    cmathach    bygender
0.4253214     0.9832162   1.7924419
> exp(coeftab)
                 coef          LL          UL
(Intercept)  0.4253214  0.4057073  0.4458837
cmathach     0.9832162  0.9812183  0.9852181
bygender     1.7924419  1.7091372  1.8798069
```

For `cmathach`, IRR = .983. It can be interpreted as follows: The incidence rate of being absent decreases by a factor of .983 for a one-unit increase in math scores when holding the other predictor constant. It also means that the expected number of absences decreases by a factor of .983 or decreases by 1.7% for a one-unit increase in math scores. For `bygender`, IRR = 1.792. It indicates that the incidence rate of being absent for male students is 1.792 times as large as that for female students when holding the other predictor constant. The incidence rate ratio for the intercept is .425, which is the overall average incidence rate of being absent when the values of the predictor variables are 0.

The same confidence intervals for the IRRs can also be obtained with the `irtab <- cbind(ir=exp(fixef(mepois.2)), IRLL=exp(LL), IRUL=exp(UL))` or `exp(confint(mepois.2, method = "Wald"))`.

```
> irtab <- cbind(ir=exp(fixef(mepois.2)), IRLL=exp(LL), IRUL=exp(UL))
> irtab
                   ir      IRLL      IRUL
(Intercept) 0.4253214 0.4057073 0.4458837
cmathach    0.9832162 0.9812183 0.9852181
bygender    1.7924419 1.7091372 1.8798069

> exp(confint(mepois.2, method = "Wald"))
                2.5 %    97.5 %
.sig01             NA        NA
(Intercept) 0.4057076 0.4458834
cmathach    0.9812184 0.9852180
bygender    1.7091386 1.8798053
```

Likelihood Ratio Test Comparing the Unconditional Model and the Random-Intercept Model

```
> anova(mepois.1, mepois.2)
Data: mecount
Models:
mepois.1: byabsen ~ 1 + (1 | SCH_ID)
mepois.2: byabsen ~ cmathach + bygender + (1 | SCH_ID)
         npar   AIC   BIC logLik deviance  Chisq Df Pr(>Chisq)
mepois.1    2 26500 26515 -13248    26496
mepois.2    4 25684 25713 -12838    25676 820.52  2  < 2.2e-16 ***
---
Signif. codes: 0 '***' 0.001 '**' 0.01 '*' 0.05 '.' 0.1 ' ' 1
```

The likelihood ratio test is used to compare the unconditional model and the random-intercept model. The `anova(mepois.1, mepois.2)` command compares the log-likelihood statistics of these two models: $\chi^2_{(2)} = 820.52$, $p < .001$, which indicates that the random-intercept model fits the data better than the unconditional model, so we are in favor of the former model with a varying intercept.

13.3.5 Random-Coefficient Model With a Level-1 Variable (Model 3)

Just like the random-coefficient model for linear mixed modeling, we can also fit a random-coefficient Poisson regression model. Although the expressions for the level 1 equation are different between the two models, both level 2 equations for the random intercept and random coefficient are the same. This model is referred to as the random-coefficient model since both the intercept and coefficient in the level 1 equation are allowed to vary across schools at level 2. This model also estimates variance–covariance components, including variances in intercepts and coefficients/slopes and the covariance between them. The random-coefficient Poisson regression model can be expressed as follows:

$$\begin{aligned} \text{Level 1}: \ln\left(\mu_{ij}\right) &= \beta_{0j} + \beta_{1j}\texttt{cmathach}_{ij} + \beta_{2j}\texttt{bygender}_{ij} \\ \text{Level 2}: \beta_{0j} &= \gamma_{00} + u_{0j} \\ \beta_{1j} &= \gamma_{10} \\ \beta_{2j} &= \gamma_{20} + u_{2j} \end{aligned}$$

The level 1 equation includes the predictor variables, math scores (`cmathach`) and gender (`bygender`), and the count response variable, number of absences (`byabsen`). The level 2 equations for the intercept (β_{0j}) and the slope for `bygender` (β_{2j}) are both specified to be random with level 2 residuals u_{0j} and u_{2j}, respectively, whereas the slope for `cmathach` (β_{1j}) is constrained to be fixed.

The `glmer()` Command and Output

The command `mepois.3 <- glmer(byabsen ~ cmathach + bygender + (bygender|SCH_ID), data = mecount, family = poisson, nAGQ = 1)` is used to fit the random-coefficient model (model 3). The fixed part of the model is the same as that of the random-intercept model. Different from the random-intercept model, the predictor variable `bygender` is specified in the random part of the model, `(bygender|SCH_ID)`, which indicates that the slope of the level 1 predictor `bygender` is allowed to vary randomly across schools. The `nAGQ = 1` argument specifies that the number of quadrature points is 1 since only the Laplace approximation is available in the random-coefficient model with more than one random effects term. The next command `summary(mepois.3)` displays the output of the fitted model.

```
> # Random-coefficient model
>mepois.3<-glmer(byabsen~cmathach+bygender+(bygender|SCH_ID),data=mecount,
family=poisson, nAGQ=1)
> summary(mepois.3)
Generalized linear mixed model fit by maximum likelihood (Laplace
  Approximation) [glmerMod]
 Family: poisson  ( log )
Formula: byabsen ~ cmathach + bygender + (bygender | SCH_ID)
   Data: mecount


    AIC       BIC    logLik  deviance  df.resid
25652.9   25697.4  -12820.5   25640.9     12177


Scaled residuals:
    Min        1Q    Median       3Q      Max
-1.3955   -0.7349  -0.5400   0.5362   6.5963
```

```
Random effects:
Groups  Name        Variance Std.Dev. Corr
SCH_ID  (Intercept) 0.1827   0.4274
        bygender    0.1235   0.3514   -0.66
Number of obs: 12183, groups: SCH_ID, 617

Fixed effects:
             Estimate  Std. Error z value Pr(>|z|)
(Intercept) -0.889891  0.028072   -31.70  <2e-16 ***
cmathach    -0.016705  0.001039   -16.08  <2e-16 ***
bygender     0.620081  0.030502    20.33  <2e-16 ***
---
Signif. codes: 0 '***' 0.001 '**' 0.01 '*' 0.05 '.' 0.1 ' ' 1

Correlation of Fixed Effects:
         (Intr) cmthch
cmathach  0.099
bygender -0.754 -0.029
convergence code: 0
```

At the time of this writing, the adaptive Gauss-Hermite quadrature method is not available for random-coefficient models with the `glmer()` function. An error message will be displayed in the output if `nAGQ` is set more than 1. For example, if we specify `nAGQ = 7` in the command, then we will see the following error message.

```
>mepois.3<-glmer(byabsen~cmathach+bygender+(bygender|SCH_ID),data=mecount,
family=poisson, nAGQ=7)
Error in updateGlmerDevfun(devfun, glmod$reTrms, nAGQ = nAGQ) :
  nAGQ > 1 is only available for models with a single, scalar random-effects term
```

Interpreting the Output

Regarding the random effects in the model, the intercept variance (τ_{00}) is .183 and the slope variance for `bygender` is .124. The correlation between the intercept and the `bygender` slope is –.66. Since the covariance of two variables equals the product of the correlation coefficient and the standard deviations of the two variables, to compute the covariance between the intercept and the `bygender` slope, we use the following equation: .427*.351*(–.66) = –.099.

The estimated log coefficients and the corresponding incidence rate ratios for the intercept, `cmathach`, and `bygender` in the output are similar to those for the previous random-intercept model. They can also be interpreted in a similar way.

The 95% confidence intervals of the estimates are computed as follows.

```
> coef <- fixef(mepois.3)
> se <- sqrt(diag(vcov(mepois.3)))
> LL <- fixef(mepois.3)-1.96*se
> UL <- fixef(mepois.3)+1.96*se
> coeftab <- cbind(coef, LL, UL)
> coeftab
                   coef          LL          UL
(Intercept) -0.88989126 -0.94491314 -0.83486938
cmathach    -0.01670476 -0.01874082 -0.01466871
bygender     0.62008124  0.56029665  0.67986583
```

The same results can be obtained with `confint(mepois.3, method = "Wald")`.

```
> confint(mepois.3, method = "Wald")
                  2.5 %      97.5 %
.sig01               NA          NA
.sig02               NA          NA
.sig03               NA          NA
(Intercept) -0.94491212 -0.83487039
cmathach    -0.01874078 -0.01466875
bygender     0.56029774  0.67986473
```

We use the `exp(fixef(mepos.3))` command to request the incidence rate ratios for the predictor variables. The confidence intervals for the incidence rate ratio can be obtained with `exp(coeftab)`.

```
> exp(fixef(mepois.3))
(Intercept)   cmathach   bygender
0.4107004   0.9834340  1.8590791
> exp(coeftab)
                 coef        LL        UL
(Intercept) 0.4107004 0.3887133 0.4339312
cmathach    0.9834340 0.9814337 0.9854384
bygender    1.8590791 1.7511919 1.9736129
```

The same confidence intervals for the IRRs can also be obtained with the `irtab <- cbind(ir=exp(fixef(mepois.3)), IRLL=exp(LL), IRUL=exp(UL))` command or the `exp(confint(mepois.3, method = "Wald"))` command.

```
> irtab <- cbind(ir=exp(fixef(mepois.3)), IRLL=exp(LL), IRUL=exp(UL))
> irtab
                   ir      IRLL      IRUL
(Intercept) 0.4107004 0.3887133 0.4339312
cmathach    0.9834340 0.9814337 0.9854384
bygender    1.8590791 1.7511919 1.9736129

> exp(confint(mepois.3, method = "Wald"))
                2.5 %     97.5 %
.sig01             NA         NA
.sig02             NA         NA
.sig03             NA         NA
(Intercept) 0.3887137  0.4339307
cmathach    0.9814337  0.9854383
bygender    1.7511938  1.9736107
```

The incidence rate ratio for `cmathach` is .983, which indicates that the incidence rate of being absent decreases by a factor of .983 for a one-unit increase in math scores.

The incidence rate ratio for `bygender` is 1.859, which indicates that the incidence rate of being absent for male students is 1.859 times as large as that for female students.

The incidence rate ratio for the intercept is .411, which is the overall average incidence rate of being absent when the values of the predictor variables are 0.

Model Comparisons Using the Likelihood Ratio Test

```
> anova(mepois.2, mepois.3)
Data: mecount
Models:
mepois.2: byabsen ~ cmathach + bygender + (1 | SCH_ID)
mepois.3: byabsen ~ cmathach + bygender + (bygender | SCH_ID)
         npar   AIC   BIC logLik deviance  Chisq Df Pr(>Chisq)
mepois.2    4 25684 25713 -12838    25676
mepois.3    6 25653 25697 -12820    25641 34.638  2   3.01e-08 ***
---
Signif. codes: 0 '***' 0.001 '**' 0.01 '*' 0.05 '.' 0.1 ' ' 1
```

The likelihood ratio test compares the random-intercept model (model 2) and the random-coefficient model (model 3). The `anova(mepois.2, mepois.3)` function compares the log-likelihood statistics of these two models: $\chi^2_{(2)} = 34.638$, $p < .001$, which indicates that we are in favor of the random-coefficient model, so having a random coefficient in model 3 is justified.

13.3.6 Contextual Model With Level-2 Variables (Model 4)

Just like the contextual model in the linear mixed modeling, the contextual model for the count response variable also includes both level 1 and level 2 predictor variables. The two level 2 equations specify the random intercept and random slope. In this example, two school-level variables `public` and `csclimat` are added to the level 2 equation for the random intercept. In the level 2 equation for the random slope, no school-level variables are specified. This model is expressed as follows:

$$\begin{aligned}\text{Level 1}: \ln\left(\mu_{ij}\right) &= \beta_{0j} + \beta_{1j}\texttt{cmathach}_{ij} + \beta_{2j}\texttt{bygender}_{ij} \\ \text{Level 2}: \beta_{0j} &= \gamma_{00} + \gamma_{01}\texttt{public}_j + \gamma_{02}\texttt{csclimat}_j + u_{0j} \\ \beta_{1j} &= \gamma_{10} \\ \beta_{2j} &= \gamma_{20} + u_{2j}\end{aligned}$$

The `glmer()` Command and Output

The command `mepois.4 <- glmer(byabsen ~ cmathach + bygender + public + csclimate + (bygender|SCH_ID), data = mecount, family = poisson, nAGQ = 1)` is used to fit the contextual model (model 4). With other parts of the equations the same as those of the random-coefficient model (model 3), the fixed part of the model includes both the level 1 predictor variables `cmathach` and `bygender` and the two school-level predictor variables `public` and `csclimat`. Since only the Laplace approximation is available for models with more than one random-effects term, we specify the `nAGQ = 1` argument. Again, the results of the fitted model are displayed with the `summary(mepois.4)` command.

```
> # Contextual model/Random-coefficient model with predictor variables in both levels
> mepois.4 <- glmer(byabsen ~ cmathach + bygender + public + csclimate +
(bygender|SCH_ID), data=mecount, family=poisson, nAGQ=1)
Warning message:
In checkConv(attr(opt, "derivs"), opt$par, ctrl = control$checkConv,  :
 Model is nearly unidentifiable: very large eigenvalue
 - Rescale variables?
> summary(mepois.4)
Generalized linear mixed model fit by maximum likelihood (Laplace
  Approximation) [glmerMod]
 Family: poisson  ( log )
Formula: byabsen ~ cmathach + bygender + public + csclimate + (bygender |
  SCH_ID)
  Data: mecount


     AIC       BIC    logLik  deviance  df.resid
 25549.2   25608.5  -12766.6   25533.2     12175


Scaled residuals:
     Min       1Q   Median      3Q     Max
-1.3725  -0.7315  -0.5412  0.5396  6.8514
```

```
Random effects:
Groups   Name         Variance  Std.Dev.  Corr
SCH_ID   (Intercept)  0.1368    0.3699
         bygender     0.1197    0.3460    -0.68
Number of obs: 12183, groups:  SCH_ID, 617

Fixed effects:
               Estimate  Std. Error  z value  Pr(>|z|)
(Intercept)   -0.548533   0.041334   -13.27   <2e-16 ***
cmathach      -0.018140   0.001049   -17.30   <2e-16 ***
bygender       0.613879   0.030252    20.29   <2e-16 ***
publicpublic  -0.428933   0.041564   -10.32   <2e-16 ***
csclimate     -0.009975   0.026241    -0.38    0.704
---
Signif. codes:  0 '***' 0.001 '**' 0.01 '*' 0.05 '.' 0.1 ' ' 1

Correlation of Fixed Effects:
            (Intr)  cmthch  bygndr   pblcpb
cmathach    -0.020
bygender    -0.514  -0.032
publicpublc -0.767   0.112   0.026
csclimate   -0.285  -0.130   0.023    0.354
convergence code: 0
Model is nearly unidentifiable: very large eigenvalue
- Rescale variables?
```

Interpreting the Output

In the random effects section, the residual variance in the intercepts (τ_{00}) is .137 and the slope variance for `bygender` is .120. The correlation between the intercept and the `bygender` slope is –.68.The covariance between the intercept and the `bygender` slope is computed as the product of the correlation coefficient and the standard deviations of the two variables: .370*.346*(–.68) = –.087.

The fixed-effects section reports the estimated log coefficients for both level 1 and level 2 predictor variables and the intercept. The log coefficients for all the predictor variables except `csclimat` are significant. The 95% confidence intervals of the estimates are computed as follows.

```
> coef <- fixef(mepois.4)
> se <- sqrt(diag(vcov(mepois.4)))
> LL <- fixef(mepois.4)-1.96*se
> UL <- fixef(mepois.4)+1.96*se
> coeftab <- cbind(coef, LL, UL)
> coeftab
```

```
                     coef          LL           UL
(Intercept)  -0.548532854  -0.62954660  -0.46751910
cmathach     -0.018140420  -0.02019591  -0.01608493
bygender      0.613879445   0.55458627   0.67317262
publicpublic -0.428932676  -0.51039883  -0.34746652
csclimate    -0.009975233  -0.06140809   0.04145763
```

The same results can be obtained with `confint(mepois.4, method= "Wald")`.

```
> confint(mepois.4, method = "Wald")
                   2.5 %      97.5 %
.sig01                NA          NA
.sig02                NA          NA
.sig03                NA          NA
(Intercept)  -0.62954511 -0.46752059
cmathach     -0.02019587 -0.01608497
bygender      0.55458736  0.67317153
publicpublic -0.51039733 -0.34746802
csclimate    -0.06140715  0.04145668
```

We use the `exp(fixef(mepois.4))` command to request the incidence rate ratios for the predictor variables. The confidence intervals for the incidence rate ratios can be obtained with the `exp(coeftab)` command. The same confidence intervals for the incidence rate ratios can also be obtained with the `irtab <- cbind(ir= exp(fixef(mepois.4)), IRLL=exp(LL), IRUL=exp(UL))` command or the `exp(confint(mepois.4, method = "Wald"))` command.

```
> exp(coeftab)
                  coef        LL        UL
(Intercept)  0.5777969 0.5328333 0.6265548
cmathach     0.9820231 0.9800067 0.9840437
bygender     1.8475851 1.7412204 1.9604472
publicpublic 0.6512038 0.6002561 0.7064757
csclimate    0.9900744 0.9404394 1.0423290

> irtab <- cbind(ir=exp(fixef(mepois.4)), IRLL=exp(LL), IRUL=exp(UL))
> irtab
                    ir      IRLL      IRUL
(Intercept)  0.5777969 0.5328333 0.6265548
cmathach     0.9820231 0.9800067 0.9840437
bygender     1.8475851 1.7412204 1.9604472
publicpublic 0.6512038 0.6002561 0.7064757
csclimate    0.9900744 0.9404394 1.0423290
> exp(confint(mepois.4, method = "Wald"))
```

```
                    2.5 %      97.5 %
.sig01                 NA          NA
.sig02                 NA          NA
.sig03                 NA          NA
(Intercept)     0.5328341   0.6265538
cmathach        0.9800067   0.9840437
bygender        1.7412223   1.9604451
publicpublic    0.6002570   0.7064746
csclimate       0.9404403   1.0423280
```

The incidence rate ratio for `cmathach` is .982, which indicates that the incidence rate of being absent or the expected count of absences decreases by a factor of .982 for a one-unit increase in math scores when holding the other predictor variables constant.

The incidence rate ratio for `bygender` is 1.848, which indicates that the incidence rate of being absent for male students is 1.848 times as large as that for female students when holding the other predictor variables constant.

The incidence rate ratio for `public` is .651, which indicates that the incidence rate of being absent for students in public schools is .651 times the incidence rate for those in private schools. In other words, being students in public schools decreases the incidence rate of being absent by 34.9%.

The incidence rate ratio for `csclimat` is .990, $p = .704$, which is not significant (see the associated p value in the coefficients table). It indicates that school climate does not impact the incidence rate of being absent.

Model Comparisons Using the Likelihood Ratio Test

```
> anova(mepois.3, mepois.4)
Data: mecount
Models:
mepois.3: byabsen ~ cmathach + bygender + (bygender | SCH_ID)
mepois.4: byabsen ~ cmathach + bygender + public + csclimate + (bygender |
mepois.4:     SCH_ID)
          npar   AIC   BIC  logLik deviance  Chisq Df Pr(>Chisq)
mepois.3     6 25653 25697  -12820    25641
mepois.4     8 25549 25609  -12767    25533 107.71  2  < 2.2e-16 ***
---
Signif. codes: 0 '***' 0.001 '**' 0.01 '*' 0.05 '.' 0.1 ' ' 1
```

The likelihood ratio test compares the contextual model with both the level 1 and level 2 predictor variables (model 4: `ranslop2`) with the random-coefficient model with level 1 predictors only (model 3: `ranslop1`). The `anova(mepois.3, mepois.4)` command compares the log-likelihood statistics of these two models. It yields a value of $\chi^2_{(2)} = 107.71$, $p < .001$, which indicates that the contextual model fits the data better.

Model Comparison Using the Likelihood Ratio Test

We use the `anova(mepois.1, mepois.2, mepois.3, mepois.4)` command to compare the log-likelihood statistics of all the four models fitted above. The output is as follows.

```
> # Model comparisons
> anova(mepois.1, mepois.2, mepois.3, mepois.4)
Data: mecount
Models:
mepois.1: byabsen ~ 1 + (1 | SCH_ID)
mepois.2: byabsen ~ cmathach + bygender + (1 | SCH_ID)
mepois.3: byabsen ~ cmathach + bygender + (bygender | SCH_ID)
mepois.4: byabsen ~ cmathach + bygender + public + csclimate + (bygender |
mepois.4:     SCH_ID)
         npar   AIC   BIC  logLik deviance    Chisq  Df  Pr(>Chisq)
mepois.1    2 26500 26515 -13248    26496
mepois.2    4 25684 25713 -12838    25676  820.517  2   < 2.2e-16 ***
mepois.3    6 25653 25697 -12820    25641   34.638  2    3.01e-08 ***
mepois.4    8 25549 25609 -12767    25533  107.710  2   < 2.2e-16 ***
---
Signif. codes: 0 '***' 0.001 '**' 0.01 '*' 0.05 '.' 0.1 ' ' 1
```

The results of the likelihood ratio test indicate that the contextual model (`melr.4`) fits the data best among all the four models. In addition, the AIC and BIC statistics of the fourth model are 25,549 and 25,609, respectively, which are also the smallest compared to those in the other three models.

13.3.7 Interpreting the Marginal Effects With the `margins()` Function in the `margins` Package

We use the `margins(mepois.4)` command to compute the average marginal effects and name the results `marg`. The `margins` package needs to be loaded with the `library(margins)` command first. The summary results are obtained with `summary(marg)` as follows.

```
> # Marginal effect with margins()
> library(margins)
> marg <- margins(mepois.4)
> summary(marg)
       factor     AME     SE        z      p   lower   upper
     bygender  0.3748 0.0198  18.9621 0.0000  0.3360  0.4135
     cmathach -0.0114 0.0007 -16.5727 0.0000 -0.0127 -0.0100
    csclimate -0.0063 0.0165  -0.3801 0.7039 -0.0385  0.0260
 publicpublic -0.3038 0.0336  -9.0292 0.0000 -0.3698 -0.2379
```

The average marginal effect for `bygender` is .375. This indicates that male students have on average .375 more absences than female students when holding all other predictors constant.

The average marginal effect for cmathach is –.011. The result indicates that there are on average .011 less absences for a one-unit increase in math scores when holding all other predictors constant.

The marginal effects of the other two predictor variable can be interpreted in a similar way.

13.3.8 Interpreting the Predicted Counts With the ggpredict() Function in the ggeffects Package

By using the ggpredict() function in the ggeffects package (Lüdecke, 2018b), we can compute the predicted number of events at specified values of the predictor variables in the multilevel Poisson regression model. We use the ggpredict() function after the contextual model (model 4: mepois.4) is fitted.

We first load the package with library(ggeffects) since it has been installed in previous chapters. In the following example, we compute the predicted counts of absences for csclimat at different values by the two groups in public when holding other variables constant at their means. The command is as follows: mepoi4.efpub <- ggpredict(mepois.4, terms = c("csclimate", "public")). In the ggpredict() function, mepois.4 is the model object and the terms = c("csclimat", "public") option specifies both csclimat and public, with the latter as the grouping variable. The output is assigned to an object named mepoi4.efpub and is plotted with plot(mepoi4.efpub). The as.data.frame() function or the sqrt(diag(vcov())) function can be used to request the standard errors, and the output is omitted here.

```
> library(ggeffects)
> mepoi4.efpub <- ggpredict(mepois.4, terms = c("csclimate", "public"))
> mepoi4.efpub

# Predicted counts of byabsen

# public = catholic or other private

csclimate  |  Predicted  |         95% CI
-----------------------------------------
    -2.80  |       0.81  |   [0.68, 0.97]
    -2.20  |       0.80  |   [0.69, 0.93]
    -1.60  |       0.80  |   [0.71, 0.90]
    -1.00  |       0.79  |   [0.72, 0.88]
    -0.20  |       0.79  |   [0.73, 0.85]
     1.00  |       0.78  |   [0.72, 0.84]

# public = public

csclimate  |  Predicted  |         95% CI
-----------------------------------------
    -2.80  |       0.53  |   [0.46, 0.61]
    -2.20  |       0.52  |   [0.47, 0.59]
    -1.60  |       0.52  |   [0.48, 0.57]
    -1.00  |       0.52  |   [0.49, 0.55]
    -0.20  |       0.51  |   [0.49, 0.53]
     1.00  |       0.51  |   [0.47, 0.54]
```

```
Adjusted for:
* cmathach = 0.00
* bygender = 0.50
*  SCH_ID = 0 (population-level)
> plot(mepoi4.efpub)
Loading required namespace: ggplot2
```

The output displays the predicted counts or incidence rate of absences for `csclimat` at different values for both Catholic or other private schools and public schools when other predictor variables are fixed at their means. The last section under the title "`Adjusted for`" lists the names and the values of the other variables.

Figure 13.1 shows the predicted counts for `csclimat` at different values by the grouping variable `public`. As shown in the graph, the predicted counts of absences do

FIGURE 13.1 ● Predicted Counts of Absences for `csclimat` by `public`

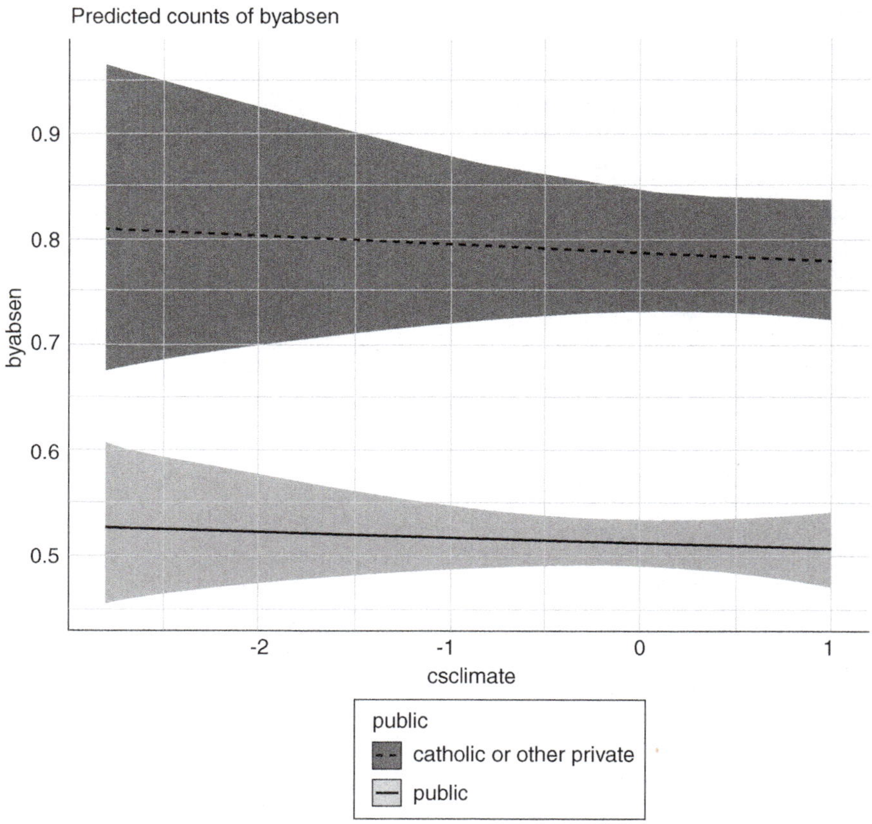

not change much with the increase of school climate, whereas the predicted counts of absences for students in Catholic or other private schools are higher than those in public schools.

In the example above, we disregard the random effects component and the predicted counts are on the population level. To account for the variability of the random effects, we need to specify the `type = "re"` argument. With this option, the predicted counts are still on the population level and stay the same, but the confidence intervals change after accounting for the variances of the random effects. The command is as follows: `mepoi4.efpub.r <- ggpredict(mepois.4, terms = c("csclimate", "public"), type = "re")`. In the `ggpredict()` function, everything stays the same exception the added `type = "re"` option. The output is assigned to an object named `mepoi4.efpub.r` and is plotted with `plot(mepoi4.efpub.r)`.

```
>mepoi4.efpub.r<-ggpredict(mepois.4, terms = c("csclimate", "public"), type = "re")
> mepoi4.efpub.r

# Predicted counts of byabsen

# public = catholic or other private

csclimate  |   Predicted  |          95% CI
---------------------------------------
    -2.80  |        0.81  |    [0.11, 5.78]
    -2.20  |        0.80  |    [0.11, 5.74]
    -1.60  |        0.80  |    [0.11, 5.69]
    -1.00  |        0.79  |    [0.11, 5.65]
    -0.20  |        0.79  |    [0.11, 5.60]
     1.00  |        0.78  |    [0.11, 5.53]

# public = public

csclimate  |   Predicted  |          95% CI
---------------------------------------
    -2.80  |        0.53  |    [0.07, 3.76]
    -2.20  |        0.52  |    [0.07, 3.73]
    -1.60  |        0.52  |    [0.07, 3.70]
    -1.00  |        0.52  |    [0.07, 3.67]
    -0.20  |        0.51  |    [0.07, 3.64]
     1.00  |        0.51  |    [0.07, 3.60]

Adjusted for:
* cmathach = 0.00
* bygender = 0.50
*   SCH_ID = 0 (population-level)
Intervals are prediction intervals.

> plot(mepoi4.efpub.r)
```

Figure 13.2 shows the predicted counts of absences for `csclimat` by the grouping variable `public` with random effects. As shown in the graph, the predicted counts are the same as those in Figure 13.2, but the confidence intervals are wider than those in Figure 13.1.

FIGURE 13.2 ● Predicted Counts of Absences for **`csclimat`** by **`public`** With Random Effects

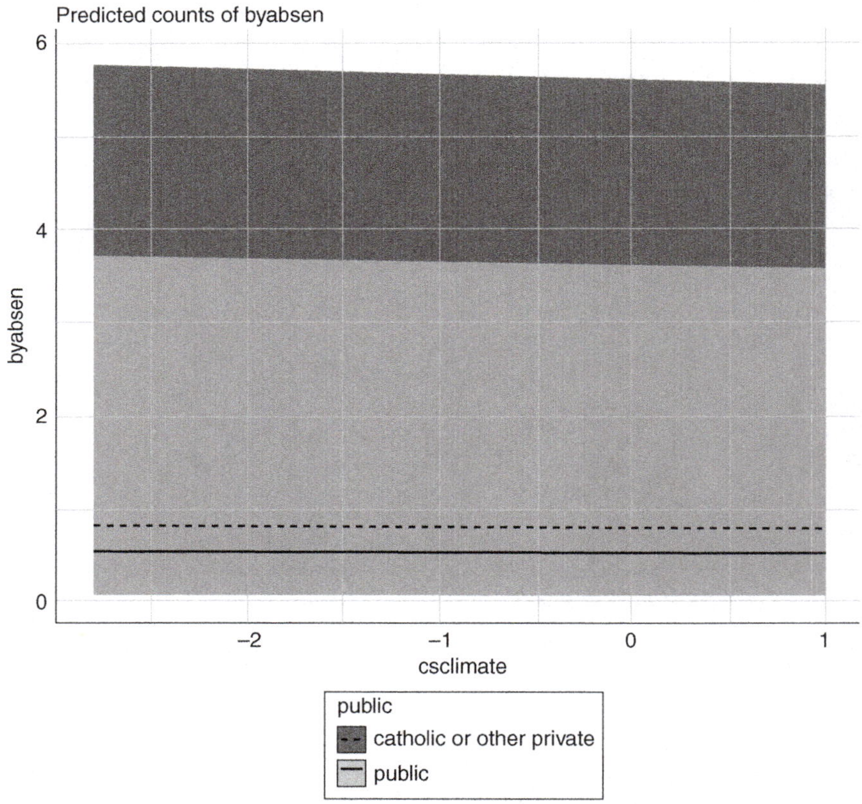

13.4 MULTILEVEL MODELING FOR COUNT RESPONSE VARIABLES WITH THE glmmTMB() FUNCTION IN THE glmmTMB PACKAGE

We can also use the glmmTMB() function in the glmmTMB package (Brooks et al., 2017) to fit multilevel models for count response variables. The glmmTMB package can be used to fit a wide variety of linear and generalized linear mixed models with broad extensions. In particular, this package is capable of fitting various multilevel models for count response variables, such as multilevel Poisson models, multilevel negative binomial models, and multilevel zero-inflated models. Since this is a user-written package, we need to install it first by typing install.packages("glmmTMB") and then load it by typing library(glmmTMB).

The syntax for glmmTMB() is similar to that for glmer(). The model formula in glmmTMB() also includes the fixed effects and random effects, which are connected by a plus (+) sign. The fixed effects part of the model specifies the dependent variable and the predictor variable(s) with the dependent variable on the left-hand side of the tilde (~). Same as the random effects part of glmer(), the random coefficients term is specified first, followed by a vertical line (|), and then, an identifier variable at a higher level as the grouping variable, which should be a factor variable. In addition to the model formula, we also need to specify the family = argument. For example, we can specify the family = poisson argument for Poisson distribution, family = nbinom1 for negative binomial distribution with a linear form for the variance (NB1 model), and family = nbinom2 for negative binomial distribution with a quadratic form (NB2 model). As of this writing, the glmmTMB() function uses the Laplace approximation rather than the adaptive Gauss-Hermite quadrature method for model estimation. For more details on how to use this function, type help(glmmTMB) in the command prompt after loading the glmmTMB package.

In the following example, the command mepois.4b <- glmmTMB(byabsen ~ cmathach + bygender + public + csclimate + (bygender|SCH_ID), data = mecount, family = poisson) tells R to fit the same contextual model (model 4) in the previous section after the two level 2 predictor variables public and csclimat are added to the model. The fitted model is named mepois.4b. The following output is displayed by the summary(mepois.4b) command.

```
> # Multilevel Poisson model with glmmTMB()
> # glmmTMB needs to be installed by using install.package()
> library(glmmTMB)
> mepois.4b <- glmmTMB(byabsen ~ cmathach + bygender + public + csclimate +
(bygender|SCH_ID), data=mecount, family=poisson)
> summary(mepois.4b)
 Family: poisson  ( log )
```

```
Formula:
byabsen ~ cmathach + bygender + public + csclimate + (bygender |   SCH_ID)
Data: mecount


     AIC       BIC     logLik  deviance  df.resid
 25549.2   25608.5   -12766.6   25533.2     12175

Random effects:


Conditional model:
Groups   Name         Variance  Std.Dev.  Corr
SCH_ID   (Intercept)  0.1370    0.3701
         bygender     0.1199    0.3463    -0.68
Number of obs: 12183, groups: SCH_ID, 617


Conditional model:
              Estimate  Std. Error  z value  Pr(>|z|)
(Intercept)  -0.548889   0.041536  -13.215   <2e-16 ***
cmathach     -0.018141   0.001055  -17.189   <2e-16 ***
bygender      0.614155   0.030469   20.157   <2e-16 ***
publicpublic -0.428997   0.041704  -10.287   <2e-16 ***
csclimate    -0.009986   0.026336   -0.379    0.705
---
Signif. codes: 0 '***' 0.001 '**' 0.01 '*' 0.05 '.' 0.1 ' ' 1
```

The R output produced by the `glmmTMB()` function is almost identical to that by the `glmer()` function introduced in the previous section. It also includes the family of the distribution, the model formula, the data, the model fit statistics such as the AIC and BIC, the log-likelihood value and the deviance with the corresponding degrees of freedom, the random effects, and the number of observations and groups. In addition, the fixed effects section at the end is titled "`Conditional model.`"

We use the `confint(mepois.4b)` command to extract the 95% confidence intervals of the parameter estimates. The output provides both the confidence intervals and the parameter estimates as follows.

```
> confint(mepois.4b)
                                 2.5 %       97.5 %      Estimate
(Intercept)                -0.63029695  -0.46748032  -0.548888639
cmathach                   -0.02020906  -0.01607209  -0.018140577
bygender                    0.55443793   0.67387249   0.614155210
publicpublic               -0.51073519  -0.34725791  -0.428996551
csclimate                  -0.06160296   0.04163031  -0.009986323
Std.Dev.(Intercept)         0.31809746   0.43056505   0.370083300
Std.Dev.bygender            0.27613499   0.43422034   0.346270743
Cor.bygender.(Intercept)   -0.77815595  -0.51022635  -0.675501615
```

The `exp(confint(mepois.4b))` command is used to obtain the incidence rate ratios and the corresponding confidence intervals. The following output is created.

```
> exp(confint(mepois.4b))
                                2.5 %       97.5 %     Estimate
(Intercept)                 0.5324337    0.6265791    0.5775914
cmathach                    0.9799938    0.9840564    0.9820230
bygender                    1.7409622    1.9618198    1.8480947
publicpublic                0.6000543    0.7066231    0.6511622
csclimate                   0.9402561    1.0425090    0.9900634
Std.Dev.(Intercept)         1.3745102    1.5381264    1.4478552
Std.Dev.bygender            1.3180258    1.5437590    1.4137853
Cor.bygender.(Intercept)    0.4592521    0.6003597    0.5089011
```

Currently we cannot use the `anova()` function to conduct the likelihood ratio test to compare models since it does not work with the `glmmTMB()` function. However, we can use the `AICtab()` function in the `bbmle` package (Bolker et al., 2021) to compare the AIC statistic between models. We need to install the package first by typing `install.packages("bbmle")` and then load it by typing `library(bbmle)`. The `AICtab(mepois.3, mepois.4b)` command compares the difference in the AIC statistic between `mepois.3` and `mepois.4b`.

```
> # install.packages("bbmle")
> library(bbmle)
> AICtab(mepois.3, mepois.4b)
            dAIC  df
mepois.4b     0.0  8
mepois.3    103.7  6
```

In the output, the AIC difference between `mepois.3` and `mepois.4b` is 103.7 with `mepois.4b` as the base model. We are in favor of the `mepois.4b` model since it has a smaller AIC. The `AICtab() function` can also be used to compare multiple models. We use the `AICtab(mepois.1, mepois.2, mepois.3, mepois.4b)` command to compare the AIC differences of all the four fitted models above. The output is as follows.

```
> AICtab(mepois.1, mepois.2, mepois.3, mepois.4b)
            dAIC  df
mepois.4b     0.0  8
mepois.3    103.7  6
mepois.2    134.4  4
mepois.1    950.9  2
```

In the output, mepois.4b is the base model. The difference in AIC between mepois.1 and mepois.4b is 950.9 and the difference between mepois.2 and mepois.4b is 134.4. The column titled df displays the degrees of freedom for the corresponding model. Since the mepois.4b model has the smallest AIC, we prefer this model to the others.

13.5 MULTILEVEL MODELING FOR COUNT RESPONSE VARIABLES WITH THE glmmPQL() FUNCTION IN THE MASS PACKAGE

In addition to the glmer() function in the lme4 package and the glmmTMB() function in the glmmTMB package, we can use the glmmPQL() function in the MASS package (Venables & Ripley, 2002) for multilevel models for count response variables. You need to install both the MASS package and the nlme package. If they are not installed, type install.packages("MASS") and then load the package with library(MASS) and also type install.packages("nlme") and then load it. Since the user-written nlme package has been installed for the previous chapter, you only need to load the package by typing library(nlme).

The syntax for glmmPQL() is similar to that for lme() for multilevel models for continuous response variables. The model formula in glmmPQL() also includes the components of the fixed effects and random effects, which are separated by a comma (,). The fixed effects part of the model specifies the dependent variable and the predictor variable(s) with the dependent variable on the left-hand side of the tilde (~). The predictor variables on the right-hand side are connected by plus (+) symbols. Next, the random effects part of the model is specified with the random = argument. Same as the random effects part of lme(), the random coefficients term is specified first, followed by a vertical line (|), and then, an identifier variable at a higher level as the grouping variable. For example, the command glmmPQL(y ~ x, random = ~ 1|schid, data = data1, family = poisson) tells R to fit a multilevel model to estimate a count outcome variable y with a predictor variable x. The random = ~ 1|schid argument specifies the random intercepts varying across schools with 1 as the random intercept on the left of the vertical line (|) and schid as the identifier variable on the right. Next, the data = data1 argument specifies the data frame data1. Finally the family = poisson argument specifies Poisson distribution. The glmer() function uses the penalized quasi-likelihood method rather than the Laplace approximation and Gauss-Hermite quadrature for model estimation although the latter two methods provide more accurate estimates. For more details on how to use this function, type help(glmmPQL) in the command prompt after loading the MASS package.

In the following example, the command mepois.4c <- glmmPQL(byabsen ~ cmathach + bygender + public + csclimate, random = ~bygender|SCH_ID, data = mecount, family = poisson) tells R to fit the same contextual model (model 4) in the previous section after the two level 2 predictor variables `public` and `csclimat` are added to the model. The fitted model is named `mepois.4c`. The following output is displayed by the `summary (mepois.4c)` command.

```
> # Multilevel Poisson model with glmmPQL()
> library(MASS)
> library(nlme)
>mepois.4c <- glmmPQL(byabsen ~ cmathach + bygender + public + csclimate, random =
~bygender|SCH_ID, data=mecount, family=poisson)
iteration 1
iteration 2
iteration 3
iteration 4
> summary(mepois.4c)
Linear mixed-effects model fit by maximum likelihood
Data: mecount
AIC  BIC  logLik
 NA   NA      NA

Random effects:
 Formula: ~bygender | SCH_ID
 Structure: General positive-definite, Log-Cholesky parametrization
              StdDev     Corr
(Intercept)  0.3366247  (Intr)
bygender     0.2940303  -0.646
Residual     1.0425759

Variance function:
 Structure: fixed weights
 Formula: ~invwt
Fixed effects: byabsen ~ cmathach + bygender + public + csclimate
                   Value  Std.Error    DF    t-value  p-value
(Intercept)   -0.5076067 0.03934589 11564 -12.901138   0.0000
cmathach      -0.0180898 0.00109048 11564 -16.588817   0.0000
bygender       0.5942889 0.02839070 11564  20.932521   0.0000
publicpublic  -0.4227360 0.04088316   614 -10.340100   0.0000
csclimate     -0.0089074 0.02605211   614  -0.341908   0.7325

 Correlation:
              (Intr)  cmthch  bygndr  pblcpb
cmathach      -0.019
bygender      -0.472  -0.042
publicpublic  -0.775   0.116  -0.005
csclimate     -0.286  -0.138   0.010   0.351

Standardized Within-Group Residuals:
       Min          Q1         Med          Q3         Max
-1.2872309  -0.7090059  -0.5335218  0.5080196  6.5034850

Number of Observations: 12183
Number of Groups: 617
```

The R output produced by the `glmmPQL()` function is similar to that by the `lme()` function introduced in the previous chapter. It includes the note for the estimation method, the data, the random effects, the fixed effects, the correlation matrix for the predictor variables, the standardized within-group residuals, the number of observations, and the number of groups.

The random effects section (labeled `Random effects:`) provides the model formula of the random effects and the standard deviations of the intercept, the random slope of `bygender`, and the residual. It also provides the correlation coefficient for `bygender` and the intercept. The results are similar to those provided by the `glmer()` function and the `glmmTMB()` function in the previous sections.

The results in the fixed effects section (labeled `Fixed effects:`) look similar to those produced by the `glmer()` function and the `glmmTMB()` function. To avoid redundancy, the interpretation of the results is omitted here.

We use the `intervals(mepois.4c)` command to extract the 95% confidence intervals of the parameter estimates for both the fixed effects and random effects. The results are displayed as follows.

```
> intervals(mepois.4c)
Approximate 95% confidence intervals

Fixed effects:
                   lower          est.         upper
(Intercept)  -0.58471549  -0.507606724  -0.43049796
cmathach     -0.02022694  -0.018089844  -0.01595275
bygender      0.53864975   0.594288896   0.64992805
publicpublic -0.50300727  -0.422735965  -0.34246466
csclimate    -0.06005898  -0.008907435   0.04224411

attr(,"label")
[1] "Fixed effects:"

 Random Effects:
  Level: SCH_ID
                               lower        est.       upper
sd((Intercept))            0.2850591   0.3366247   0.3975183
sd(bygender)               0.2190803   0.2940303   0.3946216
cor((Intercept),bygender) -0.7792886  -0.6464610  -0.4580440

 Within-group standard error:
   lower       est.     upper
1.029202   1.042576  1.056124
```

We can use the `intervals(mepois.4c, which = "fixed")` command to extract the 95% confidence intervals of the parameter estimates for the fixed effects only.

```
> intervals(mepois.4c, which = "fixed")
Approximate 95% confidence intervals


 Fixed effects:
                   lower         est.        upper
(Intercept)   -0.58471549 -0.507606724 -0.43049796
cmathach      -0.02022694 -0.018089844 -0.01595275
bygender       0.53864975  0.594288896  0.64992805
publicpublic  -0.50300727 -0.422735965 -0.34246466
csclimate     -0.06005898 -0.008907435  0.04224411
attr(,"label")
[1] "Fixed effects:"
```

In the output, the coefficients are listed in the middle with the lower and upper confidence intervals on the left and right sides. We can use the following commands to obtain the incidence rate ratios and the corresponding confidence intervals. The following output is created.

```
> coef <- fixef(mepois.4c)
> se <- sqrt(diag(vcov(mepois.4c)))
> LL <- fixef(mepois.4c)-1.96*se
> UL <- fixef(mepois.4c)+1.96*se
> coeftab <- cbind(coef, LL, UL)
> coeftab
                      coef          LL          UL
(Intercept)   -0.507606724 -0.58470884 -0.43050461
cmathach      -0.018089844 -0.02022675 -0.01595293
bygender       0.594288896  0.53865455  0.64992324
publicpublic  -0.422735965 -0.50285051 -0.34262142
csclimate     -0.008907435 -0.05995909  0.04214422

> irtab <- cbind(ir=exp(fixef(mepois.4c)), IRLL=exp(LL), IRUL=exp(UL))
> irtab
                    ir      IRLL      IRUL
(Intercept)  0.6019345 0.5572681 0.6501809
cmathach     0.9820728 0.9799764 0.9841736
bygender     1.8117421 1.7136996 1.9153938
publicpublic 0.6552516 0.6048042 0.7099069
csclimate    0.9911321 0.9418031 1.0430449
```

We use the `VarCorr(mepois.4c)` command to request the variance components of the model.

```
> VarCorr(mepois.4c)
SCH_ID = pdLogChol(bygender)
             Variance    StdDev     Corr
(Intercept)  0.11331622  0.3366247  (Intr)
bygender     0.08645384  0.2940303  -0.646
Residual     1.08696455  1.0425759
```

13.6 MULTILEVEL NEGATIVE BINOMIAL MODELS WITH THE `glmmTMB()` FUNCTION IN THE `glmmTMB` PACKAGE

We can also use the `glmmTMB()` function in the `glmmTMB` package to fit multilevel negative binomial models for count response variables when there is overdispersion, i.e., the variance of the count response variable is greater than the mean. The variance of the count response variable can be expressed either as a linear form or a quadratic form of the mean. With the linear form, when variance $(Y) = \mu(1 + \alpha)$, this negative binomial model is referred to as the NB1 model where α is the dispersion parameter. With the quadratic form, variance $(Y) = \mu + \alpha\mu^2$, so this model is also referred to as the NB2 model. The section focuses on multilevel NB2 models since they are more commonly used than NB1 models. In the following example, the command `menb.4 <- glmmTMB(byabsen ~ cmathach + bygender + public + csclimate + (bygender|SCH_ID), data = mecount, family = nbinom2)` tells R to fit the multilevel negative binomial model (NB2) with the two level 1 predictor variables `cmathach` and `bygender` and the two level 2 predictor variables `public` and `csclimat`. The random effects part of the model `(bygender|SCH_ID)` specifies the random slope of `bygender`. In addition, the `family = nbinom2` argument specifies the NB2 model. The fitted model is named `menb.4`. The following output is displayed by the `summary(menb.4)` command.

```
> # Multilevel negative binomial model with glmmTMB()
> library(glmmTMB)
> menb.4 <- glmmTMB(byabsen ~ cmathach + bygender + public + csclimate +
(bygender|SCH_ID), data=mecount, family=nbinom2)
> summary(menb.4)
 Family: nbinom2  ( log )
Formula:
byabsen ~ cmathach + bygender + public + csclimate + (bygender | SCH_ID)
Data: mecount
```

```
     AIC       BIC     logLik  deviance  df.resid
 25447.9   25514.6   -12714.9   25429.9     12174

Random effects:

Conditional model:
Groups  Name          Variance  Std.Dev.  Corr
SCH_ID  (Intercept)   0.11457   0.3385
        bygender      0.07765   0.2787    -0.68

Number of obs: 12183, groups: SCH_ID, 617

Overdispersion parameter for nbinom2 family (): 4.8

Conditional model:
              Estimate  Std. Error  z value  Pr(>|z|)
(Intercept)  -0.539578   0.041393  -13.036   <2e-16 ***
cmathach     -0.018342   0.001127  -16.273   <2e-16 ***
bygender      0.613807   0.030345   20.228   <2e-16 ***
publicpublic -0.428986   0.041684  -10.291   <2e-16 ***
csclimate    -0.009679   0.026334   -0.368    0.713
---
Signif. codes: 0 '***' 0.001 '**' 0.01 '*' 0.05 '.' 0.1 ' ' 1
```

The R output produced by the `glmmTMB()` function for the multilevel negative binomial model is similar to that by the same function for the multilevel Poisson model (Model 4) introduced in the previous section. It also includes the family of the distribution, the model formula, the data, the model fit statistics including the AIC, BIC, log-likelihood, and deviance with the corresponding degrees of freedom, the random effects, the number of observations and groups, and the fixed effects for the conditional model.

The overdispersion parameter for the `nbinom2` family is displayed between the random effects and the fixed effects sections. Recall in the NB2 model, variance $(Y) = \mu(1 + \alpha\mu) = \mu + \alpha\mu^2$. The `glmmTMB()` function uses a slightly different form, variance $(Y) = \mu + \mu^2/\theta$. Here, $\theta = 1/\alpha$. The overdispersion parameter `Theta` = 4.8. We can easily compute $\alpha = 1/4.8 = .208$.

We use the `confint(menb.4)` command to extract the 95% confidence intervals of the parameter estimates. The results are displayed as follows.

```
> confint(menb.4)
                                  2.5 %       97.5 %      Estimate
(Intercept)                 -0.62070596  -0.45844990  -0.539577929
cmathach                    -0.02055152  -0.01613302  -0.018342270
bygender                     0.55433261   0.67328114   0.613806876
publicpublic                -0.51068497  -0.34728668  -0.428985825
csclimate                   -0.06129254   0.04193412  -0.009679208
Std.Dev.(Intercept)          0.28293904   0.40493856   0.338486230
Std.Dev.bygender             0.19666450   0.39485986   0.278666314
Cor.bygender.(Intercept)    -0.80558066  -0.44619908  -0.680667869
```

We use the `exp(confint(menb.4))` command to obtain the incidence rate ratios and the corresponding confidence intervals. The following output is created.

```
> exp(confint(menb.4))
                              2.5 %      97.5 %     Estimate
(Intercept)                 0.5375648   0.6322630   0.5829943
cmathach                    0.9796582   0.9839964   0.9818249
bygender                    1.7407788   1.9606600   1.8474510
publicpublic                0.6000844   0.7066027   0.6511692
csclimate                   0.9405481   1.0428258   0.9903675
Std.Dev.(Intercept)         1.3270243   1.4992104   1.4028224
Std.Dev.bygender            1.2173356   1.4841762   1.3213663
Cor.bygender.(Intercept)    0.4468284   0.6400563   0.5062788
```

The output includes incidence rate ratios and the corresponding confidence intervals for both the fixed effects and random effects. The interpretation of the incidence rate ratios in multilevel negative binomial models is the same as that in multilevel Poisson models, so it is omitted here.

13.7 MAKING PUBLICATION-QUALITY TABLES

13.7.1 Presenting the Results Using the `stargazer` Package

We can use the `stargazer` package (Hlavac, 2018) to make a table containing the results of the fitted unconditional model and the contextual model. After fitting these two models, `mepois.1` and `mepois.4`, with the `glmer()` function, we use the command as follows: `stargazer(mepois.1, mepois.4, type = "text", align = TRUE, out = "mepoison.txt")`. In the `stargazer()` function, we first specify the two model objects to be presented and then the type of the table. The `type = "text"` argument specifies the table type and the `align = TRUE` argument aligns the results of the models. The `out = "mepoison.txt"` argument saves the output named `mepoison.txt`.

```
> library(stargazer)

Please cite as:

Hlavac, Marek (2018). stargazer: Well-Formatted Regression and Summary Statistics
Tables.
R package version 5.2.2. https://CRAN.R-project.org/package=stargazer

> stargazer(mepois.1, mepois.4, type="text", align=TRUE, out="mepoison.txt")
```

```
==================================================
                          Dependent variable:
                     -----------------------------
                               byabsen
                          (1)           (2)
--------------------------------------------------
cmathach                              -0.018***
                                      (0.001)

bygender                               0.614***
                                      (0.030)

publicpublic                          -0.429***
                                      (0.042)

csclimate                             -0.010
                                      (0.026)

Constant                -0.507***     -0.549***
                        (0.018)       (0.041)

--------------------------------------------------
Observations             12,183        12,183
Log Likelihood         -13,248.050   -12,766.610
Akaike Inf. Crit.       26,500.090    25,549.230
Bayesian Inf. Crit.     26,514.910    25,608.490
==================================================
Note:               *p<0.1;  **p<0.05;  ***p<0.01
> stargazer(mepois.1, mepois.4, type="html", align=TRUE, out="mepoison.htm")
```

We can also create the table in the HTML format and copy it into Microsoft Word. The command is as follows: `stargazer(mepois.1, mepois.4, type = "html", align = TRUE, out = "mepoison.htm")`. It produces Table 13.1, as shown here in its original format, presenting the results for the unconditional model and the contextual model.

TABLE 13.1 ● Results of the Two Multilevel Poisson Regression Models: Unconditional Model and Contextual Model (Shown in Original Format Generated by the stargazer Package)

	Dependent variable:	
	byabsen	
	(1)	(2)
cmathach		−0.018***
		(0.001)
bygender		0.614***
		(0.030)

Table 13.1 *(Continued)*

	Dependent variable:	
	byabsen	
	(1)	(2)
publicpublic		−0.429***
		(0.042)
csclimate		−0.010
		(0.026)
Constant	−0.507***	−0.549***
	(0.018)	(0.041)
Observations	12,183	12,183
Log Likelihood	−13,248.050	−12,766.610
Akaike Inf. Crit.	26,500.090	25,549.230
Bayesian Inf. Crit.	26,514.910	25,608.490

*$p < .1$
**$p < .05$
***$p < .01$

13.7.2 Presenting the Results Using the `texreg` Package

We can also use the `screenreg()` and `htmlreg()` functions from the `texreg` package (Leifeld, 2013) to display the results. Since `texreg` is a user-written package, you need to install it first by typing `install.packages("texreg")` and then load the package by typing `library(texreg)`.

The same results of the unconditional model and the contextual model can also be displayed in a table with the following command: `screenreg(list(mepois.1, mepois.4))`.

```
> # Presenting the results with texreg()
> library(texreg)
Version:     1.37.5
Date:        2020-06-17
Author:      Philip Leifeld (University of Glasgow)

Please cite the JSS article in your publications -- see citation("texreg").
> screenreg(list(mepois.1, mepois.4))
```

```
==============================================================
                                  Model 1          Model 2
--------------------------------------------------------------
(Intercept)                         -0.51 ***        -0.55 ***
                                    (0.02)           (0.04)
cmathach                                             -0.02 ***
                                                     (0.00)
bygender                                              0.61 ***
                                                     (0.03)
publicpublic                                         -0.43 ***
                                                     (0.04)
csclimate                                            -0.01
                                                     (0.03)
--------------------------------------------------------------
AIC                              26500.09         25549.23
BIC                              26514.91         25608.49
Log Likelihood                  -13248.05        -12766.61
Num. obs.                        12183            12183
Num. groups: SCH_ID                617              617
Var: SCH_ID (Intercept)              0.11             0.14
Var: SCH_ID bygender                                  0.12
Cov: SCH_ID (Intercept) bygender                     -0.09
==============================================================
*** p < 0.001, ** p < 0.01, * p < 0.05

>   htmlreg(list(mepois.1,   mepois.4),   file="mepoison.doc",   doctype=TRUE,
html.tag=TRUE, head.tag=TRUE)
The table was written to the file 'mepoison.doc'.
```

To create the table in the HTML format and copy it into Microsoft Word, we use the command as follows: `htmlreg(list(mepois.1, mepois.4), file = "mepoison.doc", doctype = TRUE, html.tag = TRUE, head.tag = TRUE)`. The original table presenting the results for the unconditional model and the contextual model is omitted. With edits on the model titles and the added IRRs and labels for the fixed and random effects, the final table is displayed as in Table 13.2.

TABLE 13.2 ● Results of the Two Multilevel Poisson Regression Models: Unconditional Model and Contextual Model (Edited)

	Unconditional Model (Model 1)	Contextual Model (Model 4)	
	Coefficient (SE)	Coefficient (SE)	IRR
Fixed Effects			
(Intercept)	−0.51***	−0.55***	.58***
	(0.02)	(0.04)	
cmathach		−0.02***	.98***
		(0.00)	

Table 13.2 *(Continued)*

	Unconditional Model (Model 1)	Contextual Model (Model 4)	
	Coefficient (SE)	Coefficient (SE)	IRR
bygender		0.61***	1.85***
		(0.03)	
public		−0.43***	.65***
		(0.04)	
csclimate		−0.01	.99
		(0.03)	
AIC	26,500.09	25,549.23	
BIC	26,514.91	25,608.49	
Log Likelihood	−13,248.05	−12,766.61	
Observations	12,183	12,183	
Number of groups: SCH_ID	617	617	
Random Effects			
Between-School Var: SCH_ID (Intercept)	0.11	0.14	
Slope Var: SCH_ID bygender		0.12	
Cov: SCH_ID (Intercept) bygender		−0.09	

Note. Standard errors are shown in parentheses.
***$p < .001$
**$p < .01$
*$p < .05$

13.8 REPORTING THE RESULTS

The same reporting guidelines for multilevel modeling for continuous response variables can be applied to the reporting for the multilevel Poisson regression modeling. The only difference is that the incidence rate ratios may need to be reported and interpreted in the latter models. The following is an example of summarizing the results of a multilevel contextual Poisson regression model.

> Multilevel Poisson regression models were fitted to predict the number of absences from four predictor variables: students' math scores, gender, the school type, and the school climate. The first predictor was a student-level

variable, and the other two were school-level variables. Multilevel modeling for count response variables was used since the outcome variable was a count variable. Four models, from the unconditional (null) model to the contextual model, were fitted. To save space, only the results for the final contextual model (model 4) were reported as follows.

Results for the Contextual Model (Model 4)

The log coefficient for `cmathach` (γ_{10}) = −.018, z = −17.30, $p < .001$. The incidence rate ratio for `cmathach` was .982, which indicated that the incidence rate of being absent or the expected count of absences decreased by a factor of .982 for a one-unit increase in math scores when holding the other predictor constant.

The log coefficient for `bygender` (γ_{20}) = .614, z = 20.29, $p < .001$. The incidence rate ratio for `bygender` was 1.848, which indicated that the incidence rate of being absent or the expected counts of absences for male students was 1.848 times as large as that for female students when holding the other predictor constant.

The log coefficient for `public` (γ_{01}) = −.429, z = −10.32, $p < .001$. The incidence rate ratio for `public` was .651, which indicated that the incidence rate of being absent for students in public schools was .651 times the incidence rate for those in private schools. In other words, being students in public schools decreased the incidence rate of being absent by 34.9%.

The log coefficient for `csclimat` (γ_{02}) = −.010, z = −.38, p = .704. The incidence rate ratio for `csclimat` was .990. Since $p > .05$, the coefficient was not significant. It indicated that school climate did not impact the incidence rate of being absent.

Regarding the variance and covariance components for the random effects, the between-group variance, the intercept variance (τ_{00}), was .14. The slope variance (τ_{10}) for `bygender` was .12. The covariance between the intercept and the slope was −.087.

The likelihood ratio test was used to compare all the four models. The results indicated that the contextual model (model 4) fit the data best. In addition, the AIC and BIC statistics of the model were 25,549 and 25,609, respectively, which were also the smallest compared to those in the other three models.

13.9 SUMMARY OF R COMMANDS IN THIS CHAPTER

```
# Chap 13 R Script

# Remove all objects
rm(list = ls(all = TRUE))

# The following user-written packages need to be installed first by using
install.packages(" ") and then by loading it with library()
# library(lme4)           # It is already installed for Chapter 10
# library(margins)        # It is already installed for Chapter 3
# library(ggeffects)      # It is already installed for Chapter 2
# library(stargazer)      # It is already installed for Chapter 2
# library(texreg)         # It is already installed for Chapter 4
# library(glmmTMB)
# library(bbmle)
# library(MASS)           # It is part of R base distribution
# library(nlme)           # It is part of R base distribution

# Import the dataset
library(foreign)
mecount <- read.dta("C:/CDA/mecount.dta")
attach(mecount)
library(sjmisc)
descr(mecount, byabsen, cmathach, bygender, public, csclimate)

library(lme4)
# Unconditional Model/Null model with glmer()
mepois.1 <- glmer(byabsen ~ 1 + (1|SCH_ID), data=mecount, family=poisson, nAGQ=1)
summary(mepois.1)
coef <- fixef(mepois.1)
se <- sqrt(diag(vcov(mepois.1)))
LL <- fixef(mepois.1)-1.96*se
UL <- fixef(mepois.1)+1.96*se
Coeftab <- cbind(coef, LL, UL)
coeftab
exp(fixef(mepois.1))
exp(coeftab)
irtab <- cbind(ir=exp(fixef(mepois.1)), IRLL=exp(LL), IRUL=exp(UL))
irtab

confint(mepois.1, method = "Wald")
exp(confint(mepois.1, method = "Wald"))

mepois.1b <- glmer(byabsen ~ 1 + (1|SCH_ID), data=mecount, family=poisson, nAGQ=7)
summary(mepois.1b)

# Random-intercept model
mepois.2 <- glmer(byabsen ~ cmathach + bygender + (1|SCH_ID), data=mecount,
family=poisson, nAGQ=1)
summary(mepois.2)
```

```
coef <- fixef(mepois.2)
se <- sqrt(diag(vcov(mepois.2)))
LL <- fixef(mepois.2)-1.96*se
UL <- fixef(mepois.2)+1.96*se
Coeftab <- cbind(coef, LL, UL)
coeftab
exp(fixef(mepois.2))
exp(coeftab)
irtab <- cbind(ir=exp(fixef(mepois.2)), IRLL=exp(LL), IRUL=exp(UL))
irtab

confint(mepois.2, method="Wald")
exp(confint(mepois.2, method="Wald"))

anova(mepois.1, mepois.2)

# Random-coefficient model
mepois.3 <- glmer(byabsen ~ cmathach + bygender + (bygender|SCH_ID), data=mecount,
family=poisson, nAGQ=1)
summary(mepois.3)
coef <- fixef(mepois.3)
se <- sqrt(diag(vcov(mepois.3)))
LL <- fixef(mepois.3)-1.96*se
UL <- fixef(mepois.3)+1.96*se
coeftab <- cbind(coef, LL, UL)
coeftab
exp(fixef(mepois.3))
exp(coeftab)
irtab <- cbind(ir=exp(fixef(mepois.3)), IRLL=exp(LL), IRUL=exp(UL))
irtab

confint(mepois.3, method="Wald")
exp(confint(mepois.3, method="Wald"))

anova(mepois.2, mepois.3)

# Contextual model/Random-coefficient model with predictor variables in both levels
mepois.4 <- glmer(byabsen ~ cmathach + bygender + public + csclimate +
(bygender|SCH_ID), data=mecount, family=poisson, nAGQ=1)
summary(mepois.4)
coef <- fixef(mepois.4)
se <- sqrt(diag(vcov(mepois.4)))
LL <- fixef(mepois.4)-1.96*se
UL <- fixef(mepois.4)+1.96*se
Coeftab <- cbind(coef, LL, UL)
coeftab
exp(fixef(mepois.4))
exp(coeftab)
irtab <- cbind(ir=exp(fixef(mepois.4)), IRLL=exp(LL), IRUL=exp(UL))
irtab

confint(mepois.4, method="Wald")
exp(confint(mepois.4, method="Wald"))

anova(mepois.3, mepois.4)
```

```
# Model comparisons
anova(mepois.1, mepois.2, mepois.3, mepois.4)

# Marginal effect with margins()
library(margins)
marg <- margins(mepois.4)
summary(marg)

# Marginal effects/Predicted counts with ggpredict() in ggeffects
library(ggeffects)
mepoi4.efpub <- ggpredict(mepois.4, terms=c("csclimate", "public"))
mepoi4.efpub
plot(mepoi4.efpub)

mepoi4.efpub.r <- ggpredict(mepois.4, terms=c("csclimate", "public"), type="re")
mepoi4.efpub.r
plot(mepoi4.efpub.r)

# Presenting the results with stargazer()
library(stargazer)
stargazer(mepois.1, mepois.4, type="text", align=TRUE, out="mepoison.txt")
stargazer(mepois.1, mepois.4, type="html", align=TRUE, out="mepoison.htm")

# Presenting the results with texreg()
library(texreg)
screenreg(list(mepois.1, mepois.4))
htmlreg(list(mepois.1, mepois.4), file="mepoison.doc", doctype=TRUE, html.tag=
TRUE, head.tag=TRUE)

# Multilevel Poisson model with glmmTMB()
# install.packages("glmmTMB")
library(glmmTMB)
mepois.4b <- glmmTMB(byabsen ~ cmathach + bygender + public + csclimate +
(bygender|SCH_ID), data=mecount, family=poisson)
summary(mepois.4b)

confint(mepois.4b)
exp(confint(mepois.4b))

# Model comparisons with AICtab()
# install.packages("bbmle")
library(bbmle)
AICtab(mepois.3, mepois.4b)
AICtab(mepois.1, mepois.2, mepois.3, mepois.4b)

# Multilevel Poisson model with glmmPQL()
library(MASS)
library(nlme)
mepois.4c <- glmmPQL(byabsen ~ cmathach + bygender + public + csclimate, random =
~bygender|SCH_ID, data=mecount, family=poisson)
summary(mepois.4c)
intervals(mepois.4c)
intervals(mepois.4c,which="fixed")
exp(intervals(mepois.4c,which="fixed"))
```

```
coef <- fixef(mepois.4c)
se <- sqrt(diag(vcov(mepois.4c)))
LL <- fixef(mepois.4c)-1.96*se
UL <- fixef(mepois.4c)+1.96*se
coeftab <- cbind(coef, LL, UL)
coeftab
exp(fixef(mepois.4c))
exp(coeftab)
irtab <- cbind(ir=exp(fixef(mepois.2c)), IRLL=exp(LL), IRUL=exp(UL))
irtab
VarCorr(mepois.4c)

# Multilevel negative binomial model with glmmTMB()
menb.4 <- glmmTMB(byabsen ~ cmathach + bygender + public + csclimate + (bygender|
SCH_ID), data=mecount, family=nbinom2)
summary(menb.4)
confint(menb.4)
exp(confint(menb.4))

detach(mecount)
```

Glossary

Cross-level interactions are interaction effects between level 1 and level 2 predictor variables.

In **a random-coefficient Poisson regression model**, both the intercept and coefficient in the level 1 equation are allowed to vary across groups at level 2.

In **a random-intercept Poisson regression model**, the intercept is allowed to vary freely across groups.

Multilevel negative binomial models for count response variables are fitted when there is overdispersion, i.e., the variance of the count response variable is greater than the mean. With the quadratic form, variance $(Y) = \mu + \alpha\mu^2$, this model is also referred to as the multilevel NB2 model.

The contextual model for a count response variable includes both level 1 and level 2 predictor variables.

The incidence rate is the expected number of events during a period of time or in a location.

The incidence rate ratios are the exponentiated log coefficients. They are interpreted as the change in the incidence rate or the expected number of events for a one-unit increase in an independent variable.

The unconditional model, or null model, for count response variables includes no predictor variable at any level. This model estimates the overall average of the log expected counts of events or the log predicted mean count and the variability in the predicted log mean count between groups (e.g., schools).

Exercises

Use the `medpar` data, a part of the U.S. Medicare data (Hilbe, 2014) available at **https://edge.sagepub.com/liu1e** for the following problems. The following variables are used for the multilevel Poisson modeling.

`los`: length of stay in a hospital

`white`: ethnicity (1 = white; 0 = others)

`hmo`: health maintenance organization (1 = hmo; 0 = not)

`age80`: age 80 or more (1 = age 80 or more; 0 = not)

`provnum`: medical provider number (hospitals, the clustering variable)

Multilevel Poisson regression models will be fitted to estimate factors related to how long a patient stayed in a hospital. The count outcome variable is `los`, and the predictor variables are `white`, `hmo`, and `age80`. Perform the analysis and answer the following questions:

1. Use the `glmer()` function to fit an unconditional model and obtain the between-group variance for the intercept.

2. Fit a two-level, random-intercept model with a random intercept and the three predictor variables `white`, `hmo` and `age80`.
3. Compute the IRRs and interpret them for the predictor variables.
4. Conduct a likelihood ratio test comparing the random-intercept model and the unconditional model. Interpret the results.
5. Write a report to summarize the results of the random-intercept model.

14

MULTILEVEL MODELING FOR NOMINAL RESPONSE VARIABLES

OBJECTIVES OF THIS CHAPTER

This chapter introduces multilevel modeling for nominal response variables. It starts with an introduction to the model specification, the odds and odds ratios in the model, and a description of the research questions and data. Then several models, from the unconditional (null) model to the random-intercept model with a level-1 predictor to the random-intercept model with predictor variables in both levels, are illustrated using R with step-by-step instructions. The `mclogit` package is used for fitting multilevel multinomial logistic regression models. The other two packages, `MCMCglmm` and `brms` are used for Bayesian multilevel modeling for nominal response variables. R commands are explained, and the output is interpreted for each model in detail. This chapter also illustrates how the results are displayed in publication-quality tables using R and reported in text. It focuses on model fitting with R, as well as on interpreting and presenting the results. After reading this chapter, you should be able to:

- Determine when multilevel modeling for nominal response variables is used.
- Formulate multilevel models.
- Conduct the multilevel modeling and Bayesian multilevel modeling for nominal response variables using R.
- Interpret the output.
- Compute and plot the predicted probabilities using the `ggeffects()` function.
- Be familiar with model fitting strategies.
- Present results in publication-quality tables using R.
- Write the results for publication.

14.1 MULTILEVEL MODELING FOR NOMINAL OUTCOME VARIABLES

14.1.1 Model Specification

In the previous chapters, we introduced multilevel models for binary and ordinal outcome variables. Multilevel modeling can be extended for categorical response variables when the nominal outcome variable is an unordered variable with more than two categories. In this chapter, multilevel models for nominal response variables (Goldstein, 2003; Hedeker, 2007; Hedeker & Gibbon, 1994; Hox, 2010; Raudenbush & Bryk, 2002) are introduced.

Recall that the single-level multinomial logistic regression model introduced in Chapter 7 is an extension of binary logistic regression models. Instead of estimating the nominal response variable directly, we estimate the probability of being in a particular category relative to that of the base category via a logistic transformation, logit$\left(\frac{P(Y=k)}{P(Y=K)}\right)$ or ln(odds). The link function transforms the original nominal outcome variable Y with K categories and connects it with the linear predictor η_k.

The multinomial logistic regression model is expressed as follows:

$$\eta_k = \ln\left(\frac{P(Y=k|x_1, x_2, \ldots, x_p)}{P(Y=K|x_1, x_2, \ldots, x_p)}\right) = \alpha_k + \beta_{k1}X_1 + \beta_{k2}X_2 + \ldots + \beta_{kp}X_p \quad (14.1)$$

where $k = 1, 2, \ldots, K-1$; K is the base category, which can be any category but is generally the highest one or the lowest one; it is also referred to as the baseline category, the reference category, or the comparison category; α_k are the intercepts; and β_{k1}, β_{k2}, …, β_{kp} are the logit coefficients for each comparison. Multinomial logistic regression can be treated as a simultaneous estimation of a series of binary logistic regression models comparing a particular category and the base category. In each binary model, being in a particular category of the nominal outcome variable is dichotomized as the binary outcome of 1 and being in the base category is coded as 0. When a nominal response variable has K levels, there are $K-1$ binary logistic models with comparisons between any other categories and the base category.

In multilevel modeling for nominal response variables, we have a nested data structure that allows us to estimate the relationships between predictor variables from different levels and the nominal response variable. Let us look at a basic two-level multinomial logistic regression model with only one predictor variable at each level of the equations. For example, researchers are interested in estimating whether students are proficient in any of the four mathematics levels from a student-level variable, student's major language, and a school-level variable, public or private. The outcome variable is assumed to be unordered four-level math proficiency. The level 1 predictor variable is whether a student's native language is English (`stlang`). The level 2 predictor is whether the school is public or private (`public`). The intercepts are allowed to vary randomly across schools.

The level 1 equation is similar to that for the single-level multinomial logistic regression model, whereas in multilevel modeling, students are nested with schools so we have a subscript i to represent students and a subscript j to represent schools. As with a single-level multinomial logistic regression model, no error term is specified in the level 1 equation:

$$\begin{aligned} \text{Level 1: } \eta_{kij} &= \ln\left(\frac{\pi_{kij}}{\pi_{Kij}}\right) = \ln\left(\frac{P(Y_{ij} = k)}{P(Y_{ij} = K)}\right) = \beta_{0jk} + \beta_{1jk}\texttt{stlang}_{ij} \\ \text{Level 2: } \beta_{0jk} &= \gamma_{00k} + \gamma_{01k}\texttt{public}_j + u_{0jk} \\ \beta_{1jk} &= \gamma_{10k} \end{aligned} \tag{14.2}$$

In the level 1 equation, $\ln\left(\frac{\pi_{kij}}{\pi_{Kij}}\right)$ is the logit link for the probability of being a particular category of the nominal response variable for the ith student in the jth school. We define the probability of being in a particular category k, $\pi_{kij} = P(Y_{kij} = k)$, and the probability of being in the base category K, $\pi_{Kij} = P(Y_{Kij} = K)$, so $\ln\left(\frac{\pi_{kij}}{\pi_{Kij}}\right) = \ln\left(\frac{P(Y_{ij} = k)}{P(Y_{ij} = K)}\right)$. Since the nominal outcome variable math proficiency has four levels, the level 1 equation can be express as three binary logistic regression models with the lowest level 1 as the base or reference category as follows.

$$\ln\left(\frac{P(Y_{ij} = 2)}{P(Y_{ij} = 1)}\right) = \beta_{0j2} + \beta_{1j2}\texttt{stlang}_{ij}$$

$$\ln\left(\frac{P(Y_{ij} = 3)}{P(Y_{ij} = 1)}\right) = \beta_{0j3} + \beta_{1j3}\texttt{stlang}_{ij}$$

$$\ln\left(\frac{P(Y_{ij} = 4)}{P(Y_{ij} = 1)}\right) = \beta_{0j4} + \beta_{1j4}\texttt{stlang}_{ij}$$

In level 2 equations, γ_{00k}, γ_{01k}, and γ_{10k} are level 2 coefficients, which are the fixed effects.

- γ_{00k} is the intercept, the average logit or log odds of being in a particular math proficiency level across schools. Since we have three binary models above with one for each model, three intercepts will be estimated.
- γ_{10k} is the average slope for `stlang`. Three coefficients for `stlang` will also be estimated for the three binary models.
- u_{0jk} is an error term associated with the random intercepts, the random variation for the intercepts. Since we have three intercepts for the binary models, three between-school random intercepts will be estimated.
- γ_{01k} is the fixed effect of the level 2 predictor `public` on the log odds of in a particular math proficiency level relative to the base category (i.e., level 1 in this example). Three coefficients for `public` will also be estimated.

We can rewrite the level 2 equations as follows.

$$
\begin{aligned}
\beta_{0j2} &= \gamma_{00(2)} + \gamma_{01(2)}\texttt{public}_j + u_{0j(2)} \\
\beta_{0j3} &= \gamma_{00(3)} + \gamma_{01(3)}\texttt{public}_j + u_{0j(3)} \\
\beta_{0j4} &= \gamma_{00(4)} + \gamma_{01(4)}\texttt{public}_j + u_{0j(4)} \\
\beta_{1j2} &= \gamma_{10(2)} \\
\beta_{1j3} &= \gamma_{10(3)} \\
\beta_{1j4} &= \gamma_{10(4)}
\end{aligned}
$$

Estimation of the level 2 random effects results in the variance and covariance matrix, which includes the intercept variances either with or without the covariances between the intercepts. When the intercepts are correlated, we estimate the unstructured variance and covariance matrix, whereas when the intercepts are unrelated, only the variances for the random intercepts are estimated.

14.1.2 Odds and Odds Ratios in Multilevel Multinomial Logistic Regression Models

As with the single-level conventional multinomial logistic regression model, the odds and odds ratios in multilevel models for nominal response variables can be interpreted in a similar way. The odds are still the ratio of the probability of being in a particular category to the probability of being in the base category.

The odds ratios are the exponentiated logit coefficients. They are interpreted as the change in the odds of being in a particular category relative to the base category for a one-unit increase in an independent variable. When an odds ratio (OR) is larger than 1, the odds of being in a particular category relative the base category for a one-unit increase in the predictor variable; when an OR is less than 1, the odds being in a particular category relative the base category decreases for a one-unit increase in the predictor variable. In addition, an OR of 1 indicates that there is no relationship between the predictor and the odds.

14.2 RESEARCH EXAMPLE AND DESCRIPTION OF THE DATA AND SAMPLE

14.2.1 Research Example and Research Questions

In this research example, using the ELS:2002, researchers are interested in predicting the four-level math proficiency from three predictor variables: whether English is students' native language, school type, and school climate. The first predictor is a student-level variable, and the other two are school-level variables. For demonstration purposes, we assume that the four-level math proficiency is a nominal categorical outcome variable. The following research questions will be addressed:

1. Can the probability of being in mathematics proficient level 1 be predicted by whether English is the students' native language?
2. Do school characteristics, such as school type and school climate, impact mathematics proficiency?
3. Does students' mathematics proficiency level 1 vary across schools?

14.2.2 Description of the Data and Sample

The ELS:2002 base-year data are used for the following analyses. The variables are listed as follows:

- `prof4`: 4-level mathematics proficiency, a recoded variable of `Profmath`
- `stlang`: student language is English (1 = yes, 0 = no)
- `public`: school type (1 = public, 0 = private and others)
- `sclimat`: school climate
- `csclimat`: school climate (grand-mean centered)

For demonstration purposes, the missing values of the variables are deleted. The descriptive statistics are summarized and the school-level variable `sclimat` is grand-mean centered.

```
> # Import the dataset
> library(foreign)
> nom <- read.dta("C:/CDA/els-nom.dta")
> # Recode the outcome variable
> library(sjmisc)
> nom$prof4 <- rec(nom$Profmath, rec="0:1=1; 2=2; 3=3; 4:5=4; else=NA")
> # Create factor variables
> nom$prof4 <- factor(nom$prof4)
> nom$SCH_ID <- factor(nom$SCH_ID)
> attach(nom)
> table(prof4)
prof4
   1      2      3      4
3612   2716   3683   2782
> table(stlang)
stlang
  No     Yes
1986   10807
> table(public)
public
catholic or other private                    public
                     2829                      9964
> mean(csclimat)                   [1] -3.948514e-05
```

14.3 MULTILEVEL MODELING FOR NOMINAL RESPONSE VARIABLES WITH R

14.3.1 Packages and Functions for Multilevel Modeling for Nominal Response Variables in R

The `mblogit()` function in the `mclogit` package (Elff, 2021) is used for fitting multilevel models with nominal outcome variables. In addition, the `MCMCglmm()` function in the `MCMCglmm` package (Hadfield, 2010) and the `brm()` function in the `brms` package (Bürkner, 2017) are used for Bayesian multilevel modeling for nominal variables.

14.3.2 The `mblogit()` Function in the `mclogit` Package

The `mblogit()` function in the `mclogit` package is used for multilevel models with nominal outcome variables or mixed-effects multinomial logistic regression models. The `mclogit` package can be used to fit both multinomial logistic regression models and mixed-effects multinomial logistic models. Since `mclogit` is a user-written package, you need to install it first by typing `install.packages("mclogit")` and then load the package by typing `library(mclogit)`. If you have installed the package, you just need to load it. The model formula syntax of `mblogit()` is similar to that of `lme()` introduced in the previous chapters. The model formula in `mblogit()` also includes two components connected by a plus (+) sign. The first component specifies the dependent variable and the predictor variable(s), which are separated by the tilde (~). This is the fixed effects part of the model. Next, the random effects part of the model is specified with the `random =` argument. Same as the random effects part of `lme()`, the random coefficients term is specified first, followed by a vertical line (|), and then, an identifier variable at a higher level as the grouping variable. The default argument is `random = NULL`, which is used for the conventional baseline category logistic model or the multinomial logistic model without random effects. For example, the command `mblogit(y ~ x, random = ~1|schid, data = data1)` tells R to fit a multilevel model to estimate a nominal outcome variable `y` on a predictor variable `x` with random intercepts varying across schools by specifying `random = ~1|schid`, where `1` indicates the random intercept on the left of the vertical line (|) and `schid` is the identifier variable on the right. Next, the `data = data1` argument specifies the data frame `data1`. The `mblogit()` function uses the penalized quasi-likelihood method (PQL) with the Laplace approximation and the marginal quasi-likelihood method (MQL) with the Solomon-Cox approximation for model estimation with the former as the default. Since both PQL and MQL approximate quasi-likelihood values, likelihood-based inferences such as the likelihood ratio test and the AIC and BIC statistics are inappropriate for model comparisons. Therefore, using the `anova()` function to compare the deviances is not recommended. For more details on how to use this function, type `help(mblogit)` in the command prompt after loading the `mclogit` package.

14.3.3 Unconditional Model or Null Model (Model 1)

The unconditional model, or null model, for nominal response variables includes no predictor variable at any level. This model estimates the average logit or log odds of being in a particular nominal category and the variability in the intercepts between groups (e.g., schools). It can be expressed as follows:

$$\text{Level 1: } \eta_{kij} = \ln\left(\frac{\pi_{kij}}{\pi_{Kij}}\right) = \ln\left(\frac{P(Y_{ij} = k)}{P(Y_{ij} = K)}\right) = \beta_{0jk} \qquad (14.4)$$

$$\text{Level 2: } \beta_{0jk} = \gamma_{00k} + u_{0jk}$$

In a multilevel multinomial logistic regression model, $\ln\left(\frac{\pi_{kij}}{\pi_{Kij}}\right)$ is the logit link for the probability of being in a particular category of the nominal response variable relative to that of being in the base category for the *i*th student in the *j*th school. When the nominal outcome variable has four categories, this model includes three binary models with each one comparing a particular category and the base category. The level 1 equation can be expressed as three binary logistic regression models with the lowest level 1 as the baseline or reference category as follows.

$$\ln\left(\frac{P(Y_{ij} = 2)}{P(Y_{ij} = 1)}\right) = \beta_{0j2}$$

$$\ln\left(\frac{P(Y_{ij} = 3)}{P(Y_{ij} = 1)}\right) = \beta_{0j3}$$

$$\ln\left(\frac{P(Y_{ij} = 4)}{P(Y_{ij} = 1)}\right) = \beta_{0j4}$$

In the level 2 equation, γ_{00k} is the overall average logit or log odds of being in a particular math proficiency level across schools and u_{0jk} is an error term associated with the random intercepts at the school level. Since we have three underlying binary models, we can rewrite the level 2 equation as follows:

$$\begin{aligned} \beta_{0j2} &= \gamma_{00(2)} + u_{0j(2)} \\ \beta_{0j3} &= \gamma_{00(3)} + u_{0j(3)} \\ \beta_{0j4} &= \gamma_{00(4)} + u_{0j(4)} \end{aligned}$$

The `mblogit()` Function and Output

The command `mb.null <- mblogit(prof4 ~ 1, random = ~1|SCH_ID, data = nom)` is used to fit the unconditional model (model 1). This command tells R to fit a mixed-effect multinomial logistic regression model for the nominal response variable `prof4`. Since this is the unconditional model, no predictors are specified on the right-hand side of the tilde (~). The random effects part of the model specifies the random effects at the school level within parentheses, `random = ~1|SCH_ID`, where `1` indicates the random intercepts and `SCH_ID` is the identifier variable. No random coefficients for any predictor variables are specified in this model. The model is assigned to an object named `mb.null`. Next, the `summary()` function is used to display the output of the model.

```
> # Unconditional model/Null model with mblogit() in mclogit
> # Install mclogit by using install.packages(); you only need to install it once
> library(mclogit)
> mb.null <- mblogit(prof4 ~ 1, random=~1|SCH_ID, data=nom)
> summary(mb.null)

Call:
mblogit(formula = prof4 ~ 1, data = nom, random = ~1 | SCH_ID)

Equation for 2 vs 1:
              Estimate   Std. Error   z value   Pr(>|z|)
(Intercept)   -0.19419     0.03884         -5   5.72e-07 ***

Equation for 3 vs 1:
              Estimate   Std. Error   z value   Pr(>|z|)
(Intercept)    0.10587     0.04854      2.181     0.0292 *

Equation for 4 vs 1:
              Estimate   Std. Error   z value   Pr(>|z|)
(Intercept)   -0.38803     0.07212      -5.38   7.44e-08 ***
---
Signif. codes: 0 '***' 0.001 '**' 0.01 '*' 0.05 '.' 0.1 ' ' 1


(Co-)Variances:
Grouping level: 1

      Estimate           Std.Err.
2~1 0.4630               0.05717
3~1 0.3538 1.0246        0.13629 0.31791
4~1 0.4979 1.2228 2.6023 0.24848 0.57786 1.04173

Null Deviance:          35470
Residual Deviance:      34020
Number of Fisher Scoring iterations: 11
Number of observations
   Groups by SCH_ID: 619
   Individual observations: 12793
```

Interpreting the Output

The R output produced by the `mblogit()` function includes the function call, the parameter estimations for each binary logistic model, the variances and covariances for the random effects, the null and residual deviances, the number of iterations, and the number of observations.

The first section, which is the fixed effects section, displays the parameter estimates for the three intercepts for the corresponding binary logistic regression models. Recall that a multinomial logistic regression model includes a series of binary logistic models comparing each category with the baseline category. This section displays the estimate for the intercept, its standard error, the *z* statistic, and the associated *p* value for each binary model. These three equations, labeled `Equation for 2 vs 1`, `Equation for 3 vs 1`, and `Equation for 4 vs 1`, compare categories 2 with 1, categories 3 with 1, and categories 4 with 1, respectively.

For the first equation comparing categories 2 with 1, the intercept $\gamma_{00(2)}$ (labeled `Intercept`) is −.195, which is significant (`z = -5, P(>|z|) = 5.72e-07`). This means that the average logit or log odds of being in math proficient level 2 versus the base level across all schools for the first equation is −.195. The estimated logit can be transformed to the odds ratio, which will be explained next.

The second equation compares category 3 and category 1. The intercept $\gamma_{00(3)}$ (labeled `Intercept`) is .106, which is also significant (`z = 2.181, P(>|z|) = 0.0292`). This means that the average logit or log odds of being in math proficient level 3 versus the base level across all schools is .106.

The third equation compares category 4 and category 1. The intercept $\gamma_{00(4)}$ (labeled `Intercept`) is −.388, which is also significant (`z = -5.38, P(>|z|) = 7.44e-08`). This means that the average logit or log odds being in math proficient level 4 versus the base level across all schools for the third equation is −.388.

Next, the random effects section (labeled `(Co-)Variances`) provides the variances and the covariances for the intercepts across all schools. In the variance and covariance matrix, the diagonal lists the variances and the off-diagonal provides the covariances. The between-school variances (τ_{00k}) for the three intercepts are .463, 1.025, and 2.602, respectively.

We use the `coef(mb.null)` command to extract the coefficients table for the three underlying binary models which compare categories 2 with 1, categories 3 with 1, and categories 4 with 1, respectively. The 95% confidence interval of the estimate is computed as follows.

```
> coef(mb.null)
2~(Intercept)    3~(Intercept)    4~(Intercept)
   -0.1941919        0.1058692       -0.3880320
> coef <- coef(mb.null)
> se <- sqrt(diag(vcov(mb.null)))
> se
2~(Intercept)    3~(Intercept)    4~(Intercept)
   0.03883627       0.04853822       0.07212120
> LL <- coef(mb.null)-1.96*se
> UL <- coef(mb.null)+1.96*se
> coeftab <- cbind(LL, UL)
> coeftab
                        LL          UL
2~(Intercept)   -0.27031095   -0.1180728
3~(Intercept)    0.01073425    0.2010041
4~(Intercept)   -0.52938960   -0.2466745
```

With the `exp(coef(mb.null))` command, we can get the odds ratios being in a particular category versus the base category for the predictor variables. Since this is the unconditional model, only the odds ratios for the intercepts are reported. The odds ratios for the three intercepts are .823, 1.112, and .678, respectively, which are the average odds of being in math proficient levels 2, 3, and 4 versus the base level 1. The confidence intervals for the odds ratios can be obtained with the `exp(coeftab)` command.

```
> exp(coef(mb.null))
2~(Intercept)    3~(Intercept)    4~(Intercept)
    0.8234999        1.1116764        0.6783906
> exp(coeftab)
                       LL          UL
2~(Intercept)   0.7631422   0.8886314
3~(Intercept)   1.0107921   1.2226298
4~(Intercept)   0.5889644   0.7813950
```

The same confidence intervals for the ORs can also be obtained with the `ortab <- cbind(or=exp(coef(mb.null)), ORLL=exp(LL), ORUL=exp(UL))` command.

```
> ortab <- cbind(or=exp(coef(mb.null)), ORLL=exp(LL), ORUL=exp(UL))
> ortab
                       or        ORLL        ORUL
2~(Intercept)   0.8234999   0.7631422   0.8886314
3~(Intercept)   1.1116764   1.0107921   1.2226298
4~(Intercept)   0.6783906   0.5889644   0.7813950
```

14.3.4 Random-Intercept Model (Model 2)

Next, we include the predictor variable, whether a student's native language is English (`stlang`), to the level 1 equation. The model is referred to as the random-intercept model since the intercept is allowed to vary freely across schools. This model can be expressed as follows:

$$\begin{aligned} \text{Level 1:}\quad \eta_{kij} &= \ln\left(\frac{\pi_{kij}}{\pi_{Kij}}\right) = \ln\left(\frac{P(Y_{ij}=k)}{P(Y_{ij}=K)}\right) = \beta_{0jk} + \beta_{1jk}\texttt{stlang}_{ij} \\ \text{Level 2:}\quad \beta_{0jk} &= \gamma_{00k} + u_{0jk} \\ \beta_{1jk} &= \gamma_{10k} \end{aligned} \tag{14.5}$$

The level 1 equation includes the predictor variable, whether student's native language is English (`stlang`), and the nominal response variable, the four-level math proficiency (`prof4`). The level 1 equation can be express as three binary logistic regression models with the lowest level 1 as the base or reference category as follows.

$$\ln\left(\frac{P(Y_{ij} = 2)}{P(Y_{ij} = 1)}\right) = \beta_{0j2} + \beta_{1j2}\texttt{stlang}_{ij}$$

$$\ln\left(\frac{P(Y_{ij} = 3)}{P(Y_{ij} = 1)}\right) = \beta_{0j3} + \beta_{1j3}\texttt{stlang}_{ij}$$

$$\ln\left(\frac{P(Y_{ij} = 4)}{P(Y_{ij} = 1)}\right) = \beta_{0j4} + \beta_{1j4}\texttt{stlang}_{ij}$$

In the level 2 equation, γ_{00k} is the overall average logit or log odds of being in a particular math proficiency level across schools and u_{0jk} is an error term associated with the random intercepts at the school level. Since we have three underlying binary models, we can rewrite the level 2 equation as follows.

$$\begin{aligned}
\beta_{0j2} &= \gamma_{00(2)} + u_{0j(2)} \\
\beta_{0j3} &= \gamma_{00(3)} + u_{0j(3)} \\
\beta_{0j4} &= \gamma_{00(4)} + u_{0j(4)} \\
\beta_{1j2} &= \gamma_{10(2)} \\
\beta_{1j3} &= \gamma_{10(3)} \\
\beta_{1j4} &= \gamma_{10(4)}
\end{aligned}$$

The level 2 equation for the intercept (β_{0jk}) is specified to be random, whereas the slope for `stlang` (β_{1jk}) is constrained to be fixed.

The `mblogit()` Command and Output

The command `mb.2 <-mblogit(prof4 ~ stlang, random = ~1 |SCH_ID, data = nom)` is used to fit the random-intercept model (model 2) after a predictor variable `stlang` is added to the fixed part of the model immediately after the tilde (~). Just like the unconditional model, no random coefficients are specified in the random part of the model. Again, the `summary()` function is used to display the output of the fitted random-intercept model named `mb.2`. The following output is displayed.

```
> # Random-intercept model
> mb.2 <- mblogit(prof4 ~ stlang, random=~1|SCH_ID, data=nom)
> summary(mb.2)
Call:
mblogit(formula = prof4 ~ stlang, data = nom, random = ~1 | SCH_ID)
Equation for 2 vs 1:

              Estimate   Std. Error   z value   Pr(>|z|)
(Intercept)   -0.52300      0.07339    -7.126   1.03e-12 ***
stlangYes      0.39012      0.07490     5.208   1.90e-07 ***
```

```
Equation for 3 vs 1:
              Estimate    Std. Error    z value    Pr(>|z|)
(Intercept)   -0.36961     0.07976      -4.634     3.59e-06 ***
stlangYes      0.56103     0.07565       7.416     1.21e-13 ***
Equation for 4 vs 1:
              Estimate    Std. Error    z value    Pr(>|z|)
(Intercept)   -0.81082     0.10319      -7.858     3.91e-15 ***
stlangYes      0.49749     0.08729       5.699     1.20e-08 ***
---
Signif. codes: 0 '***' 0.001 '**' 0.01 '*' 0.05 '.' 0.1 ' ' 1

(Co-)Variances:
Grouping level: 1
    Estimate                 Std.Err.
2~1 0.4420                   0.05063
3~1 0.3159 0.9579            0.12331 0.29304
4~1 0.4661 1.1609 2.5660     0.23420 0.55529 1.04374

Null Deviance:          35470
Residual Deviance:      33970
Number of Fisher Scoring iterations: 10
Number of observations
   Groups by SCH_ID: 619
   Individual observations: 12793
```

Interpreting the Output

The first section for the fixed effects reports the coefficients for the intercept and `stlang` for the three binary logistic models. Again, these three equations compare categories 2 with 1, categories 3 with 1, and categories 4 with 1, respectively.

The intercepts γ_{00k} (labeled `Intercept`) for the three binary models are $-.523$, $-.370$, and $-.811$, respectively. They are all significantly different from 0. The results indicate that the estimated logit or log odds of being in math proficient levels 2, 3, and 4 versus the base level 1 for students whose native language is not English (`stlang` $= 0$) in a typical school are $-.523$, $-.370$, and $-.811$, respectively.

The logit coefficients for `stlang` (γ_{10k}) for the three underlying binary models are .390, .561, and .497, respectively. They are also significantly different from 0. For example, in the equation comparing category 2 and category 1, the coefficient for `stlang` $= .390$ $z = 5.208$, $p < .001$, which indicates that the logit or log odds of being in math proficient levels 2 versus level 1 for students with English as their native language is .390 times as great as that for those students whose native language is not English. The coefficients for `stlang` in the other two binary models can be interpreted in the same way.

The second section for the random effects provides the variances and the covariances for the intercepts across all schools. Again in the variance and covariance matrix, the diagonal lists the variances and the off-diagonal lists the covariances. The between-school variances (τ_{00k}) for the three intercepts are .442, .958, and 2.566, respectively. Their corresponding standard errors are .051, .293, and 1.044, respectively.

We extract the coefficients for the fixed effects and compute the corresponding 95% confidence intervals as follows.

```
> coef(mb.2)
2~(Intercept)    3~(Intercept)    4~(Intercept)    2~stlangYes    3~stlangYes
   -0.5229954       -0.3696053       -0.8108173      0.3901155      0.5610288
  4~stlangYes
    0.4974856
> coef <- coef(mb.2)
> se <- sqrt(diag(vcov(mb.2)))
> se
2~(Intercept) 3~(Intercept) 4~(Intercept) 2~stlangYes 3~stlangYes 4~stlangYes
   0.07338934    0.07975947    0.10318736  0.07490019  0.07565431  0.08729107
> LL <- coef(mb.2)-1.96*se
> UL <- coef(mb.2)+1.96*se
> coeftab <- cbind(LL, UL)
> coeftab
                      LL         UL
2~(Intercept)  -0.6668385 -0.3791523
3~(Intercept)  -0.5259339 -0.2132768
4~(Intercept)  -1.0130645 -0.6085701
2~stlangYes     0.2433111  0.5369199
3~stlangYes     0.4127463  0.7093112
4~stlangYes     0.3263951  0.6685761
```

We use `exp(coef(mb.2))` to compute the odds ratios of being in a particular category versus the base category for the predictor variables. The confidence intervals for the odds ratios can be obtained with the `exp(coeftab)` command.

```
> exp(coef(mb.2))
2~(Intercept)    3~(Intercept)    4~(Intercept)    2~stlangYes    3~stlangYes
    0.5927424        0.6910070        0.4444946      1.4771514      1.7524745
  4~stlangYes
    1.6445809
> exp(coeftab)
                     LL        UL
2~(Intercept)  0.5133289 0.6844413
3~(Intercept)  0.5910032 0.8079325
4~(Intercept)  0.3631045 0.5441284
2~stlangYes    1.2754654 1.7107294
3~stlangYes    1.5109617 2.0325908
4~stlangYes    1.3859628 1.9514566
```

The same confidence intervals for the ORs can also be obtained with the `ortab <- cbind(or=exp(coef(mb.2)), ORLL=exp(LL), ORUL=exp(UL))` command.

```
> ortab <- cbind(or=exp(coef(mb.2)), ORLL=exp(LL), ORUL=exp(UL))
> ortab
                       or        ORLL        ORUL
2~(Intercept)   0.5927424   0.5133289   0.6844413
3~(Intercept)   0.6910070   0.5910032   0.8079325
4~(Intercept)   0.4444946   0.3631045   0.5441284
2~stlangYes     1.4771514   1.2754654   1.7107294
3~stlangYes     1.7524745   1.5109617   2.0325908
4~stlangYes     1.6445809   1.3859628   1.9514566
```

Interpretation of odds ratios in multilevel multinomial logistic regression is the same as that for multilevel logistic regression. The only difference is that the odds ratios for each predictor variable need to be interpreted across the three binary logistic models. For `stlang`, the odds ratios of being in math proficient level 2 versus level 1, level 3 versus level 1, and level 4 versus level 1 are 1.477, 1.752, and 1.645, respectively. The results indicate that the odds of being in level 2 versus the base level, level 3 versus the base level, and level 4 versus the base category for students with English as their native language are 1.477, 1.752, and 1.645 times as great as the odds for those students whose native language is not English. In other words, being students with English as their native language increases the odds by 47.7%, 75.2%, and 64.5%, respectively.

The odds ratios for the three intercepts are .593, .691, and .444, respectively, which are the average odds of being in math proficient levels 2, 3, and 4 versus level 1 when holding the predictor variable constant.

14.3.5 Random-Intercept Model With Level-2 Variables (Model 3)

The random-intercept model for the nominal response variable can also include both level 1 and level 2 predictor variables. Since random-coefficient models for nominal response variables are computationally cumbersome to fit, here we only focus on the random-intercept model with predictor variables in both levels. The two level 2 equations specify the random intercepts. Multiple random intercepts are specified since the multilevel multinomial model includes a series of binary logistic models. In this example, two school-level variables `public` and `csclimat` are added to the level 2 equation for the random intercepts. This model is expressed as follows:

$$\begin{aligned}
\text{Level 1: } \eta_{kij} &= \ln\left(\frac{\pi_{kij}}{\pi_{Kij}}\right) = \ln\left(\frac{P(Y_{ij}=k)}{P(Y_{ij}=K)}\right) = \beta_{0jk} + \beta_{1jk}\texttt{stlang}_{ij} \\
\text{Level 2: } \beta_{0jk} &= \gamma_{00k} + \gamma_{01k}\texttt{public}_j + \gamma_{02k}\texttt{csclimat}_j + u_{0jk} \\
\beta_{1jk} &= \gamma_{10k}
\end{aligned} \tag{14.7}$$

The level 1 equation includes the predictor variable, students' native language (stlang), and the nominal response variable (prof4). It can be expressed as three binary logistic regression models with the lowest level 1 as the base or reference category as follows:

$$\ln\left(\frac{P(Y_{ij} = 2)}{P(Y_{ij} = 1)}\right) = \beta_{0j2} + \beta_{1j2}\texttt{stlang}_{ij}$$

$$\ln\left(\frac{P(Y_{ij} = 3)}{P(Y_{ij} = 1)}\right) = \beta_{0j3} + \beta_{1j3}\texttt{stlang}_{ij}$$

$$\ln\left(\frac{P(Y_{ij} = 4)}{P(Y_{ij} = 1)}\right) = \beta_{0j4} + \beta_{1j4}\texttt{stlang}_{ij}$$

The level 2 equation for the intercepts includes the two school-level variables public and csclimat and the intercepts are specified to be random with level 2 residuals u_{0jk}, whereas the slope for stlang (β_{1jk}) is constrained to be fixed. Since we have three underlying binary models, the level 2 equation can also be expressed as follows:

$$\begin{aligned}
\beta_{0j2} &= \gamma_{00(2)} + \gamma_{01(2)}\texttt{public}_j + \gamma_{02(2)}\texttt{csclimat}_j + u_{0j(2)} \\
\beta_{0j3} &= \gamma_{00(3)} + \gamma_{01(3)}\texttt{public}_j + \gamma_{02(3)}\texttt{csclimat}_j + u_{0j(3)} \\
\beta_{0j4} &= \gamma_{00(4)} + \gamma_{01(4)}\texttt{public}_j + \gamma_{02(4)}\texttt{csclimat}_j + u_{0j(4)} \\
\beta_{1j2} &= \gamma_{10(2)} \\
\beta_{1j3} &= \gamma_{10(3)} \\
\beta_{1j4} &= \gamma_{10(4)}
\end{aligned}$$

The mblogit() Command and Output

The command mb.3 <- mblogit(prof4 ~ stlang + public + csclimat, random = ~1|SCH_ID, data = nom) is used to fit the contextual model (model 3). With other parts of the equations the same as those of the random-intercept model (model 2), the fixed part of the model includes both the level 1 predictor stlang and the two school-level predictor variables public and csclimat. The random part of the model, random = ~1|SCH_ID, is the same as that of the random-intercept model, where 1 indicates the random intercepts and SCH_ID is the identifier variable or the grouping factor. No random coefficients for any predictor variables are specified in this model. Again, the results of the fitted model are displayed with the summary(mb.3) command.

```
> # Contextual model/Random-intercept model with predictor variables
> mb.3 <-mblogit(prof4 ~ stlang + public + csclimat, random = ~1 |SCH_ID,
data=nom)
> summary(mb.3)
Call:
mblogit(formula = prof4 ~ stlang + public + csclimat, data = nom,
  random = ~1 | SCH_ID)
```

```
Equation for 2 vs 1:
                  Estimate    Std. Error    z value     Pr(>|z|)
(Intercept)       -0.05346     0.11189      -0.478       0.633
stlangYes          0.36278     0.07453       4.867      1.13e-06 ***
publicpublic      -0.53889     0.10212      -5.277      1.31e-07 ***
csclimat           0.27713     0.05731       4.836      1.32e-06 ***

Equation for 3 vs 1:
                  Estimate    Std. Error    z value     Pr(>|z|)
(Intercept)        0.40833     0.11842       3.448      0.000565 ***
stlangYes          0.52136     0.07481       6.969      3.18e-12 ***
publicpublic      -0.92528     0.11140      -8.306       < 2e-16 ***
csclimat           0.52301     0.06568       7.963      1.69e-15 ***

Equation for 4 vs 1:
                   Estimate     Std. Error    z value     Pr(>|z|)
(Intercept)        0.006398     0.165575      0.039        0.969
stlangYes          0.464700     0.086880      5.349      8.86e-08 ***
publicpublic      -0.985677     0.167476     -5.885      3.97e-09 ***
csclimat           0.812456     0.103825      7.825      5.07e-15 ***
---
Signif. codes: 0 '***' 0.001 '**' 0.01 '*' 0.05 '.' 0.1 ' ' 1

(Co-)Variances:
Grouping level: 1
Estimate                    Std.Err.
2~1 0.3590                  0.03058
3~1 0.1639 0.6394           0.07859 0.19280
4~1 0.3012 0.7613 1.9955    0.17519 0.43039 0.95245

Null Deviance:           35470
Residual Deviance:       33920
Number of Fisher Scoring iterations: 10
Number of observations
   Groups by SCH_ID: 619
   Individual observations: 12793
```

Interpreting the Output

The fixed-effects section reports the estimated logit coefficients for the intercept and both level 1 and level 2 predictor variables for the three binary logistic models. The logit coefficients for all the predictor variables are significant.

The estimated logit coefficients for the level 1 predictor variable `stlang` in the output are similar to those for the previous random-intercept model. They can also be interpreted in a similar way.

Regarding the level 2 predictor variables, the logit coefficients for `public` (γ_{01k}) for the three underlying binary models are −.539, −.925, and −.986, respectively. They are significantly different from 0. In addition, the logit coefficients for `csclimat` (γ_{02k}) for the three binary models are .277, .523, and .812, respectively. They are also significantly different from 0.

Regarding the random effects in the model, the between-school variances (τ_{00k}) for the three intercepts are .359, .639, and 1.996, respectively.

We extract the coefficients for the fixed effects and compute the corresponding 95% confidence intervals as follows.

```
> coef(mb.3)
2~(Intercept)     3~(Intercept)     4~(Intercept)       2~stlangYes    3~stlangYes
 -0.053459296       0.408325704       0.006397519       0.362775882    0.521364073
  4~stlangYes   2~publicpublic    3~publicpublic    4~publicpublic     2~csclimat
  0.464699650      -0.538889518      -0.925284810      -0.985677199    0.277127403
   3~csclimat        4~csclimat
  0.523012834       0.812456212
> coef<-coef(mb.3)
> se<-sqrt(diag(vcov(mb.3)))
> se
2~(Intercept)  3~(Intercept) 4~(Intercept) 2~stlangYes 3~stlangYes 4~stlangYes 2~publicpublic 3~publicpublic
   0.11188870     0.11842383    0.16557535  0.07453387  0.07480796  0.08688041     0.10211628     0.11139929
4~publicpublic    2~csclimat    3~csclimat  4~csclimat
    0.16747630    0.05730544    0.06568422  0.10382549
> LL<-coef(mb.3)-1.96*se
> UL<-coef(mb.3)+1.96*se
> coeftab<-cbind(LL, UL)
> coeftab
                         LL          UL
2~(Intercept)    -0.2727611   0.1658425
3~(Intercept)     0.1762150   0.6404364
4~(Intercept)    -0.3181302   0.3309252
2~stlangYes       0.2166895   0.5088623
3~stlangYes       0.3747405   0.6679877
4~stlangYes       0.2944140   0.6349853
2~publicpublic   -0.7390374  -0.3387416
3~publicpublic   -1.1436274  -0.7069422
4~publicpublic   -1.3139307  -0.6574237
2~csclimat        0.1648087   0.3894461
3~csclimat        0.3942718   0.6517539
4~csclimat        0.6089582   1.0159542
```

We use the exp(coef(mb.3)) command to request the odds ratios for the intercepts and predictor variables. The confidence intervals for the odds ratios can be obtained with the exp(coeftab) command.

```
> exp(coef(mb.3))
2~(Intercept)   3~(Intercept)   4~(Intercept)     2~stlangYes  3~stlangYes
    0.9479445       1.5042970       1.0064180       1.4373137    1.6843236
  4~stlangYes  2~publicpublic  3~publicpublic  4~publicpublic   2~csclimat
    1.5915361       0.5833957       0.3964185       0.3731864    1.3193344
```

```
 3~csclimat     4~csclimat
  1.6871030      2.2534361
> exp(coeftab)
                       LL        UL
2~(Intercept)   0.7612746 1.1803872
3~(Intercept)   1.1926945 1.8973087
4~(Intercept)   0.7275081 1.3922557
2~stlangYes     1.2419584 1.6633976
3~stlangYes     1.4546138 1.9503087
4~stlangYes     1.3423396 1.8869943
2~publicpublic  0.4775734 0.7126666
3~publicpublic  0.3186610 0.4931499
4~publicpublic  0.2687615 0.5181846
2~csclimat      1.1791676 1.4761629
3~csclimat      1.4833036 1.9189035
4~csclimat      1.8385151 2.7619976
```

The same confidence intervals for the ORs can also be obtained with the `ortab <-cbind(or=exp(coef(mb.3)), ORLL=exp(LL), ORUL=exp(UL))` command.

```
> ortab <- cbind(or=exp(coef(mb.3)), ORLL=exp(LL), ORUL=exp(UL))
> ortab
                       or      ORLL      ORUL
2~(Intercept)   0.9479445 0.7612746 1.1803872
3~(Intercept)   1.5042970 1.1926945 1.8973087
4~(Intercept)   1.0064180 0.7275081 1.3922557
2~stlangYes     1.4373137 1.2419584 1.6633976
3~stlangYes     1.6843236 1.4546138 1.9503087
4~stlangYes     1.5915361 1.3423396 1.8869943
2~publicpublic  0.5833957 0.4775734 0.7126666
3~publicpublic  0.3964185 0.3186610 0.4931499
4~publicpublic  0.3731864 0.2687615 0.5181846
2~csclimat      1.3193344 1.1791676 1.4761629
3~csclimat      1.6871030 1.4833036 1.9189035
4~csclimat      2.2534361 1.8385151 2.7619976
```

The odds ratios for `stlang` are 1.437, 1.684, and 1.592, respectively. The results indicate that the odds of being in level 2, level 3, and level 4 versus the base category for students with English as their native language are 1.437, 1.684, and 1.592 times, respectively, as great as the odds for those students whose native language is not English when holding all other variables constant. In other words, being students with English as their native language increases the odds by 43.7%, 68.4%, and 59.2%, respectively.

The odds ratios for `public` are .583, .396, and .373, respectively, which indicate that the odds of being in level 2, level 3, and level 4 versus the base category for students in public schools are .583, .396, .373 times the odds for those in private schools, respectively.

The odds ratios for `csclimat` are 1.319, 1.687, and 2.253, respectively, which indicate that a one-unit increase in school climate corresponds to an increase of 1.319, 1.687, and 2.253 points, respectively in the odds of being proficient in math level 2, level 3, and level 4 versus the base category. In other words, being students in schools with a more positive school climate increases the odds of being proficient in math level 2, level 3, and level 4 versus the base category.

The odds ratios for the intercepts are .948, 1.504, and 1.006, respectively, which are the average odds of being proficient in math level 2, level 3, and level 4 versus the base category when the predictor variables are held constant at zero.

14.4 BAYESIAN MULTILEVEL MODELING FOR NOMINAL OUTCOME VARIABLES WITH THE `MCMCglmm()` FUNCTION IN THE `MCMCglmm` PACKAGE

The Bayesian multilevel modeling can be used to estimate nominal response variables. We will introduce the `MCMCglmm()` function in the `MCMCglmm` package (Hadfield, 2010) and the `brm()` function in the `brms` package (Bürkner, 2017) for the Bayesian analysis of the nominal variables. The `MCMCglmm` package is introduced first.

14.4.1 The `MCMCglmm()` Function in the `MCMCglmm` Package: Syntax for Bayesian Multilevel Multinomial Regression Models

The `MCMCglmm` package is well-known for fitting various Bayesian generalized linear mixed models using the Markov chain Monte Carlo (MCMC) methods. It uses both Metropolis-Hastings sampling and Gibbs sampling for MCMC to simulate draws. Since `MCMCglmm` is a user-written package, you need to install it first by typing `install.packages("MCMCglmm")` and then load the package with `library(MCMCglmm)`.

Similar to that for `glmer()` for mixed-effects logistic regression models, the syntax for `MCMCglmm()` for multilevel multinomial regression models also includes the components for fixed effects, random effects, the `family` argument and the `data` argument, but it has more components.

First, the model formula for the fixed effects in `MCMCglmm()` specifies the dependent variable and the predictor variable(s) with the dependent variable on the left-hand side of the tilde (~). The predictor variables on the right-hand side are connected by plus (+) symbols. This component is similar to that in any other mixed-effects model.

Second, the random effects part of the model is specified with the `random = ~` argument. For example, for the random-intercept model, we specify `random = ~SCH_ID`, where `SCH_ID` is an identifier variable at a higher level as the grouping variable or grouping factor. In a random-intercept multinomial model, since we have multiple intercepts for the underlying binary models, we specify `random = ~us(trait):SCH_ID`, where `trait` is the reserved variable for indexing the categories of the nominal response variable and `us()` means that the variance-covariance structure is unstructured, which allows the random effects of the multiple intercepts to be correlated. When the random effects of the multiple intercepts are uncorrelated, we can specify `random = ~idh(trait):SCH_ID`, where `idh()` means that the random effects are independent.

Third, different from the syntax for `glmer()`, we can specify the residual variance structure (i.e., R-structure) in `MCMCglmm()` with the `rcov = ~us(trait):units` argument. Like `trait`, `units` is also a reserved variable. We specify `us()` for the unstructured variance-covariance R-structure.

Fourth, we use the `family = "categorical"` argument to specify the distribution of the nominal response variable.

Fifth, we use the `data =` argument to specify the data frame.

Finally, we use the `prior =` argument to specify the priors for the fixed effects (B), the variance and covariance matrix for the random effects (G-structure), and the residual variance structure (R-structure). Let us see an example before fitting the model below.

```
k <- length(levels(nom$prof4))
I <- diag(k-1)
J <- matrix(rep(1, (k-1)^2), c(k-1, k-1))
priors = list(R = list(fix=1, V= (1/k) * (I + J), nu = k - 1),
              G = list(G1 = list(V = diag(k - 1), nu = k - 1)))
```

- The first line computes the number of categories for the nominal outcome variable `prof4` in the `nom` data frame and assign an objected named `k`. Here `k = 4`.
- The second line creates a 3 by 3 identity matrix, a diagonal matrix with three 1s on the main diagonal.
- The third line creates a 3 by 3 matrix with all the elements as 1s.
- The fourth line defines the priors. `R` defines the residual variance structure; `V= (1/k) * (I + J)` defines the variance and covariance matrix for the R-structure; `fix=1` indicates that the whole matrix is fixed; and `nu` is the number of degrees of freedom parameters. In addition, `G` defines the random effects for the G-structure. For the random-intercept model with `random = ~us(trait):SCH_ID`, we only have one matrix for the random intercepts, so we only specify `G1`.
- Since we know that the number of categories of the nominal variable is four, the matrix for the R-structure `V = (1/k) * (I + J)` can be expressed as

V= (1/4) * (diag(3) + matrix(1, 3, 3)), where I is diag(3) and J is matrix(1, 3, 3). Similarly, the matrix for the G-structure V = diag(k - 1) can be expressed as V= diag(3). The syntax above can also be coded as follows. You can use either one.

```
IJ <- (1/4)*(diag(3) + matrix(1, 3, 3))
priors = list(R = list(fix=1, V=IJ, nu = 3),
              G = list(G1 = list(V = diag(3), nu = 3)))
```

For more details on how to use this function, type help(MCMCglmm) in the command prompt after loading the MCMCglmm package.

14.4.2 Multilevel Multinomial Regression Models With MCMCglmm(): Correlated Random Effects

In the following example, the command mb.mc.3e <-MCMCglmm(prof4 ~ -1 + trait + trait:(stlang + public + csclimat), random = ~us(trait):SCH_ID, rcov = ~us(trait):units, data = nom, family = "categorical", prior = priors)) tells R to fit the same random-intercept model with predictor variables in both levels (model 3) in the previous section after the two level 2 predictor variables public and csclimat are added to the model. In the model formula for the fixed effects, -1 means that the overall intercept is omitted. Instead, with the reserved variable, trait, the three intercepts for the underlying binary models comparing a particular category to the base category will be estimated. In addition, the formula trait:(stlang + public + csclimat) tells R that the fixed effects for the three predictor variables will be estimated across the three binary models. Without trait:(), the posterior means or the parameter estimates for the three predictor variables will be constant across the three binary models. In other words, without the interaction term, each predictor variable only has one posterior mean rather than three. The fitted model is named mb.mc.3e. The following output is displayed by the summary (mb.mc.3e) command.

```
> library(MCMCglmm)
> # Model with correlated random effects
> # use set.seed(123) to get the same results each time with the same seed
> # Set priors
> k <- length(levels(nom$prof4))
> I <- diag(k-1)
> J <- matrix(rep(1, (k-1)^2), c(k-1, k-1))
> priors = list(R = list(fix=1, V=(1/k) * (I + J), nu = k - 1),
+               G = list(G1 = list(V = diag(k - 1), nu = k - 1)))
> mb.mc.3e <- MCMCglmm(prof4 ~ -1 + trait + trait:(stlang + public + csclimat), random
= ~us(trait):SCH_ID,
+ rcov=~us(trait):units, data=nom, family = "categorical", prior = priors)
```

```
                     MCMC iteration = 0
Acceptance ratio for liability set 1 = 0.000312

                     MCMC iteration = 1000

Acceptance ratio for liability set 1 = 0.281123

... Other MCMC iterations are omitted here.
                     MCMC iteration = 13000

Acceptance ratio for liability set 1 = 0.284418
> summary(mb.mc.3e)
Iterations = 3001:12991
Thinning interval = 10
Sample size = 1000

DIC: 32717.72

G-structure: ~us(trait):SCH_ID

                                  post.mean  l-95% CI  u-95% CI  eff.samp
traitprof4.2:traitprof4.2.SCH_ID     0.3771    0.2802    0.4843     19.09
traitprof4.3:traitprof4.2.SCH_ID     0.4501    0.3387    0.5648     29.84
traitprof4.4:traitprof4.2.SCH_ID     0.6628    0.5333    0.8288     44.20
traitprof4.2:traitprof4.3.SCH_ID     0.4501    0.3387    0.5648     29.84
traitprof4.3:traitprof4.3.SCH_ID     0.7767    0.6186    0.9224     41.66
traitprof4.4:traitprof4.3.SCH_ID     1.0832    0.8976    1.2680     74.09
traitprof4.2:traitprof4.4.SCH_ID     0.6628    0.5333    0.8288     44.20
traitprof4.3:traitprof4.4.SCH_ID     1.0832    0.8976    1.2680     74.09
traitprof4.4:traitprof4.4.SCH_ID     1.8548    1.5855    2.1759     59.59


R-structure:  ~us(trait):units

                                  post.mean  l-95% CI  u-95% CI  eff.samp
traitprof4.2:traitprof4.2.units        0.50      0.50      0.50         0
traitprof4.3:traitprof4.2.units        0.25      0.25      0.25         0
traitprof4.4:traitprof4.2.units        0.25      0.25      0.25         0
traitprof4.2:traitprof4.3.units        0.25      0.25      0.25         0
traitprof4.3:traitprof4.3.units        0.50      0.50      0.50         0
traitprof4.4:traitprof4.3.units        0.25      0.25      0.25         0
traitprof4.2:traitprof4.4.units        0.25      0.25      0.25         0
traitprof4.3:traitprof4.4.units        0.25      0.25      0.25         0
traitprof4.4:traitprof4.4.units        0.50      0.50      0.50         0

Location effects: prof4 ~ -1 + trait + trait:(stlang + public + csclimat)

                          post.mean   l-95% CI   u-95% CI  eff.samp   pMCMC
traitprof4.2                0.04065   -0.22787    0.26832     32.16   0.710
traitprof4.3                0.47908    0.21729    0.73904     47.40   0.002 **
traitprof4.4                0.12945   -0.20423    0.49189     66.30   0.464
traitprof4.2:stlangYes      0.40330    0.27322    0.55361     43.23  <0.001 ***
traitprof4.3:stlangYes      0.59908    0.45015    0.75141     70.02  <0.001 ***
traitprof4.4:stlangYes      0.49397    0.32090    0.69196     59.23  <0.001 ***
traitprof4.2:publicpublic  -0.63302   -0.85411   -0.39417     34.16  <0.001 ***
traitprof4.3:publicpublic  -1.02120   -1.28024   -0.77498     59.29  <0.001 ***
traitprof4.4:publicpublic  -1.12207   -1.43718   -0.78705     98.67  <0.001 ***
traitprof4.2:csclimat       0.31884    0.20197    0.43430     78.09  <0.001 ***
traitprof4.3:csclimat       0.61642    0.47370    0.75617    104.93  <0.001 ***
traitprof4.4:csclimat       0.89417    0.70633    1.09325    112.44  <0.001 ***

---
Signif. codes:  0 '***' 0.001 '**' 0.01 '*' 0.05 '.' 0.1 ' ' 1
```

The R output produced by the `MCMCglmm()` function includes the number of iterations, the thinning interval, the sample size, and the DIC, the G structure for the random effects, the R structure for the residual variances, and the location effects for the fixed effects.

The default number of MCMC iterations is 13,000 with the number of burn-in of 3,000. In other words, the number of initial iterations of 3,000 was discarded. The number of thinning interval is 10. The sample size is 1,000, which equals the difference between the number of iterations and burn-in divided by the number of thinning interval: (13,000 − 3,000)/10 = 1,000.

First, the random effects section (labeled `G-structure: ~us(trait):SCH_ID`) reports the variances of the random intercepts and covariances between them. In the table, the first column labeled `post.mean` lists the variances and covariances of the random intercepts. The next two columns labeled `l-95% CI` and `u-95% CI` list the 95% credible intervals, which are analogous to the confidence intervals in multinomial logistic regression models. Finally, the fourth column labeled `eff.samp` reports the effective sample size, which is the number of independent samples. The first row labeled `traitprof4.2:traitprof4.2.SCH_ID` reports the variance of the random intercept for the binary model comparing category 2 with the base category level 1. In the label, `traitprof4.2` represents the category 2 of the nominal outcome variable `prof4`. Similarly, `traitprof4.3` and `traitprof4.4` represent the categories 3 and 4 of the nominal outcome variable. So the three between-school variances for the random intercepts are .377, .777, and 1.855, respectively. Regarding the covariances, the second row labeled `traitprof4.3:traitprof4.2.SCH_ID` reports the covariance of the random intercepts for categories 3 and 2. The covariances for the intercepts between categories 3 and 2, categories 4 and 2, and categories 4 and 3 are .450, .663, and 1.083, respectively.

Second, the residual variances section (labeled `R-structure: ~us(trait):SCH_ID`) reports the residual variances and covariances.

Third, the fixed effects section (labeled `Location effects: prof4 ~ -1 + trait + trait:(stlang + public + csclimat`) reports the location effects for the intercepts and predictor variables. In the table, the first column labeled `post.mean` lists the posterior means, which are the similar to the logit coefficients in the multilevel multinomial regression model (model 3) provided by the `mblogit()` function in the previous section. Recall that a multinomial logistic regression model contains a series of binary logistic models where the dichotomized outcome variables compare each category to the base category.

The posterior means for `stlang` for the three underlying binary models are .430, .599, and .494, respectively. The corresponding 95% credible intervals do not contain zeros and pMCMC <.001. Therefore, these posterior means for `stlang` are meaningful. Please note that the pMCMC is not the same as the *p*-value in frequentist statistics. It is the probability that the parameter value is either greater than or less than zero. Here, the posterior means for the predictor variables are the parameter estimates which are the estimates for the true values of the parameters.

In addition, the posterior means for `public` for the three underlying binary models are −.633, −1.021, and −1.122, respectively. Their 95% credible intervals do not contain zeros and pMCMC <.001, so they are also meaningful. Finally, the posterior means for `csclimat` for the three binary models are .319, .616, and .894, respectively. The credible intervals and pMCMC value also suggest that these posterior means are meaningful. Recall that the posterior means are on the scale of logit or log odds. We can exponentiate them to obtain the odds ratios as with those in the multilevel multinomial regression model (model 3).

14.4.3 Multilevel Multinomial Regression Models With `MCMCglmm()`: Uncorrelated Random Effects

In the G-structure for the random effects above, we allow the random effects to be correlated with the `random = ~us(trait):SCH_ID` argument, so the matrix for the variance-covariance structure is unstructured, where the main diagonals include the variance for each intercept and the off-diagonals are the covariances between intercepts. However, the random effects of the multiple intercepts can also be uncorrelated, so in the matrix for the variance-covariance structure, the main diagonals still include the variance for each intercept, whereas the off-diagonals for the covariances are set to be zeros. Researchers need to be cautious since different software packages may use a different variance-covariance structure as the default. In the following example, with everything else the same as the original syntax, we use the `random = ~idh(trait):SCH_ID` argument to specify uncorrelated random effects, where `idh()` means that the random effects are independent.

```
> # Model with uncorrelated random effects
> mb.mc.3f <- MCMCglmm(prof4 ~ -1 + trait + trait:(stlang + public + csclimat), random
= ~idh(trait):SCH_ID,
+ rcov=~us(trait):units, data=nom, family = "categorical", prior = priors)

(Acceptance ratio omitted)

> summary(mb.mc.3f)

Iterations = 3001:12991
Thinning interval  = 10
Sample size  = 1000

DIC: 33191.66

G-structure:  ~idh(trait):SCH_ID

                       post.mean   l-95% CI   u-95% CI   eff.samp
traitprof4.2.SCH_ID       0.1208     0.0835     0.1548      56.79
traitprof4.3.SCH_ID       0.2304     0.1643     0.2873      53.59
traitprof4.4.SCH_ID       0.9103     0.7498     1.0860      43.24
```

```
R-structure: ~us(trait):units

                                 post.mean  l-95% CI  u-95% CI  eff.samp
traitprof4.2:traitprof4.2.units       0.50      0.50      0.50         0
traitprof4.3:traitprof4.2.units       0.25      0.25      0.25         0
traitprof4.4:traitprof4.2.units       0.25      0.25      0.25         0
traitprof4.2:traitprof4.3.units       0.25      0.25      0.25         0
traitprof4.3:traitprof4.3.units       0.50      0.50      0.50         0
traitprof4.4:traitprof4.3.units       0.25      0.25      0.25         0
traitprof4.2:traitprof4.4.units       0.25      0.25      0.25         0
traitprof4.3:traitprof4.4.units       0.25      0.25      0.25         0
traitprof4.4:traitprof4.4.units       0.50      0.50      0.50         0

 Location effects: prof4 ~ -1 + trait + trait:(stlang + public + csclimat)

                           post.mean  l-95% CI  u-95% CI eff.samp   pMCMC
traitprof4.2                -0.15514  -0.32767   0.01480    45.03   0.092 .
traitprof4.3                 0.33977   0.14073   0.50903    69.96  <0.001 ***
traitprof4.4                -0.04847  -0.31419   0.19247   100.33   0.698
traitprof4.2:stlangYes       0.41326   0.28058   0.54630    34.62  <0.001 ***
traitprof4.3:stlangYes       0.58768   0.46830   0.73246    72.45  <0.001 ***
traitprof4.4:stlangYes       0.51742   0.34900   0.65453    62.05  <0.001 ***
traitprof4.2:publicpublic   -0.55274  -0.70054  -0.38235    45.54  <0.001 ***
traitprof4.3:publicpublic   -0.97481  -1.13978  -0.81130    78.45  <0.001 ***
traitprof4.4:publicpublic   -1.05522  -1.27735  -0.78975   149.72  <0.001 ***
traitprof4.2:csclimat        0.27659   0.17588   0.36639    36.90  <0.001 ***
traitprof4.3:csclimat        0.52371   0.42666   0.62649    57.39  <0.001 ***
traitprof4.4:csclimat        0.83952   0.66908   0.98231   115.91  <0.001 ***

---
Signif. codes: 0 '***' 0.001 '**' 0.01 '*' 0.05 '.' 0.1 ' ' 1
```

In the output, the random effects section (labeled `G-structure: ~us(trait):SCH_ID`) reports the variances of the random intercepts. The three between-school variances for the random intercepts are .121, .230, and .910, respectively. When the random intercepts are uncorrelated, their between-school variances are much smaller than those correlated random intercepts reported in the preceding output for the unstructured variance-covariance matrix.

The posterior means for the predictor variables in the fixed effects section (labeled `Location effects: prof4 ~ -1 + trait + trait:(stlang + public + csclimat)`) look similar to those in the previous output. The interpretation is omitted here.

We use the `autocorr(mb.mc.3f$Sol)` command to evaluate the autocorrelations for the fixed effects of the model. The results are displayed as follows.

```
> autocorr(mb.mc.3f$Sol)
, , traitprof4.2

         traitprof4.2  traitprof4.3  traitprof4.4  traitprof4.2:stlangYes
Lag 0      1.00000000    0.41375222    0.24065387             -0.55474787
Lag 10     0.74719702    0.33663799    0.16930562             -0.43789698
Lag 50     0.53929207    0.24941730    0.13527311             -0.29389617
Lag 100    0.34968138    0.18424904    0.12918949             -0.17574399
Lag 500   -0.04696522   -0.06762418   -0.02986882             -0.04129599
```

```
              traitprof4.3:stlangYes      traitprof4.4:stlangYes
Lag 0                    -0.21962161               -0.1267767379
Lag 10                   -0.13821235               -0.0620316279
Lag 50                   -0.09101946               -0.0319398699
Lag 100                  -0.04524355               -0.0004840206
Lag 500                   0.09106414               -0.0325478030
           traitprof4.2:publicpublic   traitprof4.3:publicpublic
Lag 0                     -0.6958593                 -0.32494463
Lag 10                    -0.4948245                 -0.30019088
Lag 50                    -0.3698108                 -0.22822017
Lag 100                   -0.2362940                 -0.18576707
Lag 500                    0.1146540                  0.02372664
         traitprof4.4:publicpublic traitprof4.2:csclimat  traitprof4.3:csclimat
Lag 0                  -0.19762835            0.001759688            -0.03012252
Lag 10                 -0.16048406            0.067320481            -0.03484957
Lag 50                 -0.14274315            0.057973782             0.01609562
Lag 100                -0.14485241           -0.005708791             0.02919098
Lag 500                 0.02707301           -0.053218364            -0.04896842
           traitprof4.4:csclimat
Lag 0                 0.06528344
Lag 10                0.06649361
Lag 50                0.09887341
Lag 100               0.05928063
Lag 500               0.07826845
```

The autocorrelations get smaller when the number of lags increases. There are no patterns for high autocorrelations.

We can use the deviance information criterion (DIC) to compare models. The DIC for the Bayesian multilevel multinomial model with the unstructured variance-covariance matrix for the random effects is 32,717.72, whereas the DIC for the model with the unrelated random effects is 33,191.66. The difference is 473.94, so we prefer the former model with the correlated random effects.

14.5 BAYESIAN MULTILEVEL MODELING FOR NOMINAL OUTCOME VARIABLES WITH THE `brm()` FUNCTION IN THE `brms` PACKAGE

14.5.1 The `brm()` Function in the `brms` Package: Syntax for Bayesian Multilevel Multinomial Regression Models

In the last section, we have introduced the `MCMCglmm` package and have used the `MCMCglmm()` function to fit the Bayesian multilevel models for nominal response variables. In this section, we introduce the `brm ()` function in the `brms` package to fit the same models.

The `brms` package (Bürkner, 2017) is another popular package for fitting a variety of Bayesian generalized linear and non-linear multilevel models using the programming language Stan. It uses Hamilton Monte Carlo (HMC) and the No-U-Turn Sampler (NUTS) in Stan rather than Metropolis-Hastings sampling and Gibbs sampling for

MCMC to simulate draws. Since `brms` is a user-written package, you need to install it first by typing `install.packages("brms")` and then load the package with `library(brms)`. To make `brms` work, you also need to install Rtools from https://cran.r-project.org/bin/windows/Rtools/. Please note that Rtools is not an add-on package, you need to install it just like other programs on your computer.

The syntax for `brm()` includes the major components for fixed effects, random effects, the `family` argument and the `data` argument. It also includes the arguments to specify the number of iterations, warm-up, chains, and cores, which will be explained below.

First, the model formula for multilevel multinomial regression models in `brm()` is similar to that for `glmer()` in the `lme4` package and is simpler than that for the `MCMCglmm()` function. It also includes two components including the fixed effects and random effects, which are connected by a plus (+) sign. The fixed effects part of the model specifies the dependent variable and the predictor variable(s) with the dependent variable on the left-hand side of the tilde (~). The predictor variables on the right-hand side are connected by plus (+) symbols. Then, the random effects part of the model is specified within parentheses. Similar to the random effects part of `glmer()`, the random coefficients term is specified first, followed by a vertical line (|), and then, an identifier variable at a higher level as the grouping variable.

Second, the `data = data1` argument specifies the data frame `data1`.

Third, the `family = categorical` argument to specify the distribution of the nominal response variable. The `brm ()` function includes various families for continuous, binary, ordinal, nominal count, and time-to-event response variables.

For example, the command `brm(y ~ x + (1|schid), data = data1, family = categorical)` tells R to fit a Bayesian multilevel model to estimate a nominal response variable `y` on a predictor variable `x` with a randomly varying intercept across schools by specifying `(1|schid)`, where `1` indicates the random intercept on the left of the vertical line (|) and `schid` is the identifier variable on the right.

We can also specify the number of iterations with `iter =`, the number of warm-up or burn-in iterations with `warmup =`, the number of Markov chain with `chain =`, and the number of cores for executing the chains. The default number of iterations is 2,000 with 1,000 warm-up iterations and the default number of chains is 4. In addition, we can specify priors with the `prior =` argument. The `set_prior()` function can be used to set up priors before model fitting. For more details on how to use this function, type `help(brm)` in the command prompt after loading the `brms` package.

14.5.2 Multilevel Multinomial Regression Models With `brm()`: Uncorrelated Random Effects

In the following example, the command `mb.br.3c <-brm(prof4 ~ stlang + public + csclimat + (1|SCH_ID), data = nom, family = categorical, warmup = 1000, iter = 3000, cores = 2, chains = 2, seed = 123)` tells R to fit the same random-intercept model with predictor variables in both levels (model 3) in the previous section after the two level 2 predictor variables `public` and `csclimat` are added to the model. In the syntax, the model formula includes the three predictor variables for the fixed effects and `(1|SCH_ID)` specifies the random effects for the intercepts only. Here only the intercepts are allowed to vary across schools. We can add a predictor variable to the random effects component if we would like to fit a random-coefficient model. Since the nominal outcome variable has four levels, the between-school variances for the three intercepts will be estimated. The default variance-covariance structure in `brm()` assumes that the between-school variances for the intercepts are independent from each other, but we can specify correlated variances for the intercepts. An example with the syntax will be provided later. Next, the `data = norm` argument specifies the data frame `norm`. Further, the `family = categorical` argument to specify the distribution of the nominal response variable. We also increase the number of iterations from the default 2,000 to 3,000 with `iter = 3000` and keep the default number of warm-up with `warmup = 1000`. Finally, we specify two cores for executing the chains with `cores = 2` but reduce the number of chains to 2 with `chains = 2` since fitting the current model with four chains is very time-consuming. It may take a couple of hours or more to fit the current model with two chains. We set the random seed so the results can be replicated. It is normal to get slightly different output due to MCMC sampling if we rerun the same analysis.

The fitted model is named `mb.br.3c`. The following output is displayed by the `summary(mb.br.3c)` command.

```
> # Model with uncorrelated random effects
> library(brms)
> mb.br.3c <-brm(prof4 ~ stlang + public + csclimat  + (1|SCH_ID), data=nom,
family=categorical,
+                  warmup = 1000, iter = 3000,
+                  cores = 2, chains = 2, seed=123)
Compiling Stan program...
Start sampling
Warning messages:
1: There were 1 divergent transitions after warmup. Increasing adapt_delta above 0.8
may help. See
http://mc-stan.org/misc/warnings.html#divergent-transitions-after-warmup
2: Examine the pairs() plot to diagnose sampling problems


> summary(mb.br.3c)
 Family: categorical
  Links: mu2 = logit; mu3 = logit; mu4 = logit
Formula: prof4 ~ stlang + public + csclimat + (1 | SCH_ID)
   Data: nom (Number of observations: 12793)
Samples: 2 chains, each with iter = 3000; warmup = 1000; thin = 1;
         total post-warmup samples = 4000

Group-Level Effects:
~SCH_ID (Number of levels: 619)
                  Estimate Est.Error l-95% CI u-95% CI Rhat Bulk_ESS Tail_ESS
sd(mu2_Intercept)     0.19      0.06     0.05     0.29 1.01      455      538
sd(mu3_Intercept)     0.42      0.03     0.35     0.49 1.00     1331     2084
sd(mu4_Intercept)     0.89      0.04     0.80     0.97 1.00     1558     2438
```

```
Population-Level Effects:
                    Estimate Est.Error l-95% CI u-95% CI Rhat Bulk_ESS Tail_ESS
mu2_Intercept          -0.11      0.10    -0.30     0.08 1.00     4402     3073
mu3_Intercept           0.33      0.10     0.13     0.52 1.00     3373     3085
mu4_Intercept          -0.03      0.13    -0.29     0.22 1.00     2128     2617
mu2_stlangYes           0.37      0.07     0.23     0.50 1.00     4416     3125
mu2_publicpublic       -0.53      0.08    -0.69    -0.38 1.00     4036     3068
mu2_csclimat            0.26      0.04     0.17     0.34 1.00     4338     2653
mu3_stlangYes           0.55      0.07     0.41     0.69 1.00     3768     3202
mu3_publicpublic       -0.93      0.08    -1.08    -0.76 1.00     3188     2845
mu3_csclimat            0.50      0.05     0.40     0.59 1.00     2786     2996
mu4_stlangYes           0.47      0.08     0.31     0.63 1.00     3566     2987
mu4_publicpublic       -1.00      0.12    -1.24    -0.76 1.00     1694     2606
mu4_csclimat            0.79      0.08     0.64     0.94 1.00     1657     2090

Samples were drawn using sampling(NUTS). For each parameter, Bulk_ESS
and Tail_ESS are effective sample size measures, and Rhat is the potential
scale reduction factor on split chains (at convergence, Rhat = 1).
Warning message:
There were 1 divergent transitions after warmup. Increasing adapt_delta above 0.8 may
help. See http://mc-stan.org/misc/warnings.html#divergent-transitions-after-warmup
```

The first section R output produced by the `brm()` function includes the model information, including the family, the link functions for the three binary models, the model formula for the fixed and random effects, the data frame with the number of observations, and the number of chains and iterations, and the total number of posterior samples. With 3,000 iterations and 1,000 warm-up for each of the two chains, the total number of posterior samples = 2 * (3,000 − 1,000) = 4,000.

The second section is the random effects section (labeled `Group-Level Effects:`). It provides the standard deviations rather than the variances for the three intercepts in the binary models comparing categories 2 with 1, categories 3 with 1, and categories 4 with 1, respectively. The standard deviations for the three intercepts are .19, .42, and .89. We can easily obtain the variances by squaring the standard deviations. The corresponding variances for the intercepts are .036, .176, and .792, respectively. The table also provides the estimated errors, 95% credible intervals, the `Rhat` values for convergence, and the effective sample sizes (labeled `Bulk_ESS` and `Tail_ESS`).

The third section is the fixed effects section (labeled `Population-Level Effects:`). It provides the posterior means for the intercepts and predictor variables for the three binary models. As with the table for the random effects section, it also provides the estimated errors, 95% credible intervals, the `Rhat` values for convergence, and the effective sample sizes. The labels `mu2`, `mu3`, and `mu4`, are the names for the three binary models, so the posterior means of `stlang` for the three models are .37, .55, and .47, respectively. They looks similar to those in the Bayesian multilevel multinomial model with uncorrelated random effects produced by the `MCMCglmm()` function.

We can extract the results of the fixed effects and the credible intervals with the `coef <-fixef(mb.br.3c, summary = TRUE)` command. We can also use the `exp(fixef(mb.br.3c)[, -2])` command to obtain the odds ratios and the corresponding credible intervals. The estimated errors should not be exponentiated, so we exclude the second column in the fixed effects table with the indexing `[, -2]`. The output is omitted here.

The odds ratios for the three predictor variables can be interpreted in a similar way to those in the multilevel multinomial model with the `mblogit()` function.

14.5.3 Multilevel Multinomial Regression Models With `brm()`: Correlated Random Effects

The default variance-covariance structure in `brm()` with the `(1|SCH_ID)` argument assumes that the between-school variances for the intercepts are independent from each other. To specify correlated variances for the intercepts or any random effects, we need to add an arbitrary letter or number in the random effects component followed by a vertical line (|). In this example, we specify `(1|p|SCH_ID)` with the arbitrary letter `p`, so the matrix for the variance-covariance structure is unstructured, where the diagonals include the variance for each intercept and the off-diagonals are the covariances between intercepts. Since `p` is an arbitrary letter here, you can replace it with any letter between the two vertical lines to specify an unstructured variance and covariance matrix for the random effects.

In the next example, the command `mb.br.4 <- brm(prof4 ~ stlang + public + csclimat + (1|p|SCH_ID), data = nom, family = categorical, warmup = 1000, iter = 3000, cores = 2, chains = 2, seed = 123)` tells R to fit the same random-intercept model (model 3) in the previous section after the two level 2 predictor variables `public` and `csclimat` are added to the model. With everything else in the syntax the same as that for the preceding model, the model formula includes the new random effects component for the correlated intercepts with `(1|p|SCH_ID)`.

The fitted model is named `mb.br.4`. The following output is displayed by the `summary(mb.br.4)` command.

```
> # Model with uncorrelated random effects
> mb.br.4 <- brm(prof4 ~ stlang + public + csclimat  + (1|p|SCH_ID), data=nom,
family=categorical,
+                 warmup = 1000, iter = 3000,
+                 cores = 2, chains = 2, seed=123)
Compiling Stan program...
Start sampling
> summary(mb.br.4)
 Family: categorical
  Links: mu2 = logit; mu3 = logit; mu4 = logit
Formula: prof4 ~ stlang + public + csclimat + (1 | p | SCH_ID)
   Data: nom (Number of observations: 12793)
Samples: 2 chains, each with iter = 3000; warmup = 1000; thin = 1;
         total post-warmup samples = 4000

Group-Level Effects:
~SCH_ID (Number of levels: 619)
```

```
                                   Estimate  Est.Error  l-95% CI  u-95% CI  Rhat
sd(mu2_Intercept)                      0.52       0.04      0.44      0.60  1.00
sd(mu3_Intercept)                      0.80       0.04      0.72      0.89  1.00
sd(mu4_Intercept)                      1.29       0.06      1.18      1.41  1.00
cor(mu2_Intercept,mu3_Intercept)       0.95       0.03      0.89      0.99  1.00
cor(mu2_Intercept,mu4_Intercept)       0.86       0.04      0.78      0.94  1.00
cor(mu3_Intercept,mu4_Intercept)       0.94       0.02      0.91      0.97  1.00
                                     Bulk_ESS  Tail_ESS
sd(mu2_Intercept)                        1791      2491
sd(mu3_Intercept)                        1825      2005
sd(mu4_Intercept)                        1349      1892
cor(mu2_Intercept,mu3_Intercept)          362       813
cor(mu2_Intercept,mu4_Intercept)          212       316
cor(mu3_Intercept,mu4_Intercept)          847      2038

Population-Level Effects:
                   Estimate  Est.Error  l-95% CI  u-95% CI  Rhat  Bulk_ESS  Tail_ESS
mu2_Intercept          0.07       0.11     -0.14      0.28  1.00      3286      3350
mu3_Intercept          0.48       0.12      0.24      0.71  1.00      2907      3280
mu4_Intercept          0.11       0.16     -0.20      0.42  1.00      2809      2651
mu2_stlangYes          0.35       0.07      0.21      0.50  1.00      5220      3200
mu2_publicpublic      -0.58       0.10     -0.77     -0.39  1.00      2788      2650
mu2_csclimat           0.31       0.06      0.21      0.42  1.00      2471      3126
mu3_stlangYes          0.57       0.08      0.42      0.71  1.00      4760      3351
mu3_publicpublic      -0.97       0.11     -1.19     -0.75  1.00      2359      2601
mu3_csclimat           0.57       0.07      0.44      0.70  1.00      2087      2649
mu4_stlangYes          0.46       0.09      0.30      0.64  1.00      4919      3234
mu4_publicpublic      -1.05       0.16     -1.35     -0.75  1.00      2444      2585
mu4_csclimat           0.85       0.10      0.66      1.04  1.00      2017      2578

Samples were drawn using sampling(NUTS). For each parameter, Bulk_ESS
and Tail_ESS are effective sample size measures, and Rhat is the potential
scale reduction factor on split chains (at convergence, Rhat = 1).
```

The random effects section (labeled `Group-Level Effects:`) reports the standard deviations for the three intercepts in the binary models. The standard deviations for the three intercepts are .52, .80, and 1.29. The corresponding variances for the intercepts can be computed as .270, .640, and 1.664, respectively. The table also provides the estimated errors, 95% credible intervals, the `Rhat` values for convergence, and the effective sample sizes (labeled `Bulk_ESS` and `Tail_ESS`). The correlation coefficients between the intercepts for categories 2 and 3 (labeled `mu2_Intercept, mu3_Intercept`), categories 2 and 4 (labeled `mu2_Intercept, mu4_Intercept`), and categories 3 and 4 (labeled `mu3_Intercept, mu4_Intercept`) are .95, .86, and .94, respectively. Recall that the covariance between two variables equals the product of the correlation coefficient and the two standard deviations, so the covariance between the intercepts for categories 3 and 2 is computed as follows:

$$\text{Covariance}(\texttt{mu2\_Intercept},\ \texttt{mu3\_Intercept}) = .52*.80*.95 = .395$$

Similarly, the covariances between the intercepts for categories 2 and 4 and for categories 3 and 4 can also be computed. They are .577, and .970, respectively.

Covariance(`mu2_Intercept`, `mu4_Intercept`) = .52*1.29*.86 = .577
Covariance(`mu3_Intercept`, `mu4_Intercept`) = .80*1.29*.94 = .970

The same variances and covariances can be easily computed with the `VarCorr (mb.br.4)` command. The output is omitted here.

The fixed effects section (labeled `Population-Level Effects:`) provides the posterior means, the estimated errors, the 95% credible intervals, the `Rhat` values for convergence, and the effective sample sizes for the intercepts and predictor variables for the three binary models. The results are similar to those in the Bayesian multilevel multinomial model with correlated random effects produced by the `MCMCglmm()` function and those by the `mblogit()` function.

We extract the results of the fixed effects and the credible intervals with the `coef <- fixef(mb.br.4, summary = TRUE)` command.

```
> coef <- fixef(mb.br.4, summary=TRUE)
> coef
                       Estimate    Est.Error         Q2.5        Q97.5
mu2_Intercept        0.07035344   0.10807173   -0.1416809    0.2826723
mu3_Intercept        0.47787899   0.11995639    0.2414931    0.7123711
mu4_Intercept        0.11208741   0.15775924   -0.1961439    0.4172594
mu2_stlangYes        0.35370009   0.07175611    0.2145659    0.4967404
mu2_publicpublic    -0.57855908   0.09865861   -0.7745356   -0.3859330
mu2_csclimat         0.31372128   0.05502020    0.2074229    0.4216058
mu3_stlangYes        0.56685064   0.07512380    0.4211559    0.7149932
mu3_publicpublic    -0.97221320   0.11269028   -1.1926967   -0.7492996
mu3_csclimat         0.56670571   0.06669238    0.4382574    0.6981288
mu4_stlangYes        0.46444062   0.08505443    0.2972842    0.6352088
mu4_publicpublic    -1.04655367   0.15668480   -1.3549974   -0.7491179
mu4_csclimat         0.85113730   0.09755534    0.6637627    1.0407936
```

We use the `exp(fixef(mb.br.4)[, -2])` command to obtain the odds ratios and the corresponding credible intervals. Again, the estimated errors are not exponentiated by excluding the second column in the fixed effects table with the indexing `[, -2]`. The following output is created.

```
> exp(fixef(mb.br.4)[, -2])
                    Estimate        Q2.5       Q97.5
mu2_Intercept      1.0728873   0.8678982   1.3266703
mu3_Intercept      1.6126503   1.2731487   2.0388197
mu4_Intercept      1.1186106   0.8218939   1.5177962
mu2_stlangYes      1.4243280   1.2393237   1.6433559
mu2_publicpublic   0.5607057   0.4609178   0.6798160
mu2_csclimat       1.3685082   1.2305029   1.5244075
mu3_stlangYes      1.7627069   1.5237218   2.0441728
mu3_publicpublic   0.3782450   0.3034020   0.4726975
mu3_csclimat       1.7624515   1.5500038   2.0099881
mu4_stlangYes      1.5911239   1.3461979   1.8874162
mu4_publicpublic   0.3511458   0.2579480   0.4727834
mu4_csclimat       2.3423093   1.9420860   2.8314631
```

14.5.4 Multilevel Multinomial Regression Models With `brm()`: Model Diagnostics With `plot()`

We use the `plot(mb.br.4)` command to visually inspect the posterior distribution and the trace plots of the fixed effect and random effects. We only show the first part of the plots here (Figure 14.1).

The posterior distribution plots and the trace plots (Figure 14.1) show good convergence.

FIGURE 14.1 ● Posterior Density Plots and Trace Plots for the Bayesian Multilevel Multinomial Model

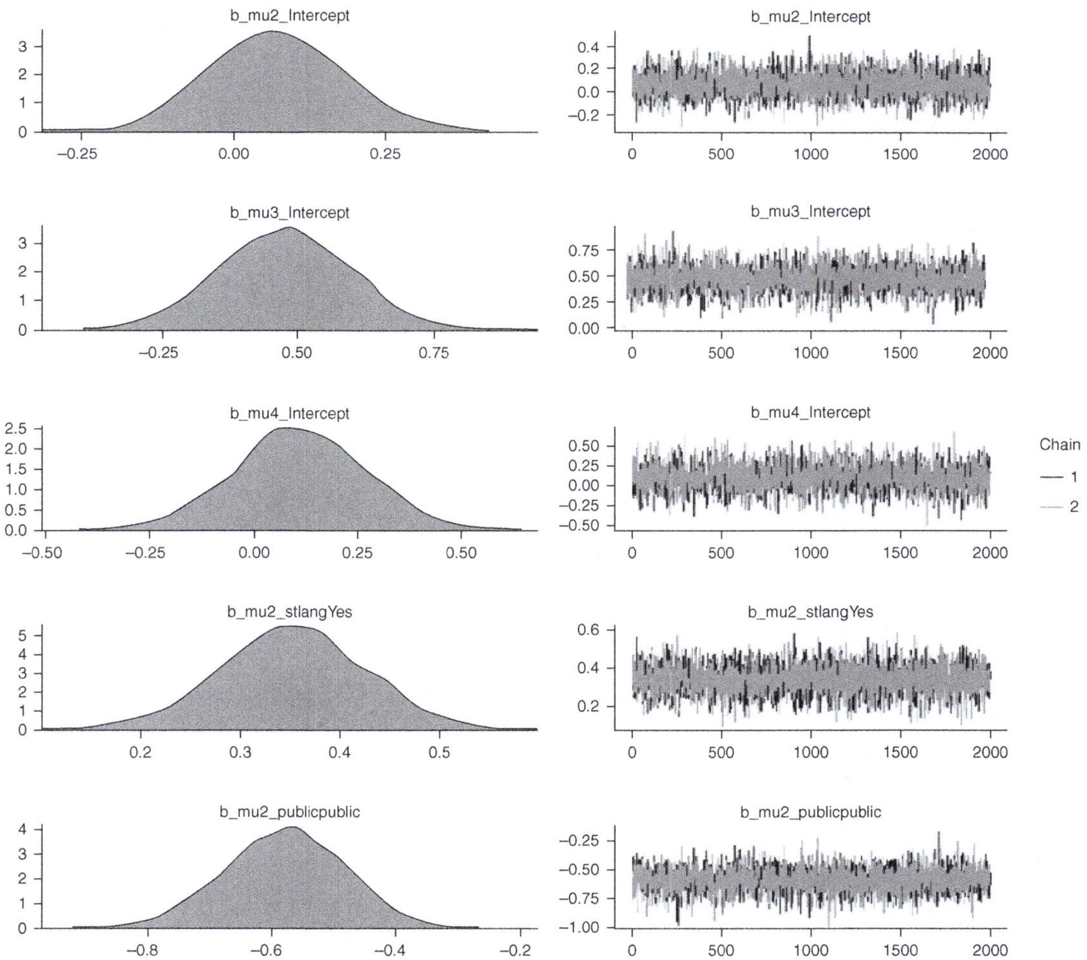

14.5.5 Conditional Effects With `conditional_effects ()`

We can use the `conditional_effects(mb.br.4, categorical = TRUE, surface = TRUE)` command to display the effects of the fixed effects, the population-level effects of the predictor variables. To save space, only the graph for the predictor variable `public` is shown here (Figure 14.2).

FIGURE 14.2 ● Predicted Probabilities of Being in Each Category of Math Proficiency for `public`

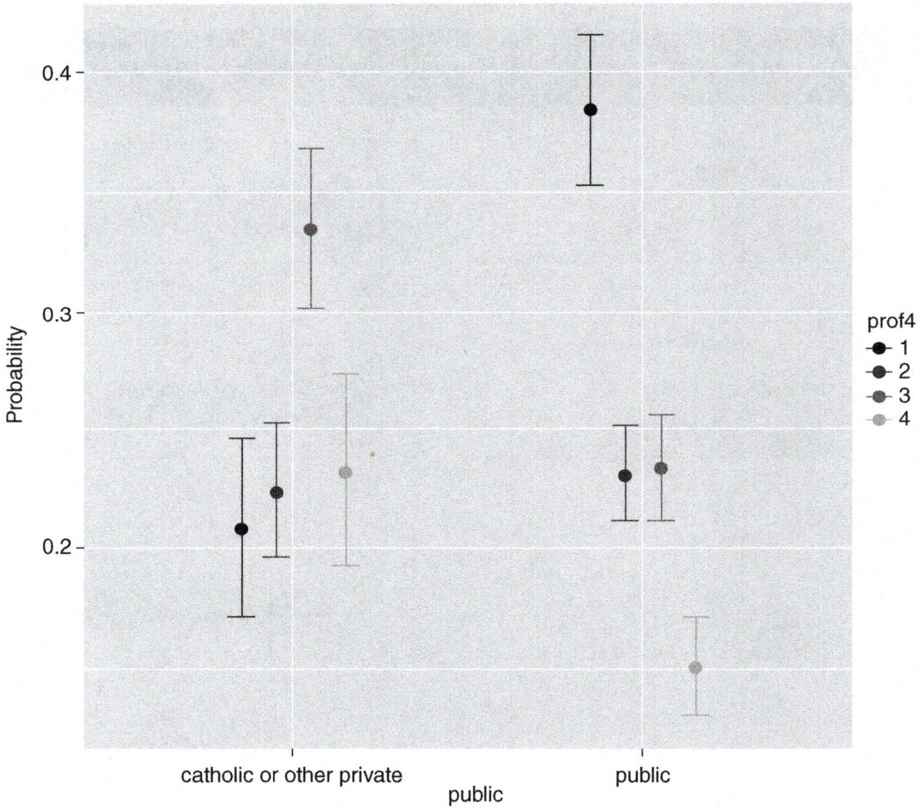

14.5.6 Interpreting the Predicted Probabilities With the `ggpredict()` Function in the `ggeffects` Package

By using the `ggpredict()` function in the `ggeffects` package (Lüdecke, 2018b), we can compute the predicted probabilities for predictor variables at specified values in the Bayesian multilevel multinomial logistic regression model. The `ggpredict()` function is executed for the random-intercept model with both level 1 and level 2 predictors (model 3).

We first load the package with `library(ggeffects)` since it has been installed in previous chapters. The command `mb.br4.pub <-ggpredict(mb.br.4, terms = "public")` tells R to compute the predicted probabilities using the `ggpredict()` function. The argument inside the function includes the estimated model, `mb.br.4`, and the `terms = "public"` option which specifies the predictor variable `public`. The other predictor variables are held constant. The output is assigned to an object named `mb.br4.pub`.

```
> # Predicted probabilities with ggpredict() in ggeffects
> library(ggeffects)
> mb.br4.pub <- ggpredict(mb.br.4, terms="public")
Note: uncertainty of error terms are not taken into account. You may want to use
`rstantools::posterior_predict()`.
Warning message:
posterior_linpred(transform = TRUE) is deprecated. Please use posterior_epred()
instead, without the 'transform' argument.
> mb.br4.pub

# Predicted probabilities of prof4

# Response Level = 1

public                    | Predicted |       95% CI
----------------------------------------------------
catholic or other private |      0.21 | [0.17, 0.25]
public                    |      0.38 | [0.35, 0.41]

# Response Level = 2

public                    | Predicted |       95% CI
----------------------------------------------------
catholic or other private |      0.22 | [0.20, 0.25]
public                    |      0.23 | [0.21, 0.25]

# Response Level = 3

public                    | Predicted |       95% CI
----------------------------------------------------
catholic or other private |      0.34 | [0.30, 0.37]
public                    |      0.23 | [0.21, 0.26]

# Response Level = 4

public                    | Predicted |       95% CI
----------------------------------------------------
catholic or other private |      0.23 | [0.19, 0.27]
public                    |      0.15 | [0.13, 0.17]

Adjusted for:
*   stlang =    No
* csclimat = -0.00
*   SCH_ID = NA (population-level)
> plot(mb.br4.pub)
```

The output displays the predicted probabilities of being in four different levels of the nominal response variable for both Catholic or other private schools and public schools when other predictor variables are fixed at zeros. The last section under the title "`Adjusted for`" lists the names and the values of the other variables. To visualize the results, the predicted probabilities are plotted using `plot(mb.br4.pub)`. Figure 14.3 shows the predicted probabilities of being proficient in four different levels for `public` at 0 and 1 with others fixed at their means or at the reference level or category.

As shown in the graph, the predicted probabilities of being in math proficiency levels 1 and 2 for students in public schools are higher than those in Catholic or other private schools, whereas students in Catholic or other private schools are more likely than

FIGURE 14.3 ● Predicted Probabilities for the Binary Predictor Variable `public`

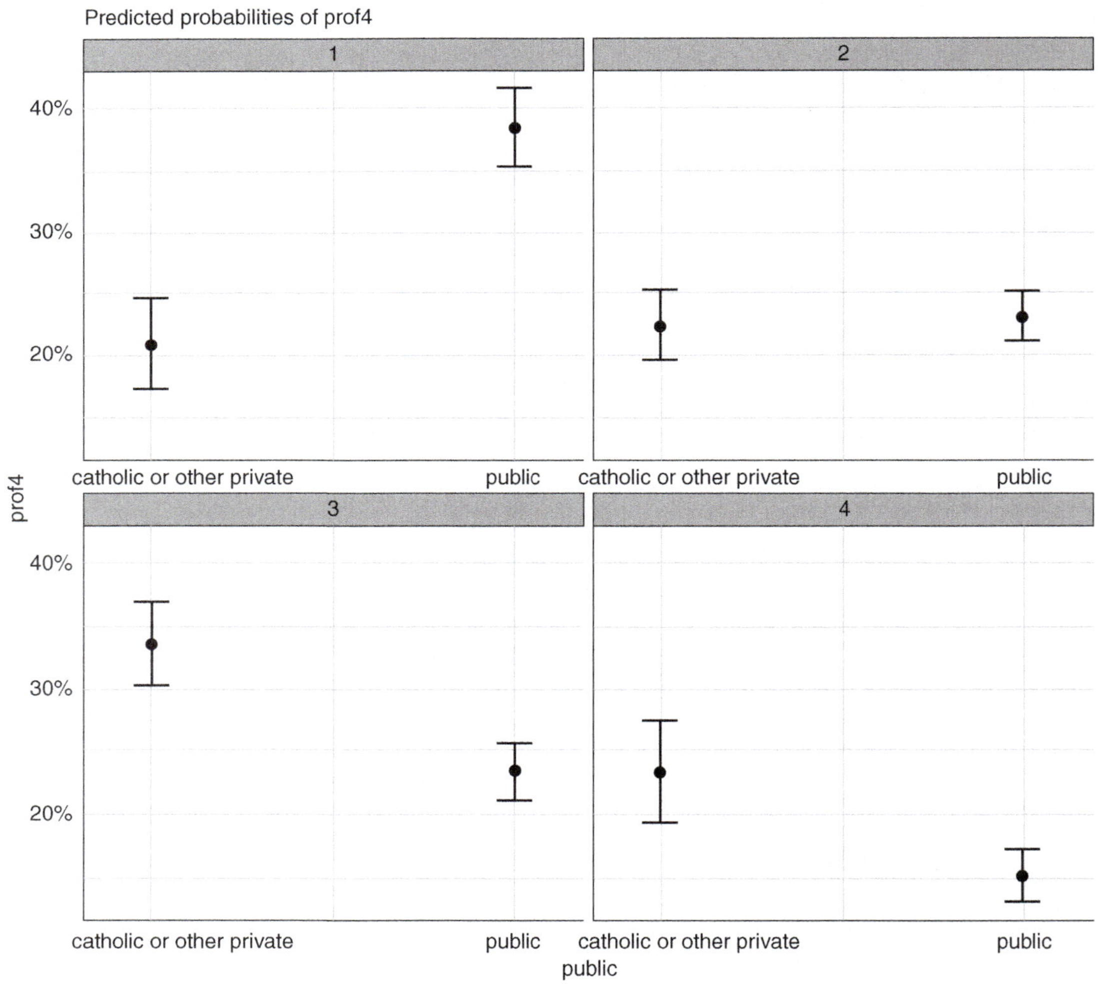

students in public schools to be proficient in levels 3 and 4. This pattern is the same as that in the plot by the `conditional_effects()` function above.

14.6 MAKING PUBLICATION-QUALITY TABLES

14.6.1 Presenting the Results Using the texreg Package

As of this writing, since the `stargazer()` function cannot directly produce the results table from the models by the `mblogit()`, `brm()` and `MCMCglmm()` functions, we use the `screenreg()` and `htmlreg()` functions from the `texreg` package (Leifeld, 2013). Since `texreg` has been installed for previous chapters, we load the package first by typing `library(texreg)`. To make the package work with the `brm()` function for the Bayesian multilevel multinomial models for nominal response variables, we need to first customize the `extract()` function and the `createTexreg()` function (Leifeld, 2013). The R code for customizing the function is included at the end of the chapter.

After using the `brm()` function to fit the two Bayesian multilevel multinomial regression models with uncorrelated and correlated random intercepts, `mb.br.3c` and `mb.br.4`, we create a table containing the results of these two models with the following command: `screenreg(list(mb.br.3c, mb.br.4))`. In the `screenreg()` function, we specify the two model objects to be presented. The output is a plain text table.

```
> library(texreg)
Version:     1.36.23
Date:        2017-03-03
Author:      Philip Leifeld (University of Glasgow)

Please cite the JSS article in your publications -- see citation("texreg").
> screenreg(list(mb.br.3c, mb.br.4))

=====================================================
                             Model 1         Model 2
-----------------------------------------------------
mu2_Intercept                 -0.11            0.07
                          [-0.30; 0.08]   [-0.14; 0.28]
mu3_Intercept                  0.33 *          0.48 *
                          [ 0.13; 0.52]   [ 0.24; 0.71]
mu4_Intercept                 -0.03            0.11
                          [-0.29; 0.22]   [-0.20; 0.42]
mu2_stlangYes                  0.37 *          0.35 *
                          [ 0.23; 0.50]   [ 0.21; 0.50]
mu2_publicpublic              -0.53 *         -0.58 *
                          [-0.69; -0.38]  [-0.77; -0.39]
mu2_csclimat                   0.26 *          0.31 *
                          [ 0.17; 0.34]   [ 0.21; 0.42]
mu3_stlangYes                  0.55 *          0.57 *
                          [ 0.41; 0.69]   [ 0.42; 0.71]
mu3_publicpublic              -0.93 *         -0.97 *
                          [-1.08; -0.76]  [-1.19; -0.75]
```

```
mu3_csclimat                                      0.50 *          0.57 *
                                             [ 0.40; 0.59]   [ 0.44; 0.70]
mu4_stlangYes                                     0.47 *          0.46 *
                                             [ 0.31; 0.63]   [ 0.30; 0.64]
mu4_publicpublic                                 -1.00 *         -1.05 *
                                             [-1.24; -0.76]  [-1.35; -0.75]
mu4_csclimat                                      0.79 *          0.85 *
                                             [ 0.64; 0.94]   [ 0.66; 1.04]
---------------------------------------------------------------------------
SD: SCH_ID: sd(mu2_Intercept)                     0.19            0.52
SD: SCH_ID: sd(mu3_Intercept)                     0.42            0.80
SD: SCH_ID: sd(mu4_Intercept)                     0.89            1.29
Num. obs.                                     12793           12793
SD: SCH_ID: cor(mu2_Intercept,mu3_Intercept)                      0.95
SD: SCH_ID: cor(mu2_Intercept,mu4_Intercept)                      0.86
SD: SCH_ID: cor(mu3_Intercept,mu4_Intercept)                      0.94
===========================================================================
* Null hypothesis value outside the confidence interval.
> htmlreg(list(mb.br.3c, mb.br.4), file="mbbr.doc", doctype=TRUE, html.tag=TRUE,
head.tag=TRUE)
The table was written to the file 'mbbr.doc'.
```

To create the table in the HTML format and copy it into Microsoft Word, we use the command as follows: `htmlreg(list(mb.br.3c, mb.br.4), file = "mbbr.doc", doctype = TRUE, html.tag = TRUE, head.tag = TRUE)`. To save space, the table with the original format is omitted.

With edits on the model titles and the added labels for the fixed and random effects, the final table is displayed as in Table 14.1.

TABLE 14.1 ● Results of the Two Bayesian Multilevel Multinomial Regression Models With `brm()`: Uncorrelated Random Intercepts and Correlated Random Intercepts (Edited)

	Random-Intercept Model (Uncorrelated Intercepts	Random-Intercept Model (Correlated Intercepts)
	Estimates (Credible Intervals)	Estimates (Credible Intervals)
Fixed Effects		
mu2_Intercept	−0.11	0.07
	[−0.30; 0.08]	[−0.14; 0.28]
mu3_Intercept	0.33[a]	0.48[a]
	[0.13; 0.52]	[0.24; 0.71]

Table 14.1 *(Continued)*

	Random-Intercept Model (Uncorrelated Intercepts	Random-Intercept Model (Correlated Intercepts)
	Estimates (Credible Intervals)	Estimates (Credible Intervals)
mu4_Intercept	−0.03	0.11
	[−0.29; 0.22]	[−0.20; 0.42]
mu2_stlang	0.37[a]	0.35[a]
	[0.23; 0.50]	[0.21; 0.50]
mu2_public	−0.53[a]	−0.58[a]
	[−0.69; −0.38]	[−0.77; −0.39]
mu2_csclimat	0.26[a]	0.31[a]
	[0.17; 0.34]	[0.21; 0.42]
mu3_stlang	0.55[a]	0.57[a]
	[0.41; 0.69]	[0.42; 0.71]
mu3_public	−0.93[a]	−0.97[a]
	[−1.08; −0.76]	[−1.19; −0.75]
mu3_csclimat	0.50[a]	0.57[a]
	[0.40; 0.59]	[0.44; 0.70]
mu4_stlang	0.47[a]	0.46[a]
	[0.31; 0.63]	[0.30; 0.64]
mu4_public	−1.00[a]	−1.05[a]
	[−1.24; −0.76]	[−1.35; −0.75]
mu4_csclimat	0.79[a]	0.85[a]
	[0.64; 0.94]	[0.66; 1.04]
Random Effects		
SD: SCH_ID: sd(mu2_Intercept)	0.19	0.52
SD: SCH_ID: sd(mu3_Intercept)	0.42	0.80
SD: SCH_ID: sd(mu4_Intercept)	0.89	1.29

Table 14.1 *(Continued)*

	Random-Intercept Model (Uncorrelated Intercepts	Random-Intercept Model (Correlated Intercepts)
	Estimates (Credible Intervals)	Estimates (Credible Intervals)
SD: SCH_ID: cor(mu2_Intercept,mu3_Intercept)		0.95
SD: SCH_ID: cor(mu2_Intercept,mu4_Intercept)		0.86
SD: SCH_ID: cor(mu3_Intercept,mu4_Intercept)		0.94
Num. obs.	12,793	12,793

[a]Null hypothesis value outside the confidence interval.

14.7 REPORTING THE RESULTS

The same reporting guidelines for multilevel modeling for binary response variables can be applied to the reporting for multilevel multinomial logistic regression models. Since a multilevel multinomial logistic regression model includes $K - 1$ binary comparisons, so the parameter estimates for the fixed effects of the predictor variables and the corresponding odds ratios need to be reported across these underlying binary models. In addition, report the variances and covariances or the standard deviations and correlations for the random effects in text. The following is an example of summarizing the results of a multilevel multinomial logistic regression model.

Multilevel multinomial logistic regression models were fitted to predict whether high-school students were proficient in each of the four math levels from three predictor variables: whether English is the students' native language, the school type, and the school climate. The first predictor was a student-level variable, and the other two were school-level variables. Multilevel modeling for nominal response variables was used since the outcome variable was nominal. Three models, from the unconditional (null) model to the random-intercept model with both level 1 and level 2 predictor variables, were fitted using the `mblogit()` function from the `mclogit` package. To save space, only the results for the final random-intercept model with predictor variables in both levels (model 3) were reported as follows.

Results for the Random-Intercept Model With Both Level 1 and Level 2 Predictor Variables (Model 3)

The logit coefficients for `stlang` (γ_{10k}) for the three underlying binary models are .363, .521, and .465, respectively. They are significantly different from 0. The odds ratios for `stlang` are 1.437, 1.684, and 1.592, respectively. The results indicate that the odds of being in level 2, level 3, and level 4 versus the base category for students with English as their native language are 1.437, 1.684, and 1.592 times, respectively, as great as the odds for those students whose native language is not English when holding all other variables constant. In other words, being students with English as their native language increases the odds by 43.7%, 68.4%, and 59.2%, respectively.

The logit coefficients for `public` (γ_{01k}) for the three underlying binary models are −.539, −.925, and −.986, respectively. They are significantly different from 0. The odds ratios for `public` are .583, .396, and .373, respectively, which indicate that the odds of being in level 2, level 3, and level 4 versus the base category for students in public schools are .583, .396, .373 times the odds for those in private schools, respectively.

In addition, the logit coefficients for `csclimat` (γ_{02k}) for the three binary models are .277, .523, and .812, respectively. They are also significantly different from 0. The odds ratios for `csclimat` are 1.319, 1.687, and 2.253, respectively, which indicate that a one-unit increase in school climate corresponds to an increase of 1.319, 1.687, and 2.253 points, respectively in the odds of being proficient in math level 2, level 3, and level 4 versus the base category. In other words, being students in schools with a more positive school climate increases the odds of being proficient in math level 2, level 3, and level 4 versus the base category.

Regarding the variance and covariance components for the random effects, the between-group variances, the three intercepts variances (τ_{00k}) are .359, .639, and 1.996, respectively. The covariances between the intercepts for categories 2 and 3, categories 2 and 4, and categories 3 and 4 are .164, .301, and .761, respectively.

14.8 SUMMARY OF R COMMANDS IN THIS CHAPTER

```
# Chap 14 R Script

# Remove all objects
rm(list = ls(all = TRUE))

# The following user-written packages need to be installed first by using
install.packages(" ") and then by loading it with library()
# library(sjmisc)          # It is already installed for Chapter 1
# library(mclogit)
# library(MCMCglmm)
# library(brms)
# library(ggeffects)       # It is already installed for Chapter 2
# library(texreg)          # It is already installed for Chapter 4


# Import the dataset
library(foreign)
nom <- read.dta("C:/CDA/els-nom.dta")
str(nom)


# Recode the outcome variable
library(sjmisc)
nom$prof4 <- rec(nom$Profmath, rec="0:1=1; 2=2; 3=3; 4:5=4; else=NA")


# Create factor variables
nom$prof4 <- factor(nom$prof4)
nom$SCH_ID <- factor(nom$SCH_ID)
attach(nom)
table(prof4)
table(stlang)
table(public)
mean(csclimat)


# Unconditional model/Null model with mblogit() in mclogit
install.packages("mclogit")
library(mclogit)
mb.null <- mblogit(prof4 ~ 1, random=~1|SCH_ID, data=nom)
summary(mb.null)


coef(mb.null)
coef <- coef(mb.null)
se <- sqrt(diag(vcov(mb.null)))
se
LL <- coef(mb.null)-1.96*se
UL <- coef(mb.null)+1.96*se
Coeftab <- cbind(LL, UL)
coeftab
exp(coef(mb.null))
exp(coeftab)
ortab <- cbind(or=exp(coef(mb.null)), ORLL=exp(LL), ORUL=exp(UL))
ortab
```

```
# Random-intercept model
mb.2 <- mblogit(prof4 ~ stlang, random=~1|SCH_ID, data=nom)
summary(mb.2)

coef(mb.2)
coef <- coef(mb.2)
se <- sqrt(diag(vcov(mb.2)))
se
LL <- coef(mb.2)-1.96*se
UL <- coef(mb.2)+1.96*se
coeftab <- cbind(LL, UL)
coeftab
exp(coef(mb.2))
exp(coeftab)
ortab <- cbind(or=exp(coef(mb.2)), ORLL=exp(LL), ORUL=exp(UL))
ortab

# Contextual model/Random-intercept model with predictor variables in both levels
mb.3 <- mblogit(prof4 ~ stlang + public + csclimat, random=~1|SCH_ID, data=nom)
summary(mb.3)

coef(mb.3)
coef <- coef(mb.3)
se <- sqrt(diag(vcov(mb.3)))
se
LL <- coef(mb.3)-1.96*se
UL <- coef(mb.3)+1.96*se
coeftab <- cbind(LL, UL)
coeftab

exp(coef(mb.3))
exp(coeftab)
ortab <- cbind(or=exp(coef(mb.3)), ORLL=exp(LL), ORUL=exp(UL))
ortab

# Bayesian multilevel multinomial models with MCMCglmm()
install.packages("MCMCglmm")
library(MCMCglmm)

# Use set.seed(123) to get replicable results each time with the same seed
set.seed(123)
# Set priors
k <- length(levels(nom$prof4))
I <- diag(k-1)
J <- matrix(rep(1, (k-1)^2), c(k-1, k-1))
priors = list(R = list(fix=1, V=(1/k) * (I + J), n = k - 1),
              G = list(G1 = list(V = diag(k - 1), n = k - 1)))

# Model with correlated random effects
# It will take around 20 minutes to fit the model
mb.mc.3e <- MCMCglmm(prof4 ~ -1 + trait + trait:(stlang + public + csclimat),
               random = ~us(trait):SCH_ID,
               rcov=~us(trait):units, data=nom,
               family = "categorical", prior = priors)
summary(mb.mc.3e)
```

```
# Model with uncorrelated random effects
mb.mc.3f <- MCMCglmm(prof4 ~ -1 + trait + trait:(stlang + public + csclimat),
                random = ~idh(trait):SCH_ID,
                rcov=~us(trait):units, data=nom,
                family = "categorical", prior = priors)
summary(mb.mc.3f)
autocorr(mb.mc.3f$Sol)

# Bayesian multilevel multinomial models with brm()
library(brms)

# Model with uncorrelated random effects
# It will take around 1 hour and 40 minutes to fit the model
mb.br.3c <- brm(prof4 ~ stlang + public + csclimat + (1|SCH_ID), data=nom,
                 family=categorical, warmup = 1000, iter = 3000,
                 cores = 2, chains = 2, seed=123)
summary(mb.br.3c)
coef<-fixef(mb.br.3c, summary=TRUE)
coef
exp(fixef(mb.br.3c)[, -2])
save(mb.br.3c, file = "mbbr3c.rda")

# Model with correlated random effects
mb.br.4 <- brm(prof4 ~ stlang + public + csclimat + (1|p|SCH_ID), data=nom,
family=categorical,
                warmup = 1000, iter = 3000,
                cores = 2, chains = 2, seed=123)
summary(mb.br.4)
coef <- fixef(mb.br.4, summary=TRUE)
coef
exp(fixef(mb.br.4)[, -2])
save(mb.br.4, file = "mbbr4.rda")

# Model diagnostics with plot()
plot(mb.br.4)

# Conditional effects with conditional_effects()
conditional_effects(mb.br.4, categorical=TRUE, surface=TRUE)

# Predicted probabilities with ggpredict() in ggeffects
library(ggeffects)
mb.br4.pub <- ggpredict(mb.br.4, terms="public")
mb.br4.pub
plot(mb.br4.pub)

# Present the results with texreg()
library(texreg)
extract.brmsfit <- function (model,
                              use.HDI = TRUE,
                              HDI.prob = 0.9,
                              include.random = TRUE,
                              include.rsquared = FALSE,
                              include.nobs = TRUE,
                              include.loo.ic = FALSE,
                              reloo = FALSE,
                              include.waic = FALSE,
                              ...) {
```

```
 sf <- summary(model, ...)$fixed
 coefnames <- rownames(sf)
 coefs <- sf[, 1]
 se <- sf[, 2]
 ci.low = sf[, 3]
 ci.up = sf[, 4]
 gof <- numeric()
 gof.names <- character()
 gof.decimal <- logical()
 if (isTRUE(include.random) & isFALSE(!nrow(model$ranef))) {
  sr <- summary(model, ...)$random
  sd.names <- character()
  sd.values <- numeric()
  for (i in 1:length(sr)) {
  sd <- sr[[i]][, 1]
  sd.names <- c(sd.names, paste0("SD: ", names(sr)[[i]], ": ", names(sd)))
  sd.values <- c(sd.values, sd)
  }
  gof <- c(gof, sd.values)
  gof.names <- c(gof.names, sd.names)
  gof.decimal <- c(gof.decimal, rep(TRUE, length(sd.values)))
 }
 if (isTRUE(include.nobs)) {
  n <- stats::nobs(model)
  gof <- c(gof, n)
  gof.names <- c(gof.names, "Num. obs.")
  gof.decimal <- c(gof.decimal, FALSE)
 }
 tr <- createTexreg(coef.names = coefnames,
                    coef = coefs,
                    se = se,
                    ci.low = ci.low,
                    ci.up = ci.up,
                    gof.names = gof.names,
                    gof = gof,
                    gof.decimal = gof.decimal)
 return(tr)
}


setMethod("extract", signature = className("brmsfit", "brm"),
          definition = extract.brmsfit)


screenreg(list(mb.br.3c, mb.br.4))
screenreg(list(mb.br.4))
htmlreg(list(mb.br.3c, mb.br.4), file="mbbr.doc", doctype=TRUE, html.tag=TRUE,
head.tag=TRUE)


detach(nom)
```

Glossary

In **a random-intercept multinomial logistic regression model**, the intercept is allowed to vary freely across groups.

Multilevel modeling for nominal outcome variables is a multilevel modeling technique for unordered response variables with more than two categories.

The `brms` package is a popular, user-friendly package for fitting a variety of Bayesian generalized linear and non-linear multilevel models using the programming language Stan. It uses Hamilton Monte Carlo (HMC) and the No-U-Turn Sampler (NUTS) in Stan rather than Metropolis-Hastings sampling and Gibbs sampling for MCMC to simulate draws.

The `MCMCglmm` package is well-known for fitting various Bayesian generalized linear mixed models using the Markov chain Monte Carlo (MCMC) methods. It uses both Metropolis-Hastings sampling and Gibbs sampling for MCMC to simulate draws.

The odds in multilevel modeling for nominal outcome variables are the ratio of the probability of being in a particular category to the probability of being in the base category.

The odds ratios are the exponentiated logit coefficients. They are interpreted as the change in the odds of being in a particular category relative to the base category for a one-unit increase in an independent variable

The unconditional model, or null model, for nominal response variables includes no predictor variable at any level. This model estimates the average logit or log odds of being in a particular nominal category and the variability in the intercepts between groups (e.g., schools).

Exercises

Use the ELS:2002 data available at **https://edge.sagepub.com/liu1e** with the following variables for the multilevel multinomial logistic regression models.

`Profmath`: mathematics proficiency level with six categories from 0 to 5

`gender`: gender (1 = female; 0 = male)

`byses`: socioeconomic status composite for base-year data

`SCH_ID`: school ID, clustering variable

This study investigates how predictor variables influence students' mathematics proficiency level and whether there are random intercepts across schools. The nominal outcome variable is `Profmath` with six levels, and the predictor variables are `gender` and `byses`. Recode `Profmath` by merging categories 0 and 1 into category 1 and merging categories 4 and 5 into category 4 and create a new nominal variable `prof4` with four levels. Conduct the multilevel modeling for the new nominal response variable and then answer the following questions:

1. Use the `mblogit()` function to fit an unconditional model and obtain the between-group variances for the intercepts.
2. Fit a two-level, random-intercept model with a random intercept and the two student-level predictor variables `gender` and `byses`. Use the grand-mean centering for `byses` before fitting the model.
3. Interpret the odds ratios of the two predictor variables.
4. Write a report to summarize the results of the random-intercept model.

15

BAYESIAN GENERALIZED LINEAR MODELS

OBJECTIVES OF THIS CHAPTER

This chapter introduces Bayesian generalized linear models. It starts with a review of generalized linear models. Then several Bayesian generalized linear models, including logistic regression models for binary response variables, ordinal logistic regression models for ordinal response variables, multinomial logistic regression models for nominal response variables, and Poisson regression and negative binomial regression models for count response variables, are illustrated using R with step-by-step instructions. The `rstanarm` and `brms` packages are used for Bayesian modeling for categorical response variables. R commands are explained, and the output is interpreted for each model in detail. This chapter also illustrates how the results are displayed in publication-quality tables using R and reported in the text. It focuses on model fitting with R, as well as on interpreting and presenting the results. After reading this chapter, you should be able to:

- Determine when Bayesian modeling for categorical response variables is used.
- Conduct the Bayesian generalized linear modeling for categorical response variables using R.
- Evaluate MCMC convergence.
- Interpret the results.
- Present the results in publication-quality tables using R.
- Write the results for publication.

15.1 BAYESIAN GENERALIZED LINEAR MODELS

15.1.1 Bayesian Inference: A Brief Introduction

So far, most models discussed in previous chapters belong to classical statistics or frequentist statistics. In the frequentist framework, we assume that data for a certain study are collected by random sampling, so the data are a random sample of the population. When we specify a statistical model, we assume that the parameters of interest are unknown but fixed. For example, in simple logistic regression, we first specify a logistic regression model and then estimate the logit coefficient for a predictor variable. The coefficient is an unknown but fixed parameter which can be estimated with the maximum likelihood estimation (MLE) approach. Once the model is fitted, we obtain the coefficient of the predictor variable, which is the point estimate of the parameter, its standard error, the Wald z statistic, and the associated p value. We then test if the coefficient of the predictor variable is zero with the Wald z statistic. If the associated p value is less than .05, then we reject the null hypothesis and claim that the predictor variable is a significant predictor of the dependent variable.

However, Bayesian statisticians have a different perspective. In Bayesian statistics, the parameters of interest are unknown, random, and have their own distributions. In other words, we are interested in the distribution of the parameters rather than one fixed parameter. Bayesian statistics focus on the uncertainty of the unknown parameters in terms of probability with prior beliefs. The three major elements in Bayesian inference include the prior distributions of the parameters, the likelihood function, and the posterior distributions of the parameters. The prior distributions of the parameters, $p(\theta)$, are the prior beliefs or the prior knowledge on the parameter distributions, which are based on the existing knowledge or the results of previous studies. The likelihood function can be expressed as the form of the probability distribution of the data given the parameters, $p(y|\theta)$. The posterior distributions of the parameters given the data, $p(\theta|y)$, are based on both the prior distributions of the parameter and the likelihood function, and are proportional to the multiplication of these two terms. In other words, the posterior distributions of the parameters given the data take account of the prior knowledge on the parameters and evidence from the data. According to the Bayes' rule, the posterior distribution of a parameter given the data is expressed in the form below.

$$p(\theta|y) \propto p(y|\theta)p(\theta)$$

where $\propto$ denotes the proportionality.

Different from MLE in frequentist statistics, Bayesian statistics approximate the posterior distribution with simulated approaches. We normally use Markov chain Monte Carlo (MCMC) approaches to repeatedly draw samples from the posterior distribution. MCMC uses algorithms such as Metropolis-Hastings sampling, Gibbs sampling, Hamilton Monte Carlo (HMC), or the No-U-Turn Sampler (NUTS) to simulate draws. Technical details of MCMC are beyond the scope of this text. For introductory

Bayesian statistics with examples, see Kruschke (2015) and McElreath (2020). In addition, consult Carlin and Louis (2009), Gelman et al. (2013), Gill (2015), Kaplan (2014), and Wakefield (2013) for more comprehensive coverage of Bayesian data analysis.

15.1.2 Specifying Priors

Specifying the prior distributions of the parameters is not an easy task and has been an obstacle for novices. To fit a regression model, we need to specify the prior distributions of the regression coefficients for the intercept and the predictor variable(s). The prior information can be based on existing knowledge or results of previous studies, and is referred to as an informative prior. For example, we can specify a prior distribution of a math test based on the scores from a previous study. When the prior distribution and the posterior distribution of a parameter come from the same family, the two distributions are conjugate, and the prior is referred to as a conjugate prior. For example, a normal prior distribution is conjugate with a normal distribution, a beta distribution is conjugate with a binomial distribution, and a Dirichlet distribution is conjugate with a multinomial distribution. Using conjugate priors is computationally convenient since the form for the posterior distribution can be solved analytically. For example, when a beta distribution is a conjugate prior for a binomial likelihood function, according to the Bayes' rule, we can easily derive the posterior distribution by multiplying the beta prior and the likelihood. The resulting posterior distribution is also a beta distribution. Sometimes, if we have little existing knowledge on the parameters, then we may specify so-called noninformative priors, which may have no or little influence on the posterior distribution. For example, a uniform prior, which has a uniform distribution, is a simple noninformative prior since all values in the range of the distribution have equal prior weight on a parameter. Another type of prior is the weakly informative prior (Gelman et al., 2008), which provides some information on the distribution of a parameter so it is helpful for model fitting. For example, we may specify a weakly informative prior for a coefficient in logistic regression with a mean of 0 and a standard deviation of 2.5 for the normal distribution, $N(0, 2.5^2)$ rather than a noninformative prior with an extremely large standard deviation of 10,000, $N(0, 10{,}000^2)$. The former is a more reasonably estimated prior for the parameter value than the latter. Both the `rstanarm` (Goodrich et al., 2020) and the `brms` (Bürkner, 2017) packages use weakly informative priors as the default priors with the option for specifying user-defined priors.

15.1.3 Number of Iterations and Warm-Up

Since MCMC uses simulated approaches to draw correlated samples from the posterior distribution, we need to draw a large number of samples so the chain does not depend on its initial state in order to form independent draws, and MCMC algorithms can converge to the stable posterior distribution or the target distribution. The warm-up iterations are the samples which are discarded before the target distribution is reached. The warm-up process is also called the burn-in period. The number of independent draws from the posterior distribution is the effective sample size. When the number of independent draws is small, we can increase the total number of iterations and use thinning to choose every *n*th draw. Ideally we would like to have more chains with a large number of iterations for each, but MCMC sampling is computationally intensive and time consuming, so it may take hours or more to fit a complex model. In either the `rstanarm` or `brms` package, the default number of Markov chains is four with 2,000 iterations and 1,000 warm-up draws for each chain.

15.1.4 Evaluating MCMC Convergence

Before we interpret the parameter estimates, we need to check convergence of MCMC. If MCMC does not converge, then the inference from the MCMC sample is invalid. There is no single magic number to evaluate MCMC convergence. The MC standard error, the Rhat statistic, and the effective sample size can be used for MCMC diagnostics. The MC standard error represents the variability of the parameter estimates associated with repeated MCMC estimations. It is the ratio of the posterior standard deviation to the square root of the effective sample size. We would like to see a small MC standard error. The Rhat statistic is the potential scale reduction factor, which compares the between-chain variation to the within-chain variation. We would like to see that Rhat is close to 1 but less than 1.1, which suggests MCMC convergence. The effective sample size is the number of independent samples from the posterior distribution. Because MCMC samples are correlated, we would like to see a large number of independent draws from the total number of iterations.

Graphic tools such as trace plots, histograms of the posterior density, and autocorrelation plots can also be used to evaluate MCMC convergence. The trace plots show the parameter estimates for the predictor variables at each iteration for the chains after warm-up draws are discarded. When the iterations have stable paths of movement and there are no substantial fluctuations, MCMC has reached convergence. The histograms of the posterior density show the distributions of the posterior density. The autocorrelation plots show the autocorrelations between iterations. We would like to see low autocorrelations over the number of iterations, which suggests that we have a large number of independent samples from the posterior distribution.

15.1.5 Point Estimates and Credible Intervals

To summarize the results of the posterior distributions of the parameters, we use the point estimates such as means, medians, or modes as the central tendency measures and uncertainty intervals such as credible intervals and highest posterior density intervals.

The posterior means are the means of the posterior distributions of the parameters or the point estimates of the parameters. They are analogous to the regression coefficients in frequentist statistics. The standard errors of the posterior means are the standard deviations of the posterior distributions. They are analogous to the standard errors of regression coefficients in the frequentist approach.

Analogous to the confidence interval in frequentist statistics, the posterior distribution of a parameter has an uncertainty interval, which displays the range of the uncertainty of the parameter. This interval is important since Bayesian statistics is all about uncertainty of parameters in terms of probability distribution. There are two types of uncertainty interval. One is the credible interval which is based on the quantiles of the posterior distribution. A credible interval has a direct probability interpretation. For example, a 95% credible interval corresponds to the 2.5% and 97.5% quantiles, respectively. It indicates that there is 95% probability that the parameter lies in that interval. The other interval is the highest posterior density (HPD) interval, which is based on the posterior density. A 95% HPD interval indicates that there is 95% probability that the parameter lies in that interval with a higher posterior density than anywhere outside the interval. One advantage of the HPD is that the posterior density does not need to be symmetric.

In Bayesian statistics, there is no significance test of the regression coefficient, as in frequentist statistics with the *p*-value hypothesis test. We do not say that the posterior mean for a predictor variable is statistically significant since we are interested in uncertainty in the posterior distributions of the parameter. Alternatively, Bayesian factors are used for Bayesian hypothesis testing for model comparison and selection. The Bayesian factor is the ratio of the likelihood of two competing models or the ratio of the posterior odds to the prior odds. Its value can be used to decide whether we accept or reject the Bayesian null hypothesis or to decide which model is preferred.

Analogous to the AIC and BIC in frequentist statistics, the widely applicable information criterion (WAIC) or Watanabe-Akaike information criterion (Watanabe, 2010), the leave-one-out (LOO) cross validation (Vehtari et al., 2017), and the deviance information criterion (DIC) can be used to compare models in Bayesian statistics. Different packages in R may provide different measures of these information criteria and LOO. Although they are computationally intensive, the WAIC and LOO are available in both the `rstanarm` and `brms` packages. Technical details of these measures are beyond the scope of this text.

15.1.6 A Review of Generalized Linear Models

As introduced in Chapter 3, generalized linear models extend the linear regression model when the response variable follows a distribution from the exponential family. The three major components of generalized linear models include the random component, the systematic component, and the link function.

First, a probability distribution of the response variable Y is the random component of the model.

Second, the systematic component is a linear combination of the predictor variables, which is the linear predictor with the form $\alpha + \beta_1 X_1 + \beta_2 X_2 + \cdots + \beta_p X_p$. If we define the linear predictor as η (*eta*), the systematic component can be expressed as the form: $\eta = \alpha + \beta_1 X_1 + \beta_2 X_2 + \cdots + \beta_p X_p$.

Third, the link function connects the linear predictor and the expected value $E(Y)$ or the conditional mean μ (mu) of the response variable. The link function is written as $g(\mu)$, which links the linear predictor η and the conditional mean μ of the response

TABLE 15.1 ● Outcome Variables, Distributions, Link Functions, Function Names, and the Corresponding Models

Outcome Variable	Distribution	Link Function	Link Function Name	Model
Continuous	Gaussian	μ	identity	Linear regression
Binary	Bernoulli or Binomial	$\ln(\mu/1-\mu)$	logit	Logistic regression
Ordinal	Multinomial	$\ln\left(\frac{P(Y \le j)}{P(Y > j)}\right)$	logit	Ordinal logistic regression (proportional odds model)
Nominal	Multinomial	$\ln\left(\frac{P(Y = j)}{P(Y = J)}\right)$	logit	Multinomial logistic regression
Count	Poisson	$\ln(\mu)$	natural log	Poisson regression
Count	Negative binomial	$\ln(\mu)$	natural log	Negative binomial regression (NB-2)

variable. For example, the logit link, $g(\mu) = \ln(\mu/1-\mu)$, is used for logistic regression models, and the natural log link, $g(\mu) = \ln(\mu)$, is used for Poisson regression and negative binomial regression models. Table 15.1 provides the types of outcome variables, the probability distributions, the link functions with names, and the corresponding models which are covered in this chapter.

15.2 BAYESIAN LOGISTIC REGRESSION WITH R

In this section, we introduce how to fit Bayesian logistic regression models for binary outcome variables with R. We use the same GSS 2016 data as those in the chapter for the classical logistic regression models with the frequentist approach.

15.2.1 Description of the Data and Sample

The data for the following analyses were the General Social Survey 2016 (GSS 2016). The following are the variables:

- `healthy`: the recoded variable of health (health status) with 1 = good health and 0 = not good health
- `wrkfull`: working full time with 1 = full time and 0 = not full time

- `maritals`: the recoded variable of marital (marital status) with 1 = currently married and 0 = not currently married
- `female`: recoded variable of sex with 1 = female and 0 = male
- `educ`: the highest education.

15.2.2 Packages and Functions for Bayesian Logistic Regression With R

The `stan_glm()` function in the `rstanarm` package (Goodrich et al., 2020) and the `brm()` function in the `brms` package (Bürkner, 2017) are used for fitting Bayesian logistic regression models for binary outcome variables.

15.2.3 The `stan_glm()` Function in the `rstanarm` Package

The `stan_glm()` function can be used to fit various generalized linear models via Stan. Since `rstanarm` is a user-written package, you need to install it first by typing `install.packages("rstanarm")` and then load the package by typing `library(rstanarm)`. If you have installed the package, you just need to load it.

We can specify the three major arguments including the model formula, the family, and the data. First, the model formula syntax of `stan_glm()` is similar to that of `glm()` introduced in the previous chapters. We can specify the dependent variable and the predictor variable(s), which are separated by the tilde (~). Second, we also need to specify the probability distribution of the outcome variable with the `family=` argument. For example, the `family = binomial(link = "logit")` argument is used for a binary outcome variable in logistic regression with the logit link and `family = poisson` is specified for a count outcome variable in Poisson regression. Third, we can specify the data with the `data=` argument. In addition to these three arguments, there are other important arguments such as the `algorithm` argument for the estimation approach and the `prior` argument for setting up the prior distribution. For more details on how to use this function, type `help(stan_glm)` in the command prompt after loading the `rstanarm` package.

We first use the `glm()` function to fit the logistic regression model introduced in Chapter 3 (model `LR.2`). Next, the `summary()` function is used to display the output of the model.

```
> # Logistic regression with glm()
> LR.2 <- glm(healthy ~ wrkfull + maritals + female + educ, data = chp3.lr, family =
binomial(link = "logit"))
> summary(LR.2)


Call:
glm(formula = healthy ~ wrkfull + maritals + female + educ, family = binomial(link =
"logit"),
    data = chp3.lr)
```

```
Deviance Residuals:
    Min         1Q     Median       3Q       Max
-2.3201    -1.1877    0.6439    0.8652    1.7620

Coefficients:
               Estimate   Std. Error   z value    Pr(>|z|)
(Intercept)   -1.82016      0.26294    -6.922    4.45e-12 ***
wrkfull        0.65505      0.11112     5.895    3.74e-09 ***
maritals       0.36738      0.10943     3.357    0.000787 ***
female         0.18217      0.10807     1.686    0.091868 .
educ           0.16184      0.01881     8.605     < 2e-16 ***
---
Signif. codes:  0 '***' 0.001 '**' 0.01 '*' 0.05 '.' 0.1 ' ' 1

(Dispersion parameter for binomial family taken to be 1)

    Null deviance: 2258.9  on 1872  degrees of freedom
Residual deviance: 2103.4  on 1868  degrees of freedom
AIC: 2113.4

Number of Fisher Scoring iterations: 4
```

We then use the Bayesian approach to fit the same model with the `stan_glm()` function. The fitted model is named `LR.2.stan`. The following output is displayed by the `summary(LR.2.stan)` command.

```
> library(rstanarm)
> LR.2.stan <- stan_glm(healthy ~ wrkfull + maritals + female + educ, data = chp3.lr,
family = binomial(link = "logit"))

SAMPLING FOR MODEL 'bernoulli' NOW (CHAIN 1).
Chain 1:
Chain 1: Gradient evaluation took 0.021 seconds
Chain 1: 1000 transitions using 10 leapfrog steps per transition would take 210
seconds.
Chain 1: Adjust your expectations accordingly!
Chain 1:
Chain 1:
Chain 1: Iteration:    1 / 2000 [  0%]  (Warmup)
Chain 1: Iteration:  200 / 2000 [ 10%]  (Warmup)
Chain 1: Iteration:  400 / 2000 [ 20%]  (Warmup)
Chain 1: Iteration:  600 / 2000 [ 30%]  (Warmup)
Chain 1: Iteration:  800 / 2000 [ 40%]  (Warmup)
Chain 1: Iteration: 1000 / 2000 [ 50%]  (Warmup)
Chain 1: Iteration: 1001 / 2000 [ 50%]  (Sampling)
Chain 1: Iteration: 1200 / 2000 [ 60%]  (Sampling)
Chain 1: Iteration: 1400 / 2000 [ 70%]  (Sampling)
Chain 1: Iteration: 1600 / 2000 [ 80%]  (Sampling)
Chain 1: Iteration: 1800 / 2000 [ 90%]  (Sampling)
Chain 1: Iteration: 2000 / 2000 [100%]  (Sampling)
Chain 1:
Chain 1:  Elapsed Time: 2.531 seconds (Warm-up)
Chain 1:                2.943 seconds (Sampling)
Chain 1:                5.474 seconds (Total)
Chain 1:
```

```
SAMPLING FOR MODEL 'bernoulli' NOW (CHAIN 2).
Chain 2:
Chain 2: Gradient evaluation took 0 seconds
Chain 2: 1000 transitions using 10 leapfrog steps per transition would take 0
seconds.
Chain 2: Adjust your expectations accordingly!
Chain 2:
Chain 2:
Chain 2: Iteration:    1 / 2000 [  0%]  (Warmup)
Chain 2: Iteration:  200 / 2000 [ 10%]  (Warmup)
Chain 2: Iteration:  400 / 2000 [ 20%]  (Warmup)
Chain 2: Iteration:  600 / 2000 [ 30%]  (Warmup)
Chain 2: Iteration:  800 / 2000 [ 40%]  (Warmup)
Chain 2: Iteration: 1000 / 2000 [ 50%]  (Warmup)
Chain 2: Iteration: 1001 / 2000 [ 50%]  (Sampling)
Chain 2: Iteration: 1200 / 2000 [ 60%]  (Sampling)
Chain 2: Iteration: 1400 / 2000 [ 70%]  (Sampling)
Chain 2: Iteration: 1600 / 2000 [ 80%]  (Sampling)
Chain 2: Iteration: 1800 / 2000 [ 90%]  (Sampling)
Chain 2: Iteration: 2000 / 2000 [100%]  (Sampling)
Chain 2:
Chain 2:  Elapsed Time: 2.557 seconds (Warm-up)
Chain 2:                2.659 seconds (Sampling)
Chain 2:                5.216 seconds (Total)
Chain 2:

SAMPLING FOR MODEL 'bernoulli' NOW (CHAIN 3).
Chain 3:
Chain 3: Gradient evaluation took 0 seconds
Chain 3: 1000 transitions using 10 leapfrog steps per transition would take 0
seconds.
Chain 3: Adjust your expectations accordingly!
Chain 3:
Chain 3:
Chain 3: Iteration:    1 / 2000 [  0%]  (Warmup)
Chain 3: Iteration:  200 / 2000 [ 10%]  (Warmup)
Chain 3: Iteration:  400 / 2000 [ 20%]  (Warmup)
Chain 3: Iteration:  600 / 2000 [ 30%]  (Warmup)
Chain 3: Iteration:  800 / 2000 [ 40%]  (Warmup)
Chain 3: Iteration: 1000 / 2000 [ 50%]  (Warmup)
Chain 3: Iteration: 1001 / 2000 [ 50%]  (Sampling)
Chain 3: Iteration: 1200 / 2000 [ 60%]  (Sampling)
Chain 3: Iteration: 1400 / 2000 [ 70%]  (Sampling)
Chain 3: Iteration: 1600 / 2000 [ 80%]  (Sampling)
Chain 3: Iteration: 1800 / 2000 [ 90%]  (Sampling)
Chain 3: Iteration: 2000 / 2000 [100%]  (Sampling)
Chain 3:
Chain 3:  Elapsed Time: 2.394 seconds (Warm-up)
Chain 3:                2.84 seconds (Sampling)
Chain 3:                5.234 seconds (Total)
Chain 3:

SAMPLING FOR MODEL 'bernoulli' NOW (CHAIN 4).
Chain 4:
Chain 4: Gradient evaluation took 0 seconds
Chain 4: 1000 transitions using 10 leapfrog steps per transition would take 0
seconds.
Chain 4: Adjust your expectations accordingly!
Chain 4:
Chain 4:
```

```
Chain 4: Iteration:    1 / 2000 [  0%]  (Warmup)
Chain 4: Iteration:  200 / 2000 [ 10%]  (Warmup)
Chain 4: Iteration:  400 / 2000 [ 20%]  (Warmup)
Chain 4: Iteration:  600 / 2000 [ 30%]  (Warmup)
Chain 4: Iteration:  800 / 2000 [ 40%]  (Warmup)
Chain 4: Iteration: 1000 / 2000 [ 50%]  (Warmup)
Chain 4: Iteration: 1001 / 2000 [ 50%]  (Sampling)
Chain 4: Iteration: 1200 / 2000 [ 60%]  (Sampling)
Chain 4: Iteration: 1400 / 2000 [ 70%]  (Sampling)
Chain 4: Iteration: 1600 / 2000 [ 80%]  (Sampling)
Chain 4: Iteration: 1800 / 2000 [ 90%]  (Sampling)
Chain 4: Iteration: 2000 / 2000 [100%]  (Sampling)
Chain 4:
Chain 4:  Elapsed Time: 2.411 seconds (Warm-up)
Chain 4:                2.706 seconds (Sampling)
Chain 4:                5.117 seconds (Total)
Chain 4:
> summary(LR.2.stan)

Model Info:
 function:     stan_glm
 family:       binomial [logit]
 formula:      healthy ~ wrkfull + maritals + female + educ
 algorithm:    sampling
 sample:       4000 (posterior sample size)
 priors:       see help('prior_summary')
 observations: 1873
 predictors:   5

Estimates:
              mean   sd   10%   50%   90%
(Intercept) -1.8    0.3 -2.2  -1.8  -1.5
wrkfull      0.7    0.1  0.5   0.7   0.8
maritals     0.4    0.1  0.2   0.4   0.5
female       0.2    0.1  0.0   0.2   0.3
educ         0.2    0.0  0.1   0.2   0.2

Fit Diagnostics:
           mean   sd   10%   50%   90%
mean_PPD 0.7    0.0  0.7   0.7   0.7

The mean_ppd is the sample average posterior predictive distribution of the outcome
variable (for details see help('summary.stanreg')).

MCMC diagnostics
              mcse  Rhat  n_eff
(Intercept)   0.0   1.0   5285
wrkfull       0.0   1.0   4997
maritals      0.0   1.0   5018
female        0.0   1.0   5176
educ          0.0   1.0   4835
mean_PPD      0.0   1.0   4513
log-posterior 0.0   1.0   1850

For each parameter, mcse is Monte Carlo standard error, n_eff is a crude measure of
effective sample size, and Rhat is the potential scale reduction factor on split
chains (at convergence Rhat=1).
```

The beginning of the output shows the log for the MCMC estimation. The default number of Markov chains is four, with 2,000 iterations and 1,000 warm-up draws for each chain. These warm-up draws are the initially discarded iterations. The posterior sample size is 4,000, which equals the difference between the number of iterations and warm-up multiplied by the number of chains: (2,000 − 1,000)*4 = 4,000.

The `summary()` function displays the summary of the fitted model including the model information, the estimates for the intercept and the predictor variables, the fit diagnostics, and the MCMC diagnostics.

The default number of digits for displaying decimal places of numbers is one and the default probabilities for specifying the quantiles of the posterior distribution are 0.1, 0.5, and 0.9 with the `prob = c(0.1, 0.5, 0.9)` argument. To request three decimal places for the numeric values and the 95% credible intervals for the estimates, we use the `summary(LR.2.stan, digits = 3, prob = c(.025, .5, .975))` command. The resulting output is shown below.

```
> summary(LR.2.stan, digits = 3, prob = c(.025, .5, .975))

Model Info:
 function:      stan_glm
 family:        binomial [logit]
 formula:       healthy ~ wrkfull + maritals + female + educ
 algorithm:     sampling
 sample:        4000 (posterior sample size)
 priors:        see help('prior_summary')
 observations:  1873
 predictors:    5

Estimates:
               mean     sd      2.5%     50%      97.5%
(Intercept)  -1.824   0.262   -2.344   -1.824   -1.314
wrkfull       0.659   0.110    0.444    0.661    0.871
maritals      0.368   0.110    0.153    0.370    0.584
female        0.183   0.111   -0.033    0.182    0.403
educ          0.162   0.019    0.126    0.162    0.200

Fit Diagnostics:
           mean   sd     2.5%  50%    97.5%
mean_PPD  0.709  0.014  0.681  0.709  0.736

The mean_ppd is the sample average posterior predictive distribution of the outcome
variable (for details see help('summary.stanreg')).

MCMC diagnostics
                mcse   Rhat   n_eff
(Intercept)     0.004  1.000   5285
wrkfull         0.002  1.000   4997
maritals        0.002  0.999   5018
female          0.002  1.000   5176
educ            0.000  1.000   4835
mean_PPD        0.000  1.001   4513
log-posterior   0.037  1.000   1850

For each parameter, mcse is Monte Carlo standard error, n_eff is a crude measure of
effective sample size, and Rhat is the potential scale reduction factor on split
chains (at convergence Rhat=1).
> prior_summary(LR.2.stan)
Priors for model 'LR.2.stan'
```

We can also use the `prior_summary()` function to check the priors.

```
> prior_summary(LR.2.stan)
Priors for model 'LR.2.stan'
------
Intercept (after predictors centered)
 ~ normal(location = 0, scale = 10)


Coefficients
  Specified prior:
    ~ normal(location = [0,0,0,...], scale = [2.5,2.5,2.5,...])
  Adjusted prior:
    ~ normal(location = [0,0,0,...], scale = [2.50,2.50,2.50,...])
------
See help('prior_summary.stanreg') for more details
```

The prior distribution for the intercept is set to be a normal distribution with a mean of 0 and a standard deviation of 10. In addition, the prior distributions for the coefficients of the predictor variables are also normal with means of 0 and standard deviations of 2.5. These are the weakly informative priors which provide prior information for regularizing inferences.

MCMC Diagnostics

Before we interpret the parameter estimates, we need to check convergence of MCMC. If MCMC does not converge, then the inference from the MCMC sample is invalid. There is no single magic number to evaluate MCMC convergence. The MCMC diagnostic section in the output provides the MC standard errors (`mcse`), the Rhat statistics (`Rhat`), and the effective sample sizes (`n_eff`). Recall that the MC standard error represents the variability of the parameter estimates associated with repeated MCMC estimations, which is the ratio of the posterior standard deviation to the square root of the number of effective sample size. Rhat is the potential scale reduction factor, which compares the between-chain variation to the within-chain variation. We would like to see that Rhat is close to 1 but less than 1.1. The effective sample size is the number of independent samples from the posterior distribution. In the output, the MC standard errors are close to zero, the Rhat statistics are 1 or close to 1, and the effective sample sizes are large enough. These values indicate the convergence of MCMC.

Graphic tools such as trace plots, autocorrelation plots, and histograms of the posterior density can also be used to evaluate MCMC convergence. We use the `plot()` function to create trace plots for the intercept and the four predictor variables as follows.

```
> plot(LR.2.stan, "trace")
```

FIGURE 15.1 ● Trace Plots for the Intercept and the Predictor Variables

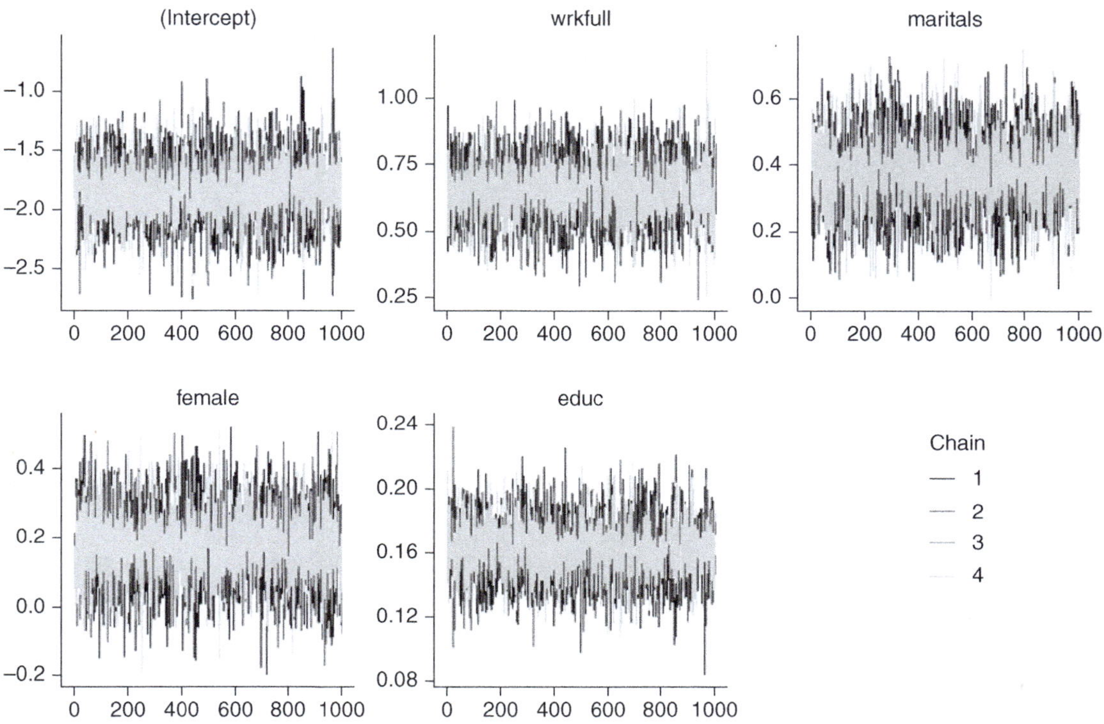

The trace plots (Figure 15.1) show the parameter estimates for the predictor variables at each iteration for all four chains after warm-up draws are discarded. All five trace plots show good convergence since the iterations have stable paths of movement and they all center around the posterior means.

We can also use the `launch_shinystan()` function in the `shinystan` package (Gabry, 2018) to explore more diagnostic tools. We need to install it first by typing `install.packages("shinystan")` and then load the package by typing `library(shinystan)`. In addition, you can use another useful tool, the `bayesplot` package, to plot parameter estimates and MCMC diagnostics. This add-on package also needs to be installed.

Interpreting the Results

The parameter estimates section (labeled `Estimates`) displays the parameter estimates for the intercept and the four predictor variables. This section is analogous to the coefficients table in the output of the frequentist logistic regression model with the `glm()` function. The column labeled `mean` displays the posterior mean values, which are the means of the posterior distributions of the parameters or the point estimates of the parameters. They are analogous to the regression coefficients in frequentist statistics. The column labeled `sd` displays the standard deviations of the posterior distributions, which are the standard errors of the posterior means, or the standard errors of the point estimates of the parameters. They are analogous to the standard errors of the regression coefficients in frequentist approach. The last three columns (labeled `2.5%`, `50%`, and `97.5%`, respectively) display the quantiles of the posterior distributions. They are the lower 95% credible interval, the median, and the upper 95% credible interval, respectively.

Interpretation of the posterior means is the same as that of the logit coefficients in logistic regression models with a frequentist approach. For example, the posterior mean for `wrkfull` is .659, indicating the log odds of having good health for the people who work full time are .659 times the log odds of having good health for those who do not work full time.

The corresponding 95% credible interval is [.444, .871], indicating that there is a 95% probability that the true parameter lies in the interval. The posterior means and the corresponding credible intervals for the other predictor variables in the model can be interpreted in a similar way.

We use the `coef()` function to extract the coefficients and use the `se()` function to extract the standard errors of the coefficients. Please note that the extracted coefficients are the posterior medians rather than posterior means, although they are almost identical here. Correspondingly, the extracted standard errors are the standard errors for the posterior medians.

```
> coef <- coef(LR.2.stan)
> coef
(Intercept)      wrkfull     maritals      female        educ
 -1.8243249    0.6608091    0.3699910   0.1817335   0.1620288
> se <- se(LR.2.stan)
> se
(Intercept)       wrkfull      maritals       female         educ
 0.26201315    0.11132204    0.11061987   0.11147501   0.01842461
> cbind(coef,se)
                  coef          se
(Intercept)   -1.8243249   0.26201315
wrkfull        0.6608091   0.11132204
maritals       0.3699910   0.11061987
female         0.1817335   0.11147501
educ           0.1620288   0.01842461
```

We can use the `posterior_interval()` function to compute the credible interval. The default credible interval is the 90% credible interval since it is more stable than the 95% credible interval. For the 90% credible interval, the columns are labeled `5%` and `95%`, respectively.

```
> posterior_interval(LR.2.stan)
                      5%           95%
(Intercept)    -2.26064240    -1.3949920
wrkfull         0.47890401     0.8386653
maritals        0.18490254     0.5491935
female         -0.00137025     0.3670291
educ            0.13221892     0.1935243
```

To request the 95% credible interval for the posterior means or the coefficients, we use the `posterior_interval(LR.2.stan, prob = 0.95)` command. In the syntax, we specify the `prob = 0.95` argument, so the resulting columns are labeled `2.5%` and `97.5%`, respectively.

```
> posterior_interval(LR.2.stan, prob = 0.95)
                    2.5%          97.5%
(Intercept)    -2.34366781    -1.3140985
wrkfull         0.44423957     0.8707925
maritals        0.15323457     0.5842631
female         -0.03315781     0.4033471
educ            0.12559879     0.1995854
```

We can use the `cbind()` function to combine the coefficients and the 95% credible intervals.

```
> ci <- posterior_interval(LR.2.stan, prob = 0.95)
> cbind(coef,ci)
                   coef           2.5%          97.5%
(Intercept)    -1.8243249    -2.34366781    -1.3140985
wrkfull         0.6608091     0.44423957     0.8707925
maritals        0.3699910     0.15323457     0.5842631
female          0.1817335    -0.03315781     0.4033471
educ            0.1620288     0.12559879     0.1995854
```

With the `exp(cbind(coef,ci))` command, we can get the odds ratios and the 95% credible intervals.

```
> exp(cbind(coef,ci))
                  coef          2.5%         97.5%
(Intercept)    0.1613265    0.09597497    0.2687165
wrkfull        1.9363585    1.55930400    2.3888032
maritals       1.4477216    1.16559837    1.7936688
female         1.1992946    0.96738588    1.4968264
educ           1.1758941    1.13382717    1.2208965
```

Interpretation of the odds ratios in Bayesian logistic regression is similar to that in logistic regression in Chapter 3. For example, for `wrkfull`, odds ratio = 1.936. The odds of having good health for the people who work full-time are 1.936 times the odds of having good health for those who do not work full-time.

15.2.4 The `brm()` Function in the `brms` Package: Syntax for Bayesian Logistic Regression Models

In the last section, we used the `stan_glm()` function in the `rstanarm` package to fit the Bayesian logistic regression model. In this section, we introduce the `brm()` function in the `brms` package to fit the same models.

The `brms` package (Bürkner, 2017) is another popular package for fitting a variety of Bayesian generalized linear models and multilevel models using the programming language Stan. You need to install the package first if it is not installed by typing `install.packages("brms")` and then load it with `library(brms)`. Since the package has been installed in the last chapter, we only need to load it by typing `library(brms)`. To make `brms` work, you also need to download and install Rtools from https://cran.r-project.org/bin/windows/Rtools/. Please note that Rtools is not an add-on package, you need to install it just like other programs on your computer.

Similar to the syntax for the `stan_glm()` function in the `rstanarm` package, the syntax for `brm()` also contains three major components, including the model formula, the `family` argument, and the `data` argument. It also includes the arguments to specify the number of iterations, warm-up, chains, and cores, which are explained below.

We specify the same model formula in `brm()` as that in `stan_glm()`. In addition, we specify the number of iterations with `iter=` , the number of warm-up or burn-in iterations with `warmup=` , the number of Markov chains with `chain=` , and the number of cores for executing the chains. In the following example, we specify 3,000 iterations with 500 warm-up iterations and two chains. We specify the random seed with `seed=` so the results can be replicable. Please note that we may still get slightly different results due to MCMC sampling if we run the same syntax. The fitted model is named `LR.2.brm`. The following output is displayed by the `summary(LR.2.brm)` command. For more details on how to use this function, type `help(brm)` in the command prompt after loading the `brms` package.

```
> library(brms)
> LR.2.brm <- brm(healthy ~ wrkfull + maritals + female + educ, data = chp3.lr, family =
bernoulli(link = "logit"),
+                          warmup = 500, iter = 3000,
+                          cores = 2, chains = 2, seed=123)
Compiling Stan program...
Start sampling
> summary(LR.2.brm)
```

```
 Family: bernoulli
  Links: mu = logit
Formula: healthy ~ wrkfull + maritals + female + educ
   Data: chp3.lr (Number of observations: 1873)
Samples: 2 chains, each with iter = 3000; warmup = 500; thin = 1;
         total post-warmup samples = 5000

Population-Level Effects:
             Estimate    Est.Error     l-95% CI     u-95% CI    Rhat    Bulk_ESS     Tail_ESS
Intercept       -1.82         0.26        -2.35        -1.32    1.00        6032         4295
wrkfull          0.66         0.11         0.44         0.88    1.00        5639         3745
maritals         0.37         0.11         0.15         0.59    1.00        6350         3797
female           0.18         0.11        -0.03         0.39    1.00        6159         4231
educ             0.16         0.02         0.13         0.20    1.00        5477         4269

Samples were drawn using sampling(NUTS). For each parameter, Bulk_ESS
and Tail_ESS are effective sample size measures, and Rhat is the potential
scale reduction factor on split chains (at convergence, Rhat = 1).
```

To request three decimal places for the numeric values, we use the `print()` function with the `digits = 3` argument.

```
> print(LR.2.brm, digits = 3)

 Family: bernoulli
  Links: mu = logit
Formula: healthy ~ wrkfull + maritals + female + educ
   Data: chp3.lr (Number of observations: 1873)
Samples: 2 chains, each with iter = 3000; warmup = 500; thin = 1;
         total post-warmup samples = 5000

Population-Level Effects:
             Estimate    Est.Error     l-95% CI     u-95% CI     Rhat    Bulk_ESS     Tail_ESS
Intercept      -1.825        0.263       -2.348       -1.318    1.000        6032         4295
wrkfull         0.656        0.111        0.442        0.879    1.001        5639         3745
maritals        0.367        0.113        0.146        0.588    1.001        6350         3797
female          0.182        0.108       -0.034        0.392    1.000        6159         4231
educ            0.162        0.019        0.126        0.199    1.000        5477         4269

Samples were drawn using sampling(NUTS). For each parameter, Bulk_ESS
and Tail_ESS are effective sample size measures, and Rhat is the potential
scale reduction factor on split chains (at convergence, Rhat = 1).
```

The estimates section (labeled `Population-Level Effects:`) provides the posterior means, the estimate errors, the 95% credible intervals, the `Rhat` values for convergence, and the effective sample sizes for the intercept and the predictor variables. The first column labeled `Estimate` displays the posterior mean values, which are analogous to the coefficients in classical generalized linear models. The second column labeled `Est.Error` displays the estimate errors, which are the standard deviations of the posterior distribution and are analogous to the standard errors of the coefficients in frequentist statistics. The next two columns labeled `l-95% CI` and `u-95% CI` list the 95% credible intervals, which are the 2.5% and 97.5% quantiles. For example, the posterior mean for `wrkfull` is .656, which is similar to those produced by the `stan_glm()` function and the `glm()` function. The corresponding 95% credible interval is [.442, .879], which is also close to the credible interval produced by the `stan_glm()` function.

The last three columns list the `Rhat` values and the effective sample sizes (labeled `Bulk_ESS` and `Tail_ESS`), which are used to evaluate MCMC convergence. An `Rhat` value close to 1 but less than 1.1 and a large effective sample size indicate good MCMC convergence. The `Rhat` values are equal to or close to 1 and the effective sample size is larger than 3,700 in the output, which indicate good MCMC convergence. The graphical evaluation of the convergence of MCMC with the `plot()` function is introduced later in this section.

We extract the results of the estimates and the 95% credible intervals with the `coef <- fixef(LR.2.brm, summary = TRUE)` command.

```
> coef <- fixef(LR.2.brm, summary = TRUE)
> coef
             Estimate   Est.Error        Q2.5      Q97.5
Intercept  -1.8249508  0.26262384 -2.34783070 -1.3177734
wrkfull     0.6561403  0.11099502  0.44164573  0.8785531
maritals    0.3669892  0.11262435  0.14613527  0.5883135
female      0.1816765  0.10840090 -0.03403228  0.3920829
educ        0.1623058  0.01894869  0.12606514  0.1990811
```

We use the `exp(fixef(LR.2.brm)[, -2])` command to obtain the odds ratios and the corresponding 95% credible intervals. The estimated errors should not be exponentiated, so we exclude the second column in the fixed effects table with the indexing `[, -2]`. The following output is created.

```
> exp(fixef(LR.2.brm)[, -2])
            Estimate       Q2.5     Q97.5
Intercept  0.1612256 0.09557627 0.2677308
wrkfull    1.9273391 1.55526466 2.4074138
maritals   1.4433824 1.15735273 1.8009485
female     1.1992262 0.96654030 1.4800605
educ       1.1762198 1.13435606 1.2202809
> save(LR.2.brm, file = "LR2brm.rda")
```

To graphically evaluate MCMC convergence, we use the `plot()` function to create the posterior density plots and the trace plots (Figure 15.2) for the intercept and the four predictor variables as follows.

```
> plot(LR.2.brm)
```

FIGURE 15.2 ● Posterior Density Plots and Trace Plots for the Intercept and the Predictor Variables

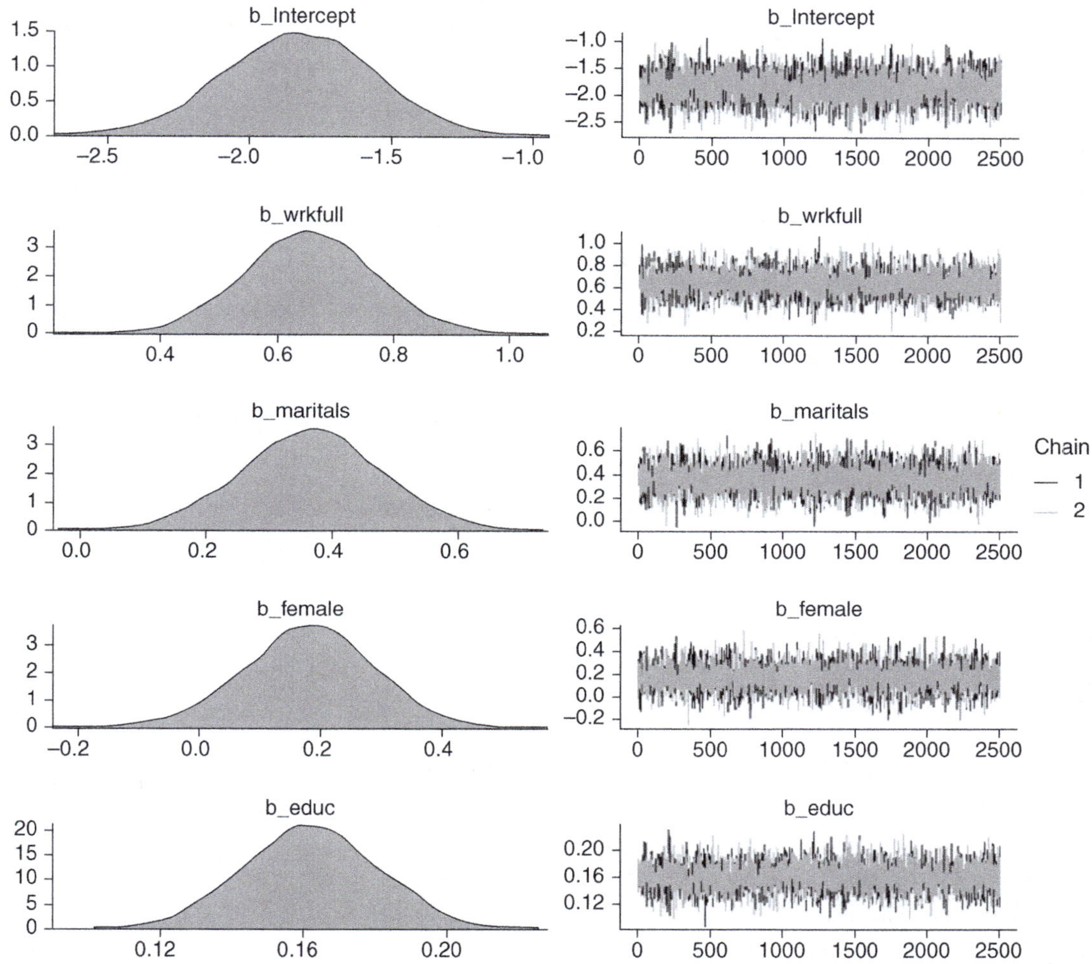

Similar to the convergence diagnostics in the preceding section, the posterior distribution plots show the posterior distributions center over the posterior means and the trace plots show smooth paths over iterations. Both plots show good convergence for the intercept and the four predictor variables.

We can also use the `stanplot()` function with the `type=` argument in the `brms` package to create other types of plot, such as the plots for Rhat (`type = "rhat"`), MC standard errors (`type = "mcse"`), effective sample sizes (`type = "ess"`), and autocorrelation (`type = "ac"`). In addition, you can use the `launch_shinystan()` function in the `shinystan` package (Gabry, 2018) to explore more diagnostic tools. We need to install it first by typing `install.packages("shinystan")` and then load the package by typing `library(shinystan)`.

15.3 BAYESIAN ORDINAL AND MULTINOMIAL LOGISTIC REGRESSION WITH R

Bayesian ordinal logistic regression allows us to estimate ordinal response variables when outcome variables are ordered with more than two categories. When an outcome variable is unordered with more than two categories, a Bayesian multinomial regression model can be fitted. When the outcome variable has only two categories, both models are equivalent to the binary logistic regression model. In this section, we briefly introduce how to fit Bayesian ordinal and multinomial logistic regression with R. The same GSS 2016 data are used for the following analyses.

15.3.1 Packages and Functions for Bayesian Ordinal Logistic Regression With R

The `stan_polr()` function in the `rstanarm` package and the `brm()` function in the `brms` package are used for fitting Bayesian ordinal logistic regression models with ordinal outcome variables. Since Bayesian ordinal logistic regression models are more complex than Bayesian logistic regression models, they are more time-consuming to fit. A good practice is to save the model object with the `save()` function after we fit a model so we can reuse it when needed with the `load()` function without rerunning the analysis.

15.3.2 The `stan_glm()` Function in the `rstanarm` Package

The `stan_polr()` function in the `rstanarm` package (Goodrich et al., 2020) is used for fitting Bayesian ordinal logistic regression models. It is the Bayesian counterpart of the `polr()` function in the `MASS` package. Similar to the `polr()` function, we can specify the three major arguments including the model formula, the method, and the data in the `stan_polr()` function. First, the model formula syntax of `stan_polr()` is similar to that of `polr()`. Second, we need to specify the link function with the `method =` argument. You can specify "`logistic`", "`probit`", or other link functions. Third, we can specify the data with the `data=` argument. In addition to these three arguments, there are other important arguments such as the `algorithm` argument for the estimation approach and the `prior` argument for setting up the prior distribution. We need to specify the prior location of R^2, which is the proportion of variance in the outcome variable explained by the predictor variables. This prior can be estimated from the results of the `polr()` function. Here we specify an R^2 of 0.2 with `prior = R2(0.2, "mean")` where the mean is specified for the beta distribution. For more details on how to use this function, type `help(stan_polr)` in the command prompt after loading the `rstanarm` package.

In this example, the ordinal response variable `healthre` is predicted from the three predictor variables `maritals`, `educ`, and `female`. The fitted model is named `olr.stan`. The following output is displayed by the `summary(olr.stan, digits = 3, prob = c(.025, .5, .975))` command.

```
> olr.stan <- stan_polr(healthre ~ maritals + educ + female, data = chp4.po, method =
"logistic",
+                  prior = R2(0.2, "mean"), seed = 123)

> summary(olr.stan, digits = 3, prob = c(.025, .5, .975))

Model Info:
 function:        stan_polr
 family:          ordered [logistic]
 formula:         healthre ~ maritals + educ + female
 algorithm:       sampling
 sample:          4000 (posterior sample size)
 priors:          see help('prior_summary')
 observations:    1873

Estimates:
            mean     sd      2.5%     50%      97.5%
maritals    0.286    0.088   0.115    0.288    0.456
educ        0.171    0.015   0.142    0.171    0.201
female      0.065    0.089  -0.116    0.066    0.239
1|2        -0.308    0.223  -0.741   -0.304    0.122
2|3         1.578    0.216   1.157    1.578    2.002
3|4         3.864    0.233   3.413    3.860    4.323

Fit Diagnostics:
              mean     sd      2.5%     50%      97.5%
mean_PPD:1    0.064    0.008   0.049    0.064    0.080
mean_PPD:2    0.229    0.014   0.202    0.229    0.257
mean_PPD:3    0.487    0.016   0.455    0.486    0.519
mean_PPD:4    0.220    0.013   0.195    0.221    0.247

The mean_ppd is the sample average posterior predictive distribution of the outcome
variable (for details see help('summary.stanreg')).

MCMC diagnostics
               mcse     Rhat     n_eff
maritals       0.001    0.999    4488
educ           0.000    1.002    2609
female         0.002    1.000    2964
1|2            0.005    1.001    2310
2|3            0.004    1.001    2563
3|4            0.004    1.001    3014
mean_PPD:1     0.000    1.000    3676
mean_PPD:2     0.000    0.999    3694
mean_PPD:3     0.000    1.000    3843
mean_PPD:4     0.000    1.000    3563
log-posterior  0.070    1.004    1004

For each parameter, mcse is Monte Carlo standard error, n_eff is a crude measure of
effective sample size, and Rhat is the potential scale reduction factor on split
chains (at convergence Rhat=1).
```

MCMC Diagnostics

The MCMC diagnostic section in the output provides the MC standard errors (`mcse`), the Rhat statistics (`Rhat`), and the number of effective sample size (`n_eff`). In the output, the MC standard errors are close to zero, the Rhat statistics are 1 or close to 1, and the effective sample sizes are large enough. Therefore, these indicators suggest good MCMC convergence. If MCMC does not converge, then we may increase the iterations and the number of chains to refit the model. As with Bayesian logistic regression, we use the `plot()` function to graphically evaluate MCMC convergence. The plots are omitted here.

Interpreting the Results

The parameter estimates section (labeled `Estimates`) displays the parameter estimates for the three predictor variables and the cutpoints or thresholds. This section is analogous to the coefficients table in the output of the frequentist proportional odds model with the `plor()` function. The column labeled `mean` displays the posterior mean values, which are the means of the posterior distributions of the parameters or the point estimates of the parameters. The column labeled `sd` displays the standard deviations of the posterior distributions, which are the standard errors of the posterior means, or the standard errors of the point estimates of the parameters. They are analogous to the standard errors of regression coefficients in the frequentist approach. The last three columns (labeled `2.5%`, `50%`, and `97.5%`, respectively) display the quantiles of the posterior distributions. They are the lower 95% credible interval, the median, and the upper 95% credible interval, respectively.

Interpretation of the posterior means is the same as that of the logit coefficients in proportional odds models with a frequentist approach. For example, the posterior means for `maritals` is .286, indicating the logit or log odds of being above a category of health status relative to being at or below that category for the married is .286 times as high as that for the unmarried.

The corresponding 95% credible interval is [.115, .456], indicating that there is a 95% probability that the true parameter lies in the interval. The posterior means and the corresponding credible intervals for the other predictor variables in the model can in interpreted in a similar way.

The fit diagnostics section provides the `mean_PPD` for each category of the ordinal response variable, which is the sample average posterior predictive distribution of the outcome variable. The column labeled `mean` displays the mean values, which are the means of each category in the posterior predictive distribution, similar to the percentage of each category in the frequency distribution of the ordinal outcome variable with a total percentage of 1.

We can use the `coef()` function to extract the coefficients and use the `exp()` function to obtain the odds ratios being above a category of health status relative to being at or below that category for the predictor variables. The output is omitted here.

15.3.3 The `brm()` Function in the `brms` Package: Syntax for Bayesian Ordinal Logistic Regression Models

To fit a Bayesian ordinal logistic regression model with `brm()`, we specify the model formula, the `family = cumulative("logit")` argument, and the `data` argument. We also specify the number of iterations, warm-up, chains, and cores. Here, we specify 3,000 iterations with 500 warm-up iterations and two chains. The fitted model is named `olr.brm`. The following output is displayed by the `print (olr.brm, digits = 3)` command. For more details on how to use this function, type `help(brm)` in the command prompt after loading the `brms` package.

```
> library(brms)
> olr.brm <- brm(healthre ~ maritals + educ + female, data = chp4.po,
+               family=cumulative("logit"),
+               warmup = 500, iter = 3000,
+               cores = 2, chains = 2, seed=123)
> print(olr.brm, digits = 3)

 Family: cumulative
  Links: mu = logit; disc = identity
Formula: healthre ~ maritals + educ + female
   Data: chp4.po (Number of observations: 1873)
Samples: 2 chains, each with iter = 3000; warmup = 500; thin = 1;
         total post-warmup samples = 5000

Population-Level Effects:
                Estimate  Est.Error  l-95% CI  u-95% CI   Rhat  Bulk_ESS  Tail_ESS
Intercept[1]      -0.262      0.226    -0.697     0.181  1.000      5084      3632
Intercept[2]       1.628      0.218     1.208     2.052  1.000      5360      3523
Intercept[3]       3.917      0.236     3.458     4.371  1.000      5005      3311
maritals           0.292      0.087     0.122     0.462  1.000      4666      3315
educ               0.174      0.015     0.144     0.204  1.001      5132      3696
female             0.069      0.091    -0.106     0.248  1.001      4255      3559

Family Specific Parameters:
       Estimate  Est.Error  l-95% CI  u-95% CI   Rhat  Bulk_ESS  Tail_ESS
disc      1.000      0.000     1.000     1.000  1.000      5000      5000

Samples were drawn using sampling(NUTS). For each parameter, Bulk_ESS
and Tail_ESS are effective sample size measures, and Rhat is the potential
scale reduction factor on split chains (at convergence, Rhat = 1).
```

The estimates section (labeled `Population-Level Effects:`) provides the posterior means, the estimate errors, the 95% credible intervals, the `Rhat` values for convergence, and the effective sample sizes for the three intercepts or cutpoints and the predictor variables. For example, the posterior mean for `maritals` is .292, which is similar to that produced by the `stan_polr()` function. The corresponding 95% credible interval is [.122, .462], which is also close to the credible interval produced by the `stan_polr()` function. The `Rhat` values are equal to or close to 1 and the effective sample sizes are larger than 3,000. These values indicate good MCMC convergence.

We can also use the `plot()` function or the `stanplot()` function to create the posterior density plots and the trace plots for the intercepts and the predictor variables. The graphs are omitted here.

15.3.4 Packages and Functions for Bayesian Multinomial Logistic Regression With R

As of the time of this writing, the `rstanarm` package has not supported Bayesian models for nominal response variables, so only the `brm()` function in the `brms` package is used for fitting Bayesian multinomial logistic regression models for nominal outcome variables.

15.3.5 The `brm()` Function in the `brms` Package: Syntax for Bayesian Multinomial Logistic Regression Models

To fit a Bayesian multinomial logistic regression model with `brm()`, we specify the model formula, the `family = categorical` argument, and the `data` argument. The link function (`link = "logit"`) is omitted since it is the default for the `categorical` family. We also specify the number of iterations, warm-up, chains, and cores. Here, we again specify 3,000 iterations with 500 warm-up iterations for two chains. The fitted model is named `mulm.brm`. The following output is displayed by the `print(mulm.brm, digits = 3)` command.

```
> library(brms)
> mulm.brm <- brm(healthre ~ educ + maritals + female + wrkfull, data = gss16,
+                 family=categorical,
+                 warmup = 500, iter = 3000,
+                 cores = 2, chains = 2, seed=123)
Compiling Stan program...
Start sampling
> print(mulm.brm, digits = 3)

 Family: categorical
  Links: mu2 = logit; mu3 = logit; mu4 = logit
Formula: healthre ~ educ + maritals + female + wrkfull
   Data: gss16 (Number of observations: 1873)
Samples: 2 chains, each with iter = 3000; warmup = 500; thin = 1;
         total post-warmup samples = 5000

Population-Level Effects:
               Estimate  Est.Error  l-95% CI  u-95% CI   Rhat  Bulk_ESS  Tail_ESS
mu2_Intercept     0.046      0.452    -0.850     0.947  1.001      3789      2958
mu3_Intercept    -1.192      0.447    -2.051    -0.308  1.000      3631      3512
mu4_Intercept    -3.641      0.509    -4.641    -2.611  1.001      3871      3882
mu2_educ          0.083      0.034     0.015     0.148  1.001      3447      3426
mu2_maritals      0.379      0.233    -0.067     0.834  1.000      3046      3417
mu2_female       -0.030      0.215    -0.449     0.407  1.000      3637      3057
mu2_wrkfull       0.326      0.232    -0.134     0.782  1.001      4059      3273
mu3_educ          0.194      0.033     0.129     0.258  1.000      3375      3277
mu3_maritals      0.685      0.222     0.257     1.113  1.000      3122      3360
mu3_female        0.165      0.206    -0.243     0.571  1.000      3629      3185
mu3_wrkfull       0.959      0.220     0.519     1.408  1.000      4027      3319
mu4_educ          0.316      0.037     0.241     0.386  1.000      3673      3356
mu4_maritals      0.642      0.237     0.191     1.099  1.001      3068      3328
mu4_female        0.151      0.221    -0.282     0.578  1.000      3636      3208
mu4_wrkfull       0.822      0.237     0.352     1.295  1.001      4026      3434

Samples were drawn using sampling(NUTS). For each parameter, Bulk_ESS
and Tail_ESS are effective sample size measures, and Rhat is the potential
scale reduction factor on split chains (at convergence, Rhat = 1).
```

The estimates section (labeled `Population-Level Effects:`) provides the posterior means for the intercepts and predictor variables for the three binary models. It also provides the estimate errors, the 95% credible intervals, the `Rhat` values for convergence, and the effective sample sizes. The labels mu2, mu3, and mu4 are the names for the three binary models. Only three binary models are estimated since the base outcome is category 1. For example, the posterior means of educ for the three models are .083, .194, and .316, respectively, which are almost identical to those estimated by the vglm() function in the VGAM package for the multinomial logistic regression model. The corresponding 95% credible intervals for the three posterior means are [.015, .148], [.129, .258], and [.241, .386], respectively. The Rhat values are equal to or close to 1 and the effective sample sizes are large, which indicate good MCMC convergence. We can also use the plot() function to graphically evaluate the convergence of MCMC. The graphs are omitted here.

15.4 BAYESIAN POISSON REGRESSION WITH R

15.4.1 Description of the Data and Sample

The GSS 2016 data are used for the following analyses. The following are the variables used for data analysis in this chapter:

- vistzoo: the recoded variable of the number of zoo visits in a year
- maritals: the recoded variable of marital (marital status) with 1 = currently married and 0 = not currently married
- educ: the highest education completed
- female: recoded variable of sex with 1 = female and 0 = male
- wrkfull: working full time or not.

15.4.2 The Poisson Regression Model With the glm() Function

Before fitting a Bayesian model, we use the glm() function to fit a multiple Poisson regression with a frequentist approach. In the glm() function, the model equation is specified as vistzoo ~ educ + maritals + female + wrkfull, the family = poisson argument specifies the Poisson family, and the data argument specifies data = count. The fitted model is named PR.2. The output is shown by the summary(PR.2) function.

```
> # Multiple-predictor Poisson regression model with glm()
> PR.2 <- glm(vistzoo ~ educ + maritals + female + wrkfull, family = poisson,
data = count)
> summary(PR.2)

Call:
glm(formula = vistzoo ~ educ + maritals + female + wrkfull, family = poisson,
  data = count)

Deviance Residuals:
    Min        1Q     Median       3Q      Max
-1.7880   -1.1659   -0.9280   0.4835   4.4634
```

```
Coefficients:
              Estimate  Std. Error  z value   Pr(>|z|)
(Intercept)   -1.34993     0.20084   -6.721   1.80e-11 ***
educ           0.05086     0.01343    3.787   0.000152 ***
maritals       0.21382     0.07494    2.853   0.004328  **
female         0.04837     0.07535    0.642   0.520896
wrkfull        0.53962     0.07792    6.925   4.35e-12 ***
---
Signif. codes:  0 '***' 0.001 '**' 0.01 '*' 0.05 '.' 0.1 ' ' 1

(Dispersion parameter for poisson family taken to be 1)

    Null deviance: 1364.6  on 901  degrees of freedom
Residual deviance: 1275.8  on 897  degrees of freedom
AIC: 2286.8

Number of Fisher Scoring iterations: 6
```

15.4.3 Packages and Functions for Bayesian Poisson Regression With R

The `stan_glm()` function in the `rstanarm` package and the `brm()` function in the `brms` package are used for fitting Bayesian Poisson regression models for count outcome variables.

15.4.4 The `stan_glm()` Function in the `rstanarm` Package

The `stan_glm()` function in the `rstanarm` package (Goodrich et al., 2020) is used for fitting Bayesian Poisson regression models within the generalized linear models framework via Stan. In the syntax, we need to specify the probability distribution of a count outcome variable with the `family = poisson` argument.

In the example below, we fit the Bayesian Poisson regression model with the `stan_glm()` function. In the function, the model equation is specified as `vistzoo ~ educ + maritals + female + wrkfull`. The `family = poisson` argument specifies that the Poisson family is used to fit the model. The data argument specifies `data = count`. The fitted model is named `LR.2.stan` The output is shown by the `summary(LR.2.stan, digits = 3, prob = c(.025, .5, .975))` command.

```
> library(rstanarm)
> PR.2.stan <- stan_glm(vistzoo ~ educ + maritals + female + wrkfull, family = poisson,
data=count)
> summary(PR.2.stan, digits = 3, prob = c(.025, .5, .975))


Model Info:
 function:        stan_glm
 family:          poisson [log]
 formula:         vistzoo ~ educ + maritals + female + wrkfull
 algorithm:       sampling
 sample:          4000 (posterior sample size)
 priors:          see help('prior_summary')
 observations:    902
 predictors:      5
```

```
Estimates:
                mean     sd      2.5%     50%     97.5%
(Intercept)    -1.353   0.202   -1.753   -1.350   -0.957
educ            0.051   0.014    0.024    0.051    0.078
maritals        0.215   0.076    0.064    0.216    0.362
female          0.049   0.076   -0.100    0.050    0.199
wrkfull         0.540   0.080    0.380    0.540    0.698

Fit Diagnostics:
            mean    sd      2.5%    50%     97.5%
mean_PPD   0.810   0.043   0.727   0.809   0.896

The mean_ppd is the sample average posterior predictive distribution of the outcome
variable (for details see help('summary.stanreg')).

MCMC diagnostics
                mcse    Rhat    n_eff
(Intercept)     0.003   1.001   4913
educ            0.000   1.001   5090
maritals        0.001   1.000   4425
female          0.001   1.001   4816
wrkfull         0.001   1.000   4794
mean_PPD        0.001   1.001   4458
log-posterior   0.038   1.001   1846

For each parameter, mcse is Monte Carlo standard error, n_eff is a crude measure of
effective sample size, and Rhat is the potential scale reduction factor on split chains (at
convergence Rhat=1).
```

The MCMC estimation log is omitted in the output. We still use the four default Markov chains with 2,000 iterations and 1,000 warm-up draws for each chain. The posterior sample size is 4,000, which equals the difference between the number of iterations and warm-up multiplied by the number of chains: (2,000 − 1,000)*4 = 4,000.

The `summary()` function displays the summary of the fitted model including the model information, the estimates for the intercept and the predictor variables, the fit diagnostics, and the MCMC diagnostics.

We use the `prior_summary()` function to check the priors.

```
> prior_summary(PR.2.stan)
Priors for model 'PR.2.stan'
------
Intercept (after predictors centered)
~ normal(location = 0, scale = 10)


Coefficients
  Specified prior:
    ~ normal(location = [0,0,0,...], scale = [2.5,2.5,2.5,...])
  Adjusted prior:
    ~ normal(location = [0,0,0,...], scale = [0.87,2.50,2.50,...])
------
See help('prior_summary.stanreg') for more details
```

The prior distribution for the intercept is set to be normal with a mean of 0 and a standard deviation of 10. In addition, the prior distributions for the coefficients of the predictor variables are originally set to be normal with means of 0 and standard deviations of 2.5 and but are later adjusted for estimation.

MCMC Diagnostics

The MCMC diagnostic section in the output provides the MC standard errors (`mcse`), the Rhat statistics (`Rhat`), and the number of effective sample sizes (`n_eff`). In the output, the MC standard errors are close to zero, the Rhat statistics are 1 or close to 1 and the effective sample sizes are large enough. These three indices suggest that MCMC has converged.

Graphic tools such as trace plots, autocorrelation plots, and histograms of the posterior density can also be used to evaluate MCMC convergence. We use the `plot()` function to create trace plots for the intercept and the four predictor variables as follows (see Figure 15.3).

```
> plot(PR.2.stan, "trace")
```

FIGURE 15.3 ● Trace Plots for the Intercept and the Predictor Variables

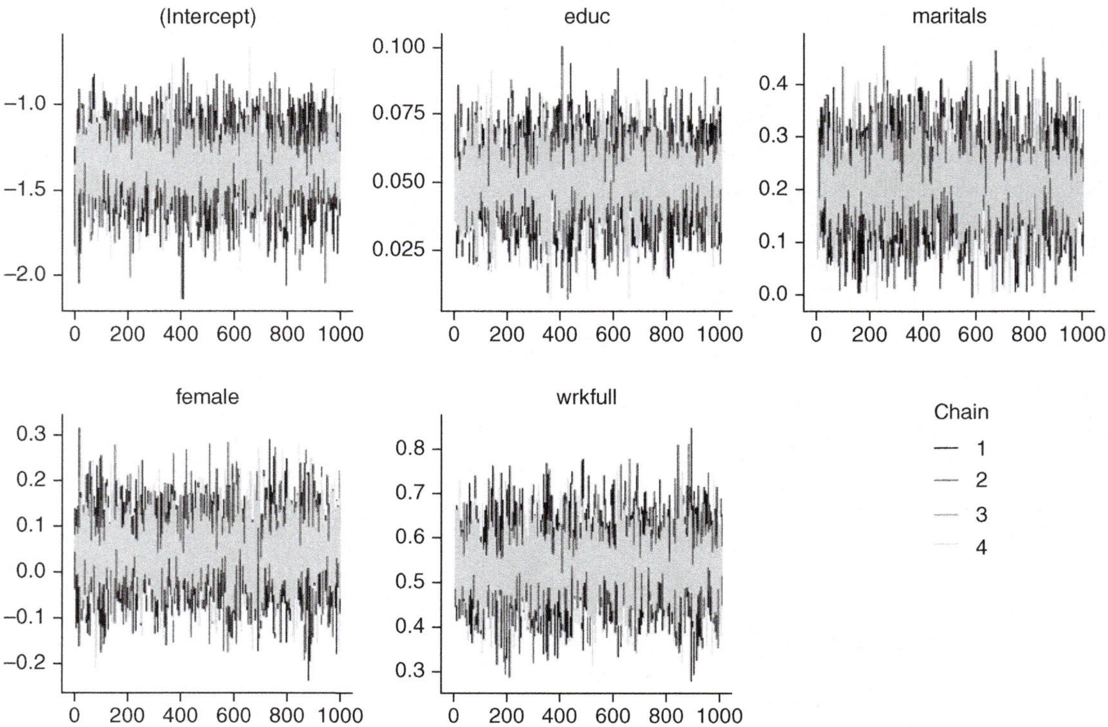

The trace plots show the parameter estimates for the intercepts and the four predictor variables at each iteration for all four chains after warm-up draws are discarded. All five trace plots show good convergence since the iterations have stable paths of movement and they all center around the posterior means.

We can also use the `launch_shinystan()` function in the `shinystan` package to explore more diagnostic tools. In addition, you can use another useful tool, the `bayesplot` package, to plot parameter estimates and MCMC diagnostics. These add-on packages need to be installed.

Interpreting the Results

The parameter estimates section (labeled `Estimates`) displays the parameter estimates for the intercept and the four predictor variables. This section is analogous to the coefficients table in the output of the frequentist Poisson regression model with the `glm()` function. The column labeled `mean` displays the posterior mean values, which are the means of the posterior distributions of the parameters or the point estimates of the parameters. They are analogous to the regression coefficients in frequentist statistics. The column labeled `sd` displays the standard deviations of the posterior distributions, which are the standard errors of the posterior means, or the standard errors of the point estimates of the parameters. They are analogous to the standard errors of the regression coefficients in the frequentist approach. The last three columns (labeled `2.5%`, `50%`, and `97.5%`, respectively) display the quantiles of the posterior distributions. They are the lower 95% credible interval, the median, and the upper 95% credible interval, respectively.

Interpretation of the posterior means is similar to that of the log coefficients in Poisson regression models with a frequentist approach. For example, the posterior means for `educ` is .051, indicating the log expected number of visits to a zoo increases by .051 for a one-unit increase in education when holding all other predictors constant.

The corresponding 95% credible interval is [.024, .078], indicating that there is a 95% probability that the true parameter lies in the interval. The posterior means and the corresponding credible intervals for the other predictor variables in the model can be interpreted in a similar way.

We use the `coef()` function to extract the coefficients and use the `se()` function to extract the standard errors of the coefficients. The extracted coefficients are the posterior medians. Correspondingly, the extracted standard errors are the standard errors for the posterior medians. The output is omitted here.

We can use the `posterior_interval()` function to compute the credible interval. The default credible interval is the 90% credible interval and the columns are labeled `5%` and `95%`, respectively.

```
> posterior_interval(PR.2.stan)
                     5%          95%
(Intercept)   -1.69558893  -1.03034111
educ           0.02847363   0.07290907
maritals       0.08889707   0.33675331
female        -0.07473204   0.17366945
wrkfull        0.41002587   0.67316088
```

We can use the `cbind()` function to combine the coefficients and the 90% credible intervals.

```
> ci<-posterior_interval(PR.2.stan)
> cbind(coef,ci)
                    coef           5%          95%
(Intercept)  -1.34967292  -1.69558893  -1.03034111
educ          0.05063225   0.02847363   0.07290907
maritals      0.21643925   0.08889707   0.33675331
female        0.05019143  -0.07473204   0.17366945
wrkfull       0.54047238   0.41002587   0.67316088
```

To request the 95% credible interval of the estimates, we use the `posterior_interval(PR.2.stan, prob = 0.95)` command. The output is omitted here.

We can compute the incidence rate ratios (IRRs) by exponentiating the log coefficients with `exp(cbind(coef,ci))`. The following are the IRRs with 90% credible intervals.

```
> exp(cbind(coef,ci))
                 coef         5%        95%
(Intercept) 0.2593251  0.1834911  0.3568852
educ        1.0519360  1.0288829  1.0756327
maritals    1.2416477  1.0929682  1.4003936
female      1.0514724  0.9279921  1.1896623
wrkfull     1.7168177  1.5068568  1.9604242
```

The incidence rate ratio for a predictor can be interpreted as the change in the incidence rate or the expected number of events for a one-unit increase in the predictor variable when holding other predictors constant. For example, for `educ`, the incidence rate ratio is 1.052, which indicates that the incidence rate or the expected number of visits to a zoo increases by a factor of 1.052 for a one-unit increase in education when holding all other predictors constant.

15.4.5 The `brm()` Function in the `brms` Package: Syntax for Bayesian Poisson Regression Models

To fit a Bayesian Poisson logistic regression model with `brm()`, we specify the model formula, the `family = poisson` argument, and the `data` argument. The link function `(link = "log")` is omitted since it is the default for the `poisson` family. We also specify the number of iterations, warm-up, chains, and cores. Here, we again specify 3,000 iterations with 500 warm-up iterations for two chains. The fitted model is named `PR.2.brm`. The following output is displayed by the `print(PR.2.brm, digits = 3)` command.

```
> library(brms)
> PR.2.brm <- brm(vistzoo ~ educ + maritals + female + wrkfull, family = poisson, data=count,
+                 warmup = 500, iter = 3000,
+                 cores = 2, chains = 2, seed=123)
Compiling Stan program...
Start sampling
> print(PR.2.brm, digits = 3)

 Family: poisson
  Links: mu = log
Formula: vistzoo ~ educ + maritals + female + wrkfull
   Data: count (Number of observations: 902)
Samples: 2 chains, each with iter = 3000; warmup = 500; thin = 1;
         total post-warmup samples = 5000

Population-Level Effects:
             Estimate  Est.Error  l-95% CI  u-95% CI   Rhat  Bulk_ESS  Tail_ESS
Intercept      -1.349      0.204    -1.759    -0.955  1.001      6518      3758
educ            0.050      0.013     0.024     0.078  1.000      6468      3996
maritals        0.213      0.073     0.070     0.357  1.000      6340      3609
female          0.048      0.076    -0.099     0.199  1.000      5372      3146
wrkfull         0.541      0.076     0.394     0.689  1.000      6061      4245

Samples were drawn using sampling(NUTS). For each parameter, Bulk_ESS
and Tail_ESS are effective sample size measures, and Rhat is the potential
scale reduction factor on split chains (at convergence, Rhat = 1).
```

The estimates section (labeled `Population-Level Effects:`) provides the posterior means (labeled `Estimate`), the estimate errors, the 95% credible intervals, the `Rhat` values for convergence, and the effective sample sizes for the intercept and the predictor variables. The results are similar to those produced by the `stan_glm()` function. The `Rhat` values are equal to or close to 1 and the effective sample sizes are very large, which indicate good MCMC convergence. The graphical evaluation of the convergence of MCMC with the `plot()` function is introduced later in this section.

We can extract the results of the estimates and the credible intervals with the `coef <- fixef(PR.2.brm, summary = TRUE)` command and use the `exp (fixef(PR.2.brm)[, -2])` command to obtain the incidence rate ratios and the 95% corresponding credible intervals. The estimated errors should not be exponentiated, so we can exclude the second column in the fixed effects table with the indexing `[, -2]`. The output is omitted.

Interpretation of the incidence rate ratios for the predictor variables is illustrated in the preceding section, so it is omitted here.

We save the model object with the `save()` function so we can reuse it when needed with the `load()` function.

```
> save(PR.2.brm, file = "PR2brm.rda")
```

To graphically evaluate MCMC convergence, we can use the `plot()` function to create the posterior density plots and the trace plots for the intercepts and the predictor.

We can also use the `stanplot()` function with the `type=` argument in the `brms` package to create additional plots for MCMC diagnostics. In addition, you can use the `launch_shinystan()` function in the `shinystan` package to explore more diagnostic tools.

We can use the `conditional_effects()` function to display the marginal effects of the predictor variables on the predicted number of visits to a zoo. The graphs are omitted.

15.5 BAYESIAN NEGATIVE BINOMIAL REGRESSION WITH R

15.5.1 Description of the Data and Sample

The GSS 2016 data are used for the following analyses. The same variables from the preceding section for the Bayesian Poisson regression models are used for data analysis in this section.

15.5.2 The Negative Binomial Regression Model With the `glm.nb()` Function

We first use the `glm.nb()` function in the `MASS` package to fit the negative binomial regression model with the frequentist approach. In the following example, the `glm.nb(vistzoo ~ educ + maritals + female + wrkfull, data = count)` command tells R to predict the count response variable `vistzoo` from the four independent variables. In the `glm.nb()` function, the model equation is specified as `vistzoo ~ educ + maritals + female + wrkfull`. The data argument specifies `data = count`. The `family` argument is not needed. The fitted model is named `nbr`. The `summary(nbr)` function displays the output of the fitted model.

```
> # Negative binomial regression model with glm.nb() in MASS
> library(MASS)
> nbr <- glm.nb(vistzoo ~ educ + maritals + female + wrkfull, data=count)
> summary(nbr)

Call:
glm.nb(formula = vistzoo ~ educ + maritals + female + wrkfull,
    data = count, init.theta = 1.58331383, link = log)

Deviance Residuals:
    Min        1Q     Median       3Q      Max
-1.4702   -1.0592   -0.8788   0.3811   3.3187

Coefficients:
              Estimate  Std. Error  z value   Pr(>|z|)
(Intercept)   -1.30862     0.24565   -5.327   9.98e-08 ***
educ           0.04982     0.01665    2.992    0.00277  **
maritals       0.19126     0.09337    2.048    0.04052   *
female         0.02412     0.09386    0.257    0.79719
wrkfull        0.53499     0.09525    5.617   1.94e-08 ***
---
Signif. codes:  0 '***' 0.001 '**' 0.01 '*' 0.05 '.' 0.1 ' ' 1
```

```
(Dispersion parameter for Negative Binomial(1.5833) family taken to be 1)

    Null deviance: 931.01  on 901  degrees of freedom
Residual deviance: 873.58  on 897  degrees of freedom
AIC: 2200

Number of Fisher Scoring iterations: 1


              Theta:  1.583
          Std. Err.:  0.251

 2 x log-likelihood:  -2188.013
```

The dispersion parameter labeled `Theta`, its standard error, and the 2 log likelihood value are shown at the bottom. In the NB2 model, Variance (Y) $= \mu(1 + \alpha\mu) = \mu + \alpha\mu^2$. The `glm.nb()` function uses a slightly different form, Variance (Y) $= \mu + \mu^2/\theta$. As can be seen, $\theta = 1/\alpha$. `Theta` = 1.583 and its standard error is .251. We can easily compute $\alpha = 1/1.583 = .632$.

```
> alpha <- 1/1.583
> alpha
[1] 0.6317119
```

15.5.3 Packages and Functions for Bayesian Negative Binomial Regression With R

We then use the `stan_glm()` function in the `rstanarm` package and the `brm()` function in the `brms` package to fit Bayesian negative binomial regression models for count outcome variables. The `stan_glm.nb()` function in the `rstanarm` package can also be used to fit negative binomial regression models. Since it is equivalent to the `stan_glm()` function with the `family = neg_binomial_2` argument, illustration of this function is omitted here.

15.5.4 The `stan_glm()` Function in the `rstanarm` Package

The `stan_glm()` function in the `rstanarm` package (Goodrich et al., 2020) is used for fitting Bayesian negative binomial regression models. To fit a Bayesian negative binomial regression model, we need to specify the probability distribution of a count outcome variable with the `family = neg_binomial_2` argument.

In the function of the example, the model equation is specified as `vistzoo ~ educ + maritals + female + wrkfull`. The `family = neg_binomial_2` argument specifies that the negative binomial family is used to fit the model. The data argument specifies `data = count`. The fitted model is named `nbr.stan` The output is shown by the `summary(nbr.stan, digits = 3, prob = c(.025, .5, .975))` command.

```
> library(rstanarm)
> nbr.stan <- stan_glm(vistzoo ~ educ + maritals + female + wrkfull, family =
neg_binomial_2, data=count)

> summary(nbr.stan, digits = 3, prob = c(.025, .5, .975))

Model Info:
 function:       stan_glm
 family:         neg_binomial_2 [log]
 formula:        vistzoo ~ educ + maritals + female + wrkfull
 algorithm:      sampling
 sample:         4000 (posterior sample size)
 priors:         see help('prior_summary')
 observations:   902
 predictors:     5

Estimates:
                          mean     sd      2.5%     50%      97.5%
(Intercept)              -1.310   0.247   -1.797   -1.308   -0.829
educ                      0.050   0.017    0.016    0.050    0.084
maritals                  0.191   0.091    0.010    0.190    0.366
female                    0.026   0.093   -0.155    0.026    0.205
wrkfull                   0.535   0.097    0.342    0.536    0.726
reciprocal_dispersion     1.578   0.256    1.148    1.554    2.152

Fit Diagnostics:
           mean    sd     2.5%   50%    97.5%
mean_PPD  0.811  0.053  0.706  0.810  0.915

The mean_ppd is the sample average posterior predictive distribution of the outcome
variable (for details see help('summary.stanreg')).

MCMC diagnostics
                         mcse    Rhat    n_eff
(Intercept)              0.003   1.000   5452
educ                     0.000   1.001   5560
maritals                 0.001   1.001   4926
female                   0.001   1.000   6125
wrkfull                  0.001   1.001   4710
reciprocal_dispersion    0.004   1.000   4902
mean_PPD                 0.001   1.000   4705
log-posterior            0.045   1.002   1606

For each parameter, mcse is Monte Carlo standard error, n_eff is a crude measure of
effective sample size, and Rhat is the potential scale reduction factor on split
chains (at convergence Rhat=1)
```

The dispersion parameter labeled `reciprocal_dispersion` is provided in the table for the estimates. The posterior mean for the dispersion parameter is 1.578. We can easily compute $\alpha = 1/1.578 = .633$, which is almost identical to the dispersion parameter produced by the `glm.nb()` function with the frequentist approach. Please note that the dispersion parameter in your output will be slightly different from mine, so when computing α, you need to adjust it accordingly.

```
> alpha <- 1/1.578
> alpha
[1] 0.6337136
```

We use the `prior_summary()` function to check the priors.

```
Priors for model 'nbr.stan'
------
Intercept (after predictors centered)
  ~ normal(location = 0, scale = 10)

Coefficients
  Specified prior:
    ~ normal(location = [0,0,0,...], scale = [2.5,2.5,2.5,...])
  Adjusted prior:
    ~ normal(location = [0,0,0,...], scale = [0.87,2.50,2.50,...])

Auxiliary (reciprocal_dispersion)
  ~ exponential(rate = 1)
------
See help('prior_summary.stanreg') for more details
```

The prior distributions are the same as those in the Bayesian Poisson regression model in the preceding section.

MCMC Diagnostics

The MCMC diagnostic section in the output provides the MC standard errors (`mcse`), the Rhat statistics (`Rhat`), and the effective sample sizes (`n_eff`). In the output, the MC standard errors are close to zero, the Rhat statistics are 1 or close to 1 and the effective sample sizes are large enough. These three indices suggest that MCMC has converged.

Graphic tools such as trace plots, autocorrelation plots, and histograms of the posterior density can also be used to evaluate MCMC convergence. We use the `plot()` function to create trace plots for the intercept and the four predictor variables as follows.

```
> plot(nbr.stan, "trace")
```

FIGURE 15.4 ● Trace Plots for the Intercept, the Predictor Variables, and the Dispersion Parameter

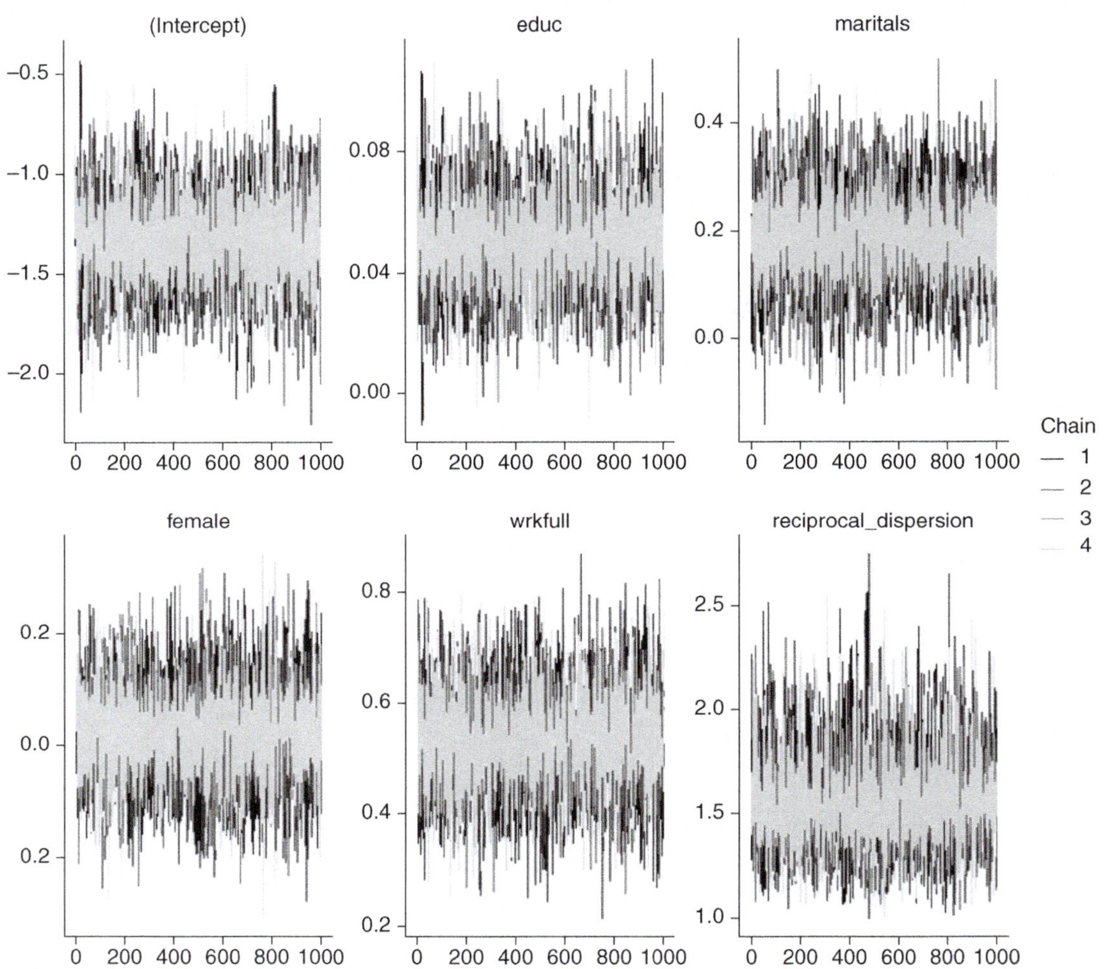

The trace plots (Figure 15.4) show the parameter estimates for the intercepts, the four predictor variables, and the dispersion parameter at each iteration for all four chains after warm-up draws are discarded. Overall the trace plots show good convergence since the iterations have stable paths of movement and they all center around the posterior means.

Interpreting the Results

The parameter estimates section (labeled `Estimates`) displays the parameter estimates (labeled `mean`) for the intercept, the four predictor variables, and the dispersion parameter. This section is analogous to the coefficients table in the output of the frequentist negative binomial model with the `glm.nb()` function. The column labeled `sd` displays the standard deviations of the posterior distributions, which are the standard errors of the posterior means. They are analogous to the standard errors of regression coefficients in the frequentist approach. The last three columns (labeled `2.5%`, `50%`, and `97.5%`, respectively) display the quantiles of the posterior distributions. They are the lower 95% credible interval, the median, and the upper 95% credible interval, respectively.

Interpretation of the posterior means is similar to that of the log coefficients in negative binomial regression models with a frequentist approach. For example, the posterior means for `educ` is .050, indicating the log expected number of visits to a zoo increases by .050 for a one-unit increase in education when holding all other predictors constant.

The corresponding 95% credible interval is [.016, .084], indicating that there is a 95% probability that the true parameter lies in the interval. The posterior means and the corresponding credible intervals for the other predictor variables in the model can be interpreted in a similar way.

We use the `coef()` function to extract the coefficients and use the `se()` function to extract the standard errors of the coefficients. The extracted coefficients are the posterior medians and the extracted standard errors are the standard errors for the posterior medians.

```
> coef <- coef(nbr.stan)
> coef
(Intercept)        educ    maritals      female     wrkfull
-1.30827614  0.05010345  0.19024463  0.02587501  0.53570773
> se<-se(nbr.stan)
> se
(Intercept)        educ    maritals      female     wrkfull
0.24614434   0.01695037  0.08938872  0.09281749  0.09632988
```

We use the `cbind(coef,se)` function to combine the posterior medians and the standard errors.

```
> cbind(coef,se)
                   coef          se
(Intercept)  -1.30827614  0.24614434
educ          0.05010345  0.01695037
maritals      0.19024463  0.08938872
female        0.02587501  0.09281749
wrkfull       0.53570773  0.09632988
```

We can use the `posterior_interval()` function to compute the credible interval. The default credible interval is the 90% credible interval and the columns are labeled `5%` and `95%`, respectively.

```
> posterior_interval(nbr.stan)
                              5%          95%
(Intercept)           -1.71938628  -0.91483156
educ                   0.02148192   0.07752313
maritals               0.03967449   0.33892837
female                -0.12443782   0.17736931
wrkfull                0.37620061   0.69159242
reciprocal_dispersion  1.20893210   2.03430449
```

15.5.5 The `brm()` Function in the `brms` Package: Syntax for Bayesian Negative Binomial Regression Models

To fit a Bayesian negative binomial regression model with `brm()`, we specify the model formula, the `family = negbinomial` argument, and the `data` argument. The link function `(link = "log")` is omitted since it is the default for the `negbinomial` family. We also specify the number of iterations, warm-up, chains, and cores. Here, we again specify 3,000 iterations with 500 warm-up iterations for two chains. The fitted model is named `nbr.brm`. The following output is displayed by the `print(nbr.brm, digits = 3)` command.

```
> library(brms)
> nbr.brm <- brm(vistzoo ~ educ + maritals + female + wrkfull, family = negbinomial,
data=count,
+               warmup = 500, iter = 3000,
+               cores = 2, chains = 2, seed=123)
Compiling Stan program...
Start sampling

> print(nbr.brm, digits = 3)

 Family: negbinomial
  Links: mu = log; shape = identity
Formula: vistzoo ~ educ + maritals + female + wrkfull
   Data: count (Number of observations: 902)
Samples: 2 chains, each with iter = 3000; warmup = 500; thin = 1;
         total post-warmup samples = 5000

Population-Level Effects:
           Estimate  Est.Error   l-95% CI   u-95% CI    Rhat  Bulk_ESS  Tail_ESS
Intercept    -1.310      0.247     -1.797     -0.831   1.000      7060      4083
educ          0.050      0.017      0.017      0.082   1.000      6798      3942
maritals      0.192      0.094      0.008      0.377   1.000      7022      4109
female        0.025      0.095     -0.159      0.208   1.001      6597      3750
wrkfull       0.537      0.097      0.349      0.727   1.000      6841      4196
```

```
Family Specific Parameters:
      Estimate  Est.Error  l-95% CI  u-95% CI   Rhat  Bulk_ESS  Tail_ESS
shape    1.595      0.262     1.171     2.182  1.000      6502      3704

Samples were drawn using sampling(NUTS). For each parameter, Bulk_ESS
and Tail_ESS are effective sample size measures, and Rhat is the potential
scale reduction factor on split chains (at convergence, Rhat = 1).
```

The estimates section (labeled `Population-Level Effects:`) provides the posterior means (labeled `Estimate`), the estimate errors, the 95% credible intervals, the `Rhat` values for convergence, and the effective sample sizes for the intercept and the predictor variables. The results are similar to those produced by the `stan_glm()` function.

We can extract the results of the estimates and the credible intervals with the `coef <- fixef(PR.2.brm, summary = TRUE)` command. We can also use the `exp(fixef(PR.2.brm)[, -2])` command to obtain the incidence rate ratios and the corresponding 95% credible intervals. The output is omitted.

To graphically evaluate MCMC convergence, we use the `plot()` function to create the posterior density plots and the trace plots for the intercepts, the predictor variables, and the dispersion parameter. In addition, we can use the `conditional_effects()` function to display the marginal effects of the predictor variables on the predicted number of visits to a zoo.

15.6 MAKING PUBLICATION-QUALITY TABLES

15.6.1 Presenting the Results Using the texreg Package

As of the time of this writing, since the `stargazer()` function cannot directly produce the results table from the models by the `rstanarm()` and `brm()` functions, we use the `screenreg()` and `htmlreg()` functions from the `texreg` package (Leifeld, 2013). You need to install `texreg` first by typing `install.packages("texreg")` if you have not done so, and then load the package by typing `library(texreg)`. This package needs to be installed only once.

To save space, we only demonstrate how to use the `texreg` package to display the results for the Bayesian logistic regression model. After using the `brm()` function to fit the Bayesian logistic regression model, `LR.2.brm`, we create a table containing the results with the following command: `screenreg(list(LR.2.brm))`. In the `screenreg()` function, we specify the model object to be presented. The output is a plain text table.

```
> library(texreg)
> screenreg(list(LR.2.brm))


===========================
                 Model 1
---------------------------
Intercept         -1.82 *
                 [-2.24; -1.38]
wrkfull            0.66 *
                 [ 0.48;  0.84]
maritals           0.37 *
                 [ 0.18;  0.55]
female             0.18 *
                 [ 0.00;  0.36]
educ               0.16 *
                 [ 0.13;  0.19]
---------------------------
R^2                0.08
Num. obs.        1873
loo IC           2113.64
WAIC             2113.63
===========================
* 0 outside the confidence interval.
> htmlreg(list(LR.2.brm), file="LR2br.doc", doctype=TRUE, html.tag=TRUE,
head.tag=TRUE)
The table was written to the file 'LR2br.doc'.
```

To create the table in the HTML format and copy it into Microsoft Word, we use the command as follows: `htmlreg(list(LR.2.brm), file = "LR2br.doc", doctype = TRUE, html.tag = TRUE, head.tag = TRUE)`. The table presenting the results for the Bayesian logistic regression model in its original format is omitted here. With edits on the model title, the final table is displayed as in Table 15.2.

TABLE 15.2 ● Results of the Bayesian Logistic Regression Model With `brm()` (Edited)

	Bayesian Logistic Regression Model
	Estimates (Credible Intervals)
Intercept	−1.82[a]
	[−2.24; −1.38]
Wrkfull	0.66[a]
	[0.48; 0.84]
maritals	0.37[a]
	[0.18; 0.55]

Table 15.2 *(Continued)*

	Bayesian Logistic Regression Model
	Estimates (Credible Intervals)
female	0.18[a]
	[0.00; 0.36]
educ	0.16[a]
	[0.13; 0.19]
R^2	0.08
Num. obs.	1873
loo IC	2,113.64
WAIC	2,113.63

[a]0 outside the confidence interval.

15.7 REPORTING THE RESULTS

Reporting the results for Bayesian models is similar to that for classical generalized linear models for categorical response variables.

First, describe the research purpose and explain why the Bayesian modeling for categorical response variables is needed for the analysis.

Second, evaluate MCMC convergence based on available measures, such as the Rhat, the effective sample sizes, and the autocorrelations, and available graphical tools, such as trace plots and density plots.

Third, report the parameter estimates, such as the posterior means, medians, or modes, for the predictor variables in a table or in the text. Interpret odds ratios or probabilities in logistic regression models and interpret incidence rate ratios in Poisson and negative binomial models.

The following is an example of summarizing the results of a Bayesian logistic regression model.

> The Bayesian logistic regression analysis with the `brm()` function was conducted to estimate the probability of having good health from four predictor variables. The dependent variable was having good health or not, and the independent variables were working full-time, marital status, years of education, and gender. The Bayesian logistic regression model was fitted since the outcome variable was dichotomous.

Evaluation of MCMC Convergence

The posterior sample size is 3,000, which equals the difference between the number of iterations and warm-up multiplied by the number of chains: (3,000 − 500)*2 = 3,000. The Rhat values were equal to or close to 1 and the effective sample sizes were large, which indicate good MCMC convergence. The graphical evaluation of the convergence of MCMC with the `plot()` function also shows that MCMC had converged in the posterior density plots and the trace plots for the intercepts and the predictor variables.

Results of the Bayesian Logistic Regression Model

Table 15.2 presents the posterior means or the logit coefficients and the credible intervals for the predictor variables of the model. For `wrkfull`, the logit coefficient = .656, and the 95% credible interval is [.442, .879]. Odds ratio = 1.927. The odds of having good health for the people who work full time are 1.927 times the odds of having good health for those who do not work full time.

For `maritals`, the logit coefficient = .367, and the 95% credible interval is [.146, .588]. Odds ratio = 1.443, which is larger than 1. This indicates that the odds of having good health for the married were 1.443 times the odds for the unmarried.

For `educ`, the logit coefficient = .162, and the 95% credible interval is [.126, .199]. Odds ratio = 1.176, indicating for each one-unit increase in education, the odds of having good health increase by a factor of 1.176.

For `female`, the logit coefficient = .182, and the 95% credible interval is [-.034, .039], which contains zero. Odds ratio = 1.199. This indicates that being a female or not does not influence whether that person has good health.

15.8 SUMMARY OF R COMMANDS IN THIS CHAPTER

```
# Chap 15 R Script

# Remove all objects
rm(list = ls(all = TRUE))

# The following user-written packages need to be installed first
# Use install.packages(" ") and then load it with library()
# library(rstanarm)
# library(brms)              # It is already installed for Chapter 14
# library(texreg)            # It is already installed for Chapter 4
```

```
# Import GSS 2016 Stata data file
library(foreign)
chp3.lr <- read.dta("C:/CDA/gss2016.dta")
attach(chp3.lr)

# Logistic regression with glm()
LR.2 <- glm(healthy ~ wrkfull + maritals + female + educ, data = chp3.lr, family =
binomial(link = "logit"))
summary(LR.2)

# Bayesian logistic regression with rstanarm
library(rstanarm)
LR.2.stan <- stan_glm(healthy ~ wrkfull + maritals + female + educ, data =
chp3.lr, family = binomial(link = "logit"))
summary(LR.2.stan)
summary(LR.2.stan, digits=3, prob=c(.025, .5, .975))
prior_summary(LR.2.stan)
coef <- coef(LR.2.stan)
coef
se <- se(LR.2.stan)
se
cbind(coef,se)
posterior_interval(LR.2.stan)
ci <- posterior_interval(LR.2.stan)
cbind(coef,ci)
exp(cbind(coef,ci))
plot(LR.2.stan, "trace")

# Bayesian logistic regression with brms
# It may take several minutes to compile the Stan program and fit the model
library(brms)
LR.2.brm <- brm(healthy ~ wrkfull + maritals + female + educ, data = chp3.lr,
family = bernoulli(link = "logit"),
                              warmup = 500, iter = 3000,
                              cores = 2, chains = 2, seed=123)
summary(LR.2.brm)
print(LR.2.brm, digits=3)
coef <- fixef(LR.2.brm, summary=TRUE)
coef
exp(fixef(LR.2.brm)[, -2])
save(LR.2.brm, file = "LR2brm.rda")
plot(LR.2.brm)

library(texreg)
screenreg(list(LR.2.brm))
htmlreg(list(LR.2.brm), file="LR2br.doc", doctype=TRUE, html.tag=TRUE,
head.tag=TRUE)

detach(chp3.lr)

# Bayesian ordinal logistic regression with rstanarm
# Remove all objects
rm(list = ls(all = TRUE))
```

```
# Import GSS 2016 Stata data file
library(foreign)
chp4.po <- read.dta("C:/CDA/gss2016.dta")
chp4.po$healthre <- factor(chp4.po$healthre, ordered=TRUE)
attach(chp4.po)

# rstanrarm and brms
# It may take several minutes to fit the model with stan_polr
library(rstanarm)
olr.stan <- stan_polr(healthre ~ maritals + educ + female, data = chp4.po,
                method = "logistic", prior = R2(0.2, "mean"), seed = 12345)
summary(olr.stan)
summary(olr.stan, digits=3, prob=c(.025, .5, .975))
prior_summary(olr.stan)
coef <- coef(olr.stan)
coef
se <- se(olr.stan)
se
cbind(coef,se)
posterior_interval(olr.stan)
ci <- posterior_interval(olr.stan)
cbind(coef,ci)
exp(cbind(coef,ci))
save(olr.stan, file = "olrstan.rda")

# Bayesian ordinal logistic regression with brms
# It may take several minutes to compile Stan and fit the model
library(brms)
olr.brm <- brm(healthre ~ maritals + educ + female, data = chp4.po,
               family=cumulative("logit"),
               warmup = 500, iter = 3000,
               cores = 2, chains = 2, seed=123)
print(olr.brm, digits=3)
coef <- fixef(olr.brm, summary=TRUE)
coef
exp(fixef(olr.brm)[, -2])
save(olr.brm, file = "olrbrm.rda")

detach(chp4.po)

# Bayesian multinomial logistic regression with brms
# It may take several minutes to compile Stan and fit the model
rm(list=ls())
library(foreign)
gss16 <- read.dta("C:/CDA/gss2016-chap2.dta")
gss16$healthre <- factor(gss16$healthre)
attach(gss16)
library(brms)
mulm.brm <- brm(healthre ~ educ + maritals + female + wrkfull, data = gss16,
               family=categorical,
               warmup = 500, iter = 3000,
               cores = 2, chains = 2, seed=123)
summary(mulm.brm)
print(mulm.brm, digits=3)
coef <- fixef(mulm.brm, summary=TRUE)
coef
```

```
exp(fixef(mulm.brm)[, -2])
save(mulm.brm, file = "mulmbrm.rda")
plot(mulm.brm)
detach(gss16)

# Bayesian Poisson regression with rstanarm
# Remove all objects
rm(list=ls(all = TRUE))

# Import the count dataset
library(foreign)
count <- read.dta("C:/CDA/count.dta")

# Convert variables from integer to numeric so they will work well with ggpredict()
count$educ <- as.numeric(count$educ)
count$wrkfull <- as.numeric(count$wrkfull)
count$maritals <- as.numeric(count$maritals)

attach(count)

# Multiple-predictor Poisson regression model with glm()
PR.2 <- glm(vistzoo ~ educ + maritals + female + wrkfull, family = poisson,
data=count)
summary(PR.2)

# stan_glm() in rstanarm
library(rstanarm)
PR.2.stan <- stan_glm(vistzoo ~ educ + maritals + female + wrkfull,
family = poisson, data=count)
summary(PR.2.stan)
summary(PR.2.stan, digits=3, prob=c(.025, .5, .975))
prior_summary(PR.2.stan)
coef <- coef(PR.2.stan)
coef
se <- se(PR.2.stan)
se
cbind(coef,se)
posterior_interval(PR.2.stan)
ci <- posterior_interval(PR.2.stan)
cbind(coef,ci)
exp(cbind(coef,ci))
plot(PR.2.stan, "trace")

# Bayesian Poisson regression with brms
# It may take a couple of minutes to compile Stan and fit the model.
library(brms)
PR.2.brm <- brm(vistzoo ~ educ + maritals + female + wrkfull, family = poisson,
data=count,
                warmup = 500, iter = 3000,
                cores = 2, chains = 2, seed=123)
summary(PR.2.brm)
print(PR.2.brm, digits=3)
coef <- fixef(PR.2.brm, summary=TRUE)
coef
exp(fixef(PR.2.brm)[, -2])
save(PR.2.brm, file = "PR2brm.rda")
plot(PR.2.brm)
conditional_effects(PR.2.brm)
```

```
# Negative binomial regression model with glm.nb() in MASS
library(MASS)
nbr <- glm.nb(vistzoo ~ educ + maritals + female + wrkfull, data=count)
summary(nbr)


# Bayesian negative binomial models with rstanarm
library(rstanarm)
nbr.stan <- stan_glm(vistzoo ~ educ + maritals + female + wrkfull, family =
neg_binomial_2, data=count)
summary(nbr.stan)
summary(nbr.stan, digits=3, prob=c(.025, .5, .975))
prior_summary(nbr.stan)
coef <- coef(nbr.stan)
coef
se <- se(nbr.stan)
se
cbind(coef,se)
posterior_interval(nbr.stan)
ci <- posterior_interval(nbr.stan)
plot(nbr.stan, "trace")


# Bayesian negative binomial models with brms
# It may take a couple of minutes to compile Stan and fit the model.
library(brms)
nbr.brm <- brm(vistzoo ~ educ + maritals + female + wrkfull, family = negbinomial,
data=count,
               warmup = 500, iter = 3000,
               cores = 2, chains = 2, seed=123)
summary(nbr.brm)
print(nbr.brm, digits=3)
coef <- fixef(nbr.brm, summary=TRUE)
coef
exp(fixef(nbr.brm)[, -2])
save(nbr.brm, file = "nbrbrm.rda")
plot(nbr.brm)
detach(count)
```

Glossary

A uniform prior, which has a uniform distribution, is a simple noninformative prior since all values in the range of the distribution have equal prior weight on a parameter.

A weakly informative prior provides some information on the distributions of the parameters, so it is helpful for model fitting.

An uncertainty interval displays the range of the uncertainty of the parameter.

If we have little existing knowledge on the parameters, we may specify so-called **noninformative priors**, which may have no or little influence on the posterior distribution.

In Bayesian statistics the parameters of interest are unknown, random, and have their own distributions. We are interested in the distribution of the parameters rather than one fixed parameter. Bayesian statistics focus on the uncertainty of the unknown parameters in terms of probability with prior beliefs.

The autocorrelation plots show the autocorrelations between iterations.

The Bayesian factor is the ratio of the likelihood of two competing models or the ratio of the posterior odds to the prior odds.

The credible interval is based on the quantiles of the posterior distribution.

The effective sample size is the number of independent draws from the posterior distribution.

The highest posterior density (HPD) interval is based on the posterior density.

The histograms of the posterior density show the distributions of the posterior density.

The likelihood function can be expressed as the form of the probability distribution of the data given the parameters, $p(y|\theta)$.

The MC standard error represents the variability of the parameter estimates associated with repeated MCMC estimations. It is the ratio of the posterior standard deviation to the square root of the effective sample size.

The posterior distributions of the parameters given the data, $p(\theta|y)$, are based on both the prior distributions of the parameter and the likelihood function, and are proportional to the multiplication of these two terms.

The posterior means are the means of the posterior distributions of the parameters or the point estimates of the parameters. They are analogous to the regression coefficients in frequentist statistics.

The prior distributions of the parameters, $p(\theta)$, are the prior beliefs or the prior knowledge on the parameter distributions, which are based on the existing knowledge or the results of previous studies.

The prior information can be based on existing knowledge or results of previous studies, which is referred to as **an informative prior**.

The Rhat statistic is the potential scale reduction factor, which compares the between-chain variation with the within-chain variation.

The standard errors of the posterior means are the standard deviations of the posterior distributions. They are analogous to the standard errors of regression coefficients in the frequentist approach.

The three major elements in Bayesian inference include the prior distributions of the parameters, the likelihood function, and the posterior distributions of the parameters.

The trace plots show the parameter estimates for the predictor variables at each iteration for the chains after warm-up draws are discarded. When the iterations have stable paths of movement and there are no substantial fluctuations, MCMC has reached convergence.

The **warm-up iterations** are the samples which are discarded before the target distribution is reached. The warm-up process is also called the burn-in period.

When the prior distribution and the posterior distribution of a parameter come from the same family, the two distributions are **conjugate**.

Exercises

Use the GSS 2016 data available at **https://edge.sagepub.com/liu1e** for the following problems. The following variables are used for the Bayesian logistic regression modeling.

`gunlaw`: favor or oppose gun permit (1 = favoring gun permit; 2 = opposing gun permit)

`sex`: gender (1 = male; 2 = female)

`educ`: highest year of school completed

`age`: age.

1. Before conducting the analysis, recode `gunlaw` into a new variable named `gun` so that 1 = favoring gun permit and 0 = opposing gun permit. Also, recode `sex` into a new variable named `female` so that 1 = female and 0 = male.
2. Use the `brm()` function or the `stan_glm()` function to conduct a Bayesian logistic regression analysis to examine the relationship between the outcome variable `gunlaw` and the three predictor variables. If using the `brm()` function, specify 3,000 iterations with 500 warm-up iterations and two chains. You may specify more iterations with more chains.
3. Evaluate MCMC convergence.
4. Identify the posterior mean and the 95% credible intervals for the predictor variable `educ`.
5. Computer the odds ratios for `educ` and `age`.
6. Write a report to summarize the results from the output.

16

BAYESIAN MULTILEVEL MODELING OF CATEGORICAL RESPONSE VARIABLES

OBJECTIVES OF THIS CHAPTER

This chapter introduces Bayesian multilevel modeling of categorical response variables or Bayesian generalized linear mixed models. Specifically, this chapter covers Bayesian multilevel logistic regression, ordinal logistic regression, Poisson regression, and negative binomial regression models. Since Bayesian multilevel multinomial logistic regression models have been introduced in Chapter 14, they are omitted here. The random-coefficient models for each type of categorical response variable with predictor variables in both levels are illustrated using R with step-by-step instructions. The `brms` and `MCMCglmm` packages are used for the Bayesian multilevel analyses. R commands are explained, and the output is interpreted for each model in detail. This chapter also illustrates how the results are displayed in publication-quality tables using R and reported in the text. It focuses on model fitting with R, as well as on interpreting and presenting the results. After reading this chapter, you should be able to:

- Determine when multilevel modeling for categorical response variables is used.
- Formulate multilevel models.
- Conduct Bayesian multilevel modeling of categorical response variables using R.
- Evaluate MCMC convergence.
- Interpret the results.
- Present the results in publication-quality tables using R.
- Write the results for publication.

16.1 BAYESIAN MULTILEVEL LOGISTIC REGRESSION WITH R

As discussed in previous chapters, we can use multilevel modeling when data are nested or hierarchical. Just like Bayesian generalized linear models for single-level data structure, Bayesian methods can also be applied to multilevel modeling to estimate Bayesian multilevel or hierarchical models. Since multilevel models include both fixed effects and random effects, priors need to be specified for both effects. The MCMC approaches are used to draw samples from the posterior distributions of parameters. This chapter presents Bayesian multilevel modeling for categorical and count response variables. Specifically, it introduces Bayesian multilevel logistic regression, ordinal logistic regression, Poisson regression, and negative binomial regression models. Since Bayesian multilevel multinomial logistic regression models have been introduced in Chapter 14, they are omitted here. We start with an introduction to Bayesian multilevel modeling for binary outcome variables, which is referred to as Bayesian multilevel logistic regression when Bayesian methods are applied to multilevel logistic regression.

16.1.1 Model Specification

For illustration purposes, the same random-coefficient model for the binary response variable (i.e., the contextual model 4 in Chapter 11) is demonstrated here. The level 1 predictor variable is whether the student's native language is English (`stlang`). The level 2 predictor variables include whether the school is public or private (`public`) and school climate (`csclimat`). The level 2 equations for the intercept and the slope for `stlang` are both specified to be random with level 2 residuals u_{0j} and u_{1j}, respectively. In the level 2 equation for the random slope, no school-level variables are specified. This model is expressed as follows:

$$\begin{aligned} \text{Level 1}: \ \text{logit}\left[\pi\left(x_{ij}\right)\right] &= \beta_{0j} + \beta_{1j}\texttt{stlang}_{ij} \\ \text{Level 2}: \ \beta_{0j} &= \gamma_{00} + \gamma_{01}\texttt{public}_j + \gamma_{02}\texttt{csclimat}_j + u_{0j} \\ \beta_{1j} &= \gamma_{10} + u_{1j} \end{aligned} \quad (16.1)$$

16.1.2 Description of the Data and Sample

The ELS:2002 base-year data are used for the following analyses. The variables are listed as follows:

- `Profmath1`: mathematics proficiency level 1 (i.e., doing simple arithmetical operations on whole numbers or not)
- `stlang`: student language is English (1 = yes, 0 = no)
- `public`: school type (1 = public, 0 = private and others)
- `sclimat`: school climate
- `csclimat`: school climate (grand-mean centered).

For demonstration purposes, the missing values of the variables are deleted. The descriptive statistics are summarized and the school-level variable `sclimat` is grand-mean centered.

16.1.3 Packages and Functions for Bayesian Multilevel Modeling for Binary Response Variables in R

We introduce the `brm()` function in the `brms` package (Bürkner, 2017) and the `MCMCglmm()` function in the `MCMCglmm` package (Hadfield, 2010) for the Bayesian multilevel analysis of binary response variables. The `brms` package is introduced first.

16.1.4 The `brm()` Function in the `brms` Package: Syntax for Bayesian Multilevel Logistic Regression Models

The `brms` package is well known for fitting a variety of Bayesian generalized linear mixed models using the programming language Stan. Since the package was installed in the last chapter, we only need to load the package by typing `library(brms)`.

As introduced in the preceding chapters, the syntax for `brm()` is similar to that for `glmer()` in the `lme4` package. To fit a Bayesian multilevel binary logistic regression model with `brm()`, we specify the model formula for the fixed and random effects, the `family = bernoulli` argument, and the `data` argument. First, the model formula includes both the fixed effects and the random effects. The fixed-effects part of the model specifies the outcome variable and the predictor variable(s) with the outcome variable on the left-hand side of the tilde (~). Then, the random-effects part of the model is specified within parentheses. The random coefficients term is specified first, followed by a vertical line (|), and then, an identifier variable at a higher level as the grouping variable. Second, we also need to specify the `family = bernoulli` argument since the outcome variable is dichotomous. Third, the `data =` argument specifies the data frame.

We also specify the number of total iterations, warm-up iterations chains, and cores. The default number of iterations is 2,000 with 1,000 warm-up iterations and the default number of chains is four. Since estimation for Bayesian multilevel modeling is computationally intensive and time consuming, to reduce estimation time, here, we specify 2,000 iterations with 500 warm-up iterations and two chains. Even though we reduce the number of chains, it may still take hours to fit the model. We specify the random seed with `seed = 123` so the results can be reproducible. If we rerun the same syntax, we may still get slightly different results due to MCMC sampling. The fitted model is named `melr.brm`. The following output is displayed by the `summary(melr.brm)` command. For more details on how to use this function, type `help(brm)` in the command prompt after loading the `brms` package.

```
> library(brms)
> melr.brm <- brm(Profmath1~ stlang + csclimat + public + (stlang|SCH_ID),
+               data=els, family=bernoulli,
+               warmup = 500, iter = 2000,
+               cores = 2, chains = 2, seed=123)
Compiling Stan program...
Start sampling
> summary(melr.brm)
 Family: bernoulli
  Links: mu = logit
Formula: Profmath1 ~ stlang + csclimat + public + (stlang | SCH_ID)
   Data: els (Number of observations: 12793)
Samples: 2 chains, each with iter = 2000; warmup = 500; thin = 1;
         total post-warmup samples = 3000


Group-Level Effects:
~SCH_ID (Number of levels: 619)
                          Estimate Est.Error l-95% CI u-95% CI Rhat Bulk_ESS
sd(Intercept)             1.07          0.17     0.76     1.41 1.00      772
sd(stlangYes)             0.83          0.23     0.33     1.25 1.01      242
cor(Intercept,stlangYes) -0.49          0.22    -0.77     0.07 1.00      436
                          Tail_ESS
sd(Intercept)             1559
sd(stlangYes)             229
cor(Intercept,stlangYes)  533

Population-Level Effects:
                Estimate Est.Error l-95% CI u-95% CI Rhat Bulk_ESS Tail_ESS
Intercept           4.00      0.25     3.53     4.49 1.00     2910     2377
stlangYes           0.69      0.15     0.37     0.98 1.00     2066     2383
csclimat            0.60      0.10     0.41     0.79 1.00     2885     2365
publicpublic       -1.31      0.22    -1.75    -0.88 1.00     4909     2345

Samples were drawn using sampling(NUTS). For each parameter, Bulk_ESS
and Tail_ESS are effective sample size measures, and Rhat is the potential
scale reduction factor on split chains (at convergence, Rhat = 1).
```

We use the `digits = 3` argument in the `print()` function to display three decimal places for the numerical values in the output.

```
> print(melr.brm, digits=3)
 Family: bernoulli
  Links: mu = logit
Formula: Profmath1 ~ stlang + csclimat + public + (stlang | SCH_ID)
   Data: els (Number of observations: 12793)
Samples: 2 chains, each with iter = 2000; warmup = 500; thin = 1;
         total post-warmup samples = 3000


Group-Level Effects:
~SCH_ID (Number of levels: 619)
                          Estimate Est.Error l-95% CI u-95% CI  Rhat Bulk_ESS
sd(Intercept)                1.068     0.165    0.764    1.406 1.001      772
sd(stlangYes)                0.834     0.234    0.333    1.255 1.007      242
cor(Intercept,stlangYes)    -0.492     0.219   -0.773    0.070 1.003      436
```

```
                              Tail_ESS
sd(Intercept)                     1559
sd(stlangYes)                      229
cor(Intercept,stlangYes)           533


Population-Level Effects:
              Estimate Est.Error l-95% CI u-95% CI  Rhat Bulk_ESS Tail_ESS
Intercept        4.000     0.247    3.528    4.492 1.000     2910     2377
stlangYes        0.690     0.155    0.368    0.981 1.000     2066     2383
csclimat         0.599     0.095    0.413    0.788 1.002     2885     2365
publicpublic    -1.308     0.223   -1.751   -0.877 1.000     4909     2345

Samples were drawn using sampling(NUTS). For each parameter, Bulk_ESS
and Tail_ESS are effective sample size measures, and Rhat is the potential
scale reduction factor on split chains (at convergence, Rhat = 1).
```

The first section R output produced by the `brm()` function includes the model information, including the family, the link function, the model formula for the fixed and random effects, the data frame with the number of observations, and the number of chains and iterations, and the total number of posterior samples. With 2,000 iterations and 500 warm-up iterations for each of the two chains, the total number of posterior samples = 2 * (2,000 − 500) = 3,000.

The second section is the random-effects section (labeled `Group-Level Effects:`). It provides the standard deviations rather than the variances for the intercept and the random coefficient. They are 1.068, and .834, respectively, which are analogous to the standard deviations for the random effects in the output produced by the `glmer()` function. We can easily obtain the variances by squaring the standard deviations. The corresponding variances for the intercept and the random coefficient are 1.141, and .696, respectively. Therefore, the residual variance in the intercept (τ_{00}) is 1.141 and the slope variance for `stlang` is .696. The covariance between the intercept and the `stlang` slope equals the product of the correlation coefficient and the two standard deviations and is computed as follows: 1.068 * .834 * (−.492) = −.438. The table also provides the estimate errors, the 95% credible intervals, the `Rhat` values for convergence, and the effective sample sizes (labeled `Bulk_ESS` and `Tail_ESS`).

The third section is the fixed-effects section (labeled `Population-Level Effects:`). It provides the estimates, the estimate errors, the 95% credible intervals, the `Rhat` values for convergence, and the effective sample sizes for the intercept and predictor variables. The column labeled `Estimate` displays the posterior mean values, which are analogous to the coefficients for the fixed effects in the output produced by the `glmer()` function. Next, the column labeled `Est.Error` displays the estimate errors, the standard deviations of the posterior distribution, which are analogous to the standard errors of the coefficients for the fixed effects in frequentist statistics. In addition, the following two columns labeled `l-95% CI` and `u-95% CI` list the 95% credible intervals, which are the 2.5% and 97.5% quantiles. Finally, the last three columns list the Rhat values for convergence and the effective sample sizes (labeled Bulk_ESS and Tail_ESS). We use these three values to evaluate MCMC convergence. An Rhat value close to 1 but less than 1.1 and a large effective sample size indicate MCMC convergence. The Rhat values and the effective sample sizes in the output indicate good MCMC convergence. The graphical evaluation of the convergence of MCMC with the plot() function is introduced later in this section.

We extract the results of the fixed effects and the 95% credible intervals with the coef <- fixef(melr.brm, summary = TRUE) command.

```
> coef <- fixef(melr.brm, summary = TRUE)
> coef
                Estimate Est.Error       Q2.5      Q97.5
Intercept      3.9998062 0.2465845  3.5279264  4.4919482
stlangYes      0.6903831 0.1548180  0.3677357  0.9808220
csclimat       0.5986294 0.0954939  0.4127223  0.7883733
publicpublic  -1.3077028 0.2230087 -1.7513564 -0.8770598
```

We use the exp(fixef(melr.brm)[, -2]) command to obtain the odds ratios and the corresponding credible intervals. The estimated errors should not be exponentiated, so we exclude the second column in the fixed-effects table with the indexing [, -2]. The following output is created.

```
> exp(fixef(melr.brm)[, -2])
                Estimate       Q2.5      Q97.5
Intercept     54.5875719 34.0532807 89.2952418
stlangYes      1.9944795  1.4444602  2.6666474
csclimat       1.8196231  1.5109253  2.1998151
publicpublic   0.2704406  0.1735384  0.4160042
```

The odds ratio for stlang is 1.994, which indicates that the odds of being proficient in math for students whose native language is English are 1.994 times as high as the odds for those whose native language is not English when holding the other predictor variables constant.

The odds ratio for csclimat is 1.820, which indicates that a one-unit increase in school climate corresponds to an increase by a factor of 1.820 points in the odds of being proficient in math. In other words, being students in schools with a more positive school climate increases the odds of being proficient in math.

The odds ratio for public is .270, which indicates that the odds of being proficient in math for students in public schools are .270 times the odds for those in private schools.

Fitting a Bayesian multilevel model is time consuming. It may take hours or more to fit a model, so it is always good practice to save the model object with the save()

function so we can reuse it when needed with the `load()` function without refitting the model.

```
> save(melr.brm, file = "melrbrm.rda")
```

To graphically evaluate MCMC convergence, we use the `plot()` function to create the posterior density plots and the trace plots for the fixed effects and the random effects as follows.

```
> plot(melr.brm)
```

The posterior distribution plots and the trace plots (Figure 16.1) show good convergence for the intercept and the four predictor variables.

We use the `conditional_effects()` function to display the marginal effects of the predictor variables on the predicted probabilities of being proficient in doing simple arithmetical operations on whole numbers. Figure 16.2 shows the conditional effect for `stlang`.

```
> conditional_effects(melr.brm)
```

Similar plots for the predicted probabilities for predictor variables at specified values can be created using the `ggpredict()` function and the `plot()` function in the `ggeffects` package. See details on how to use the `ggeffects` package in previous chapters.

16.1.5 The `MCMCglmm()` Function in the `MCMCglmm` Package: Syntax for Bayesian Multilevel Logistic Regression Models

The `MCMCglmm` package is also capable of fitting Bayesian generalized linear mixed models with the Markov chain Monte Carlo (MCMC) methods. It uses both Metropolis-Hastings sampling and Gibbs sampling for MCMC to simulate draws. Since the package has been installed in the preceding chapters, we only need to load it by typing `library(MCMCglmm)`.

Similar to that for `glmer()` for mixed-effects logistic regression models, the syntax for `MCMCglmm()` for multilevel multinomial regression models also includes the components for the fixed effects and random effects, the `family` argument, and the `data` argument, but it has more components.

First, the model formula for the fixed effects in `MCMCglmm()` specifies the outcome variable and the predictor variable(s) with the outcome variable on the left-hand side of the tilde (~).

FIGURE 16.1 ● Posterior Density Plots and Trace Plots for the Fixed Effects and the Random Effects

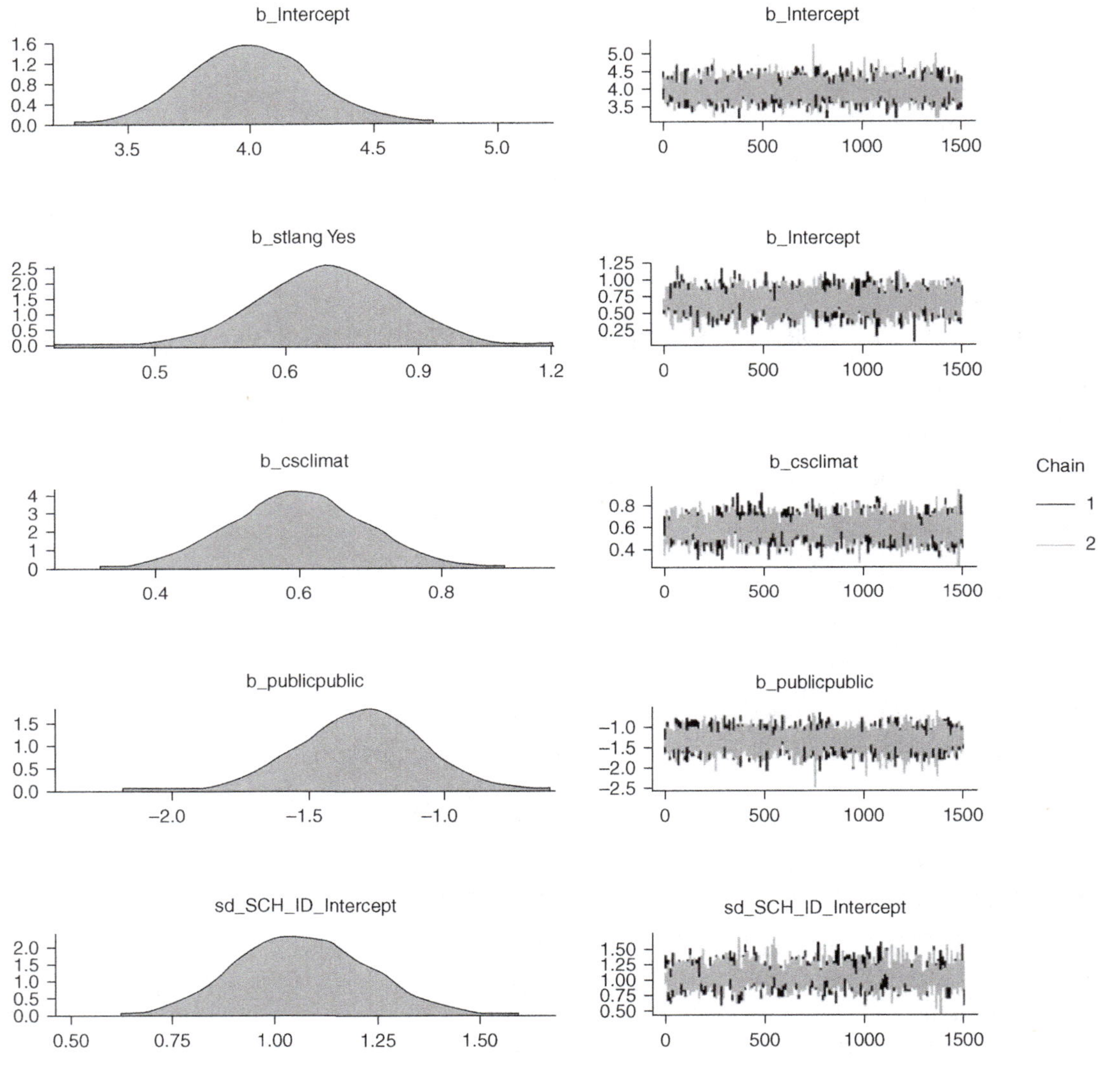

Second, the random-effects part of the model is specified with the `random = ~` argument. For example, for the random-intercept model, we specify `random = ~SCH_ID`, where `SCH_ID` is an identifier variable at a higher level as the grouping variable or grouping factor. In a random-coefficient logistic regression model, since we have both a random intercept and a random slope, we specify `random = ~us(1 + stlang):SCH_ID`, where `(1 + stlang)` represents the random intercept and the

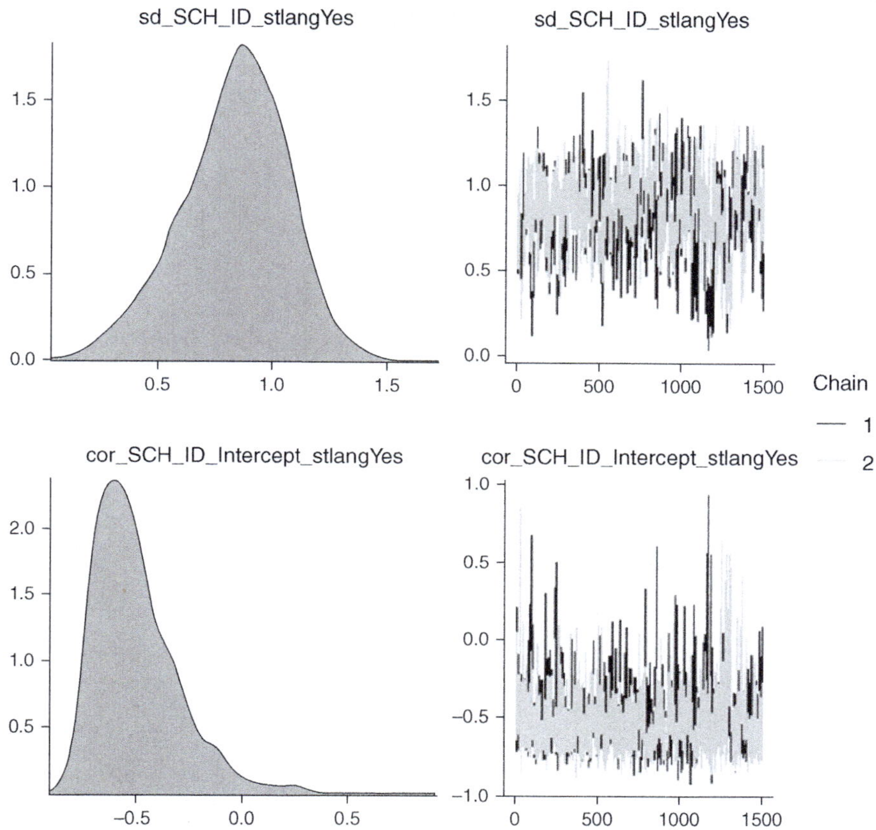

Figure 16.1 *(Continued).*

random slope for `stlang` and `us()` means that the variance–covariance structure is unstructured, which allows the random effects to be correlated.

Third, we use the `family = "categorical"` argument to specify the distribution of the binary response variable. This argument can also be used for nominal response variables.

Fourth, we use the `data =` argument to specify the data frame.

Finally, we use the `prior =` argument to specify the priors for the fixed effects (B), the variance and covariance matrix for the random effects (G-structure), and the residual variance structure (R-structure). Let us see an example before fitting the model below.

```
priors = list(R = list(V = 1, nu = 0.002, fix = 1),
              G = list(G1 = list(V = diag(2), nu = 2)))
```

FIGURE 16.2 ● Conditional Effect Plot for the Predictor Variable `stlang`

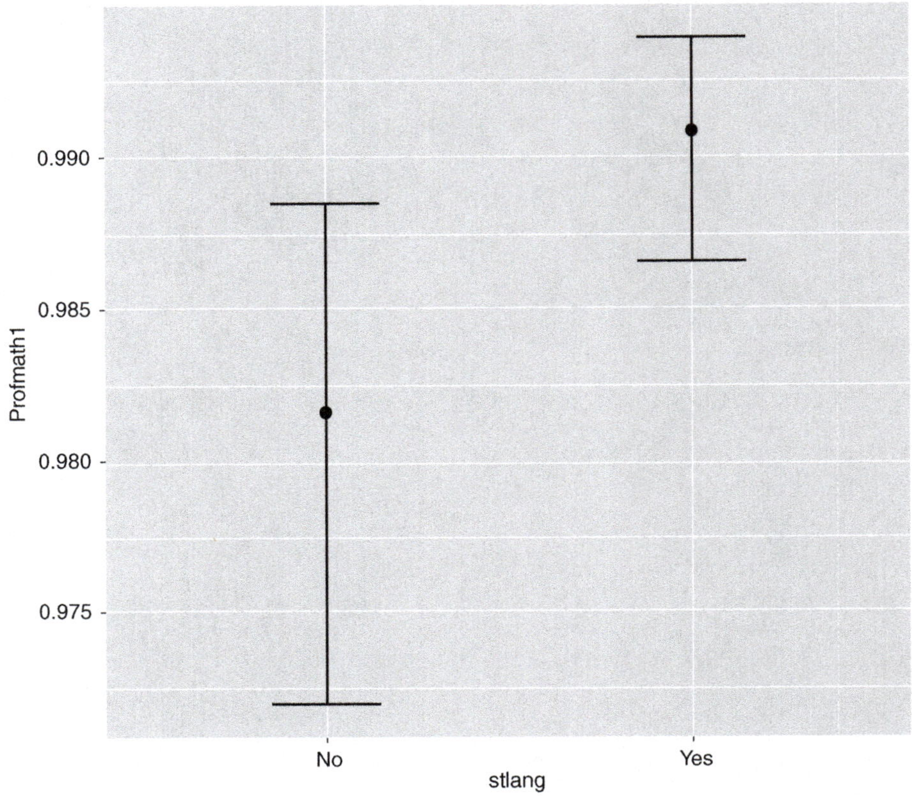

- R defines the residual variance structure; V = 1 defines that the residual variance for the Bayesian multilevel logistic regression model is 1 since it is not estimated from the data; fix = 1 indicates that the residual variance is fixed; and nu is the number of degrees of freedom parameter which is normally specified at .002.

- In addition, G defines the random effects for the G-structure. For the random-coefficient model with `random = ~us(1 + stlang):SCH_ID`, we have a variance matrix for the random intercept and the random slope, so we specify G1. This matrix is a 2 by 2 diagonal matrix for both the random intercept and the random slope rather than a single variance, so it is expressed as `V = diag(2)`. The number of degrees of freedom parameter nu is specified at 2 since the dimension of the variance matrix is 2. We normally specify nu to be equal to or greater than the number of the dimension of the variance matrix; nu is normally specified at .002 when V is a scalar.

```
> library(MCMCglmm)
> prior = list(R = list(V = 1, nu = 0.002, fix = 1),
+              G = list(G1 = list(V = diag(2), nu = 2)))
> melr.mc <- MCMCglmm(Profmath1 ~ stlang + csclimat + public,
+                      random = ~ us(1 + stlang):SCH_ID,
+                      data=els, family = "categorical", prior=prior)
> summary(melr.mc)

Iterations = 3001:12991
  Thinning interval  = 10
  Sample size  = 1000

  DIC: 4470.909

  G-structure:  ~us(1 + stlang):SCH_ID

                                   post.mean  l-95% CI  u-95% CI  eff.samp
(Intercept):(Intercept).SCH_ID        1.3801    0.6254    2.1712     26.60
stlangYes:(Intercept).SCH_ID         -0.6252   -1.3572    0.0806     29.10
(Intercept):stlangYes.SCH_ID         -0.6252   -1.3572    0.0806     29.10
stlangYes:stlangYes.SCH_ID            0.9310    0.2603    1.7969     30.39

  R-structure:  ~units

      post.mean  l-95% CI  u-95% CI  eff.samp
units         1         1         1         0

  Location effects: Profmath1 ~ stlang + csclimat + public

               post.mean  l-95% CI  u-95% CI  eff.samp   pMCMC
(Intercept)       4.4314    3.9361    5.0111     40.62  <0.001 ***
stlangYes         0.7650    0.4320    1.0520     64.60  <0.001 ***
csclimat          0.6535    0.4441    0.8365     99.46  <0.001 ***
publicpublic     -1.3704   -1.7885   -0.8912     28.58  <0.001 ***
---
Signif. codes:  0 '***' 0.001 '**' 0.01 '*' 0.05 '.' 0.1 ' ' 1
```

The results are similar to those produced by the `brm()` function and the interpretation is omitted here.

16.2 BAYESIAN MULTILEVEL ORDINAL LOGISTIC REGRESSION WITH R

16.2.1 Model Specification

Bayesian multilevel ordinal logistic regression is used when Bayesian methods are applied to multilevel modeling for ordinal response variables. The same random-coefficient model with both level 1 and level 2 predictor variables for the ordinal response variable (i.e., the contextual model 4 in Chapter 12) is demonstrated here. The level 1 equation includes the predictor variable, the student's native language (stlang), and the ordinal response variable (Profmath). The level 2 predictor variables include whether the school is public or private (public) and school climate (csclimat). The level 2 equations for the intercept and the slope for stlang are both specified to be random with level 2 residuals u_{0j} and u_{1j}, respectively. In addition, in level 2 equations, two school-level predictor variables are added to the equation for the level 1 random intercept, whereas the equation for the random slope contains no predictor variables. This model is expressed as follows:

$$\text{Level 1}: \text{logit}\left[\pi_{kij}(Y \leq k)\right] = \ln\left(\frac{\pi\left(Y_{ij} \leq k\right)}{\pi\left(Y_{ij} > k\right)}\right)$$

$$= \alpha_k - \left(\beta_{0j} + \beta_{1j}\texttt{stlang}_{ij}\right) \quad (16.2)$$

$$\text{Level 2}: \beta_{0j} = \gamma_{00} + \gamma_{01}\texttt{public}_j + \gamma_{02}\texttt{csclimat}_j + u_{0j}$$

$$\beta_{1j} = \gamma_{10} + u_{1j}$$

16.2.2 Description of the Data and Sample

The ELS:2002 base-year data are used for the following analyses. The variables are listed as follows:

- Profmath: mathematics proficiency with six levels from 0 to 5
- stlang: whether English is the student's native language
- public: school type (1= public, 0 = private and others)
- sclimat: school climate
- csclimat: school climate (grand-mean centered).

For demonstration purposes, the missing values of the variables are deleted. The descriptive statistics are summarized, and the school-level variable sclimat is grand-mean centered.

16.2.3 Packages and Functions for Bayesian Multilevel Ordinal Logistic Regression With R

As of the time of writing, the rstanarm package has not supported Bayesian multilevel modeling for ordinal and nominal response variables. In addition, the MCMCglmm package can be used to fit ordinal probit regression models rather than ordinal logistic regression models. Therefore, in the following example, we only use the brm() function in the brms package (Bürkner, 2017) to fit Bayesian ordinal logistic regression models for ordinal outcome variables.

16.2.4 The `brm()` Function in the `brms` Package: Syntax for Bayesian Multilevel Ordinal Logistic Regression Models

To fit a Bayesian multilevel ordinal logistic regression model with `brm()`, we specify the model formula for the fixed and random effects, the `family = cumulative` argument, and the `data` argument. In the model formula, `Profmath ~ stlang + public + csclimat + (stlang|SCH_ID)`, the fixed part of the model includes one level 1 predictor, `stlang`, and two level 2 predictor variables, `public` and `csclimat`. The predictor variable `stlang` is specified in the random part of the model `(stlang|SCH_ID)`, which indicates that the slope of the level 1 predictor `stlang` and the intercept are allowed to vary randomly across schools. The random part of the model can also be specified as `(1 + stlang|SCH_ID)`, where `1` represents the intercept.

We also specify the number of iterations warm-up iterations, chains, and cores. Here, we specify 2,000 iterations with 500 warm-up iterations and two chains. The fitted model is named `mpo.brm`. The `summary(mpo.brm)` command only shows the output with two decimal places, which is omitted here. To display three decimal places for the numbers, we use the `print(mpo.brm, digits = 3)` command as follows.

Please note that the model fitting may take a couple of hours, so be prepared. For more details on how to use this function, type `help(brm)` in the command prompt after loading the `brms` package.

```
> library(brms)
> mpo.brm <- brm(Profmath ~ stlang + public + csclimat + (stlang|SCH_ID),
+                data=els, family=cumulative,
+                warmup = 500, iter = 2000,
+                cores = 2, chains = 2, seed=123)
Compiling Stan program...
Start sampling
> print(mpo.brm, digits=3)
 Family: cumulative
  Links: mu = logit; disc = identity
Formula: Profmath ~ stlang + public + csclimat + (stlang | SCH_ID)
   Data: els (Number of observations: 12793)
Samples: 2 chains, each with iter = 2000; warmup = 500; thin = 1;
         total post-warmup samples = 3000


Group-Level Effects:
~SCH_ID (Number of levels: 619)
                        Estimate Est.Error l-95% CI u-95% CI  Rhat Bulk_ESS
sd(Intercept)              1.046     0.071    0.915    1.194 1.001      601
sd(stlangYes)              0.638     0.103    0.434    0.842 1.002      234
cor(Intercept,stlangYes)  -0.705     0.057   -0.804   -0.580 1.000     1206
                        Tail_ESS
sd(Intercept)               1325
sd(stlangYes)                583
cor(Intercept,stlangYes)    1693
```

```
Population-Level Effects:
              Estimate Est.Error l-95% CI u-95% CI  Rhat Bulk_ESS Tail_ESS
Intercept[1]    -3.415     0.109   -3.625   -3.198 1.001     1553     1982
Intercept[2]    -1.170     0.103   -1.371   -0.963 1.000     1410     1774
Intercept[3]    -0.091     0.103   -0.289    0.114 1.000     1370     1995
Intercept[4]     1.434     0.103    1.231    1.639 1.000     1359     2246
Intercept[5]     5.248     0.146    4.957    5.534 1.000     1829     2356
stlangYes        0.404     0.066    0.276    0.528 1.000     1860     2383
publicpublic    -0.575     0.090   -0.755   -0.404 1.000     1384     1910
csclimat         0.533     0.057    0.423    0.645 1.000     1251     1857

Family Specific Parameters:
     Estimate Est.Error l-95% CI u-95% CI  Rhat Bulk_ESS Tail_ESS
disc    1.000     0.000    1.000    1.000 1.000     3000     3000

Samples were drawn using sampling(NUTS). For each parameter, Bulk_ESS
and Tail_ESS are effective sample size measures, and Rhat is the potential
scale reduction factor on split chains (at convergence, Rhat = 1).
```

The first section R output produced by the `brm()` function includes the model information, including the family, the link function, the model formula for the fixed and random effects, the data frame with the number of observations, and the number of chains and iterations, and the total number of posterior samples. With 2,000 iterations and 500 warm-up iterations for each of the two chains, the total number of posterior samples = 2 * (2,000 − 500) = 3,000.

The second section is the random-effects section (labeled `Group-Level Effects:`). This provides the standard deviations for the random intercept and the random coefficient, which are analogous to the standard deviations for the random effects in the output produced by the `clmm()` function. They are 1.046 and .638, respectively. By squaring the standard deviations, we obtain the corresponding variances for the intercept and the random coefficient, which are 1.094 and .407, respectively. Therefore, the residual variance in the intercepts (τ_{00}) is 1.094 and the slope variance for `stlang` is .407. The table also provides the estimate errors, the 95% credible intervals, the `Rhat` values for convergence, and the effective sample sizes (labeled `Bulk_ESS` and `Tail_ESS`).

The third section is the fixed-effects section (labeled `Population-Level Effects:`). This provides the posterior means, the estimate errors, the 95% credible intervals, the `Rhat` values for convergence, and the effective sample sizes for the intercepts or the thresholds and the predictor variables. The first column labeled `Estimate` displays the posterior mean values, which are analogous to the coefficients for the fixed effects in the output produced by the `clmm()` function. The next column labeled `Est.Error` displays the estimate errors, which are the standard errors of the estimates or the posterior means. In addition, the third and fourth columns labeled `l-95% CI` and `u-95% CI` list the 95% credible intervals, which are the 2.5% and 97.5% quantiles. For example, the posterior mean for `stlang` is .404, which is similar to that produced by the `clmm()` function or by the `mixor()` function. The corresponding 95% credible interval is [.276, .528], which indicates that there is a 95%

chance that this interval contains the true parameter. Finally, the last three columns list the `Rhat` values for convergence and the effective sample sizes (labeled `Bulk_ESS` and `Tail_ESS`). We use these three values to evaluate MCMC convergence. The `Rhat` values are equal to or close to 1 and the effective sample sizes are large, which indicate good MCMC convergence. The graphical evaluation of the convergence of MCMC with the `plot()` function is introduced later in this section.

We extract the results of the fixed effects and the 95% credible intervals with the `coef <- fixef(mpo.brm, summary = TRUE)` command.

```
> coef <- fixef(mpo.brm, summary = TRUE)
> coef
                 Estimate  Est.Error       Q2.5      Q97.5
Intercept[1]  -3.41468613 0.10876679 -3.6245815 -3.1977546
Intercept[2]  -1.16953358 0.10313021 -1.3714806 -0.9632970
Intercept[3]  -0.09096858 0.10293136 -0.2892965  0.1139944
Intercept[4]   1.43354402 0.10281256  1.2305756  1.6393936
Intercept[5]   5.24753241 0.14558961  4.9573883  5.5340500
stlangYes      0.40359949 0.06609772  0.2760322  0.5280242
publicpublic  -0.57502261 0.08990415 -0.7547871 -0.4038712
csclimat       0.53342269 0.05723615  0.4230438  0.6446725
```

We use the `exp(fixef(mpo.brm)[, -2])` command to obtain the odds ratios and the corresponding credible intervals. The estimated errors should not be exponentiated, so we exclude the second column in the fixed-effects table with the indexing `[, -2]`. The following output is created.

```
> exp(fixef(mpo.brm)[, -2])
                  Estimate         Q2.5        Q97.5
Intercept[1]    0.03288673   0.02666025   0.04085383
Intercept[2]    0.31051174   0.25373100   0.38163255
Intercept[3]    0.91304640   0.74879013   1.12074587
Intercept[4]    4.19353484   3.42319927   5.15204443
Intercept[5]  190.09660854 142.22187419 253.16715396
stlangYes       1.49720418   1.31789029   1.69557883
publicpublic    0.56269214   0.47011071   0.66773015
csclimat        1.70475720   1.52660110   1.90536295
```

We save the model object with the `save()` function so we can reuse it when needed with the `load()` function without refitting the model.

```
> save(mpo.brm, file = "mpo.rda")
```

To graphically evaluate MCMC convergence, we use the `plot()` function to create the posterior density plots and the trace plots for the intercepts and the predictor variables and the random effects as follows (Figure 16.3).

FIGURE 16.3 ● Posterior Density Plots and Trace Plots for the Fixed Effects and Random Effects

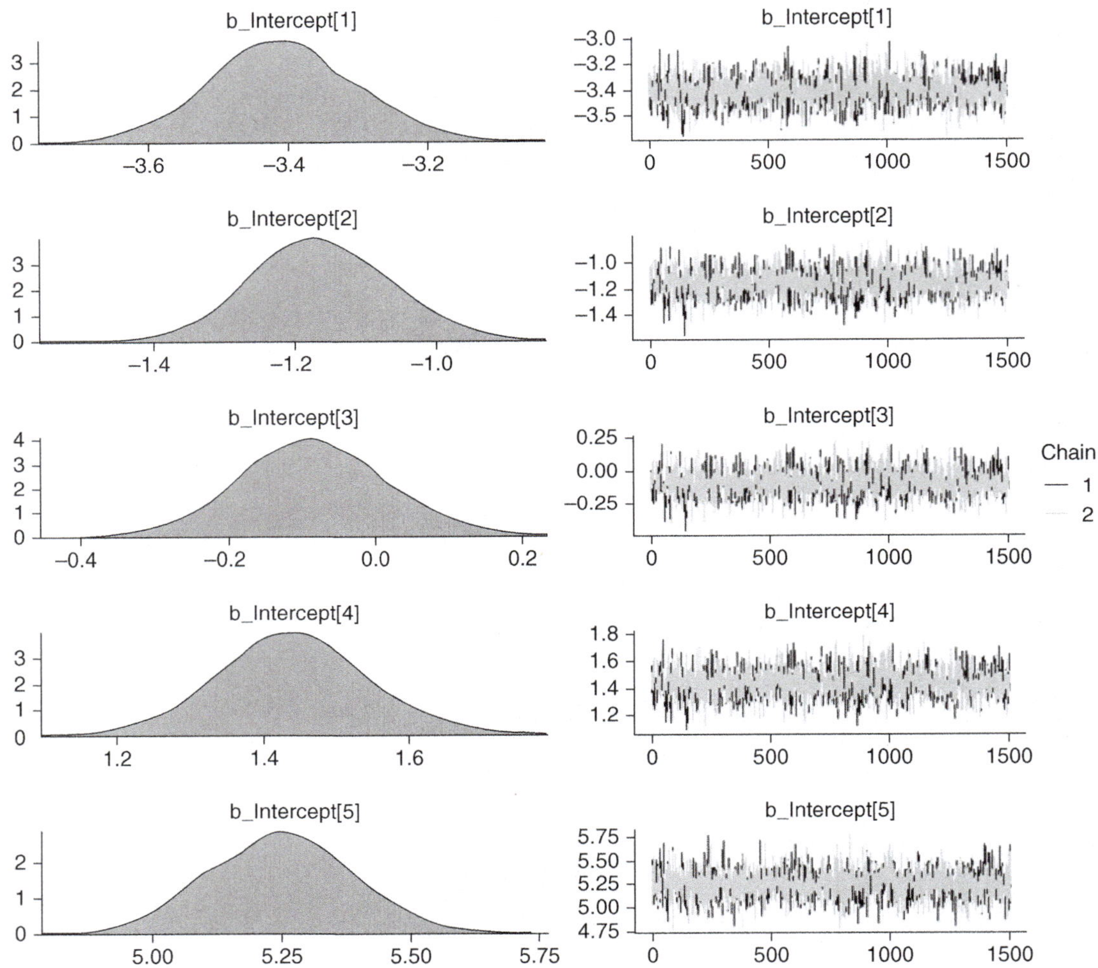


```
> plot(mpo.brm)
```

Please note that the disc parameter is the inverse of the standard deviation of the latent outcome variable, which is fixed to 1 in this model.

We use the `conditional_effects()` function to plot the marginal effect of each predictor variable on the estimated probabilities of being in the six categories of

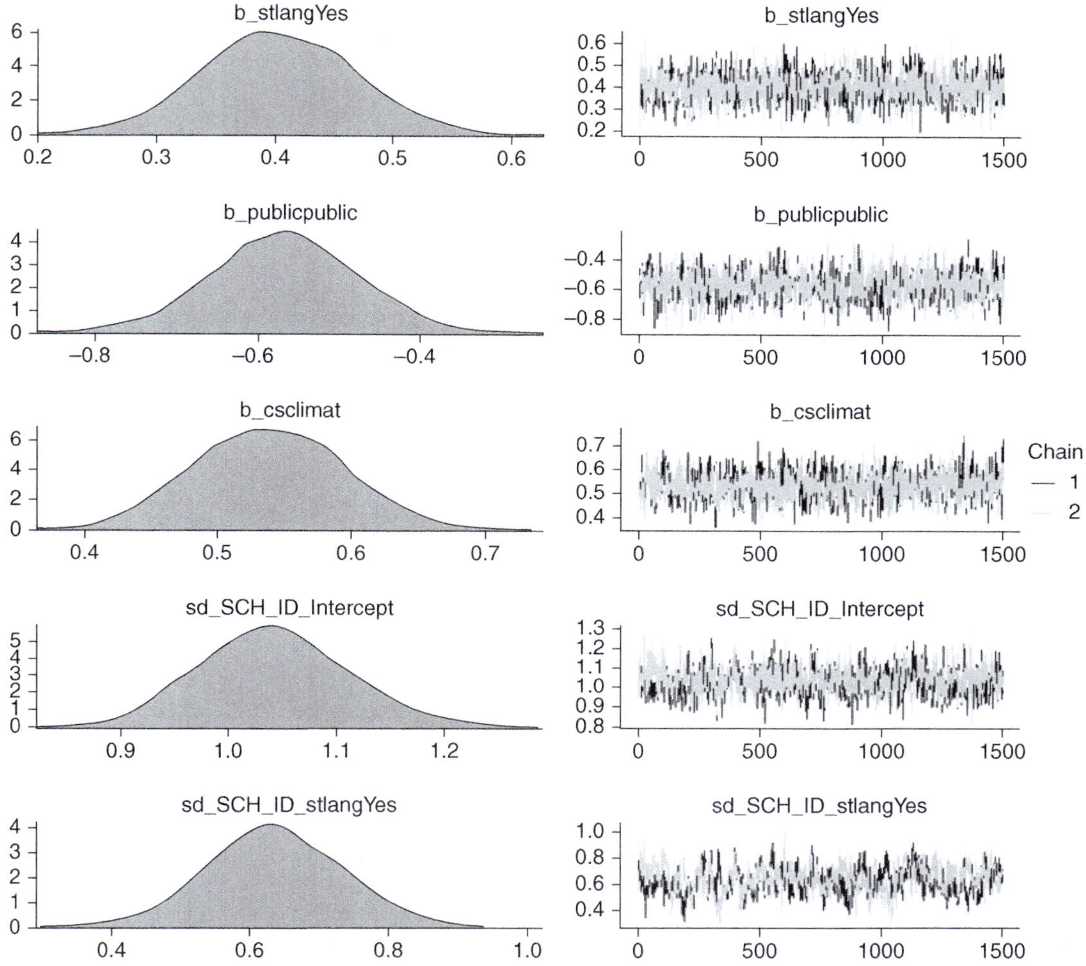

Figure 16.3 *(Continued).*

the outcome variable. In the function, the `categorical = TRUE` argument needs to be specified since the ordinal outcome variable has more than two categories. Figure 16.4 displays the conditional effect plot for predictor variable `csclimat`. The plots for the other two predictor variables are omitted here.

```
> conditional_effects(mpo.brm, categorical = TRUE)
```

In this section, we have used the `brm()` function in the `brms` package to fit the Bayesian ordinal logistic regression model for an ordinal outcome variable. For nominal

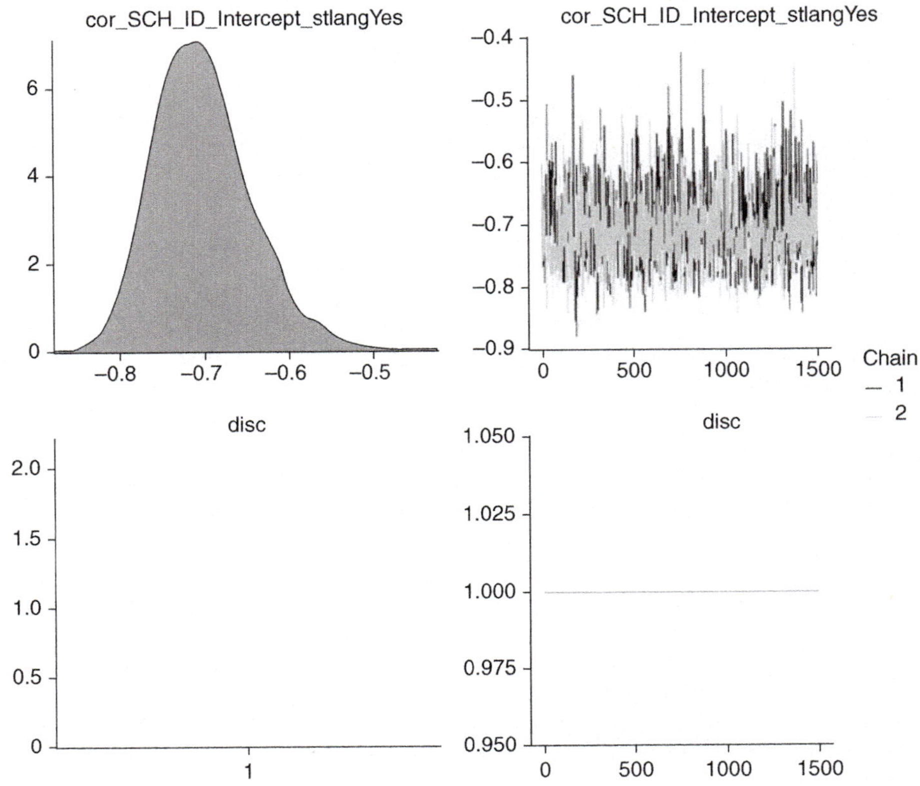

Figure 16.3 *(Continued).*

response variables, refer to Chapter 14 on how to fit Bayesian multilevel multinomial logistic regression models with `MCMCglmm` and `brms`.

16.3 BAYESIAN MULTILEVEL POISSON REGRESSION WITH R

16.3.1 Model Specification

Bayesian multilevel Poisson regression is used when Bayesian methods are applied to multilevel modeling for count response variables. The same random-coefficient Poisson regression model with both level 1 and level 2 predictor variables for the count response variable (i.e., the contextual model 4 in Chapter 13) is demonstrated here. The level 1 equation includes the predictor variables, math scores (`cmathach`) and gender (`bygender`), and the count response variable, number of absences (`byabsen`). The

FIGURE 16.4 ● Conditional Effect Plot for the Predictor Variable `csclimat`

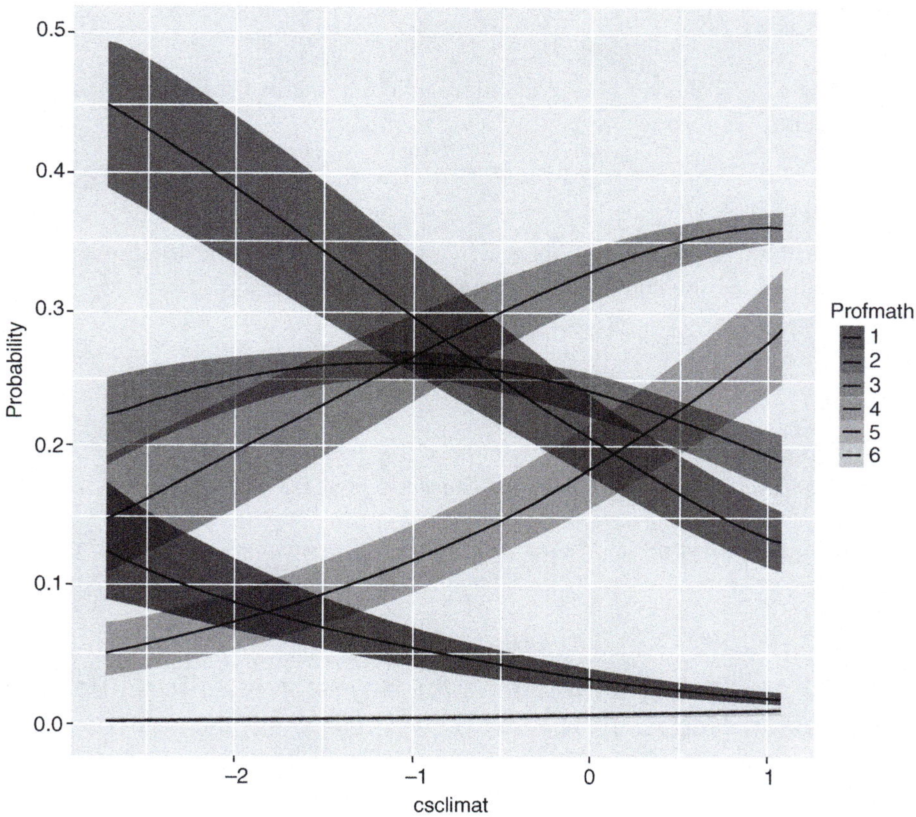

level 2 predictor variables include whether the school is public or private (`public`) and school climate (`csclimat`). The level 2 equations for the intercept (β_{0j}) and the slope for `bygender` (β_{2j}) are both specified to be random with level 2 residuals u_{0j} and u_{2j}, respectively, whereas the slope for `cmathach` (β_{1j}) is constrained to be fixed. This model is expressed as follows:

$$\begin{aligned}\text{Level 1}: \ln(\mu_{ij}) &= \beta_{0j} + \beta_{1j}\texttt{cmathach}_{ij} + \beta_{2j}\texttt{bygender}_{ij}\\ \text{Level 2}: \beta_{0j} &= \gamma_{00} + \gamma_{01}\texttt{public}_j + \gamma_{02}\texttt{csclimat}_j + u_{0j}\\ \beta_{1j} &= \gamma_{10}\\ \beta_{2j} &= \gamma_{20} + u_{2j}\end{aligned}$$

16.3.2 Description of the Data and Sample

The ELS:2002 base-year data are used for the following analyses. The variables are listed as follows:

- `byabsen`: number of absences from school in the first semester of ninth grade
- `mathach`: mathematics item response theory (IRT) estimated scores of high-school students
- `cmathach`: mathematics item response theory (IRT) estimated scores of high-school students (grand-mean centered)
- `bygender`: gender (1 = male, 0 = female)
- `public`: school type (1 = public, 0 = private and others)
- `sclimat`: school climate
- `csclimat`: school climate (grand-mean centered).

For demonstration purposes, the missing values of the variables are deleted. The student-level variable `mathach` and the school-level variable `sclimat` are grand-mean centered.

16.3.3 Packages and Functions for Bayesian Multilevel Poisson Regression Models in R

Both the `brm()` function in the `brms` package and the `MCMCglmm()` function in the `MCMCglmm` package are used for the Bayesian multilevel Poisson regression analysis of count response variables. The `brms` package is introduced first.

16.3.4 The `brm()` Function in the `brms` Package: Syntax for Bayesian Multilevel Poisson Regression Models

To fit a Bayesian multilevel Poisson regression model with `brm()`, we specify the model formula for the fixed and random effects, the `family = poisson` argument, and the `data` argument. First, the model formula for Bayesian multilevel Poisson regression models in `brm()` is similar to that for `glmer()` in the `lme4` package. Second, we also need to specify the `family = poisson` argument since the outcome variable is a count variable. Third, the `data =` argument specifies the data frame.

We also specify the number of iterations, warm-up iterations chains, and cores. Here, we specify 3,000 iterations with 500 warm-up iterations and two chains. The fitted model is named `mepois.brm`. The following output is displayed by the `summary(mepois.brm)` command. For more details on how to use this function, type `help(brm)` in the command prompt after loading the `brms` package.

```
> library(brms)
> mepois.brm <- brm(byabsen ~ cmathach + bygender + public + csclimate + (bygender|
SCH_ID),
+             family = poisson, data=mecount,
+             warmup = 500, iter = 3000,
+             cores = 2, chains = 2, seed=123)
Compiling Stan program...
Start sampling
> summary(mepois.brm)
 Family: poisson
  Links: mu = log
Formula: byabsen ~ cmathach + bygender + public + csclimate + (bygender | SCH_ID)
   Data: mecount (Number of observations: 12183)
Samples: 2 chains, each with iter = 3000; warmup = 500; thin = 1;
         total post-warmup samples = 5000

Group-Level Effects:
~SCH_ID (Number of levels: 617)
                         Estimate Est.Error l-95% CI u-95% CI Rhat Bulk_ESS
sd(Intercept)                0.37      0.03     0.31     0.42 1.00     1986
sd(bygender)                 0.33      0.04     0.25     0.41 1.00     1139
cor(Intercept,bygender)     -0.65      0.07    -0.78    -0.49 1.00     2256
                         Tail_ESS
sd(Intercept)                2677
sd(bygender)                 1449
cor(Intercept,bygender)      3080

Population-Level Effects:
             Estimate Est.Error l-95% CI u-95% CI Rhat Bulk_ESS Tail_ESS
Intercept       -0.55      0.04    -0.63    -0.47 1.00     6165     4068
cmathach        -0.02      0.00    -0.02    -0.02 1.00     5341     4322
bygender         0.61      0.03     0.55     0.67 1.00     6675     4216
publicpublic    -0.43      0.04    -0.51    -0.35 1.00     5748     3587
csclimate       -0.01      0.03    -0.06     0.04 1.00     6617     4302

Samples were drawn using sampling(NUTS). For each parameter, Bulk_ESS
and Tail_ESS are effective sample size measures, and Rhat is the potential
scale reduction factor on split chains (at convergence, Rhat = 1).
```

We again use the `digits = 3` argument in the `print()` function to display three decimal places for the numerical values in the output.

```
> print(mepois.brm, digits = 3)
 Family: poisson
  Links: mu = log
Formula: byabsen ~ cmathach + bygender + public + csclimate + (bygender | SCH_ID)
   Data: mecount (Number of observations: 12183)
Samples: 2 chains, each with iter = 3000; warmup = 500; thin = 1;
         total post-warmup samples = 5000
```

```
Group-Level Effects:
~SCH_ID (Number of levels: 617)
                          Estimate Est.Error l-95% CI u-95% CI  Rhat Bulk_ESS
sd(Intercept)                0.368     0.029    0.312    0.425 1.000     1986
sd(bygender)                 0.335     0.043    0.249    0.414 1.001     1139
cor(Intercept,bygender)     -0.652     0.075   -0.776   -0.487 1.001     2256
                          Tail_ESS
sd(Intercept)                 2677
sd(bygender)                  1449
cor(Intercept,bygender)       3080

Population-Level Effects:
              Estimate Est.Error l-95% CI u-95% CI  Rhat Bulk_ESS Tail_ESS
Intercept       -0.548     0.042   -0.631   -0.468 1.000     6165     4068
cmathach        -0.018     0.001   -0.020   -0.016 1.000     5341     4322
bygender         0.612     0.030    0.554    0.671 1.000     6675     4216
publicpublic    -0.429     0.042   -0.512   -0.346 1.000     5748     3587
csclimate       -0.010     0.026   -0.061    0.041 1.000     6617     4302

Samples were drawn using sampling(NUTS). For each parameter, Bulk_ESS
and Tail_ESS are effective sample size measures, and Rhat is the potential
scale reduction factor on split chains (at convergence, Rhat = 1).
```

The random-effects section (labeled `Group-Level Effects:`) displays the standard deviations for the random intercept and the random coefficient, which are analogous to the standard deviations for the random effects in the output produced by the `glmer()` function. The standard deviations for the random intercept and coefficient are .368, and .335, respectively, so the corresponding variances are .135, and .112, respectively. Therefore, the residual variance in the intercepts (τ_{00}) is .135 and the slope variance for `bygender` is .112. The table also provides the estimate errors, the 95% credible intervals, the `Rhat` values, and the effective sample sizes (labeled `Bulk_ESS` and `Tail_ESS`).

The fixed-effects section (labeled `Population-Level Effects:`) provides the posterior means (labeled `Estimate`), the estimate errors, the 95% credible intervals, the `Rhat` values, and the effective sample sizes for the intercept and the predictor variables. The results are similar to those produced by the `glmer()` function in the `lme4` package. The `Rhat` values are equal to 1 and the effective sample sizes are large, which indicate MCMC convergence. The graphical evaluation of the convergence of MCMC with the `plot()` function is introduced later in this section.

We extract the results of the estimates and the 95% credible intervals with the `coef <- fixef(mepois.brm, summary = TRUE)` command.

```
> coef <- fixef(mepois.brm, summary = TRUE)
> coef
                Estimate  Est.Error        Q2.5       Q97.5
Intercept    -0.54796478 0.041717016 -0.63103985 -0.46789359
cmathach     -0.01815312 0.001060119 -0.02024348 -0.01608214
bygender      0.61242901 0.029952278  0.55371547  0.67055258
publicpublic -0.42945798 0.042210377 -0.51151656 -0.34569013
csclimate    -0.01017263 0.026226567 -0.06110574  0.04107305
```

We use the `exp(fixef(mepois.brm)[, -2])` command to obtain the incidence rate ratios and the corresponding credible intervals. The estimated errors should not be exponentiated, so we exclude the second column in the fixed-effects table with the indexing `[, -2]`. The following output is created.

```
> exp(fixef(mepois.brm)[, -2])
              Estimate      Q2.5     Q97.5
Intercept     0.5781252 0.5320383 0.6263202
cmathach      0.9820107 0.9799600 0.9840465
bygender      1.8449073 1.7397048 1.9553175
publicpublic  0.6508618 0.5995856 0.7077318
csclimate     0.9898789 0.9407238 1.0419282
```

The incidence rate ratio for `cmathach` is .982, which indicates that the incidence rate of being absent or the expected count of absences decreases by a factor of .982 for a one-unit increase in math scores when holding the other predictor variables constant.

The incidence rate ratio for `bygender` is 1.845, which indicates that the incidence rate of being absent for male students is 1.845 times as high as that for female students when holding the other predictor variables constant.

The incidence rate ratio for `public` is .651, which indicates that the incidence rate of being absent for students in public schools is .651 times the incidence rate for those in private schools. In other words, being in public schools decreases the incidence rate of being absent by 34.9%.

The incidence rate ratio for `csclimat` is .990. The 95% credible interval for the incidence rate ratio is [.941, 1.042], which contains 1. In addition, the 95% credible interval for the log coefficient for `csclimat` is [−.061, .041], which contains zero. This indicates that school climate does not impact the incidence rate of being absent.

We save the model object with the `save()` function so we can reuse it when needed with the `load()` function without refitting the model.

```
> save(mepois.brm, file = "mepoisbrm.rda")
```

To graphically evaluate MCMC convergence, we use the `plot()` function to create the posterior density plots and the trace plots for the fixed effects and the random effects as follows.

```
> plot(mepois.brm)
```

The posterior distribution plots and the trace plots (Figure 16.5) show good convergence for the intercept and the four predictor variables. Those for the random effects also look fine.

FIGURE 16.5 ● Posterior Density Plots and Trace Plots for the Fixed Effects and the Random Effects

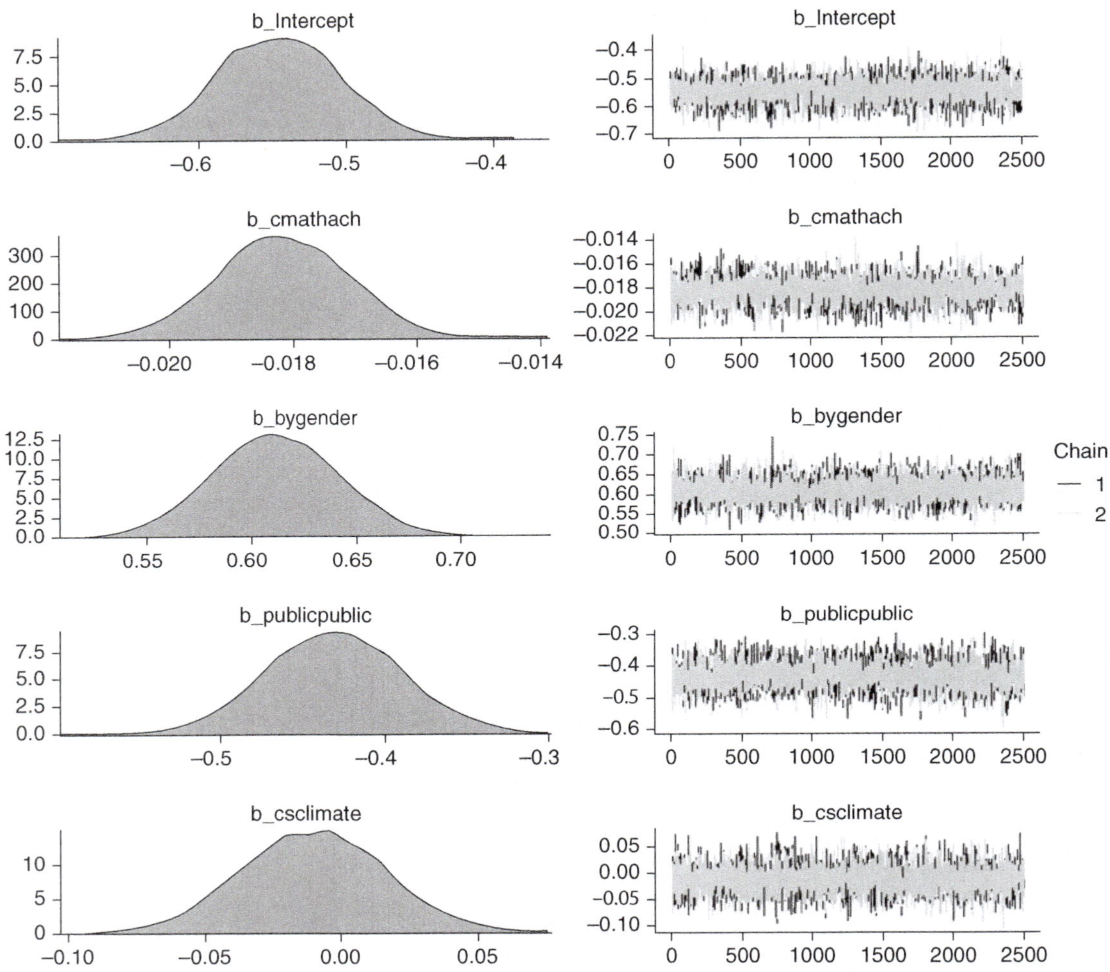

Although the graphs are not shown here, we can use the `conditional_effects()` function to display the marginal effects of the predictor variables.

16.3.5 The `MCMCglmm()` Function in the `MCMCglmm` Package: Syntax for Bayesian Multilevel Poisson Regression Models

To fit a Bayesian multilevel Poisson regression model with `MCMCglmm()`, we specify the model formula for the fixed and random effects, the `family = poisson` argument, and the `data` argument.

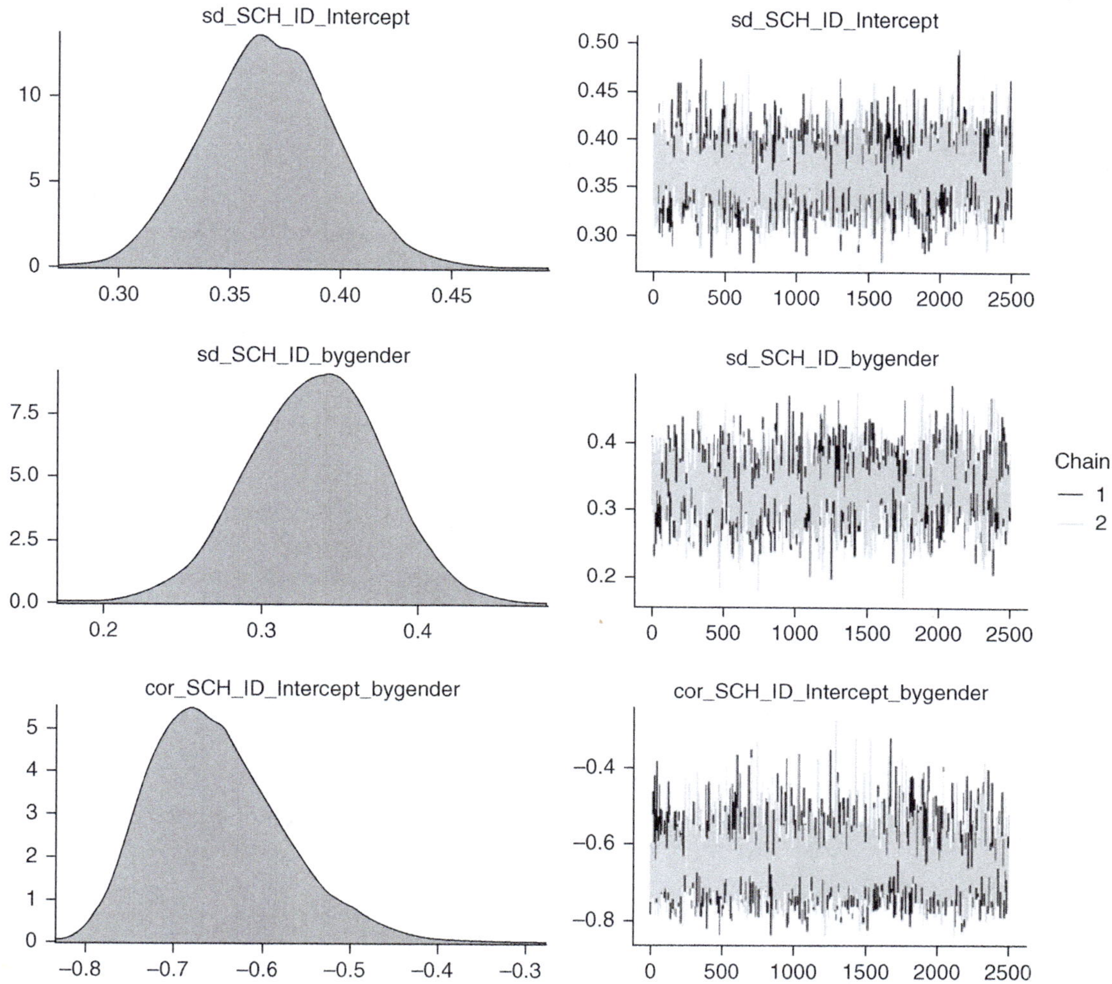

Figure 16.5 *(Continued).*

First, the model formula for the fixed effects in `MCMCglmm()` specifies the dependent variable and the predictor variable(s) with the dependent variable on the left-hand side of the tilde (~). In the model formula, `byabsen ~ cmathach + bygender + public + csclimate`, the fixed part of the model includes two level 1 predictors `cmathach` and `bygender` and two level 2 predictor variables `public` and `csclimat`.

Second, the random effects part of the model is specified with the `random = ~` argument. In a random-coefficient Poisson regression model, since we have both a random intercept and a random slope, we specify `random = ~ us(1 + bygender):`

SCH_ID, where (1 + bygender) represents the random intercept and the random slope for bygender and us() represents the unstructured variance–covariance matrix.

Third, we use the family = " poisson" argument to specify the distribution of the count response variable.

Fourth, we use the data = argument to specify the data frame.

Finally, we use the prior = argument to specify the priors for the fixed effects (B), the variance and covariance matrix for the random effects (G-structure), and the residual variance structure (R-structure).

```
priors = list(R = list(V = 1, nu = 0.002),
              G = list(G1 = list(V = diag(2), nu = 2)))
```

- R defines the residual variance structure; V = 1 defines that the residual variance for the Bayesian multilevel Poisson regression model is 1; and nu is the number of degrees of freedom parameter, which is normally specified at .002.
- In addition, G defines the random effects for the G-structure. For the random-coefficient model with random = ~us(1 + bygender): SCH_ID, we have a variance matrix for the random intercept and the random slope. This matrix is expressed as V = diag(2) for both the random intercept and the random slope. Since we only have one G-structure in this example, we specify G1. We specify the number of degrees of freedom parameter nu = 2 since the dimension of the variance matrix is 2.

```
> library(MCMCglmm)
>
> prior = list(R = list(V = 1, nu = 0.002),
+              G = list(G1 = list(V = diag(2), nu = 2)))
>
> mepois.mc <- MCMCglmm(byabsen ~ cmathach + bygender + public + csclimate,
+                   random =  ~ us(1 + bygender):SCH_ID,
+                   data=mecount, family = "poisson", prior=prior)

Iterations = 3001:12991
 Thinning interval  = 10
 Sample size  = 1000

DIC: 25132.53

G-structure:  ~us(1 + bygender):SCH_ID

                              post.mean  l-95% CI  u-95% CI eff.samp
(Intercept):(Intercept).SCH_ID  0.13363  0.09782  0.17588    62.03
bygender:(Intercept).SCH_ID    -0.07745 -0.11773 -0.04138    82.04
(Intercept):bygender.SCH_ID    -0.07745 -0.11773 -0.04138    82.04
bygender:bygender.SCH_ID        0.12698  0.08316  0.17703   111.26
```

```
 R-structure:  ~units

        post.mean  l-95% CI  u-95% CI  eff.samp
units      0.1945    0.1521    0.2355     24.78

 Location effects: byabsen ~ cmathach + bygender + public + csclimate

               post.mean  l-95% CI  u-95% CI  eff.samp  pMCMC
(Intercept)     -0.63336  -0.71332  -0.54701     290.4 <0.001 ***
cmathach        -0.01841  -0.02048  -0.01631     266.9 <0.001 ***
bygender         0.61174   0.55096   0.67682     161.1 <0.001 ***
publicpublic    -0.44505  -0.52575  -0.36032     403.3 <0.001 ***
csclimate       -0.01585  -0.06555   0.03864     430.6    0.58
---
Signif. codes:  0 '***' 0.001 '**' 0.01 '*' 0.05 '.' 0.1 ' ' 1
```

The posterior means for the predictor variables in the fixed-effects section (labeled `Location effects:`) look similar to those in the output produced by the `brm()` function in the preceding section. However, the effective sample sizes for the random effects are small.

We use the `autocorr(mepois.mc$Sol)` command to evaluate the autocorrelations for the fixed effects of the model. The results are displayed as follows.

```
> autocorr(mepois.mc$Sol)
, , (Intercept)

        (Intercept)    cmathach    bygender publicpublic   csclimate
Lag 0    1.00000000 -0.01192754 -0.50507336  -0.76092175 -0.31535627
Lag 10   0.36136731 -0.04279086 -0.32579328  -0.17681823 -0.11538117
Lag 50   0.13327929 -0.05710942 -0.11998823  -0.03237095 -0.01432434
Lag 100  0.12091052  0.02751925 -0.13355716  -0.03872956 -0.07738828
Lag 500  0.03536815 -0.03031891  0.03674749  -0.04795475 -0.04190946

, , cmathach

          (Intercept)   cmathach     bygender publicpublic   csclimate
Lag 0    -0.011927536 1.00000000 -0.092486058  0.115769990 -0.10045813
Lag 10    0.006483509 0.57834160 -0.064798840  0.056683479 -0.03645979
Lag 50    0.012108655 0.09431153 -0.051565125  0.029495276 -0.05980826
Lag 100   0.034083067 0.01169285 -0.060534665 -0.034201477 -0.01709894
Lag 500   0.034629087 0.02455739  0.004291025 -0.008330142 -0.03163264

, , bygender

         (Intercept)       cmathach    bygender publicpublic  csclimate
Lag 0   -0.505073363 -0.0924860576  1.00000000  0.047084962 0.05947034
Lag 10  -0.291078091 -0.0625766489  0.55538557  0.004872649 0.06382096
Lag 50  -0.128168820 -0.0930914688  0.21482103  0.013442891 0.07782780
Lag 100 -0.060865612 -0.0004303986  0.10224504  0.022196399 0.07260894
Lag 500  0.008794306 -0.0179060238 -0.02505884  0.012557418 0.06102620
```

```
, , publicpublic

         (Intercept)    cmathach    bygender publicpublic   csclimate
Lag 0    -0.76092175 0.115769990 0.047084962   1.00000000  0.36246204
Lag 10   -0.19508044 0.121204771 0.057814562   0.27332676  0.11847660
Lag 50   -0.04655967 0.121604401 0.006209438   0.08253207 -0.02663148
Lag 100  -0.05239676 0.006941609 0.074696530   0.04764762  0.05943240
Lag 500  -0.05210277 0.052177471 -0.017791706  0.03718589  0.02982081

, , csclimate

         (Intercept)    cmathach    bygender publicpublic    csclimate
Lag 0    -0.31535627 -0.10045813  0.05947034  0.362462041  1.000000000
Lag 10   -0.13606194 -0.03647130  0.07999367  0.153737944  0.329684022
Lag 50   -0.06059069  0.04314702  0.06779800  0.065021605  0.039198970
Lag 100  -0.01229294  0.02744985  0.05513630 -0.008904553 -0.004629702
Lag 500   0.03994165  0.04811244 -0.02461449 -0.007435269 -0.017548666
```

The output shows that the autocorrelations are small after lag 10 for all the predictor variables.

To increase the effective sample sizes, we increase the number of MCMC iterations from the default 13,000–30,000, set the number of burn-in iterations to 5,000, and rerun the model. The resulting sample size is 2,500, which equals the difference between the number of iterations and burn-in divided by the number of thinning interval: (30,000 − 5,000)/10 = 2,500.

```
> mepois.mc.2 <- MCMCglmm(byabsen ~ cmathach + bygender + public + csclimate,
+                   random =  ~ us(1 + bygender):SCH_ID,
+                   data=mecount, family = "poisson", prior=prior,
+                   burnin = 5000, nitt = 30000, thin = 10)

Acceptance ratio for liability set 1 = 0.452769
> summary(mepois.mc.2)

Iterations = 5001:29991
 Thinning interval  = 10
 Sample size  = 2500

DIC: 25131.96

 G-structure:  ~us(1 + bygender):SCH_ID

                              post.mean  l-95% CI  u-95% CI  eff.samp
(Intercept):(Intercept).SCH_ID   0.13206   0.09175   0.17403    191.6
bygender:(Intercept).SCH_ID     -0.07594  -0.11696  -0.03407    177.3
(Intercept):bygender.SCH_ID     -0.07594  -0.11696  -0.03407    177.3
bygender:bygender.SCH_ID         0.12587   0.08089   0.17396    248.6

 R-structure:  ~units

      post.mean l-95% CI u-95% CI eff.samp
units    0.1982   0.1625   0.2366    95.69
```

```
 Location effects: byabsen ~ cmathach + bygender + public + csclimate

               post.mean  l-95% CI  u-95% CI eff.samp   pMCMC
(Intercept)     -0.63277 -0.72443 -0.55315    588.8 <4e-04 ***
cmathach        -0.01839 -0.02045 -0.01609    590.0 <4e-04 ***
bygender         0.60855  0.54555  0.67076    448.1 <4e-04 ***
publicpublic    -0.44502 -0.52971 -0.36191    779.0 <4e-04 ***
csclimate       -0.01285 -0.06383  0.04301    762.0   0.645
---
Signif. codes:  0 '***' 0.001 '**' 0.01 '*' 0.05 '.' 0.1 ' ' 1
```

The effective sample sizes do increase for the random effects and the fixed effects, but the posterior means and the corresponding credible intervals are almost identical to those in the output with smaller effective sample sizes.

16.4 BAYESIAN MULTILEVEL NEGATIVE BINOMIAL REGRESSION WITH R

16.4.1 The brm() Function in the brms Package: Syntax for Bayesian Multilevel Negative Binomial Regression Models

Bayesian multilevel negative binomial regression is used when Bayesian methods are applied to multilevel modeling for a count response variable with overdispersion, where the variance of the response variable is a function of the mean μ and a dispersion parameter α. As of the time of writing, the `MCMCglmm()` function in the `MCMCglmm` package has not supported the negative binomial family, so we only introduce the `brm()` function in the `brms` package for the Bayesian multilevel negative binomial regression analysis of count response variables.

To fit a Bayesian multilevel negative binomial regression model with `brm()`, with the model formula for the fixed and random effects and the `data` argument the same as those for the Poisson model in the last section, we specify the `family = negbinomial` argument. The fitted model is named `menb.brm`. We use the `digits = 3` argument in the `print()` function to display three decimal places for the numerical values in the output.

```
> menb.brm <- brm(byabsen ~ cmathach + bygender + public + csclimate + (bygender|
SCH_ID),
+                  family = negbinomial, data=mecount,
+                  warmup = 500, iter = 3000,
+                  cores = 2, chains = 2, seed=123)
Compiling Stan program...
Start sampling
> print(menb.brm, digits = 3)
 Family: negbinomial
  Links: mu = log; shape = identity
Formula: byabsen ~ cmathach + bygender + public + csclimate + (bygender | SCH_ID)
   Data: mecount (Number of observations: 12183)
Samples: 2 chains, each with iter = 3000; warmup = 500; thin = 1;
         total post-warmup samples = 5000
```

```
Group-Level Effects:
~SCH_ID (Number of levels: 617)
                         Estimate Est.Error l-95% CI u-95% CI  Rhat Bulk_ESS
sd(Intercept)               0.334     0.032    0.272    0.398 1.001     1452
sd(bygender)                0.259     0.057    0.131    0.359 1.002      677
cor(Intercept,bygender)    -0.647     0.110   -0.821   -0.391 1.001     2443
                         Tail_ESS
sd(Intercept)                2633
sd(bygender)                  876
cor(Intercept,bygender)      2470

Population-Level Effects:
             Estimate Est.Error l-95% CI u-95% CI  Rhat Bulk_ESS Tail_ESS
Intercept      -0.539     0.041   -0.620   -0.458 1.001     3910     3433
cmathach       -0.018     0.001   -0.021   -0.016 1.000     5956     3896
bygender        0.613     0.030    0.555    0.670 1.002     4290     3808
publicpublic   -0.429     0.041   -0.509   -0.350 1.001     4638     3729
csclimate      -0.010     0.027   -0.062    0.043 1.000     4351     3878

Family Specific Parameters:
      Estimate Est.Error l-95% CI u-95% CI  Rhat Bulk_ESS Tail_ESS
shape    4.838     0.588    3.860    6.180 1.001     4467     3454

Samples were drawn using sampling(NUTS). For each parameter, Bulk_ESS
and Tail_ESS are effective sample size measures, and Rhat is the potential
scale reduction factor on split chains (at convergence, Rhat = 1).
```

The estimates or the posterior means for the random effects and the fixed effects are similar to those in the Bayesian multilevel Poisson model with the function in the preceding section, so interpretation of the results is omitted here. The `Rhat` values are equal to or close to 1 and the effective sample sizes are large, which indicate good MCMC convergence. The graphical evaluation of the convergence of MCMC with the `plot()` function is introduced later in this section.

The dispersion parameter labeled `shape` is provided in the section titled `Family Specific Parameters`. The posterior mean for the dispersion parameter is 4.838. Recall that in NB2 models, variance $(Y) = \mu(1 + \alpha\mu) = \mu + \alpha\mu^2$. We use a slightly different form for the variance of the count response variable here. Variance $(Y) = \mu + \mu^2/\theta$. Since $\theta = 4.838$, we can easily compute $\alpha = 1/4.838 = .207$. Please note that the dispersion parameter in your output will be slightly different from mine due to MCMC sampling, so when computing α, you need to adjust it accordingly.

```
> alpha <- 1/4.838
> alpha
[1] 0.206697
```

We extract the results of the estimates and the credible intervals with the `coef <- fixef(menb.brm, summary = TRUE)` command.

```
> coef <- fixef(menb.brm, summary = TRUE)
> coef
               Estimate   Est.Error        Q2.5       Q97.5
Intercept   -0.53909650 0.041071364 -0.61989856 -0.45761428
cmathach    -0.01836346 0.001141298 -0.02060532 -0.01617960
bygender     0.61250078 0.030072949  0.55451274  0.67047984
publicpublic -0.42914785 0.040821148 -0.50924746 -0.35014540
csclimate   -0.01009620 0.026671473 -0.06226952  0.04281903
```

We use the `exp(fixef(menb.brm)[, -2])` command to obtain the incidence rate ratios and the 95% corresponding credible intervals. The estimated errors should not be exponentiated, so we exclude the second column in the fixed-effects table with the indexing `[, -2]`. The following output is created.

```
> exp(fixef(menb.brm)[, -2])
              Estimate      Q2.5     Q97.5
Intercept    0.5832750 0.5379990 0.6327915
cmathach     0.9818041 0.9796055 0.9839506
bygender     1.8450397 1.7410924 1.9551753
publicpublic 0.6510637 0.6009476 0.7045856
csclimate    0.9899546 0.9396296 1.0437490
```

Interpretation of the incidence rate ratios for the predictor variables in multilevel negative binomial models is similar to that in multilevel Poisson regression models which were introduced in the preceding section, so it is omitted here.

We save the model object with the `save()` function so we can reuse it next time when needed with the `load()` function without refitting the model.

```
> save(menb.brm, file = "menbbrm.rda")
```

To graphically evaluate MCMC convergence, we use the `plot()` function to create the posterior density plots and the trace plots for the intercepts, the predictor variables, the dispersion parameter, and the random effects as follows.

```
> plot(menb.brm)
```

The posterior distribution plots and the trace plots (Figure 16.6) show good convergence for the intercept and the four predictor variables, the three random effects, and the dispersion parameter.

FIGURE 16.6 ● Posterior Density Plots and Trace Plots for the Fixed Effects, the Random Effects, and the Dispersion Parameter

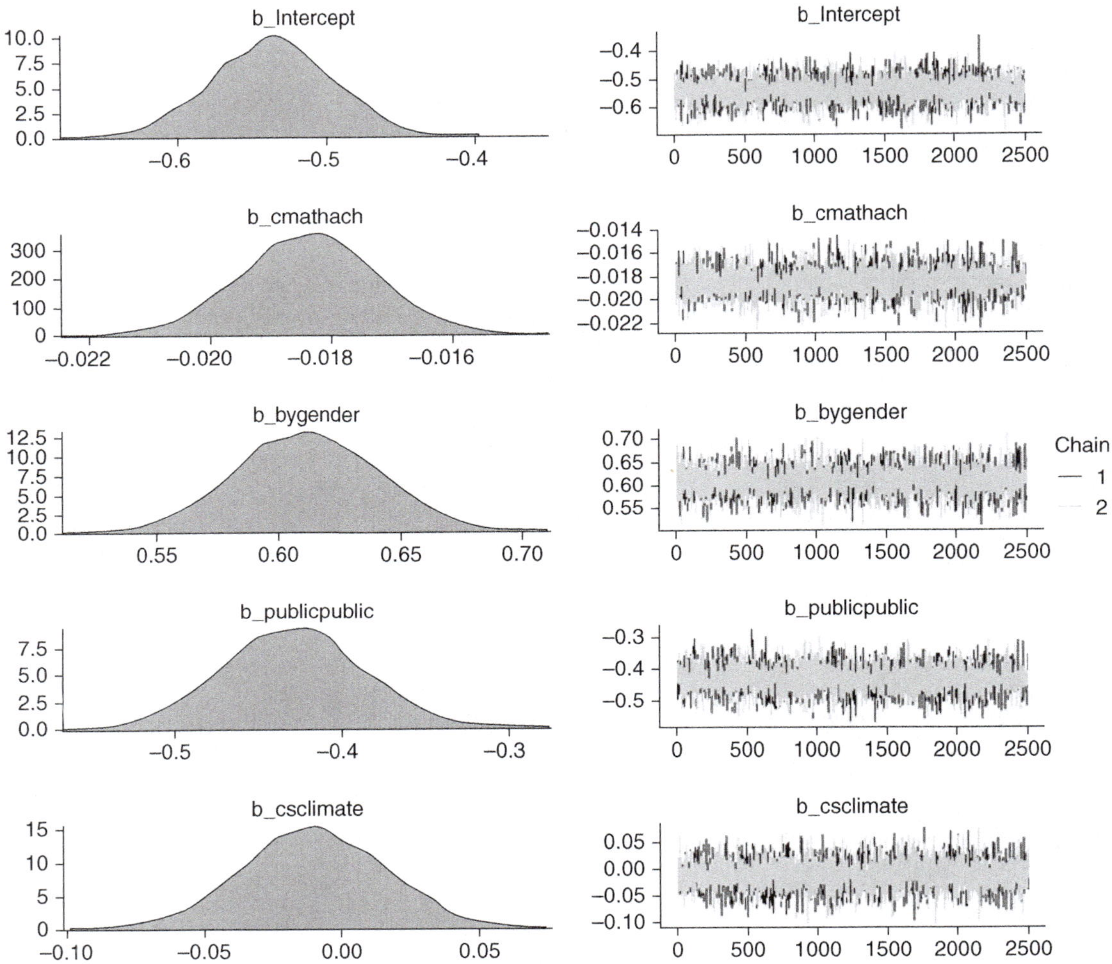

We use the `conditional_effects()` function to display the marginal effects of the predictor variables. The plots are omitted here.

```
> conditional_effects(menb.brm)
```

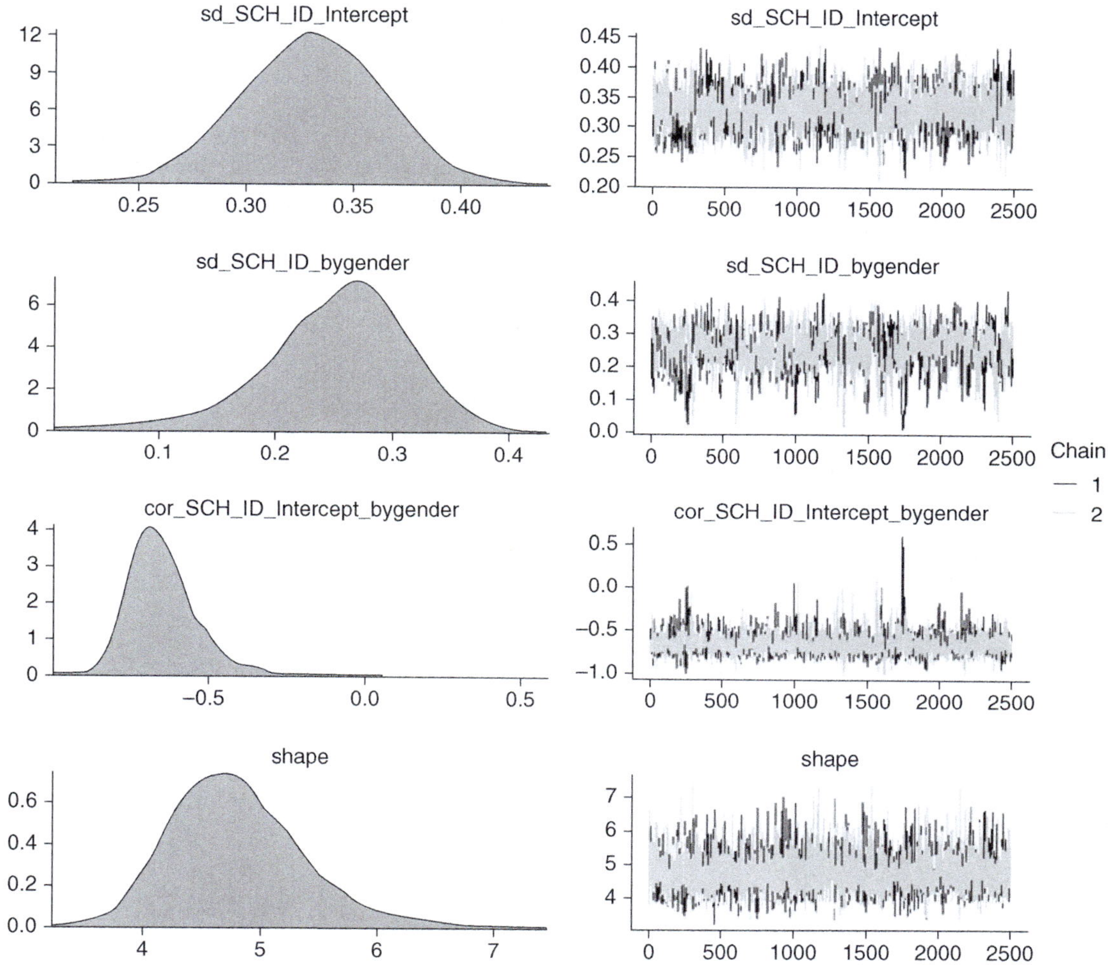

Figure 16.6 *(Continued).*

16.5 MAKING PUBLICATION-QUALITY TABLES

16.5.1 Presenting the Results Using the `texreg` Package

At the time of this writing, the `stargazer()` function cannot directly produce the results table from the models by the `brm()` and `MCMCglmm()` functions, so we use the `screenreg()` and `htmlreg()` functions from the `texreg` package (Leifeld, 2013). You need to install `texreg` first by typing `install.packages` ("texreg") if you have not done so and then load the package by typing library(texreg). This package needs to be installed only once.

Although several Bayesian multilevel models for categorical outcome variables are introduced in this chapter, due to space limitations, we only demonstrate how to use the texreg package to display the results for the Bayesian multilevel logistic regression model. The texreg package works with the brm() function, but it may be slow to create the results table since calculating the LOO information criterion (LOOIC) and the Watanabe-Akaike information criterion (WAIC) is computationally intensive. We need to first customize the extract() function and the createTexreg() function to make the texreg package work with the brms package (Leifeld, 2013). The R code for customizing the function is included at the end of the chapter.

After using the brm() function to fit the two Bayesian multilevel logistic regression model, melr.brm, we create a table containing the results of this model with the following command: screenreg(list(melr.brm)). In the screenreg() function, we specify the model object to be presented. The output is a plain text table.

```
> library(texreg)
> screenreg(list(melr.brm))
=============================================
                                     Model 1
---------------------------------------------
Intercept                             4.00 *
                                   [ 3.53; 4.49]
stlangYes                             0.69 *
                                   [ 0.37; 0.98]
csclimat                              0.60 *
                                   [ 0.41; 0.79]
publicpublic                         -1.31 *
                                   [-1.75; -0.88]
---------------------------------------------
SD: SCH_ID: sd(Intercept)             1.07
SD: SCH_ID: sd(stlangYes)             0.83
SD: SCH_ID: cor(Intercept,stlangYes) -0.49
Num. obs.                            12793
=============================================
* 0 outside the confidence interval.
> htmlreg(list(melr.brm), file="melrbr.doc", doctype=TRUE, html.tag=TRUE,
head.tag=TRUE)
The table was written to the file 'melrbr.doc'.
```

To create the table in the HTML format and copy it into Microsoft Word, we use the command as follows: htmlreg(list(melr.brm), file = "melrbr.doc", doctype = TRUE, html.tag = TRUE, head.tag = TRUE). The table presenting the results for the Bayesian multilevel logistic regression model in its original format is omitted here. With edits on the model title and the added labels for the fixed and random effects, the final table is displayed as in Table 16.1.

16.6 REPORTING THE RESULTS

Similar reporting guidelines for multilevel modeling for categorical response variables can be applied to the reporting for Bayesian multilevel models. The following are the generic guidelines which may vary across disciplines and journals.

First, as with other research examples, describe the purpose of your study and explain why the Bayesian multilevel modeling is needed for the analysis.

TABLE 16.1 ● Results of the Bayesian Multilevel Logistic Regression Models With `brm()` (Edited)

	Random-Coefficient Logistic Regression Model
	Estimates (Credible Intervals)
Fixed Effects	
`Intercept`	4.00[a]
	[3.53; 4.49]
`stlang`	0.69[a]
	[0.37; 0.98]
`csclimat`	0.60[a]
	[0.41; 0.79]
`public`	−1.31[a]
	[−1.75; −0.88]
Random Effects	
`SD: SCH_ID: sd(Intercept)`	1.07
`SD: SCH_ID: sd(stlang)`	0.83
`SD: SCH_ID: cor(Intercept,stlang)`	−0.49
Num. obs.	12,793

[a]0 outside the confidence interval.

Second, evaluate MCMC convergence based on available measures, such as the Rhat, the effective sample sizes, and the autocorrelations. In addition, use available graphical tools, such as trace plots and density plots.

Third, report the results of the fitted models in a table including both the parameter estimates for the fixed effects and the variances and covariances or the standard deviations and correlations for the random effects.

The following is an example of summarizing the results of a Bayesian multilevel logistic regression model.

Bayesian multilevel logistic regression models were fitted with the `brm()` function to predict whether high-school students were capable of doing simple arithmetical operations on whole numbers from three predictor variables: whether English is the student's native language, the school type, and the school climate. The first predictor was a student-level variable and the other two were school-level variables. Bayesian multilevel modeling for binary response variables was used since the outcome variable was dichotomous.

Evaluation of MCMC Convergence

The Rhat values were equal to or close to 1 and the effective sample sizes were large, which indicate good MCMC convergence. The graphical evaluation of the convergence of MCMC with the `plot()` function also shows that MCMC had converged in the posterior density plots and the trace plots for the intercepts and the predictor variables.

Results for the Random-Coefficient Model With Both Level 1 and Level 2 Predictor Variables

The logit coefficient for `stlang` (γ_{10}) = .618, and the 95% credible interval is [.368, .981]. The odds ratio for `stlang` is 1.994, which indicates that the odds of being proficient in math for students whose native language is English are 1.994 times as high as the odds for those whose native language is not English.

The logit coefficient for `csclimat` (γ_{02}) = .599 and the 95% credible interval is [.413, .788]. The odds ratio for `csclimat` is 1.820, which indicates that a one-unit increase in school climate corresponds to an increase by a factor of 1.820 points in the odds of being proficient in math. In other words, being students in schools with a more positive school climate increases the odds of being proficient in math.

The logit coefficient for `public` (γ_{01}) = −1.308 and the 95% credible interval is [−1.751, −0.877]. The odds ratio for `public` is .270, which indicates that the odds of being proficient in math for students in public schools are .270 times the odds for those in private schools.

Regarding the variance and covariance components for the random effects, the residual variance in the intercept (τ_{00}) is 1.141 and the slope variance for `stlang` is .696. The covariance between the intercept and the slope was −.438.

16.7 SUMMARY OF R COMMANDS IN THIS CHAPTER

```
# Chap 16 R Script

# Remove all objects
rm(list = ls(all = TRUE))
# The following user-written packages need to be installed first
# Use install.packages(" ") and then load it with library()
# library(brms)                  # It is already installed for Chapter 14
# library(MCMCglmm)              # It is already installed for Chapter 14
# library(texreg)                # It is already installed for Chapter 4

# Import the dataset
library(foreign)
els <- read.dta("C:/CDA/els2002.dta")
els <- els[!is.na(els$Profmath1)&!is.na(els$stlang)&!is.na(els$public)&!is.
na(els$sclimate), ]
attach(els)
els$csclimat <- sclimate-mean(sclimate, na.rm=TRUE)
summary(els$csclimat)

# Bayesian multilevel logistic regression models
# It may take several minutes to fit the model
library(brms)
melr.brm <- brm(Profmath1~ stlang + csclimat + public + (stlang|SCH_ID),
               data=els, family=bernoulli,
               warmup = 500, iter = 2000,
               cores = 2, chains = 2, seed=123)
summary(melr.brm)
print(melr.brm, digits=3)
coef <- fixef(melr.brm, summary=TRUE)
coef
exp(fixef(melr.brm)[, -2])
save(melr.brm, file = "melrbrm.rda")
plot(melr.brm)
conditional_effects(melr.brm)
```

```
# Present the results with texreg()
library(texreg)
# Customize the extract() function and the createTexreg() function
extract.brmsfit <- function (model,
                             use.HDI = TRUE,
                             HDI.prob = 0.9,
                             include.random = TRUE,
                             include.rsquared = FALSE,
                             include.nobs = TRUE,
                             include.loo.ic = FALSE,
                             reloo = FALSE,
                             include.waic = FALSE,
                             ...) {
 sf <- summary(model, ...)$fixed
 coefnames <- rownames(sf)
 coefs <- sf[, 1]
 se <- sf[, 2]
 ci.low = sf[, 3]
 ci.up = sf[, 4]
 gof <- numeric()
 gof.names <- character()
 gof.decimal <- logical()
 if (isTRUE(include.random) & isFALSE(!nrow(model$ranef))) {
   sr <- summary(model, ...)$random
   sd.names <- character()
   sd.values <- numeric()
   for (i in 1:length(sr)) {
     sd <- sr[[i]][, 1]
     sd.names <- c(sd.names, paste0("SD: ", names(sr)[[i]], ": ", names(sd)))
     sd.values <- c(sd.values, sd)
   }
   gof <- c(gof, sd.values)
   gof.names <- c(gof.names, sd.names)
   gof.decimal <- c(gof.decimal, rep(TRUE, length(sd.values)))
 }
 if (isTRUE(include.nobs)) {
   n <- stats::nobs(model)
   gof <- c(gof, n)
   gof.names <- c(gof.names, "Num. obs.")
   gof.decimal <- c(gof.decimal, FALSE)
 }
 tr <- createTexreg(coef.names = coefnames,
                    coef = coefs,
                    se = se,
                    ci.low = ci.low,
                    ci.up = ci.up,
                    gof.names = gof.names,
                    gof = gof,
                    gof.decimal = gof.decimal)
   return(tr)
}
# Register the function with setMethod()
setMethod("extract", signature = className("brmsfit", "brm"),
          definition = extract.brmsfit)
screenreg(list(melr.brm))
htmlreg(list(melr.brm), file="melrbr.doc", doctype=TRUE, html.tag=TRUE, head.
 tag=TRUE)
```

```
# It may take a couple of minutes to fit the model with MCMCglmm()
library(MCMCglmm)
set.seed(123)
prior = list(R = list(V = 1, nu = 0.002, fix = 1),
             G = list(G1 = list(V = diag(2), nu = 2)))
melr.mc <- MCMCglmm(Profmath1 ~ stlang + csclimat + public,
                    random =  ~ us(1 + stlang):SCH_ID,
                    data=els, family = "categorical", prior=prior)
summary(melr.mc)
save(melr.mc, file = "melrmc.rda")
detach(els)


# Bayesian multilevel ordinal logistic regression models
# It may take around 1 hour or more to fit the model with brm()
library(foreign)
els <- read.dta("C:/CDA/els2002.dta")
attach(els)
els <- els[!is.na(Profmath)&!is.na(stlang)&!is.na(public)&!is.na(sclimate), ]
els$Profmath <- factor(els$Profmath, ordered=TRUE)
els$SCH_ID <- factor(els$SCH_ID)
els$csclimat <- els$sclimate-mean(els$sclimate, na.rm=TRUE)
summary(els$csclimat)


library(brms)
mpo.brm <- brm(Profmath ~ stlang + public + csclimat + (stlang|SCH_ID),
               data=els, family=cumulative,
               warmup = 500, iter = 2000,
               cores = 2, chains = 2, seed=123)
summary(mpo.brm)
print(mpo.brm, digits=3)
coef <- fixef(mpo.brm, summary=TRUE)
coef
exp(fixef(mpo.brm)[, -2])
save(mpo.brm, file = "mpo.rda")
plot(mpo.brm)
conditional_effects(mpo.brm, categorical=TRUE)
detach(els)

# Bayesian multilevel multinomial logistic regression models (see Chapter 14)

# Bayesian multilevel Poisson regression models
library(foreign)
mecount <- read.dta("C:/CDA/mecount.dta")
attach(mecount)
library(sjmisc)
descr(mecount, byabsen, cmathach, bygender, public, csclimate)


# It may take around 10 minutes or more to compile Stan and fit the model with brm()
library(brms)
mepois.brm <- brm(byabsen ~ cmathach + bygender + public + csclimate + (bygender|
SCH_ID),
               family = poisson, data=mecount,
               warmup = 500, iter = 3000,
               cores = 2, chains = 2, seed=123)
summary(mepois.brm)
print(mepois.brm, digits=3)
coef <- fixef(mepois.brm, summary=TRUE)
coef
```

```
exp(fixef(mepois.brm)[, -2])
save(mepois.brm, file = "mepoisbrm.rda")
plot(mepois.brm)
conditional_effects(mepois.brm)


# It may take several minutes to fit the model with MCMCglmm()
library(MCMCglmm)
set.seed(123)
prior = list(R = list(V = 1, nu = 0.002),
             G = list(G1 = list(V = diag(2), nu = 2)))
mepois.mc<-MCMCglmm(byabsen ~ cmathach + bygender + public + csclimate,
                    random =  ~ us(1 + bygender):SCH_ID,
                    data=mecount, family = "poisson", prior=prior)
summary(mepois.mc)
autocorr(mepois.mc$Sol)
# It may take several minutes to fit the following model with MCMCglmm()
mepois.mc.2 <- MCMCglmm(byabsen ~ cmathach + bygender + public + csclimate,
                        random =  ~ us(1 + bygender):SCH_ID,
                        data=mecount, family = "poisson", prior=prior,
                        burnin = 5000, nitt = 30000, thin = 10)
summary(mepois.mc.2)


# Bayesian multilevel negative binomial regression models
# It may take 20 minutes or more to fit the model with brm()
menb.brm <- brm(byabsen ~ cmathach + bygender + public + csclimate + (bygender|
SCH_ID),
                family = negbinomial, data=mecount,
                warmup = 500, iter = 3000,
                cores = 2, chains = 2, seed=123)
summary(menb.brm)
print(menb.brm, digits=3)
coef <- fixef(menb.brm, summary=TRUE)
coef
exp(fixef(menb.brm)[, -2])
save(menb.brm, file = "menbbrm.rda")
plot(menb.brm)
conditional_effects(menb.brm)
detach(mecount)
```

Glossary

Bayesian multilevel modeling for binary outcome variables is referred to as Bayesian multilevel logistic regression when Bayesian methods are applied to multilevel logistic regression.

Bayesian multilevel negative binomial regression is used when Bayesian methods are applied to multilevel modeling for a count response variable with overdispersion, where the variance of the response variable is a function of the mean μ and a dispersion parameter α.

Bayesian multilevel ordinal logistic regression is used when Bayesian methods are applied to multilevel modeling for ordinal response variables.

Bayesian multilevel Poisson regression is used when Bayesian methods are applied to multilevel modeling for count response variables.

Exercises

Use the ELS:2002 data available at **https://edge.sagepub.com/liu1e** to conduct Bayesian multilevel modeling for a binary response variable. The following variables are used for the analysis.
`Profmath2`: math proficiency level 2 (1 = capable of doing simple operations with decimals, fractions, powers, and root; 0 = not achieving this level)
`gender`: gender (1 = female; 0 = male)
`byses`: socioeconomic status composite for base-year data
`urban`: urban schools or not (1 = urban; 0 = suburban or rural areas) (school-level variable).

Use a Bayesian multilevel logistic regression model to estimate whether students are capable of doing simple operations with decimals, fractions, powers, and root. The binary outcome variable is `Profmath2`, and the predictor variables are `gender`, `byses`, and `urban`. Perform the analysis and answer the following questions:

1. Use the `brm()` function to fit a two-level, random-intercept model with `gender` and `byses` as the student-level predictor variables and `urban` as the school-level predictor variable. Use the grand-mean centering for `byses` before fitting the model. No cross-level interactions are included. Also, specify 2,000 iterations with 500 warm-up iterations and two chains.
2. Evaluate MCMC convergence.
3. Use the `exp()` function to obtain the odds ratios and the corresponding 95% credible intervals for the fixed effects.
4. Interpret the odds ratio of `urban` in the model.
5. Make a publication-quality table containing the fixed effects and random effects for the model.
6. Write a report to summarize the results of the model.

REFERENCES

Agresti, A. (2010). *Analysis of ordinal categorical data* (2nd ed.). John Wiley & Sons.

Agresti, A. (2013). *Categorical data analysis* (3rd ed.). John Wiley & Sons.

Agresti, A. (2015). *Foundations of linear and generalized linear models*. John Wiley & Sons.

Agresti, A. (2019). *An introduction to categorical data analysis* (3rd ed.). John Wiley & Sons.

Akaike, H. (1974). A new look at the statistical model identification. *IEEE Transactions on Automatic Control*, AC-19, 716–723.

Allison, P. D. (2012). *Logistic regression using the SAS system: Theory and application* (2nd ed.). SAS Institute.

American Psychological Association (APA). (2020). *Publication manual of the American Psychological Association: The official guide to APA style* (7th ed.). American Psychological Association (APA).

Ananth, C. V., & Kleinbaum, D. G. (1997). Regression models for ordinal responses: A review of methods and applications. *International Journal of Epidemiology*, 26, 1323–1333.

Anderson, J. A. (1984). Regression and ordered categorical variables. *Journal of Royal Statistical Society, Series B*, 46, 1–30.

Archer, K. J., Hedeker, D., Nordgren, R., & Gibbons, R. D. (2018). *mixor: An R package for longitudinal and clustered ordinal response modeling*. (R package version 1.0.4) [Computer software]. https://cran.r-project.org/web/packages/mixor/vignettes/mixor.pdf

Armstrong, B. B., & Sloan, M. (1989). Ordinal regression models for epidemiological data. *American Journal of Epidemiology*, 129(1), 191–204.

Bates, D., Mächler, M., Bolker, B., & Walker, S. (2015). Fitting linear mixed-effects models using lme4. *Journal of Statistical Software*, 67(1), 1–48. https://doi.org/10.18637/jss.v067.i01

Bilder, C. R., & Loughin, T. M. (2015). *Analysis of categorical data with R*. Chapman & Hall/CRC.

Bolker, B., R Development Core Team, & Giné-Vázquez, I. (2021). *bbmle: Tools for general maximum likelihood estimation*. (R package version 1.0.24) [Computer software]. https://cran.r-project.org/web/packages/bbmle/index.html

Borooah, V. K. (2002). *Logit and probit: Ordered and multinomial models*. Sage.

Brooks, M. E., Kristensen, K., van Benthem, K. J., Magnusson, A., Berg, C. W., Nielsen, A., Skaug, H. J., Maechler, M., & Bolker, B. M. (2017). glmmTMB balances speed and flexibility among packages for zero-inflated generalized linear mixed modeling. *The R Journal*, 9(2), 378–400. https://journal.r-project.org/archive/2017/RJ-2017-066/index.html

Buis, M. L. (2007). *seqlogit: Stata module to fit a sequential logit model*. Statistical Software Components. Boston College Department of Economics. http://econpapers.repec.org/software/bocbocode/s456843.htm

Bürkner, P. (2017). brms: An R package for Bayesian multilevel models using Stan. *Journal of Statistical Software*, 80(1), 1–28.

Cameron, A. C., & Trivedi, P. K. (2013). *Regression analysis of count data* (2nd ed.). Cambridge University Press.

Carlin, B. P., & Louis, T. A. (2009). *Bayesian methods for data analysis* (3rd ed.). Chapman & Hall/CRC.

Christensen, R. H. B. (2019). *Ordinal: Regression models for ordinal data*. (R package version 2019.12-10)

[Computer software]. http://www.cran.r-project.org/package=ordinal/

Clogg, C. C., & Shihadeh, E. S. (1994). *Statistical models for ordinal variables*. Sage.

Cole, S. R., & Ananth, C. V. (2001). Regression models for unconstrained, partially or fully constrained continuation odds ratios. *International Journal of Epidemiology*, 30, 1379–1382.

Croissant, Y. (2020). Estimation of random utility models in R: The mlogit package. *Journal of Statistical Software*, 95(11), 1–41. https://doi.org/10.18637/jss.v095.i11

Elff, M. (2021). *mclogit: Multinomial logit models with or without random effects or overdispersion*. (R package version 0.8.7.3) [Computer software]. https://cran.r-project.org/web/packages/mclogit/index.html

Enders, C. K., & Tofighi, D. (2007). Centering predictor variables in cross-sectional multilevel models: A new look at an old issue. *Psychological Methods*, 12, 121–138.

Fienberg, S. E. (1980). *The analysis of cross-classified categorical data*. The MIT Press.

Fox, J. (2005). The R commander: A basic statistics graphical user interface to R. *Journal of Statistical Software*, 14(9), 1–42. https://www.jstatsoft.org/article/view/v014i09

Fox, J. (2016). *Applied regression analysis and generalized linear models* (3rd ed.). Sage.

Fox, J. (2017). *Using the R commander: A point-and-click interface for R*. Chapman & Hall/CRC.

Fox, J., & Weisberg, S. (2019). *An R companion to applied regression* (3rd ed.). Sage.

Friendly, M., & Meyer, D. (2016). *Discrete data analysis with R: Visualization and modeling techniques for categorical and count data*. Chapman & Hall/CRC.

Fu, V. (1998). Estimating generalized ordered logit models. *Stata Technical Bulletin*, 44, 27–30.

Fullerton, A. S. (2009). A conceptual framework for ordered logistic regression models. *Sociological Methods and Research*, 38, 306–347.

Fullerton, A. S., & Xu, J. (2016). *Ordered regression models: Parallel, partial, and non-parallel alternatives*. Chapman & Hall/CRC.

Gabry, J. (2018). *shinystan: Interactive visual and numerical diagnostics and posterior analysis for Bayesian models*. (R package version 2.5.0) [Computer software]. http://mc-stan.org/shinystan/

Garson, G. D. (Ed.) (2013). *Hierarchical linear modeling: Guide and applications*. Sage.

Garson, G. D. (2020). *Multilevel modeling: Applications in STATA, IBM SPSS, SAS, R, & HLM*. Sage.

Gelman, A., Carlin, J. B., Stern, H. S., Dunson, D. B., Vehtari, A., & Rubin, D. B. (2013). *Bayesian data analysis* (3rd ed.). Chapman & Hall/CRC.

Gelman, A., Jakulin, A., Pittau, M. G., & Su, Y. S. (2008). A weakly informative default prior distribution for logistic and other regression models. *Annals of Applied Statistics*, 2, 1360–1383.

Gill, J. (2015). *Bayesian methods: A social and behavioral sciences approach* (3rd ed.). Chapman & Hall/CRC.

Goldstein, H. (2003). *Multilevel statistical models* (3rd ed.). Edward Arnold.

Goodman, L. A. (1983). The analysis of dependence in cross-classifications having ordered categories, using loglinear models for dependencies and log-linear models for odds. *Biometrics*, 39, 149–160.

Goodrich, B., Gabry, J., Ali, I., & Brilleman, S. (2020). *rstanarm: Bayesian applied regression modeling via Stan*. (R package version 2.21.1) [Computer software]. https://mc-stan.org/rstanarm

Greenland, S. (1994). Alternative models for ordinal logistic regression. *Statistics in Medicine*, 13(16), 1665–1677.

Grosjean, P., Ibanez, F., & Etienne, M. (2018). *pastecs: Package for analysis of space-time ecological series*. (R package version 1.3.21) [Computer software]. https://cran.r-project.org/web/packages/pastecs/index.html

Hadfield, J. D. (2010). MCMC methods for multi-response generalized linear mixed models: The MCMCglmm R package. *Journal of Statistical Software*, 33(2), 1–22.

Hardin, J. W., & Hilbe, J. M. (2018). *Generalized linear models and extensions* (4th ed.). Stata Press.

Heck, R. H., & Thomas, S. L. (2015). *An introduction to multilevel modeling techniques: MLM and SEM approaches using Mplus* (3rd ed.). Routledge.

Heck, R. H., Thomas, S. L., & Tabata, L. N. (2010). *Multilevel and longitudinal modeling with IBM SPSS*. Routledge.

Heck, R. H., Thomas, S. L., & Tabata, L. N. (2012). *Multilevel modeling of categorical outcomes using IBM SPSS*. Routledge.

Hedeker, D. (2007). Multilevel models for ordinal and nominal variables. In J. de Leeuw, & E. Meijer (Eds.), *Handbook of multilevel analysis* (pp. 237–274). Springer.

Hedeker, D., & Gibbon, R. D. (1994). A random-effects ordinal regression model for multilevel analysis. *Biometrics*, 50(4), 933–944.

Hilbe, J. M. (2009). *Logistic regression models*. Chapman & Hall/CRC.

Hilbe, J. M. (2011). *Negative binomial regression* (2nd ed.). Cambridge University Press.

Hilbe, J. M. (2014). *Modeling count data*. Cambridge University Press.

Hlavac, M. (2018). *stargazer: Well-formatted regression and summary statistics tables*. (R package version 5.2.1) [Computer software]. https://CRAN.R-project.org/package=stargazer

Hoffmann, J. P. (2016). *Regression models for categorical, count, and related variables: An applied approach*. University of California Press.

Hofmann, D. A., & Gavin, M. B. (1998). Centering decisions in hierarchical linear models: Implications for research in organizations. *Journal of Management*, 24(5), 623–641.

Hosmer, D. W., Lemeshow, S., & Sturdivant, R. X. (2013). *Applied logistic regression* (3rd ed.). John Wiley & Sons.

Hox, J. J. (2010). *Multilevel analysis: Techniques and applications* (2nd ed.). Routledge.

Huck, S. W. (2012). *Reading statistics and research* (6th ed.). Pearson.

Jackman, S. (2020). *pscl: Classes and methods for R developed in the political science computational laboratory*. (R package version 1.5.5) [Computer software]. https://github.com/atahk/pscl/

Kaplan, D. (2014). *Bayesian statistics for the social sciences*. The Guilford Press.

Kaufman, R. L. (2019). *Interaction effects in linear and generalized linear models: Examples and applications using Stata*. Sage.

Kline, R. B. (2013). *Beyond significance testing: Statistics reform in the behavioral sciences* (2nd ed.). American Psychological Association.

Kreft, I. G. G., & de Leeuw, J. (1998). *Introducing multilevel modeling*. Sage.

Kreft, I. G. G., de Leeuw, J., & Aiken, L. (1995). The effect of different forms of centering in hierarchical linear models. *Multivariate Behavioral Research*, 30, 1–22.

Kruschke, J. K. (2015). *Doing Bayesian data analysis: A tutorial with R, JAGS, and Stan* (2nd ed.). Academic Press.

Kuss, O. (2006). On the estimation of the stereotype regression model. *Computational Statistics & Data Analysis*, 50, 1877–1890.

Kuznetsova, A., Brockhoff, P. B., & Christensen, R. H. B. (2017). lmerTest package: Tests in linear mixed effects models. *Journal of Statistical Software*, 82(13), 1–26. https://doi.org/10.18637/jss.v082.i13

Lambert, D. (1992). Zero-inflated Poisson regression with an application to defects in manufacturing. *Technometrics*, 34, 1–14.

Leeper, T. J. (2021). *margins: Marginal effects for model objects*. (R package version 0.3.26) [Computer software]. https://cran.r-project.org/web/packages/margins/citation.html

Leifeld, P. (2013). texreg: Conversion of statistical model output in R to LaTeX and HTML tables. *Journal of Statistical Software*, 55(8), 1–24. https://doi.org/10.18637/jss.v055.i08

Lele, S. R., Keim, J. L., & Solymos, P. (2019). *ResourceSelection: Resource selection (probability) functions for use-availability data*. (R package version 0.3-5) [Computer software]. https://cran.r-project.org/web/packages/ResourceSelection/index.html

Liao, T. F. (1994). *Interpreting probability models: Logit, probit, and other generalized linear models*. Sage.

Liu, X. (2009). Ordinal regression analysis: Fitting the proportional odds model using Stata, SAS and SPSS. *Journal of Modern Applied Statistical Methods*, 8(2), 632–645.

Liu, X. (2014). Fitting stereotype logistic regression models for ordinal response variables in educational research. *Journal of Modern Applied Statistical Methods*, 13(2), 528–543.

Liu, X. (2016a). *Applied ordinal logistic regression using Stata: From single-level to multilevel modeling*. Sage.

Liu, X. (2016b). Proportional odds models for ordinal response variables with complex sampling designs using SAS, IBM SPSS and Stata. *General Linear Model Journal*, 42, 30–39.

Liu, X., & Bai, H. (2019). Forward and backward continuation ratio models for ordinal response variables. *Journal of Modern Applied Statistical Methods*, 18(1), 1–16.

Liu, X., Bai, H., & Koirala, H. (2018). Multiple imputation for missing data analysis in proportional odds models for ordinal response variables. *General Linear Model Journal*, 44, 1–10.

Liu, X., & Koirala, H. (2012). Ordinal regression analysis: Using generalized ordinal logistic regression models to estimate educational data. *Journal of Modern Applied Statistical Methods*, 11(1), 242–254.

Liu, X., & Koirala, H. (2013). Fitting proportional odds models to educational data with complex sampling designs in ordinal logistic regression. *Journal of Modern Applied Statistical Methods*, 12(1), 235–248.

Liu, X., O'Connell, A. A., & Koirala, H. (2011). Ordinal regression analysis: Predicting mathematics proficiency using the continuation ratio model. *Journal of Modern Applied Statistical Methods*, 10(2), 513–527.

Long, J. S. (1997). *Regression models for categorical and limited dependent variables*. Sage.

Long, J. S., & Freese, J. (2014). *Regression models for categorical dependent variables using Stata* (3rd ed.). Stata Press.

Lüdecke, D. (2018a). sjmisc: Data and variable transformation functions. *Journal of Open Source Software*, 3(26), 754. https://doi.org/10.21105/joss.00754

Lüdecke, D. (2018b). ggeffects: Tidy data frames of marginal effects from regression models. *Journal of Open Source Software*, 3(26), 772. https://doi.org/10.21105/joss.00772

Lüdecke, D. (2021). *sjstats: Statistical functions for regression models*. (R package version 0.18.1) [Computer software]. https://CRAN.R-project.org/package=sjstats

Luke, D. A. (2004). *Multilevel modeling*. Sage.

Lunt, M. (2001). Stereotype ordinal regression. *Stata Technical Bulletin*, 61, 12–18.

Ma, X., Ma, L., & Bradley, K. D. (2008). Using multilevel modeling to investigate school effects. In A. A. O'Connell, & D. B. McCoach (Eds.), *Multilevel modeling of educational data* (pp. 59–110). Information Age.

Maas, C. J. M., & Hox, J. J. (2005). Sufficient sample sizes for multilevel modeling. *Methodology: European Journal of Research Methods for the Behavioral and Social Sciences*, 1(3), 86–92.

Mangiafico, S. (2021) *rcompanion: Functions to support extension education program evaluation*. (R package version 2.4.1) [Computer software]. https://cran.r-project.org/web/packages/rcompanion/index.html

McCoach, D. B. (2010). Hierarchical linear modeling. In G. R. Hancock, & R. O. Mueller (Eds.), *The reviewers' guide to quantitative methods in the social sciences* (pp. 123–140). Routledge.

McCullagh, P. (1980). Regression models for ordinal data [with discussion]. *Journal of the Royal Statistical Society Series B*, 42, 109–142.

McCullagh, P., & Nelder, J. A. (1989). *Generalized linear models* (2nd ed.). Chapman and Hall.

McElreath, R. (2020). *Bayesian rethinking: A Bayesian course with examples in R and Stan* (2nd ed.). Chapman & Hall/CRC.

Menard, S. (2002). *Applied logistic regression analysis* (2nd ed.). Sage.

Menard, S. (2010). *Logistic regression: From introductory to advanced concepts and applications*. Sage.

Meyer, D., Zeileis, A., & Hornik, K. (2020). *vcd: Visualizing categorical data*. (R package version 1.4-8) [Computer software]. https://cran.r-project.org/web/packages/vcd/

O'Connell, A. A. (2000). Methods for modeling ordinal outcome variables. *Measurement and Evaluation in Counseling and Development*, 33(3), 170–193.

O'Connell, A. A. (2006). *Logistic regression models for ordinal response variables*. Sage.

O'Connell, A. A., & Amico, K. R. (2010). Logistic regression. In G. R. Hancock, & R. O. Mueller (Eds.), *The reviewers' guide to quantitative methods in the social sciences* (pp. 221–239). Routledge.

O'Connell, A. A., Goldstein, J., Rogers, H. J., & Peng, C. Y. J. (2008). Multilevel logistic models for dichotomous and ordinal data. In A. A. O'Connell, & D. B. McCoach (Eds.), *Multilevel modeling of educational data* (pp. 199–242). Information Age.

Paccagnella, O. (2006). Centering or not centering in multilevel models? *Evaluation Review*, 30, 66–85.

Peng, C. Y. J., Lee, K. L., & Ingersoll, G. M. (2002). An introduction to logistic regression and reporting. *The Journal of Educational Research*, 96(1), 3–14.

Peterson, B., & Harrell, F. E. (1990). Partial proportional odds models for ordinal response variables. *Applied Statistics*, 39(2), 205–217.

Pinheiro, J., Bates, D., DebRoy, S., Sarkar, D., & R Core Team (2021). *nlme: Linear and nonlinear mixed effects models*. (R package version 3.1-152) [Computer software]. https://CRAN.R-project.org/package=nlme

Powers, D. A., & Xie, Y. (2008). *Statistical models for categorical data analysis* (2nd ed.). Emerald.

R Core Team (2021). R: A language and environment for statistical computing. *R Foundation for Statistical Computing*. https://www.R-project.org/

Rabe-Hesketh, S., & Skrondal, A. (2012). *Multilevel and longitudinal modeling using Stata* (3rd ed.). Stata Press.

Raudenbush, S. W., & Bryk, A. S. (2002). *Hierarchical linear models: Applications and data analysis methods* (2nd ed.). Sage.

RStudio Team (2020). RStudio: Integrated development for R. *RStudio, Inc*. http://www.rstudio.com/

Schwarz, G. (1978). Estimating the dimension of a model. *Annals of Statistics*, 6, 461–464.

Singer, J. D., & Willett, J. B. (2003). *Applied longitudinal data analysis: Modeling change and event occurrence*. Oxford University Press.

Smithson, M., & Merkle, E. C. (2014). *Generalized linear models for categorical and continuous limited dependent variables*. Chapman & Hall/CRC.

Snijders, T. A., & Bosker, R. (2012). *Multilevel analysis: An introduction to basic and advanced multilevel modeling* (2nd ed.). Sage.

Tang, W., He, H., & Tu, X. M. (2013). *Applied categorical and count data analysis*. Chapman & Hall/CRC.

Tutz, G. (1991). Sequential models in categorical regression. *Computational Statistics & Data Analysis*, 11(3), 275–295.

Tutz, G. (2012). *Regression for categorical data*. Cambridge University Press.

Vehtari, A., Gelman, A., & Gabry, J. (2017). Practical Bayesian model evaluation using leave-one-out cross-validation and WAIC. *Statistics and Computing*, 27(5), 1413–1432.

Venables, W. N., & Ripley, B. D. (2002). *Modern applied statistics with S* (4th ed.). Springer.

Wakefield, J. (2013). *Bayesian and frequentist regression models*. Springer.

Warnes, G. R., Bolker, B., Lumley, T., & Johnson, R. C. (2018). *gmodels: Various R programming tools for model fitting*. (R package version 2.18.1) [Computer software]. https://cran.r-project.org/web/packages/gmodels/index.html

Watanabe, S. (2010). Asymptotic equivalence of Bayes cross validation and widely applicable information criterion in singular learning theory. *Journal of Machine Learning Research*, 11, 3571–3594.

West, B. T., Welch, K. B., & Gałecki, A. T. (2014). *Linear mixed models: A practical guide using statistical software* (2nd ed.). Chapman & Hall/CRC.

Wickham, H. (2007). Reshaping data with the reshape package. *Journal of Statistical Software*, 21(12), 1–20. http://www.jstatsoft.org/v21/i12/

Wickham, H. (2016). *ggplot2: Elegant graphics for data analysis*. Springer-Verlag.

Wickham, H., Averick, M., Bryan, J., Chang, W., McGowan, L. D., François, R., Grolemund, G., Hayes, A., Henry, L., Hester, J., Kuhn, M., Pedersen, T., Miller, E., Bache, S., Müller, K., Ooms, J., Robinson, D., Seidel, D., Spinu, V., & Yutani, H. (2019). Welcome to the tidyverse. *Journal of Open Source Software*, 4(43), 1686. https://doi.org/10.21105/joss.01686

Wickham, H., Francois, R., Henry, L., & Müller, K. (2021). *dplyr: A grammar of data manipulation*. (R package version 1.0.6) [Computer software]. http://CRAN.R-project.org/package=dplyr

Williams, R. (2006). Generalized ordered logit/partial proportional odds models for ordinal dependent variables. *The Stata Journal*, 6(1), 58–82.

Williams, R. (2016). Understanding and interpreting generalized ordered logit models. *The Journal of Mathematical Sociology*, 40(1), 7–20.

Xie, Y. (2015). *Dynamic documents with R and knitr* (2nd ed.). Chapman & Hall/CRC.

Xie, Y. (2021). *knitr: A general-purpose package for dynamic report generation in R*. (R package version 1.33) [Computer software]. https://yihui.org/knitr/

Yee, T. W. (2010). The VGAM package for categorical data analysis. *Journal of Statistical Software*, 32(10), 1–34. https://www.jstatsoft.org/v32/i10/

Yee, T. W. (2015). *Vector generalized linear and additive models: With an implementation in R*. Springer.

Yee, T. W. (2021). *VGAM: Vector generalized linear and additive models*. (R package version 1.1-5) [Computer software]. https://CRAN.R-project.org/package=VGAM

Zeileis, A., & Hothorn, T. (2002). Diagnostic checking in regression relationships. *R News*, 2(3), 7–10.

Zeileis, A., Kleiber, C., & Jackman, S. (2008). Regression models for count data in R. *Journal of Statistical Software*, 27(8), 1–25. http://www.jstatsoft.org/v27/i08/

INDEX

S

T

V

Printed in Dunstable, United Kingdom

88190710R00420